Distributed parameter systems encompass a broad range of engineering applications from musical instruments and home stereo systems to satellites and space structures. The dynamic behavior of these systems is usually governed by one or more partial differential equations, which may accurately represent the physical system but are often difficult to solve exactly.

Research in the dynamics and control of distributed parameter structural systems has grown dramatically in recent years, owing in part to the increasing complexity of these systems. Emerging technologies such as smart structures, structronics, and mechatronics have also contributed to the growth in interest.

The purpose of this book is to document recent progress in both theory and practical applications. Specific chapters discuss new modeling, simulation, and analysis techniques used to investigate a variety of distributed elastic, electromechanical, acoustic, and control systems.

With contributions by leading authorities, this book will serve as an up-to-date resource for researchers and graduate students and also as a useful reference for practicing engineers working in the area of dynamics and control of distributed systems.

DYNAMICS AND CONTROL
OF DISTRIBUTED SYSTEMS

DYNAMICS AND CONTROL OF DISTRIBUTED SYSTEMS

Edited by

H. S. TZOU and L. A. BERGMAN

CAMBRIDGE
UNIVERSITY PRESS

CAMBRIDGE UNIVERSITY PRESS
Cambridge, New York, Melbourne, Madrid, Cape Town, Singapore, São Paulo

Cambridge University Press
The Edinburgh Building, Cambridge CB2 2RU, UK

Published in the United States of America by Cambridge University Press, New York

www.cambridge.org
Information on this title: www.cambridge.org/9780521550741

First published 1998
This digitally printed first paperback version 2006

A catalogue record for this publication is available from the British Library

Library of Congress Cataloguing in Publication data

Dynamics and control of distributed systems / edited by H.S. Tzou and L.A. Bergman.
p. cm.
Includes bibliographical references and index.
ISBN 0 521 55074 2 hardback
1. Vibration – Mathematical models. 2. Distributed parameter systems. 3. Structural control (Engineering) I. Tzou, H.S. (Horn S.) II. Bergman, Lawrence A.
TA355.D968 1998
620.3 – dc21 98-14977
CIP

ISBN-13 978-0-521-55074-1 hardback
ISBN-10 0-521-55074-2 hardback

ISBN-13 978-0-521-03374-9 paperback
ISBN-10 0-521-03374-8 paperback

Contents

Contributors

A. V. Balakrishnan
Department of Electrical Engineering, University of California, Los Angeles, CA 90095

H. T. Banks
Center for Research in Scientific Computation, North Carolina State University, Raleigh, NC 26795

L. A. Bergman
Department of Aeronautical and Astronautical Engineering, University of Illinois at Urbana-Champaign, Urbana, IL 61801

J. J. Hollkamp
Flight Dynamics Directorate, Wright Laboratory, Wright-Patterson Air Force Base, OH 45433

C. D. Mote Jr.
Department of Mechanical Engineering, University of California at Berkeley, Berkeley, CA 94720

V. N. Pilipchuk
Department of Applied Mathematics, State Chemical and Technological University of Ukraine, Dniepropetrovsk, Ukraine

R. C. Smith
Department of Mathematics, Iowa State University, Ames, IA 50011

C. A. Tan
Department of Mechanical Engineering, Wayne State University, Detroit, MI 48202

H.-S. Tzou
Department of Mechanical Engineering, University of Kentucky, Lexington, KY 40506

A. F. Vakakis
Department of Mechanical and Industrial Engineering, University of Illinois at Urbana-Champaign, Urbana, IL 61801

V. B. Venkayya
Flight Dynamics Directorate, Wright Laboratory, Wright-Patterson Air Force Base, OH 45433

B. Yang
Department of Mechanical Engineering, University of Southern California, Los Angeles, CA 90089-1453

J. Zhou
Department of Astronautics, National University of Defense Technology, Changsha, Hunan 410073, China

W. D. Zhu
Department of Mechanical Engineering, University of North Dakota, Grand Forks, ND 58202

Preface

Distributed (parameter) systems (DPSs) are the most natural and generic systems existing today. Dynamic characteristics of most natural structures, manufacturing processes, fluids, heat transfer, control, etc., all fall into this DPS category. In general, their dynamics or responses are functions of spatial and time variables, and the systems are usually modeled by partial differential equations. Common practice often discretizes these distributed systems, and their lumped approximations (discrete systems modeled by ordinary differential equations) are then analyzed and evaluated. Original (distributed) behavior can only be observed at these discrete reference locations.

The dynamics and control of distributed systems have traditionally posed many challenging issues investigated by researchers and scientists for decades. However, new R&D activities and findings on DPSs have not been systematically reported for a long while. This book aims to document recent progress on the subject and to bring these technical advances to the engineering community.

Distributed structures are often coupled to external discrete components in engineering applications (e.g., disk heads and tape drives). A new transient analysis technique is developed to investigate the dynamics of coupled distributed-discrete systems in Chapter 1. Transient responses of time-varying systems and constrained translating strings are investigated. Transient behaviors of cables transporting dynamic payloads and translating (e.g., magnetic tape-head systems) are thoroughly studied.

The problem of an elastic distributed system coupled with a moving oscillator, often referred to as the "moving-oscillator" problem, is studied in Chapter 2. The moving-oscillator problem is formulated using a relative displacement model. It is shown that, in the limiting case, the moving-mass problem is recovered. The coupled equations of motion are recast into an integral equation, which is amenable to solution by both iterative and direct numerical procedures.

Nonlinear localized modes and wave transmission of periodic distributed systems are investigated in Chapter 3. Two methodologies for studying the periodic oscillations are proposed and demonstrated. The first methodology is based on a nonsmooth transformation of the spatial variable, and it eliminates the singularities from the governing partial differential equations. The second involves the piecewise transformation of the time variable. The scattering of the primary pulse at the nonlinear stiffnesses is reduced to solving a set of strongly nonlinear first-order ordinary differential equations. Approximate analytical and exact numerical solutions are presented.

The use of continuum models in control design for stabilizing flexible structures is demonstrated in Chapter 4. Dynamics and control of a six-degree-of-freedom anisotropic flexible beam with discrete nodes attached to lumped mass actuators are studied. The Timoshenko model dynamics are translated into a canonical equation in a Hilbert space. The solution is shown to require the use of an energy norm that is no more than the total energy: potential plus kinetic. Under an appropriate extension of the notion of controllability, rate feedback with a collocated sensor can stabilize the structure such that all modes are damped and the energy decays to zero.

A model for a 3-D structural acoustic system is introduced and numerical techniques for simulation, parameter estimation, and noise control in structural acoustic systems are presented in Chapter 5. The 3-D model consists of a hard-walled cylinder with a flexible circular plate at one end. An exterior noise source causes vibrations in the plate which in turn leads to unwanted noise inside the cylinder. Control is implemented through the excitation of piezoceramic patches. Numerical techniques for analyzing the coupled distributed structural-acoustic system are proposed and numerical examples evaluated.

In Chapter 6, a new distributed transfer function method is introduced for modeling and analysis of stepped cylindrical shells and cylindrical shells stiffened by circumferential rings. Various static and dynamic problems of cylindrical shells are systemically formulated and the static/dynamic responses, natural frequencies/modes, and buckling loads of general stiffened cylindrical shells under arbitrary external excitations and boundary conditions are analyzed in exact and closed form.

Chapter 7 focuses on distributed sensing and control of a generic double-curvature elastic shell and its derived geometries laminated with distributed piezoelectric transducers. Generic distributed orthogonal sensing and actuation of shells and continua are proposed. Spatially distributed orthogonal sensors/actuators and self-sensing actuators are presented. Collocated independent modal control with self-sensing orthogonal actuators is demonstrated and its control effectiveness evaluated. Spatially distributed orthogonal piezoelectric sensors/actuators for circular ring shells are designed and their modal sensing and control are investigated. Membrane and bending contributions in sensing and control responses are also studied.

This book serves not only as an updated resource book for researchers and scientists, but also as a useful reference book for practicing engineers with advanced research and development functions. Professors can use this book in teaching distributed systems, partial differential equations, dynamics and control of distributed systems, etc. Students will also find the material essential in their advanced research in distributed systems and advanced dynamics and control.

The editors are very grateful to the distinguished contributors who deserve full credit for the success of the book and also to Cambridge University Press editor Florence Padgett who has continuously and patiently cultivated the project for over two years. The natural rule usually leads to "everything takes longer than it should"; compiling this book has also gone through the usual and unusual obstacles. Finally, it is here. The editors would just like to say thank you to all concerned.

May 14, 1997

Horn-Sen Tzou
Lawrence A. Bergman

1

On the Transient Response of Distributed Structures Interacting with Discrete Components

W. D. ZHU and C. D. MOTE, JR.

Abstract

Distributed structures are often coupled to external components in engineering applications: The recording head in disk and tape drives, the guide bearing in circular and band saws, and the payload in cable transport systems are a few examples. Because the transient response of a distributed structure is characterized by complex multiple wave scattering at an external component, this behavior has rarely been explored. Classical transient analyses by finite difference, finite element, and modal expansion approaches use spatial discretization. As a result, some discontinuities in the interaction force between a distributed structure and an external component, a characteristic of multiple wave scattering, are not predicted in the normal application of these methods. A new transient analysis is developed in this chapter for the response of distributed structures interacting with discrete components. The transient response of a time-varying, cable transport system is analyzed first. The transient and steady-state responses of constrained translating strings are obtained next. Transient behaviors of cables transporting dynamic payloads and translating, magnetic tape-head systems are thoroughly investigated. Application of the method to the classical piano string response under a hammer strike avoids key limitations of the existing standing and traveling wave methods.

1.1 Introduction

Distributed structures are usually coupled to external dynamic components in engineering applications. In magnetic tape and disk drives data transfer is accomplished through contact of recording media with a recording head. The guide bearing used in wood machining to minimize the circular and band saw vibration is another example. In cable transport systems, such as tramways and ski lifts, the payload is attached to a cable for transport across a span.

In all these cases, a wave propagating along a distributed structure will partly transmit through, and partly reflect from, an attached external system. Hence the transient response of a distributed structure coupled to an external component is characterized by complex, multiple wave scattering.

Although research interest in the dynamic analysis of constrained distributed structures has increased in recent years, most studies (Butkovskiy, 1983; Bergman and Nicholson, 1985; Yang, 1992; Tan and Chung, 1993) investigate natural frequencies, modes of vibration, and steady harmonically forced response. The transient behavior has rarely been explored. Developing an understanding of the transient behavior is important for a number of reasons. First, some constrained distributed systems do not have a steady-state response. For instance, the vibration of a cable transport system is inherently a transient response problem. Its eigensolutions and steady-state response do not exist because of the time-varying inertia and stiffness of the system as the payload moves along the span (Zhu and Mote, 1994a). Second, a steady-state analysis usually seeks response to a sinusoidal, or periodic, excitation. Naturally, many other loading conditions require response solution too. The wave motion generated in a magnetic tape when a fast moving recording head suddenly contacts or leaves the tape is of interest in magnetic recording because it is desirable to minimize transient effects (Lacey and Talke, 1990; Sundaram and Benson, 1990). A transient analysis, however, can solve for the response to arbitrary loadings and initial conditions. Third, even for the case of a sinusoidal excitation, the transient response of a constrained distributed structure is characterized by multiple wave scattering and differs significantly from its steady-state response. For a lightly damped system, the transient wave motion can dominate the response within the time interval of interest. Hence in machines that operate satisfactorily at steady state, problems can arise during transients. The design of high-precision and high-productivity machines benefits from a fundamental understanding of the transient dynamic behavior.

Transient analyses commonly use spatial discretization of the governing partial differential equation for a distributed structure through finite difference, finite element, or modal expansion approaches. Stahl et al. (1974) used finite differences in both the temporal and spatial domains to calculate the transient air pressure between the moving tape and a stationary recording head. By Galerkin discretization, Ulsoy and Mote (1982) determined the transient response of a traveling plate under transverse load. When a distributed system is approximated by spatially discretized models, some dynamic characteristics associated with the distributed model are lost. In control of a translating string vibration, use of a finite number of modes leads to control and observation spill-overs (Ulsoy, 1984). For transient analysis of a distributed structure coupled to external dynamic components, the principal task is to determine the force of interaction between the structure and the external component (Wickert and Mote, 1991). Once the force is known, the response of each subsystem can then be predicted. Characteristic of the multiple wave scattering phenomenon, the time history of the interaction force between the structure and an external component can exhibit discontinuities when the propagating wavefronts in the structure have nonzero slopes (Zhu and Mote, 1994a, 1995), and these discontinuities are not predicted by the spatially discretized models. This behavior is illustrated in the following simple system.

Consider a pulse of length l propagating in an infinite, stationary, taut string with a discrete, rigid, or flexible constraint at the origin $x = 0$, as shown in Fig. 1.1. The slopes of the leading and trailing edges of pulse are given by $\Theta'(0)$ and $\Theta'(l)$ respectively (see Appendix A). The pulse is reflected from the rigid support in Fig. 1.1(a) and is partly reflected from, and partly transmitted through, the flexible constraint in Fig. 1.1(b,c). In the first two cases, when the leading and trailing edges of the pulse collide with the constraint, the dimensionless interaction force between the string and the constraint is discontinuous, with magnitudes $-2\Theta'(0)$ and $2\Theta'(l)$, respectively. For the flexible massless constraint in Fig. 1.1(c), the discontinuities in the dimensionless interaction force are $-2c/(c+2)\Theta'(0)$ and $2c/(c+2)\Theta'(l)$ when the pulse enters and

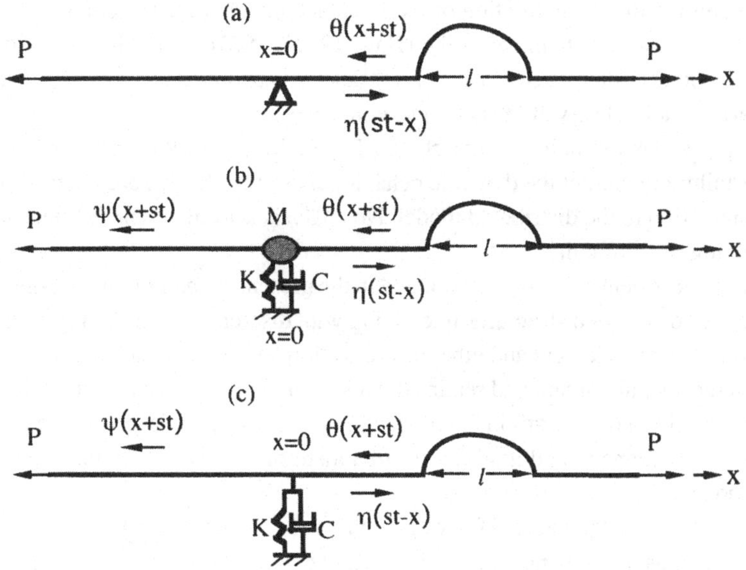

Fig. 1.1. Propagating pulses in constrained stationary strings.

exits the constraint. Here c is the dimensionless viscous damping coefficient. Hence if the pulse has nonzero slopes at either the leading or trailing edges, the interaction force will be discontinuous. The interaction force is always continuous only when the constraint is modeled by a stiffness element.

In a previous investigation, Wickert and Mote (1990) derived the general response of an axially moving material to arbitrary initial conditions and excitations in terms of the Green's function. The transient response of a traveling string with an attached inertia was subsequently studied (Wickert and Mote, 1991). Though abrupt variations in the force of interaction between the string and inertia occur at the instants when propagating wavefronts coincide with the particle, the interaction force is continuous in this case because the wavefront has vanishing slope. For steady-state response of a constrained distributed structure subject to a harmonic excitation, discontinuities can occur in the constraint force during the initial transient. When distributed or lumped damping exists in the structure, as shown in an example in Section 1.3.4, the discontinuous constraint force during the initial transient approaches, at the steady state, the sinusoidal force predicted by the steady forced analysis.

Existence of the discontinuity in the interaction force has implications to both practice and theory. Discontinuities in the tape-head contact force cause signal strength variations in magnetic recording. Discontinuities in the interaction force between the cable and payload cause discontinuous acceleration of the payload. Discontinuities in the contact force between the piano string and hammer affect the duration of their contact and alter the harmonic spectra of the string after a hammer strike (Hall, 1987a, 1987b). The discontinuities in the contact force between the string and a hard hammer have also been experimentally observed (Yanagisawa et al., 1981).

A particular limitation of the classical spatial discretization approaches is their failure to predict the dynamic behavior caused by the discontinuous force of interaction in the distributed model. No analysis is available in the literature resolving this difficulty.

A new transient analysis is developed in the present chapter predicting the response of distributed structures interacting with discrete dynamic components. Unlike the finite element and other discretization approaches, the method is exact in the spatial domain and retains the distributed property of the structure. It captures all discontinuities in the interaction force between the structure and an external component, and it allows for treatment of nonlinearity in the external components.

In Section 1.2, the free and forced responses of a cable transport system, modeled as a translating string transporting a damped, linear oscillator, are predicted for arbitrary initial conditions, external forces, and boundary disturbances.

Analogous to the mechanism for the simple case depicted in Fig. 1.1, the time history of the force of interaction between the cable and the suspension is in general discontinuous for two reasons: 1. The presence of the suspension damping causes discontinuity in the force of interaction and 2. for all initial conditions, except those satisfying the special equation (1.45), all propagating wavefronts in the string will have nonzero slopes and lead to discontinuity in the force of interaction.

The delay-integral equation describing the force of interaction between the cable and the suspension is derived using the Green's function formulation, following Wickert and Mote (1991). A difficulty arises in the analysis of the governing integral equation in which both the kernel and the dependent variable are discontinuous. The solution technique is developed to eliminate the discontinuities in the kernel using the theory of distributions. The forcing integrals are evaluated by both the distribution and series expansion methods. The limitation for the use of the series expansion of the Green's function in evaluation of the forcing integrals is indicated. Exact in the spatial domain, a finite difference algorithm in the temporal domain is presented for the solution of the resulting delay-integro-differential equation with a discontinuous dependent variable. The interaction force for the special case of harmonic support excitation at the entrance to the span is calculated in the example in Section 1.2.4.

The transient and steady-state responses of a translating string in contact with discrete, rigid, and flexible constraints are determined in Section 1.3. A translating magnetic tape-head system or a guided band saw system are modeled. The contact force between the string and the one-sided constraint is composed of an equilibrium preload component and a dynamic component.

The transient analysis developed in Section 1.3.2 predicts the response to arbitrary initial conditions, external forces, and boundary disturbances. The exact expression describing the dynamic contact force component is derived using the Green's function formulation. The eigensolutions and steady harmonic response of the constrained translating string are determined in closed form in Section 1.3.3. The transient and steady-state analyses of a magnetic tape-head system model are presented in Section 1.3.4.

Section 1.4 addresses the application of the discontinuous contact force methodology to the classical problem of the piano string response under a hammer strike. The method of solution applies to linear as well as nonlinear hammer models and handles nicely the multiple contacts occurring between the string and hammer. Particular sets of parameters were selected in the examples in Section 1.4.3 to compare the predictions with available results found in the literature and to obtain results that could not have been obtained by other existing methods.

1.2 Free and Forced Response of a Time-varying, Cable Transport System

1.2.1 Model and Equations of Motion

Consider first a cable transport system modeled as a translating string of length L and mass per unit length ρ, transporting a suspended mass M with suspension stiffness and damping constants K and C, respectively (see Fig. 1.2). A constant, subcritical transport speed V is prescribed. The string is subjected to gravitational force ρg, distributed external force $F(X, T)$, and a prescribed transverse motion $E(T)$ at the left, span entrance, end. The boundary excitation at the right, span exit, end can be included in a similar manner. The payload mass is subjected to an external force $Q(T)$.

The natural response of the above system is not adequately modeled by the solution of an eigenvalue problem, because the inertia and stiffness properties in the model depend on the position of the payload along the span. During the free transit of the payload through the span, the transient response problem under the gravity and initial conditions is referred to as the *free vibration problem*. Under external excitation $F(X, T)$, $Q(T)$, and $E(T)$, the transient response problem is the *forced vibration problem*. The procedure developed in this section determines the forced response, and the free response is obtained as a special case of it.

Application of the model is limited by the following conditions. The string transverse displacement $U(X, T)$ is small and planar. The initial tension P in

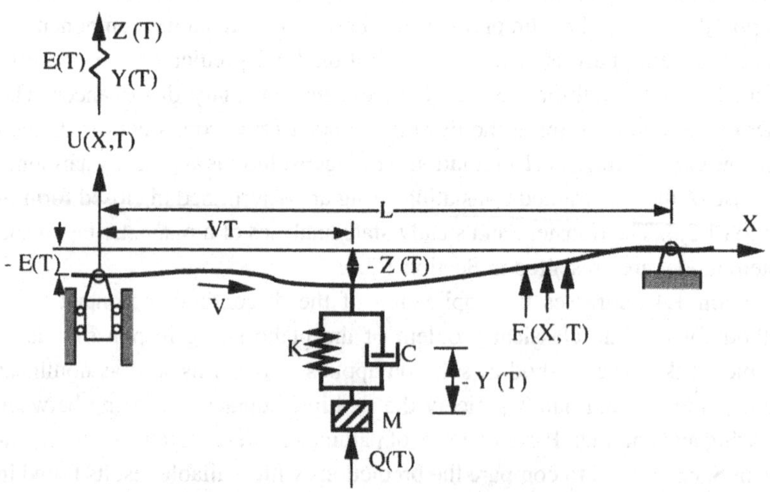

Fig. 1.2. Schematic of the cable trasport system under external forces $F(X, T)$ and $Q(T)$, and boundary excitation $E(T)$ at $x = 0$. All displacement variables are positive in the upward direction.

the string is sufficiently large that changes in tension caused by the payload have a negligible effect on the system response. The payload mass undergoes transverse vibration relative to the string without rotation.

Let $U(X, T)$ be the displacement of the string relative to the horizontal X axis. Assuming that the payload is positioned at $X = 0$ at time $T = 0$, $Z(T) = U(VT, T)$ is the displacement of its attachment point to the string. Also let $Y(T)$ be the displacement of the payload mass measured from the unstretched length of the spring at displacement $Z(T) = 0$. The interaction force $R(T)$ between the string and the payload suspension is the variable to be determined. Introduce the following dimensionless variables:

$$\begin{aligned} &x = X/L, \quad u = U/L, \quad y = Y/L, \quad z = Z/L, \quad e = E/L, \\ &t = T(P/\rho L^2)^{1/2}, \quad m = M/\rho L, \quad k = KL/P, \\ &c = C/(P\rho)^{1/2}, \quad v = V(\rho/P)^{1/2}, \quad w = \rho g L/P, \\ &f = FL/P, \quad q = Q/P, \quad r = R/P. \end{aligned} \tag{1.1}$$

The dimensionless equation of motion for a uniform traveling string is

$$u_{tt}(x,t) + 2vu_{xt}(x,t) + (v^2 - 1)u_{xx}(x,t) = r(t)\delta(x - vt) + f(x,t) - w \tag{1.2}$$

with the boundary and initial conditions given by

$$u(0,t) = e(t), \qquad u(1,t) = 0, \tag{1.3}$$

$$u(x,0) = a(x) - \frac{wx(1-x)}{2(1-v^2)}, \qquad u_t(x,0) = b(x). \tag{1.4}$$

In Eq. (1.4) $a(x)$ is the initial displacement from the static deflection $-wx(1 - x)/2(1 - v^2)$ of the traveling string under weight w, and $b(x)$ is the initial transverse velocity. The terms $a(x)$ and $b(x)$ are dimensionless. The dimensionless equation of motion and initial conditions for the payload are

$$m\ddot{y}(t) = -r(t) + q(t) - mw \tag{1.5}$$

and

$$y(0) = y_0, \qquad \dot{y}(0) = \dot{y}_0, \tag{1.6}$$

where y_0 and $\dot{y}_0$ are specified. The interaction force $r(t)$ between the suspension and the string is

$$r(t) = k(y(t) - z(t)) + c(\dot{y}(t) - \dot{z}(t)) \tag{1.7}$$

and the attachment point displacement and velocity are

$$z(t) = u(vt,t), \qquad \dot{z}(t) = u_t(vt,t) + vu_x(vt,t). \tag{1.8}$$

Equations (1.2)–(1.8) can be reduced to a single integral equation governing the interaction force $r(t)$.

1.2.2 Integral Formulation

1.2.2.1 Delay-integral Equation of Volterra Type

The solution to Eqs. (1.2)–(1.4) proceeds by first applying the transformation (Meirovitch, 1967)

$$u(x,t) = \hat{u}(x,t) + (1-x)e(t) \tag{1.9}$$

to render the boundary condition (1.3) homogeneous. The transformed equations for $\hat{u}(x,t)$ are

$$\hat{u}_{tt}(x,t) + 2v\hat{u}_{xt}(x,t) + (v^2-1)\hat{u}_{xx}(x,t)$$
$$= r(t)\delta(x-vt) + f(x,t) - w - (1-x)\ddot{e}(t) + 2v\dot{e}(t), \tag{1.10}$$

$$\hat{u}(0,t) = \hat{u}(1,t) = 0, \tag{1.11}$$

$$\hat{u}(x,0) = a(x) - \frac{wx(1-x)}{2(1-v^2)} - (1-x)e(0),$$
$$\hat{u}_t(x,0) = b(x) - (1-x)\dot{e}(0). \tag{1.12}$$

The Green's function (Wickert and Mote, 1990) can then be used to determine $\hat{u}(x,t)$. With inclusion of the parabolic static deflection of the string under its own weight, the transverse displacement of the string becomes

$$\hat{u}(x,t) = -\frac{wx(1-x)}{2(1-v^2)} + \int_0^t\int_0^1 [f(\xi,\tau) - (1-\xi)\ddot{e}(\tau) + 2v\dot{e}(\tau)]$$
$$\times g(x,\xi;t-\tau)\,d\xi\,d\tau + \int_0^t r(\tau)g(x,v\tau;t-\tau)\,d\tau$$
$$+ \int_0^1 [b(\xi) - (1-\xi)\dot{e}(0)]g(x,\xi;t)\,d\xi + \int_0^1 [a(\xi) - (1-\xi)e(0)]$$
$$\times [g_t(x,\xi;t) - 2vg_\xi(x,\xi;t)]\,d\xi, \tag{1.13}$$

where the Green's function for a fixed-fixed traveling string (Wickert and Mote, 1991),

$$g(x,\xi;t-\tau) = \sum_{n=1}^{\infty} \frac{2}{n\pi}\sin[n\pi(1-v^2)(t-\tau)$$
$$+ n\pi v(x-\xi)]\sin n\pi x \sin n\pi\xi$$
$$= \phi[(1-v^2)(t-\tau) - (1-v)(x-\xi)]$$
$$+ \phi[(1-v^2)(t-\tau) + (1+v)(x-\xi)]$$
$$- \phi[(1-v^2)(t-\tau) - (1-v)x - (1+v)\xi]$$
$$- \phi[(1-v^2)(t-\tau) + (1+v)x + (1-v)\xi], \tag{1.14}$$

is piecewise continuous, because $\phi(\cdot)$ is a Fourier sine series whose sum is the piecewise continuous sawtooth wave:

$$\phi(s) = \sum_{n=1}^{\infty} \frac{\sin(n\pi s)}{2n\pi}$$
$$= \begin{cases} \sum_{n=-\infty}^{\infty} [H(s-2n) - H(s-2(n+1))] \frac{2n+1-s}{4}, & s \neq 0, \pm 2, \pm 4, \ldots \\ 0, & s = 0, \pm 2, \pm 4, \ldots \end{cases} \tag{1.15}$$

where $H(\cdot)$ is the Heaviside function. The first two derivatives of $\phi(s)$, which can be differentiated distributionally (Stakgold, 1979), are

$$\phi'(s) = \frac{1}{2}\sum_{n=1}^{\infty} \cos(n\pi s) = -\frac{1}{4} + \frac{1}{2}\sum_{n=-\infty}^{\infty} \delta(s-2n) \tag{1.16}$$

and

$$\phi''(s) = -\frac{1}{2}\sum_{n=1}^{\infty} n\pi \sin(n\pi s) = \frac{1}{2}\sum_{n=-\infty}^{\infty} \delta'(s-2n), \tag{1.17}$$

where $\delta(\cdot)$ is the Dirac delta function. The series in Eqs. (1.16) and (1.17), which are divergent in the classical sense, converge to the generalized functions. The first partial derivatives of the Green's function are thus obtained from Eqs. (1.14) and (1.16):

$$\begin{aligned} g_x(x,\xi;t-\tau) = &-\frac{1-v}{2}\sum_{n=-\infty}^{\infty} \{\delta[(1-v^2)(t-\tau) - (1-v)(x-\xi) - 2n] \\ &+ \delta[(1-v^2)(t-\tau) - (1-v)x - (1+v)\xi - 2n]\} \\ &+ \frac{1+v}{2}\sum_{n=-\infty}^{\infty} \{\delta[(1-v^2)(t-\tau) + (1+v)(x-\xi) - 2n] \\ &- \delta[(1-v^2)(t-\tau) + (1+v)x + (1-v)\xi - 2n]\}, \end{aligned} \tag{1.18}$$

$$\begin{aligned} g_t(x,\xi;t-\tau) = &\frac{1-v^2}{2}\sum_{n=-\infty}^{\infty} \{\delta[(1-v^2)(t-\tau) - (1-v)(x-\xi) - 2n] \\ &+ \delta[(1-v^2)(t-\tau) + (1+v)(x-\xi) - 2n] \\ &- \delta[(1-v^2)(t-\tau) - (1-v)x - (1+v)\xi - 2n] \\ &- \delta[(1-v^2)(t-\tau) + (1+v)x + (1-v)\xi - 2n]\}. \end{aligned} \tag{1.19}$$

Through use of Leibnitz's rule, $g(x,\xi;0)=0$, and Eq. (1.13), Eq. (1.9) is differentiated to give

$$u_x(x,t) = -e(t) - \frac{w(1-2x)}{2(1-v^2)} + \int_0^t\int_0^1 [f(\xi,\tau) - (1-\xi)\ddot{e}(\tau) + 2v\dot{e}(\tau)] \times g_x(x,\xi;t-\tau)\,d\xi\,d\tau + \int_0^t r(\tau)g_x(x,v\tau;t-\tau)\,d\tau + \int_0^1 [b(\xi) - (1-\xi)\dot{e}(0)]g_x(x,\xi;t)\,d\xi + \int_0^1 [a(\xi) - (1-\xi)e(0)][g_{tx}(x,\xi;t) - 2vg_{\xi x}(x,\xi;t)]\,d\xi, \tag{1.20}$$

$$u_t(x,t) = (1-x)\dot{e}(t) + \int_0^t\int_0^1 [f(\xi,\tau) - (1-\xi)\ddot{e}(\tau) + 2v\dot{e}(\tau)] \times g_t(x,\xi;t-\tau)\,d\xi\,d\tau + \int_0^1 r(\tau)g_t(x,v\tau;t-\tau)\,d\tau + \int_0^1 [b(\xi) - (1-\xi)\dot{e}(0)]g_t(x,\xi;t)\,d\xi + \int_0^1 [a(\xi) - (1-\xi)e(0)][g_{tt}(x,\xi;t) - 2vg_{\xi t}(x,\xi;t)]\,d\xi. \tag{1.21}$$

The solutions of Eqs. (1.5) and (1.6) are expressed in the integral forms

$$y(t) = y_0 + \dot{y}_0 t - \frac{1}{2}wt^2 + \frac{1}{m}\int_0^t (t-\tau)q(\tau)\,d\tau - \frac{1}{m}\int_0^t (t-\tau)r(\tau)\,d\tau \tag{1.22}$$

and

$$\dot{y}(t) = \dot{y}_0 - wt + \frac{1}{m}\int_0^t q(\tau)\,d\tau - \frac{1}{m}\int_0^t r(\tau)\,d\tau. \tag{1.23}$$

Substitution of Eqs. (1.20) and (1.21) into Eq. (1.8) gives expressions for $z(t)$ and $\dot{z}(t)$. These results, along with Eqs. (1.22) and (1.23), are substituted into Eq. (1.7) to yield a Volterra integral equation of the second kind describing the interaction force $r(t)$:

$$r(t) = p(t) + \int_0^t \kappa(t,\tau)r(\tau)\,d\tau, \tag{1.24}$$

where

$$\kappa(t,\tau) = -\frac{c}{m} - \frac{k}{m}(t-\tau) - kg(vt,v\tau;t-\tau) - c\frac{\partial}{\partial t}g(vt,v\tau;t-\tau), \tag{1.25}$$

$$p(t) = ky_0 + c\dot{y}_0 + (k\dot{y}_0 - cw)t - \frac{1}{2}kwt^2 + \frac{kwvt(1-vt)}{2(1-v^2)}$$
$$+ \frac{cwv(1-2vt)}{2(1-v^2)} + [-k(1-vt) + cv]e(t) - c(1-vt)\dot{e}(t)$$
$$+ \frac{k}{m}\int_0^t (t-\tau)q(\tau)\,d\tau + \frac{c}{m}\int_0^t q(\tau)\,d\tau - \int_0^t\int_0^1 [f(\xi,\tau)$$
$$- (1-\xi)\ddot{e}(\tau) + 2v\dot{e}(\tau)]\left(k + c\frac{\partial}{\partial t}\right)g(vt,\xi;t-\tau)\,d\xi\,d\tau$$
$$- \int_0^1 [b(\xi) - (1-\xi)\dot{e}(0)]\left(k + c\frac{\partial}{\partial t}\right)g(vt,\xi;t)\,d\xi$$
$$- \int_0^1 [a(\xi) - (1-\xi)e(0)]\left(k + c\frac{\partial}{\partial t}\right)[g_t(vt,\xi;t)$$
$$- 2vg_\xi(vt,\xi;t)]\,d\xi. \tag{1.26}$$

Equations (1.2)–(1.8), governing the response of the coupled system, have thus been reduced to a single integral equation (1.24) for $r(t)$. The kernel $\kappa(t,\tau)$ in Eq. (1.25) contains delta functions that come out of the integral in Eq. (1.24). Noting that

$$\frac{\partial}{\partial t}g(vt, v\tau; t-\tau) = g_t(vt, v\tau; t-\tau) + vg_x(vt, v\tau; t-\tau) \tag{1.27}$$

and using Eqs. (1.18) and (1.19) evaluated at $x = vt$, $\xi = v\tau$ gives

$$\frac{\partial}{\partial t}g(vt, v\tau; t-\tau) = \frac{1}{2}\sum_{n=-\infty}^{\infty}\{(1-v)\delta[(1-v)(t-\tau) - 2n]$$
$$+ (1+v)\delta[(1+v)(t-\tau) - 2n]$$
$$- (1-v)\delta[(1-v)t - (1+v)\tau - 2n]$$
$$- (1+v)\delta[(1+v)t - (1-v)\tau - 2n]\}. \tag{1.28}$$

Use of Eq. (1.28) in Eq. (1.24) gives a series of interaction forces occurring at delayed times. The Volterra equation of the second kind (1.24) hence reduces to a Volterra integral equation with delay similar to that in Wickert and Mote (1991):

$$r(t) = p(t) + \int_0^t \kappa_1(t,\tau)r(\tau)\,d\tau - \frac{c}{2}\sum_{j=1}^{N}\left[r\left(t - \frac{2(j-1)}{1-v}\right)\right.$$
$$\left. - \frac{1-v}{1+v}r\left(\frac{1-v}{1+v}t - \frac{2(j-1)}{1+v}\right)\right]$$

$$\times H\left(t - \frac{2(j-1)}{1-v}\right) - \frac{c}{2}\sum_{j=1}^{N}\left[r\left(t - \frac{2j}{1+v}\right)\right.$$

$$\left. - \frac{1+v}{1-v} r\left(\frac{1+v}{1-v}t - \frac{2j}{1-v}\right)\right] H\left(t - \frac{2j}{1+v}\right), \tag{1.29}$$

where the kernel

$$\kappa_1(t, \tau) = -\frac{c}{m} - \frac{k}{m}(t - \tau) - kg(vt, v\tau; t - \tau) \tag{1.30}$$

is piecewise continuous and $N = \mathrm{Int}[(1+v)/2v]$. As shown in Section 1.2.2.4, N is the number of interactions of a downstream propagating wave with the payload during its transport across the span. Note that $N \geq 1$ for the subcritical transport speed $v < 1$. All external excitation is included in $p(t)$ in Eq. (1.29). The first ten terms in $p(t)$ in Eq. (1.26) are continuous in time, as long as $e(t)$ is continuously differentiable and $q(t)$ is piecewise continuous. The eleventh double integral term is continuous if $f(x, t)$ is also piecewise continuous. The last two terms are possibly piecewise continuous because they are single integrals containing delta functions in their integrands. The last integrand contains derivatives of delta functions, but the singular boundary terms containing the delta functions after integration by parts vanish by $a(1^-) = 0$ and continuity of the initial displacement at the left, span entrance, end (i.e., $a(0^+) = e(0)$). The last term will be shown to be continuous if $a(0^+) = e(0)$. Hence $p(t)$ is a piecewise continuous function of time.

1.2.2.2 Discontinuity in $p(t)$ and $r(t)$

The potentially discontinuous parts of $p(t)$ are denoted by $I_1(t)$ and $I_2(t)$ here and are evaluated by calculus of generalized functions (Hoskins, 1979). The expressions are

$$I_1(t) = \int_0^1 [b(\xi) - (1-\xi)\dot{e}(0)] c \frac{\partial}{\partial t} g(vt, \xi; t)\, d\xi$$

$$= \frac{c}{2}\left\{ \sum_{l_1 \leq i \leq r_1} coef_1(i) \left[b\left(-t + \frac{2i}{1-v}\right) - \left(1 + t - \frac{2i}{1-v}\right)\dot{e}(0)\right]\right.$$

$$- \frac{1-v}{1+v} \sum_{l_2 \leq i \leq r_2} coef_2(i) \left[b\left(\frac{1-v}{1+v}t - \frac{2i}{1+v}\right)\right.$$

$$\left. - \left(1 - \frac{1-v}{1+v}t + \frac{2i}{1+v}\right)\dot{e}(0)\right]$$

$$+ \sum_{l_3 \leq i \leq r_3} coef_3(i) \left[b\left(t - \frac{2i}{1+v}\right) - \left(1 - t + \frac{2i}{1+v}\right)\dot{e}(0)\right]$$

$$-\frac{1+v}{1-v}\sum_{l_4\le i\le r_4} coef_4(i)\left[b\left(-\frac{1+v}{1-v}t+\frac{2i}{1-v}\right)\right.$$

$$\left.\left.-\left(1+\frac{1+v}{1-v}t-\frac{2i}{1-v}\right)\dot{e}(0)\right]\right\} \tag{1.31}$$

and

$$I_2(t)=\int_0^1[a(\xi)-(1-\xi)e(0)]\left(k+c\frac{\partial}{\partial t}\right)[g_t(vt,\xi;t)-2vg_\xi(vt,\xi;t)]\,d\xi$$

$$=\sum_{l_1\le i\le r_1} coef_1(i)\frac{1-v}{2}\left\{k\left[a\left(-t+\frac{2i}{1-v}\right)-\left(1+t-\frac{2i}{1-v}\right)e(0)\right]\right.$$

$$\left.-c\left[a'\left(-t+\frac{2i}{1-v}\right)+e(0)\right]\right\}-\sum_{l_2\le i\le r_2} coef_2(i)\left\{\frac{1+v}{2}k\right.$$

$$\times\left[a\left(\frac{1-v}{1+v}t-\frac{2i}{1+v}\right)-\left(1-\frac{1-v}{1+v}t+\frac{2i}{1+v}\right)e(0)\right]+\frac{1-v}{2}c$$

$$\left.\times\left[a'\left(\frac{1-v}{1+v}t-\frac{2i}{1+v}\right)+e(0)\right]\right\}+\sum_{l_3\le i\le r_3} coef_3(i)\frac{1+v}{2}$$

$$\times\left\{k\left[a\left(t-\frac{2i}{1+v}\right)-\left(1-t+\frac{2i}{1+v}\right)e(0)\right]\right.$$

$$\left.+c\left[a'\left(t-\frac{2i}{1+v}\right)+e(0)\right]\right\}+\sum_{l_4\le i\le r_4} coef_4(i)\left\{-\frac{1-v}{2}k\right.$$

$$\times\left[a\left(-\frac{1+v}{1-v}t+\frac{2i}{1-v}\right)-\left(1+\frac{1+v}{1-v}t-\frac{2i}{1-v}\right)e(0)\right]$$

$$\left.+\frac{1+v}{2}c\left[a'\left(-\frac{1+v}{1-v}t+\frac{2i}{1-v}\right)+e(0)\right]\right\}, \tag{1.32}$$

where

$$l_1=\frac{1-v}{2}t,\qquad r_1=\frac{1-v}{2}(t+1),\qquad l_2=\frac{1-v}{2}t-\frac{1+v}{2},$$

$$r_2=\frac{1-v}{2}t,\qquad l_3=\frac{1+v}{2}(t-1),\qquad r_3=\frac{1+v}{2}t,\qquad l_4=\frac{1+v}{2}t,$$

$$r_4=\frac{1+v}{2}t+\frac{1-v}{2},$$

$$coef_m(i)=$$

$$\begin{cases}1, & \text{if } i \text{ is the integer in the region } (l_m, r_m),\\ 0.5, & \text{if the integer } i=l_m \text{ or } i=r_m,\\ 0, & \text{if the integer } i \text{ is not in the region } [l_m, r_m].\end{cases}\qquad (m=1,2,3,4)$$

The initial values of $I_1(t)$ and $I_2(t)$ at $t = 0^+$ are obtained from Eqs. (1.31) and (1.32):

$$I_1(0^+) = \frac{cv}{1+v}[b(0^+) - \dot{e}(0)], \qquad I_2(0^+) = cv[a'(0^+) + e(0)]. \quad (1.33)$$

The discontinuities in $I_1(t)$ and $I_2(t)$ during the payload transit time $t \in [0, 1/v]$ are calculated explicitly from Eqs. (1.31) and (1.32).

At $t = t_-^{(j)} = \dfrac{2j}{1+v}, \qquad j = 1, 2, \ldots, N$

$$I_1\left[\left(t_-^{(j)}\right)^+\right] - I_1\left[\left(t_-^{(j)}\right)^-\right] = \frac{c}{1-v}[b(0^+) - \dot{e}(0)], \quad (1.34)$$

$$I_2\left[\left(t_-^{(j)}\right)^+\right] - I_2\left[\left(t_-^{(j)}\right)^-\right] = k[a(0^+) - e(0)] = 0. \quad (1.35)$$

At $t = t_+^{(j)} = \dfrac{2j}{1-v}, \qquad j = 1, 2, \ldots, N-1$

$$I_1\left[\left(t_+^{(j)}\right)^+\right] - I_1\left[\left(t_+^{(j)}\right)^-\right] = -\frac{c}{1+v}[b(0^+) - \dot{e}(0)], \quad (1.36)$$

$$I_2\left[\left(t_+^{(j)}\right)^+\right] - I_2\left[\left(t_+^{(j)}\right)^-\right] = -k[a(0^+) - e(0)] = 0, \quad (1.37)$$

where $t_-^{(j)}$ and $t_+^{(j)}$ are the instants when the main transmitted wave component interacts with the payload, as shown in Fig. 1.3. It is seen that the discontinuities in $I_2(t)$ at $t_-^{(j)}$ and $t_+^{(j)}$ vanish if the continuity of the initial displacement of the string at the left end (payload entrance) is prescribed. The initial velocity of the string at the left end will be discontinuous (i.e., $b(0^+) \neq \dot{e}(0)$) unless it is appropriately prescribed as well. Hence the discontinuities of $I_1(t)$ are normally nonzero if c is nonzero. The initial value and discontinuities of $p(t)$ are obtained from Eq. (1.26) and Eqs. (1.33)–(1.37) as follows:

$$p(0^+) = ky_0 + c\dot{y}_0 + \frac{cwv}{2(1-v^2)} - ke(0) - c\dot{e}(0) - \frac{cv}{1+v}[b(0^+) - \dot{e}(0)] - cva'(0^+), \quad (1.38)$$

$$p\left[\left(t_-^{(j)}\right)^+\right] - p\left[\left(t_-^{(j)}\right)^-\right] = -\frac{c}{1-v}[b(0^+) - \dot{e}(0)], \quad j = 1, 2, \ldots, N, \quad (1.39)$$

$$p\left[\left(t_+^{(j)}\right)^+\right] - p\left[\left(t_+^{(j)}\right)^-\right] = \frac{c}{1+v}[b(0^+) - \dot{e}(0)], \quad j = 1, 2, \ldots, N-1. \quad (1.40)$$

The initial value of the interaction force $r(t)$ can be calculated by applying

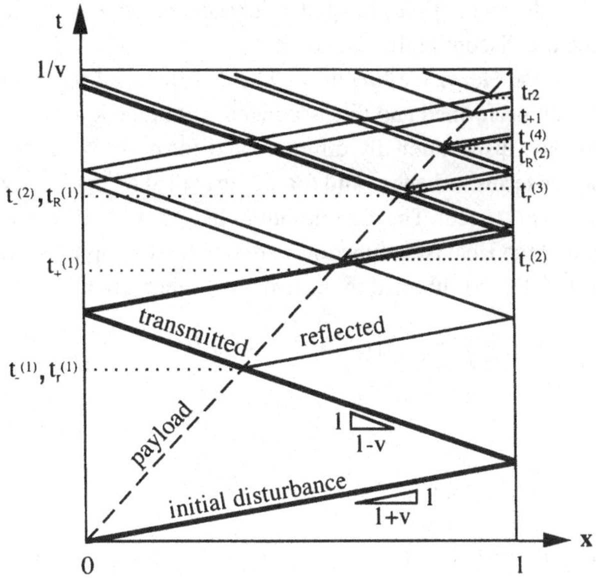

Fig. 1.3. Multiple wave scattering in the cable transport system. Intersections of the characteristic lines with the dashed line $t = x/v$ are the instants when propagating wavefronts in the string interact with the payload. The heavy line is the main transmitted wave component.

Eq. (1.29) at $t = 0^+$ and using Eq. (1.38), yielding

$$
\begin{aligned}
r(0^+) &= \frac{1+v}{1+v+cv} p(0^+) \\
&= \frac{1+v}{1+v+cv} \left[ky_0 + c\dot{y}_0 + \frac{cwv}{2(1-v^2)} - ke(0) - c\dot{e}(0) \right] \\
&\quad - \frac{cv}{1+v+cv} [b(0^+) - \dot{e}(0)] - \frac{cv(1+v)}{1+v+cv} a'(0^+). \qquad (1.41)
\end{aligned}
$$

For $c = 0$, $p(t)$ is continuous by Eqs. (1.39) and (1.40), and the Heaviside functions in Eq. (1.29) vanish. The integral in Eq. (1.29) is always continuous because the kernel $\kappa_1(t, \tau)$ is piecewise continuous. Hence $r(t)$ is continuous. As a result, Eq. (1.29) can be differentiated distributionally to eliminate the piecewise continuous $g(vt, v\tau; t - \tau)$ in the kernel for subsequent numerical solution. When c is nonzero, the discontinuities of $p(t)$ at $t_-^{(j)}$ and $t_+^{(j)}$ are nonzero by Eqs. (1.39) and (1.40). The Heaviside functions are also discontinuous at $t_-^{(j)}$ and $t_+^{(j)}$ if $r(0^+)$ is nonzero. Hence, $r(t)$ is discontinuous at $t_-^{(j)}$ and $t_+^{(j)}$ for nonzero c. Equation (1.29) is thus a discontinuous integral equation, where both the dependent variable $r(t)$ and the kernel $\kappa_1(t, \tau)$ are piecewise continuous. The differentiation strategy for subsequent numerical solution is to first determine the discontinuities in $r(t)$ at $t_-^{(j)}$ and $t_+^{(j)}$ due to the discontinuities of $p(t)$ and the Heaviside functions. The distributional differentiation of Eq. (1.29) is then performed and the associated delta functions at $t_-^{(j)}$ and $t_+^{(j)}$ are discarded. The distributionally differentiated form of Eq. (1.29), without the delta function terms, is hence valid for all t except at $t_-^{(j)}$ and $t_+^{(j)}$, where the discontinuities in $r(t)$ occur. The discontinuities in $r(t)$ at $t_-^{(j)}$, due to the discontinuities in p and the Heaviside functions, are obtained by applying Eq. (1.29) at $[t_-^{(j)}]^+$ and $[t_-^{(j)}]^-$ and subtracting one from the other, giving

$$r\left[\left(t_-^{(j)}\right)^+\right] - r\left[\left(t_-^{(j)}\right)^-\right] = p\left[\left(t_-^{(j)}\right)^+\right] - p\left[\left(t_-^{(j)}\right)^-\right] - \frac{c}{2}\left\{r\left[\left(t_-^{(j)}\right)^+\right] - r\left[\left(t_-^{(j)}\right)^-\right] + r(0^+) - \frac{1+v}{1-v}r(0^+)\right\}. \quad (1.42)$$

By use of Eq. (1.39) we obtain, for $j = 1, 2, \ldots, N$,

$$r\left[\left(t_-^{(j)}\right)^+\right] - r\left[\left(t_-^{(j)}\right)^-\right] = -\frac{2c}{(c+2)(1-v)}[b(0^+) - \dot{e}(0)] + \frac{2cv}{(c+2)(1-v)}r(0^+) \quad (1.43)$$

and likewise, for $j = 1, 2, \ldots, N-1$,

$$r\left[\left(t_+^{(j)}\right)^+\right] - r\left[\left(t_+^{(j)}\right)^-\right] = \frac{2c}{(c+2)(1+v)}[b(0^+) - \dot{e}(0)] - \frac{2cv}{(c+2)(1+v)}r(0^+) \quad (1.44)$$

Two cases of continuity in $r(t)$ are evident from Eqs. (1.43) and (1.44). If the suspension damping is absent, $c = 0$ and $r(t)$ is continuous. If the initial states satisfy

$$(1+cv)b(0^+) - \dot{e}(0) + cv^2 a'(0^+) = v\left[ky_0 + c\dot{y}_0 + \frac{cwv}{2(1-v^2)}\right] \quad (1.45)$$

then $r(t)$ is continuous since the discontinuities in $r(t)$ due to the initial conditions of the string and the payload mass cancel each other. Satisfaction of Eq. (1.45) indicates the propagating wavefronts in the string have vanishing slopes.

1.2.2.3 Delay-integro-differential Equation

A direct numerical solution of Eq. (1.29) is not adopted because the discontinuous kernel (1.30) causes oscillatory solutions. As shown in Wickert and Mote (1991), differentiation of Eq. (1.29) eliminates the discontinuity in the kernel. In the present case, however, the dependent variable $r(t)$ in Eq. (1.29) is also discontinuous in general. Hence, the delta functions arise from differentiating Eq. (1.29) distributionally due to the discontinuities of $p(t)$, $r(t)$, and the Heaviside functions at $t_-^{(j)}$ and $t_+^{(j)}$. By dropping the delta function terms, the resulting delay-integro-differential equation, which is valid for all t except at the finite number of $t_-^{(j)}$ and $t_+^{(j)}$, is

$$\begin{aligned}\frac{d}{dt}r(t) &= \frac{d}{dt}p(t) - \frac{c}{m}r(t) - \frac{k}{m}\int_0^t r(\tau)\,d\tau - \frac{k}{2}\sum_{j=1}^{N}\left[r\left(t - \frac{2(j-1)}{1-v}\right)\right.\\ &\quad \left. - \frac{1-v}{1+v}r\left(\frac{1-v}{1+v}t - \frac{2(j-1)}{1+v}\right)\right]H\left(t - \frac{2(j-1)}{1-v}\right)\\ &\quad - \frac{k}{2}\sum_{j=1}^{N}\left[r\left(t - \frac{2j}{1+v}\right) - \frac{1+v}{1-v}r\left(\frac{1+v}{1-v}t - \frac{2j}{1-v}\right)\right]\\ &\quad \times H\left(t - \frac{2j}{1+v}\right) - \frac{c}{2}\sum_{j=1}^{N}\left[\frac{d}{dt}r\left(t - \frac{2(j-1)}{1-v}\right) - \frac{1-v}{1+v}\right.\\ &\quad \left. \times \frac{d}{dt}r\left(\frac{1-v}{1+v}t - \frac{2(j-1)}{1+v}\right)\right]H\left(t - \frac{2(j-1)}{1-v}\right)\\ &\quad - \frac{c}{2}\sum_{j=1}^{N}\left[\frac{d}{dt}r\left(t - \frac{2j}{1+v}\right) - \frac{1+v}{1-v}\right.\\ &\quad \left. \times \frac{d}{dt}r\left(\frac{1+v}{1-v}t - \frac{2j}{1-v}\right)\right]H\left(t - \frac{2j}{1+v}\right), \end{aligned} \tag{1.46}$$

where d/dt denotes the distributional derivative without regard to the discontinuities at $t_-^{(j)}$ and $t_+^{(j)}$ (Stakgold, 1979). At $t_-^{(j)}$ and $t_+^{(j)}$, Eq. (1.46) is supplemented with the discontinuity relations Eqs. (1.43) and (1.44). It has been shown that all terms in $p(t)$ are continuous at $t_-^{(j)}$ and $t_+^{(j)}$ except $I_1(t)$. Hence the contributions to the delta function from differentiation of $p(t)$ distributionally arise from $I_1(t)$. The term $dp(t)/dt$ is obtained from Eq. (1.26) by Leibnitz's rule and the use of $g(vt, \xi; 0) = 0$:

$$\begin{aligned}\frac{d}{dt}p(t) &= k\dot{y}_0 - cw + \frac{kw(v-2t)}{2(1-v^2)} - \frac{cwv^2}{1-v^2} + kve(t)\\ &\quad + [-k(1-vt) + 2cv]\dot{e}(t) - c(1-vt)\ddot{e}(t)\\ &\quad + \frac{k}{m}\int_0^t q(\tau)\,d\tau + \frac{c}{m}q(\tau) - \int_0^t\!\!\int_0^1 [f(\xi,\tau) - (1-\xi)\ddot{e}(\tau)\\ &\quad + 2v\dot{e}(\tau)]k\frac{\partial}{\partial t}g(vt,\xi;t-\tau)\,d\xi\,d\tau - \frac{d}{dt}\int_0^t\!\!\int_0^1 [f(\xi,\tau)\end{aligned}$$

$$- (1 - \xi)\ddot{e}(\tau) + 2v\dot{e}(\tau)]c\frac{\partial}{\partial t}g(vt, \xi; t - \tau)\, d\xi\, d\tau$$
$$- \int_0^1 [b(\xi) - (1 - \xi)\dot{e}(0)]k\frac{\partial}{\partial t}g(vt, \xi; t)\, d\xi + \frac{d}{dt}I_1(t)$$
$$- \int_0^1 [a(\xi) - (1 - \xi)e(0)]\left(k\frac{\partial}{\partial t} + c\frac{\partial^2}{\partial t^2}\right)[g_t(vt, \xi; t)$$
$$- 2vg_\xi(vt, \xi; t)]\, d\xi. \tag{1.47}$$

The distributional derivative of $I_1(t)$, neglecting the delta functions at $t_-^{(j)}$ and $t_+^{(j)}$, is

$$\frac{d}{dt}I_1(t) = -\frac{c}{2}\Bigg\{\sum_{l_1 \le i \le r_1} coef_1(i)\left[b'\left(-t + \frac{2i}{1 - v}\right) + \dot{e}(0)\right] + \left(\frac{1 - v}{1 + v}\right)^2$$
$$\times \sum_{l_2 \le i \le r_2} coef_2(i)\left[b'\left(\frac{1 - v}{1 + v}t - \frac{2i}{1 + v}\right) + \dot{e}(0)\right]$$
$$- \sum_{l_3 \le i \le r_3} coef_3(i)\left[b'\left(t - \frac{2i}{1 + v}\right) + \dot{e}(0)\right] - \left(\frac{1 + v}{1 - v}\right)^2$$
$$\times \sum_{l_4 \le i \le r_4} coef_4(i)\left[b'\left(-\frac{1 + v}{1 - v}t + \frac{2i}{1 - v}\right) + \dot{e}(0)\right]\Bigg\}. \tag{1.48}$$

The term $\int_0^1 [b(\xi) - (1 - \xi)\dot{e}(0)]k(\partial/\partial t)g(vt, \xi; t)\, d\xi$ in Eq. (1.47) is obtained from Eq. (1.31) by replacing c by k. The last term in Eq. (1.47) is determined either by numerically differentiating $I_2(t)$, which is obtained from Eq. (1.32), or by following the method for Eq. (1.32) with integration by parts twice and noting that the singular boundary terms vanish because $I_2(t)$ is continuous. A computational algorithm for

$$I_3(t) = \int_0^t \int_0^1 [f(\xi, \tau) - (1 - \xi)\ddot{e}(\tau) + 2v\dot{e}(\tau)]k\frac{\partial}{\partial t}g(vt, \xi; t - \tau)\, d\xi\, d\tau \tag{1.49}$$

is given in Section 1.2.3. The term

$$I_4(t) = \frac{d}{dt}\int_0^t \int_0^1 [f(\xi, \tau) - (1 - \xi)\ddot{e}(\tau) + 2v\dot{e}(\tau)]c\frac{\partial}{\partial t}g(vt, \xi; t - \tau)\, d\xi\, d\tau$$

can be computed in two ways. The easier, which is used in the example in Section 1.2.4, is to evaluate the double integral by replacing k in $I_3(t)$ by c and replacing the derivative by a difference operator. The other approach uses Leibnitz's rule with the initial values of the Green's function, $g_t(vt, \xi; 0) = \delta(vt - \xi)$ and

$g_x(vt, \xi; 0) = 0$, giving

$$I_4(t) = cf(vt, t) - c(1 - vt)\ddot{e}(t) + 2cv\dot{e}(t)$$
$$+ \int_0^t \int_0^1 [f(\xi, \tau) - (1 - \xi)\ddot{e}(\tau) + 2v\dot{e}(\tau)]c \frac{\partial^2}{\partial t^2} g(vt, \xi; t - \tau)\, d\xi\, d\tau. \tag{1.50}$$

The last term in Eq. (1.50) can be calculated either by the series expansion of the Green's function (1.14) or by computing the double integral with derivatives of delta functions similar to the algorithm given in Section 1.2.3 for $I_3(t)$.

1.2.2.4 Wave–payload Interaction

We demonstrate the discontinuities in the interaction force by examination of propagating waves in a string transporting a payload.

The linear transverse vibration of the subcritical speed axially moving string is characterized by forward waves propagating downstream at speed $1 + v$ and backward waves propagating upstream at speed $1 - v$. For this nonuniform string, the incident wave is partly reflected from, and partly transmitted through, the discontinuity in the string from the attached payload. The payload enters the span ($x = 0$) at time $t = 0$ and the initial disturbance in the string originates at $x = 0$. The instants at which the propagating wavefronts interact with the payload are predicted from the characteristic lines in the t–x plane in Fig. 1.3. The reflected and transmitted waveforms are functions of m, k, and c, but the times of interaction are independent of the payload parameters.

An initial disturbance from the left end propagating downstream at $t = 0$ is reflected by the right support and then interacts with the payload at $t_-^{(1)}$. A branch of the *transmitted wave component* passes through the payload and subsequent interactions occur at $t_-^{(j)}$ ($j = 2, 3, \ldots, N$) and $t_+^{(j)}$ ($j = 1, 2, \ldots, N - 1$), where the subscripts denote the direction of the transmitted wave: $+$ downstream and $-$ upstream relative to the string velocity. N is the number of interactions of the downstream and upstream transmitted waves with the payload. A branch of the *reflected wave component* also reflects from the payload at subsequent interactions occurring at an infinite number of instants $t_r^{(n)}$:

$$t_r^{(n)} = \frac{2}{1+v}, \frac{4}{(1+v)^2}, \frac{2v^2+6}{(1+v)^3}, \ldots, \left[1 - \left(\frac{1-v}{1+v}\right)^n\right]\frac{1}{v}, \ldots \tag{1.51}$$

for $n \geq 1$. For $1/3 < v < 1$, the aforementioned times include all wave–payload interactions. For $\sqrt{5} - 2 < v < 1/3$, an additional branch of the transmitted wave component transmits through the payload at the second wave–payload interaction and then reflects from the payload in all subsequent interactions at an infinite number of instants $t_R^{(n)}$ with $t_R^{(1)} = t_-^{(2)}$ and

$$t_R^{(n)} = \frac{2(3-v)}{(1+v)^2}, \frac{4(v-2)}{(1+v)^2}, \ldots, \left[1 - \frac{1-3v}{1+v}\left(\frac{1-v}{1+v}\right)^{n-1}\right]\frac{1}{v}, \ldots, v < \frac{1}{3} \tag{1.52}$$

for $n \geq 2$. For $1/5 < v < \sqrt{5} - 2$, an additional branch of the transmitted wave component reflects from the payload at the next interaction and can interact with the payload again at $t_{+1} = 2(1+v)/(1-v)^2$. There is also a branch of the reflected wave component that is transmitted through the payload at the next wave–payload interaction. It interacts with the payload again at $t_{r2} = 4/(1-v^2)$ after reflection from the $x = 0$ support. A branch of each transmitted component of these two waves can also continue to reflect from the payload at an infinite number of instants afterwards. For $v < 1/5$, additional wave–payload interactions can be determined similarly.

The first discontinuity of $r(t)$ at $t_-^{(1)}$ is given by Eq. (1.43). The discontinuities of $r(t)$ at $t_-^{(j)}$ ($j \geq 2$) and $t_+^{(j)}$ are not those given by Eq. (1.43) and Eq. (1.44) because additional discontinuities are implicitly accounted for by the delay terms. For example, at $t_+^{(1)}$, the second delay term in Eq. (1.29) for $j = 1$ causes an additional discontinuity in $r(t)$ because r is discontinuous at $(1-v)t_+^{(1)}/(1+v) = t_-^{(1)}$. Observing that

$$\frac{1+v}{1-v}t_r^{(n)} - \frac{2}{1-v} = t_r^{(n-1)}, \quad \text{with } t_r^{(1)} = t_-^{(1)}, \tag{1.53}$$

and using Eq. (1.29), the discontinuities of $r(t)$ at $t_r^{(n)}$ are derived to be

$$r\left[\left(t_r^{(n)}\right)^+\right] - r\left[\left(t_r^{(n)}\right)^-\right] = \left[\frac{c(1+v)}{(c+2)(1-v)}\right]^{n-1} \times \left\{r\left[\left(t_-^{(1)}\right)^+\right] - r\left[\left(t_-^{(1)}\right)^-\right]\right\}, \quad n \geq 1. \tag{1.54}$$

Likewise, the discontinuities of $r(t)$ at $t_R^{(n)}$ are

$$r\left[\left(t_R^{(n)}\right)^+\right] - r\left[\left(t_R^{(n)}\right)^-\right] = \left[\frac{c(1+v)}{(c+2)(1-v)}\right]^{n-1} \times \left\{r\left[\left(t_-^{(2)}\right)^+\right] - r\left[\left(t_-^{(2)}\right)^-\right]\right\}, \quad n \geq 1. \tag{1.55}$$

From the relations

$$\frac{1-v}{1+v}t_{r2} = t_r^{(2)}, \qquad \frac{1-v}{1+v}t_{+1} = t_+^{(1)} \tag{1.56}$$

and the second delay term in Eq. (1.29) for $j = 1$, we conclude that $r(t)$ is discontinuous at t_{r2} and t_{+1}. Discontinuities of $r(t)$ at t_{r2} and t_{+1} and other instants of wave–payload interaction for slower transport speed can be obtained similarly. In short, the discontinuities in the interaction force occur at all instants when propagating waves interact with the payload if the suspension damping of the payload is nonzero and if the initial state does not satisfy Eq. (1.45).

1.2.3 Numerical Solution

Equation (1.46), in conjunction with the discontinuity conditions (1.43) and (1.44), is solved numerically for $r(t)$ by finite difference. The algorithm of Linz (1985) for Volterra equations of the second kind is extended for solution of the delay-integro-differential equation (1.46) with a discontinuous dependent variable. The time of transit $[0, 1/v]$ is divided uniformly by $n_p + 1$ points with step size $\Delta t = 1/vn_p$. The backward finite difference is used to approximate the derivative of $r(t)$ and trapezoidal integration replaces the integral in Eq. (1.46). The delay terms are evaluated by linear interpolation. Finally, the time derivatives of the delay terms are approximated by linear interpolations and then by backward finite differences. Given dp/dt in Eq. (1.47), the resulting delay-difference equation is solved recursively for $r(t)$ from the initial value in Eq. (1.41). At each $t_-^{(j)}$ or $t_+^{(j)}$, the discontinuity of $r(t)$ given by Eqs. (1.43) or (1.44) is included. The discontinuities at other times are implicitly accounted for by the delay terms.

Integration of $I_3(t)$ in Eq. (1.49) in the ξ variable yields

$$
\begin{aligned}
I_3(t) = \frac{k}{2}\int_0^t d\tau \Bigg\{ & \sum_{L_1 \le i \le R_1} \left[f\left(-t + (1+v)\tau + \frac{2i}{1-v}, \tau\right) \right. \\
& \left. - \left(1 + t - (1+v)\tau - \frac{2i}{1-v}\right)\ddot{e}(\tau) + 2v\dot{e}(\tau) \right] \\
& - \frac{1-v}{1+v} \sum_{L_2 \le i \le R_2} \left[f\left(\frac{1-v}{1+v}t - (1-v)\tau - \frac{2i}{1+v}, \tau\right) \right. \\
& \left. - \left(1 - \frac{1-v}{1+v}t + (1-v)\tau + \frac{2i}{1+v}\right)\ddot{e}(\tau) + 2v\dot{e}(\tau) \right] \\
& + \sum_{L_3 \le i \le R_3} \left[f\left(t - (1-v)\tau - \frac{2i}{1+v}, \tau\right) \right. \\
& \left. - \left(1 - t + (1-v)\tau + \frac{2i}{1+v}\right)\ddot{e}(\tau) + 2v\dot{e}(\tau) \right] \\
& - \frac{1+v}{1-v} \sum_{L_4 \le i \le R_4} \left[f\left(-\frac{1+v}{1-v}t + (1+v)\tau + \frac{2i}{1-v}, \tau\right) \right. \\
& \left. - \left(1 + \frac{1+v}{1-v}t - (1+v)\tau - \frac{2i}{1-v}\right)\ddot{e}(\tau) + 2v\dot{e}(\tau) \right] \Bigg\}, \qquad (1.57)
\end{aligned}
$$

where

$$L_1 = \frac{1-v}{2}[t-(1+v)\tau], \qquad R_1 = \frac{1-v}{2}[1+t-(1+v)\tau],$$
$$L_2 = \frac{1-v}{2}t - \frac{1+v}{2}[1+(1-v)\tau], \qquad R_2 = \frac{1-v}{2}t - \frac{1+v}{2}(1-v)\tau,$$
$$L_3 = \frac{1+v}{2}[-1+t-(1-v)\tau], \qquad R_3 = \frac{1+v}{2}[t-(1-v)\tau],$$
$$L_4 = \frac{1+v}{2}t - \frac{1-v}{2}(1+v)\tau, \qquad R_4 = \frac{1+v}{2}t + \frac{1-v}{2}[1-(1+v)\tau]. \tag{1.58}$$

Note from Eq. (1.58) that for each of the four summations in Eq. (1.57) there exists at most one integer i for $0 < v < 1$. A difficulty in evaluating the right side of Eq. (1.57) occurs because the integer i in each of the four summations is dependent on τ for specified t. The methodology used divides the domain of the integration $[0, t]$ for each summation into several subintervals, each with or without a specific integer i, so that exact integration can be performed. The separation points of these subintervals are found by noting that the corresponding integers i equal either L_m or R_m there for every $m = 1, 2, 3, 4$. The flow diagram for the algorithm is shown in Fig. 1.4. If we integrate $I_3(t)$ in the τ

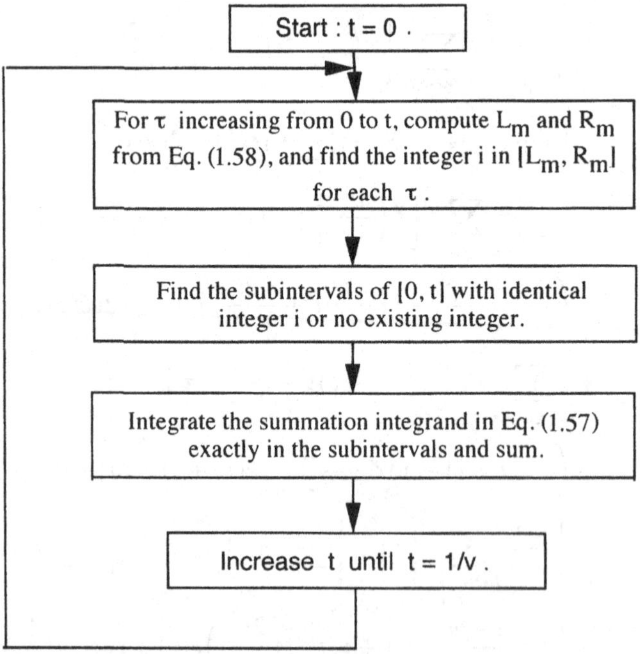

Fig. 1.4. Flowchart for integrating each of the four summation integrands in Eq. (1.57).

variable first, the number of integers in each summation can be more than one. The algorithm is essentially the same and the result obtained is identical.

Once the interaction force $r(t)$ is determined, the system response can be computed. The payload displacement and velocity are obtained from Eqs. (1.22) and (1.23), where the integrals with $r(t)$ are calculated by trapezoidal integration. The string displacement is given by Eqs. (1.9) and (1.13). The displacement and velocity of the payload attachment point are obtained from Eqs. (1.8), (1.9), (1.13), (1.20), and (1.21). The velocity $\dot{z}(t)$ is

$$\begin{aligned}\dot{z}(t) &= (1-vt)\dot{e}(t) - ve(t) - \frac{wv(1-2vt)}{2(1-v^2)}\\ &+ \int_0^t\int_0^1 [f(\xi,\tau) - (1-\xi)\ddot{e}(\tau) + 2v\dot{e}(\tau)]\frac{\partial}{\partial t}g(vt,\xi;t-\tau)\,d\xi\,d\tau\\ &+\frac{1}{2}\sum_{j=1}^{N}\left[r\left(t-\frac{2(j-1)}{1-v}\right) - \frac{1-v}{1+v}r\left(\frac{1-v}{1+v}t - \frac{2(j-1)}{1+v}\right)\right]\\ &\times H\left(t-\frac{2(j-1)}{1-v}\right) + \frac{1}{2}\sum_{j=1}^{N}\left[r\left(t-\frac{2j}{1+v}\right)\right.\\ &\left.-\frac{1+v}{1-v}r\left(\frac{1+v}{1-v}t-\frac{2j}{1-v}\right)\right]H\left(t-\frac{2j}{1+v}\right)\\ &+\int_0^1[b(\xi)-(1-\xi)\dot{e}(0)]\frac{\partial}{\partial t}g(vt,\xi;t)\,d\xi\\ &+\int_0^1[a(\xi)-(1-\xi)e(0)]\frac{\partial}{\partial t}[g_t(vt,\xi;t)-2vg_\xi(vt,\xi;t)]\,d\xi,\end{aligned} \tag{1.59}$$

where the double integral and two single integrals are calculated from Eqs. (1.49), (1.31), and (1.32), respectively, by assigning values of k and/or c there. The initial value $\dot{z}(0)$ is

$$\dot{z}(0^+) = \frac{1}{1+v}\dot{e}(0) - \frac{wv}{2(1-v^2)} + \frac{v}{1+v}r(0^+) + \frac{v}{1+v}b(0^+) + va'(0^+) \tag{1.60}$$

by Eqs. (1.59) and (1.33), where $r(0^+)$ is given by Eq. (1.41). The displacement $z(t)$ can be predicted by numerical integration of $\dot{z}(t)$ with $z(0) = e(0)$. The velocity $\dot{z}(t)$ has discontinuities at all the instants when propagating wavefronts interact with the payload following analysis of $r(t)$.

1.2.4 Examples and Discussion

The method developed is illustrated on a sample problem. The transport system is excited sinusoidally, $e(t) = e_0 \sin \Omega t$, with $\Omega = 10$. The initial conditions of

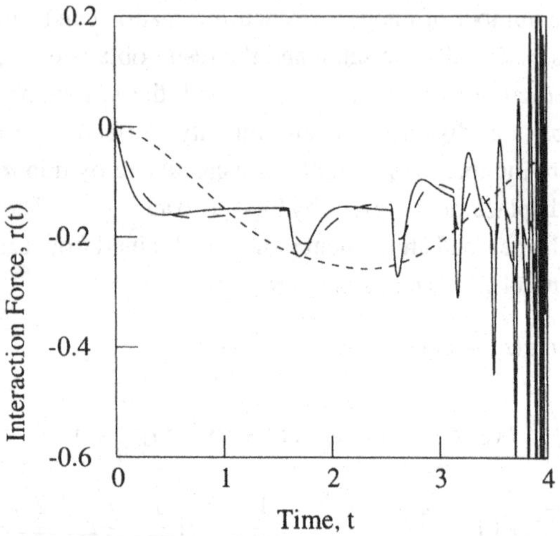

Fig. 1.5. Interaction forces in the free vibration problem during the transit of the payload across the span for different payload suspension parameters with $n_p = 1{,}000$. —, $k = 600$, $c = 0$; – – –, $k = 10$, $c = 5$; - - -, $k = 1$, $c = 0.1$.

the string are $a(x) = 0$ and $b(x) = 0$. The external forces are $f(x, t) = 0$ and $q(t) = 0$. The payload mass is $m = 3/8$, and $w = 3/8$. The transport speed is $v = 0.25$; hence $N = 2$. Since $\sqrt{5} - 2 < v < 1/3$, $t_-^{(j)}$ $(j = 1, 2)$, $t_+^{(1)}$, $t_r^{(n)}$ $(n = 2, 3, \ldots)$, and $t_R^{(n)}$ $(n = 2, 3, \ldots)$ include all instants of wave–payload interaction.

The force $r(t)$ in the free vibration problem ($e_0 = 0$) for different suspension stiffnesses and damping coefficients is shown in Fig. 1.5 for the initial conditions of the payload mass identical to those of the payload attachment point on the string (i.e., $y_0 = z(0) = 0$, $\dot{y}_0 = \dot{z}(0) = -wv/2(1-v^2)$). The initial conditions satisfy Eq. (1.45) so that the discontinuities in the interaction force are absent. As expected, $r(t)$ for large stiffness (solid line) approaches the prediction for the string–particle system (Wickert and Mote, 1991). The magnitude of the changes in $r(t)$ at the wave–payload interactions depend on the magnitudes of k and c.

As the payload approaches the exit from the span, there exist infinitely many wave reflections from the payload with decreasing intervals between successive interactions, as given by Eqs. (1.51) and (1.52). A remarkably fine temporal finite difference mesh is required to accurately predict $r(t)$ as t approaches $1/v$, and $r(t)$ cannot be determined at the limiting time $1/v$. A similar singularity is reported by Smith (1964) and in Wickert and Mote (1991). However, the $r(t)$ near the right end has a small effect on the string displacement because of its closeness to the fixed support, the small impulse associated with it due to the diminishing energy in the reflected wave, and the alternating direction of successive impulses.

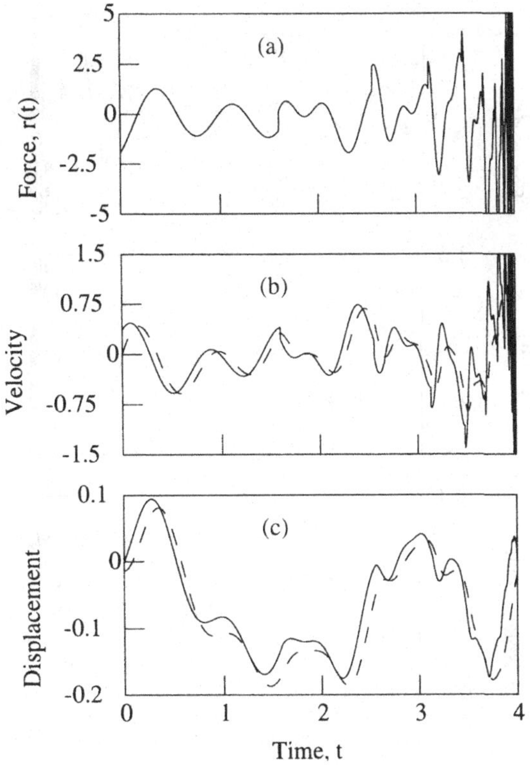

Fig. 1.6. The interaction force and response for $k = 10$, $c = 5$, $e_0 = 0.1$, $\Omega = 10$, $y = -mw/k$, $\dot{y}_0 = 0$, and $n_p = 1{,}000$. (*a*) Interaction force $r(t)$; (*b*) velocity (——, $\dot{z}(t)$; - - -, $\dot{y}(t)$); (*c*) displacement (——, $z(t)$; - - -, $y(t)$).

The functions $r(t)$, $y(t)$, $z(t)$, $\dot{y}(t)$, and $\dot{z}(t)$ with $e_0 = 0.1$ are shown in Fig. 1.6. The initial displacement of the payload mass is the static deflection of k and the initial velocity is zero. Because $c > 0$ and Eq. (1.45) is not satisfied, $r(t)$ is discontinuous. As given by Eq. (1.43), the first jump in $r(t)$ occurring at $t_-^{(1)} = 1.6$ is 0.978. The jumps of $r(t)$ at $t_+^{(1)} = 2.67$ and $t_-^{(2)} = 3.2$ are not those given by Eqs. (1.44) and (1.43) because of the additional discontinuities accounted for by the delay terms. The positive jumps of $r(t)$ at $t_r^{(n)}$ (for $n \geq 2$) $= 2.56, 3.136, 3.4816$, etc., which are implicitly accommodated by the delay terms, are also observed. These jump amplitudes agree with the analytical prediction, Eq. (1.54). The jump amplitudes of $r(t)$ at these reflection instants $t_r^{(n)}$ $(n \geq 2)$ are slowly growing from 0.978 at $t_r^{(1)}$, since $c(1+v)/(c+2)(1-v) \approx 1.19 > 1$ here. The jumps at $t_R^{(n)}$ $(n \geq 1) = 3.2, 3.52, 3.712, 3.8272$, etc. are not captured by the finite difference mesh used.

The velocity history of the attachment point $\dot{z}(t)$, shown in Fig. 1.6(b), is computed from Eqs. (1.59) and (1.60), once $r(t)$ is known. The corresponding

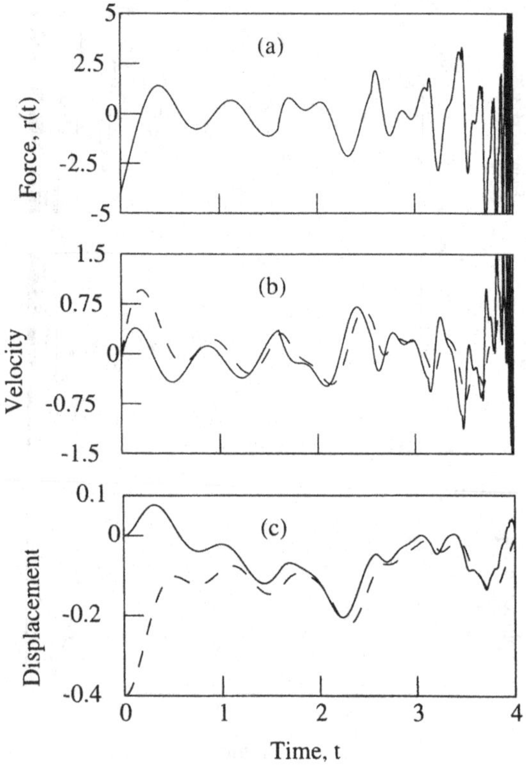

Fig. 1.7. The interaction force and response for the same parameters used in Fig.1.6 except that $y_0 = -e_0\omega/kv$ and $\dot{y}_0 = -vw/2(1 - v^2)$ such that the initial states satisfy Eq. (1.45). (*a*) Interaction force $r(t)$; (*b*) velocity (——, $\dot{z}(t)$; – – –, $\dot{y}(t)$); (*c*) displacement (——, $z(t)$; – – –, $y(t)$).

discontinuities in $\dot{z}(t)$ occur at the wave–payload interactions. The velocity of the attachment point changes rapidly and dramatically as the payload approaches the exit to the span because of the alternating impulses from $r(t)$. Consequently, "ripples" in $z(t)$ are seen as the payload approaches the right support.

In Fig. 1.7, the initial conditions of the payload mass are $y_0 = -e_0\omega/kv$ and $\dot{y}_0 = -vw/2(1 - v^2)$, satisfying Eq. (1.45). It is seen that the discontinuities in $r(t)$ and, as a consequence, $\dot{z}(t)$ vanish.

The forcing integrals $I_1(t)$, $dI_1(t)/dt$ (the distributional derivative of $I_1(t)$ without regard to the discontinuities at $t_-^{(j)}$ and $t_+^{(j)}$), $I_3(t)$, and $I_4(t)$, computed by the generalized function approach along with the algorithm developed, are shown in Fig. 1.8 as solid lines. The integral $I_2(t) = 0$. A series expansion of the Green's function in Eq. (1.14) using 100 terms gives the dash-line results for $I_1(t)$, $I_3(t)$, and $I_4(t)$ that are indistinguishable from the solid lines, but the series

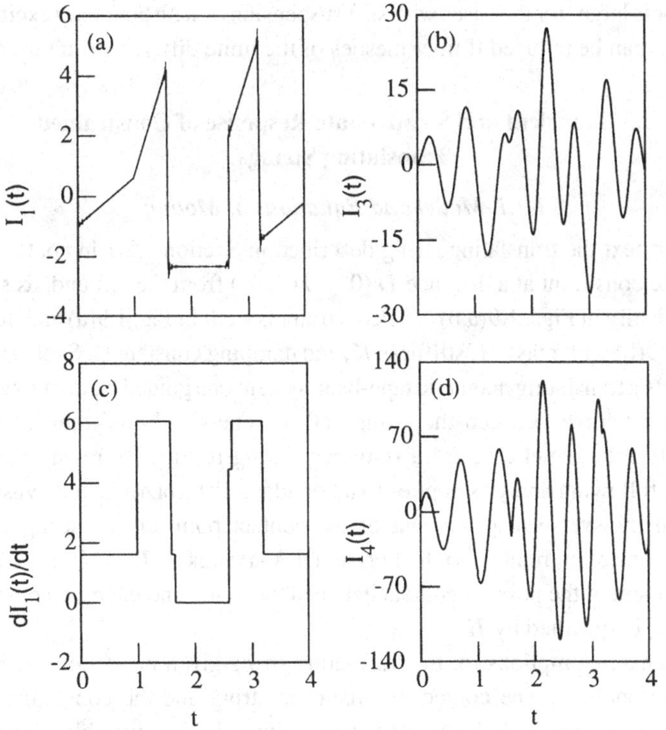

Fig. 1.8. Forcing integrals computed by the generalized function method along with the algorithm developed (——) and the series expansion method (– – –). *(a)* $I_1(t)$; *(b)* $I_3(t)$; *(c)* $dI_1(t)/dt$ (the series expansion method for $dI_1(t)/dt$ is divergent and not shown in the figure); *(d)* $I_4(t)$.

in $dI_1(t)/dt$ becomes divergent (not shown in the figure). Although the series expansion method displays some Gibb's phenomenon at the discontinuities of $I_1(t)$ (Stakgold, 1979), it gives results virtually identical to the generalized function approach for the continuous functions $I_3(t)$ and $I_4(t)$ within the scale of the plots except for tiny wiggles in $I_4(t)$ at small t.

The accuracy of the numerical solutions is evaluated by the error in satisfying Eq. (1.7) using the computed $r(t)$, $z(t)$, $y(t)$, $\dot{z}(t)$, and $\dot{y}(t)$. The error measure, defined by

$$\text{error} = \frac{\|r(t) - k(y(t) - z(t)) - c(\dot{y}(t) - \dot{z}(t))\|}{\|r(t)\|} \times 100\% \tag{1.61}$$

is calculated, where $\|\cdot\|$ is the L_2-norm evaluated from $t = 0$ to $t = 0.99/v$. With $n_p = 1{,}000$, the error in Fig. 1.5 is 1.5% (solid line), 0.17% (long dashed line), and 0.0002% (short dashed line); 0.67% in Fig. 1.6; and 1.4% in Fig. 1.7.

The error is larger for the cases with stiff suspension or with boundary excitation. The error can be reduced if finer meshes of the finite differences are used.

1.3 Transient and Steady-state Response of Constrained Translating Strings

1.3.1 Model and Equations of Motion

Consider next the translating string described in Section 1.2.1 in contact with a discrete constraint at a distance D ($0 < D < L$) from the left end, as shown schematically in Figs. 1.9(a,b). The constraint is rigid in Fig. 1.9(a) and flexible in Fig. 1.9(b) with mass M, stiffness K, and damping constant C. Such a model represents a translating magnetic tape-head system or a guided band saw system. The contact force between the string and the one-sided constraint is $R(T)$. At $V = 0$, the special case for a stationary string results. By replacing V by $-V$, the following analysis applies to a boundary disturbance at a downstream end. The transverse displacement of the contact point on the string with a flexible constraint, relative to the horizontal X-axis, is $Z(T) = U(D,T)$. The displacement of the point of contact between the string and the rigid constraint, $U(D,T)$, is specified by H.

The same assumptions for the translating string given in Section 1.2.1 constrain the analysis. The contact between the string and the constraint is assumed frictionless. With the corresponding dimensionless variables defined in Eq. (1.1), we introduce two additional dimensionless parameters: $d = D/L$ and

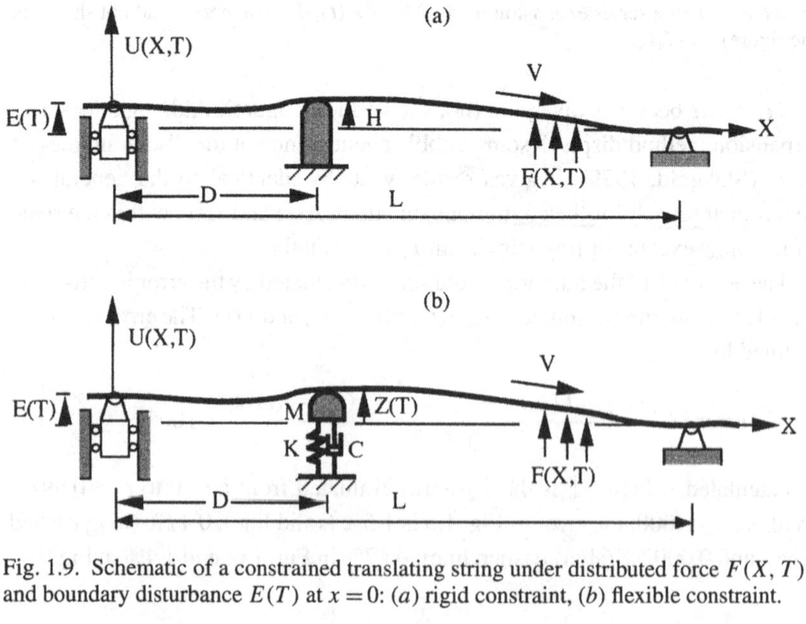

Fig. 1.9. Schematic of a constrained translating string under distributed force $F(X,T)$ and boundary disturbance $E(T)$ at $x = 0$: (*a*) rigid constraint, (*b*) flexible constraint.

$h = H/L$. The equation governing transverse motion of the translating string is

$$u_{tt}(x,t) + 2vu_{xt}(x,t) + (v^2 - 1)u_{xx}(x,t) = r(t)\delta(x-d) + f(x,t) - w \tag{1.62}$$

with the boundary and initial conditions given by

$$u(0,t) = e(t), \qquad u(1,t) = 0, \quad 0 \le t \le t_{\mathrm{f}}, \tag{1.63}$$

$$u(x,0) = a(x), \qquad u_t(x,0) = 0, \quad 0 < x < 1, \tag{1.64}$$

where $a(x)$ is the initial displacement of the string and t_{f} is an arbitrarily specified final time ($0 < t_{\mathrm{f}} < \infty$). Continuity of the initial displacement of the string at the left end $x = 0$ (i.e., $e(0) = a(0^+)$), limits the analysis to continuous displacement $u(x,t)$. The initial velocity of the string at $x = 0$, however, may be discontinuous (i.e., $\dot{e}(0) \neq 0$), the consequence of which is shown in what follows. The equation of motion and initial conditions for the flexible constraint in contact with the string are

$$-r(t) + k[z_0 + z^* - z(t)] - c\dot{z}(t) - mw = m\ddot{z}(t), \tag{1.65}$$

$$z(0) = a(d), \qquad \dot{z}(0) = 0, \tag{1.66}$$

where z^* is the equilibrium displacement of the constraint mass m to be determined, and z_0 is the compression of the linear spring k at equilibrium. The contact conditions are

$$u(d,t) = h \text{ for a rigid constraint}; \quad u(d,t) = z(t) \text{ for a flexible constraint.} \tag{1.67}$$

At equilibrium, the string displacement $u^*(x)$ and the static preload r^* are derived from the static counterparts to Eqs. (1.62), (1.63), (1.65), and (1.67):

$$(v^2 - 1)u^*_{xx}(x) = r^*\delta(x-d) + f^*(x) - w, \tag{1.68}$$

$$u^*(0) = e(0), \qquad u^*(1) = 0, \tag{1.69}$$

where

$$r^* = kz_0 - mw, \qquad z^* = u^*(d) \text{ for a flexible constraint,} \tag{1.70a}$$

$$u^*(d) = h \text{ for a rigid constraint,} \tag{1.70b}$$

and $f^*(x)$ is the static part of $f(x,t)$. The equilibrium solution for $f^*(x) = 0$ and $e(0) = 0$ is

$$u^*(x) = \begin{cases} wx^2/2(1-v^2) + Ax, & 0 < x < d, \\ w(1-x)^2/2(1-v^2) + B(1-x), & d < x < 1, \end{cases} \tag{1.71}$$

where $A = h/d - wd/2(1-v^2)$, $B = h/(1-d) - w(1-d)/2(1-v^2)$ for the rigid constraint and $A = -w/2(1-v^2) + (1-d)(kz_0 - mw)/(1-v^2)$,

$B = d(kz_0 - mw - w/2d)/(1 - v^2)$ for the flexible constraint. In addition $r^* = w/2+(1-v^2)h/d(1-d)$ for the rigid constraint; r^* is given by Eq. (1.70a) and $z^* = d(1-d)(kz_0 - mw - w/2)/(1 - v^2)$ for the flexible constraint.

Substitution of $u(x,t) = u^*(x)+\tilde{u}(x,t)$, $z(t) = z^*+\tilde{z}(t)$, $r(t) = r^*+\tilde{r}(t)$, and $f(x,t) = f^*(x)+\tilde{f}(x,t)$ into Eqs. (1.62)–(1.66), and use of Eqs. (1.68)–(1.69), yield the equations describing small-amplitude motion of the string and the flexible constraint around the system equilibrium:

$$\tilde{u}_{tt}(x,t) + 2v\tilde{u}_{xt}(x,t) + (v^2-1)\tilde{u}_{xx}(x,t) = \tilde{r}(t)\delta(x-d) + \tilde{f}(x,t), \quad (1.72)$$

$$\tilde{u}(0,t) = e(t) - e(0), \qquad \tilde{u}(1,t) = 0, \quad 0 \le t \le t_f, \quad (1.73)$$

$$\tilde{u}(x,0) = a(x) - u^*(x), \qquad \tilde{u}_t(x,0) = 0, \quad 0 < x < 1, \quad (1.74)$$

$$\tilde{r}(t) = -k\tilde{z}(t) - c\dot{\tilde{z}}(t) - m\ddot{\tilde{z}}(t), \quad (1.75)$$

$$\tilde{z}(0) = a(d) - u^*(d), \qquad \dot{\tilde{z}}(0) = 0. \quad (1.76)$$

The equation of motion for a flexible massless constraint around the equilibrium is given by Eq. (1.75) when $m = 0$. The contact conditions (1.77a) and (1.77b) become

$$\tilde{u}(d,t) = 0 \text{ for a rigid constraint} \quad (1.77a)$$

and

$$\tilde{u}(d,t) = \tilde{z}(t) \text{ for a flexible constraint.} \quad (1.77b)$$

1.3.2 Transient Response

1.3.2.1 Method of Analysis

The transient wave propagation problem in Eqs. (1.72)–(1.77b) is studied, without loss of generality, for a string initially at rest in its equilibrium configuration (i.e., $a(x) = u^*(x)$). Equations (1.72)–(1.74) are solved by first applying the transformation $\tilde{u}(x,t) = \hat{u}(x,t) + (1-x)[e(t) - e(0)]$ to render the boundary condition Eq. (1.73) homogeneous for the new variable $\hat{u}(x,t)$. The function $\hat{u}(x,t)$ is then determined from the transformed Eqs. (1.72)–(1.74) in terms of the Green's function for a fixed-fixed translating string. The final expression for $\tilde{u}(x,t)$ is

$$\tilde{u}(x,t) = (1-x)[e(t)-e(0)] + \int_0^t\int_0^1 [\tilde{f}(\xi,\tau) - (1-\xi)\ddot{e}(\tau) + 2v\dot{e}(\tau)]$$
$$\times g(x,\xi;t-\tau)\,d\xi\,d\tau + \int_0^t \tilde{r}(\tau)g(x,d;t-\tau)\,d\tau$$
$$-\dot{e}(0)\int_0^1 (1-\xi)g(x,\xi;t)\,d\xi, \quad (1.78)$$

where the Green's function is piecewise continuous, as given by Eqs. (1.14) and (1.15) with $g(x, \xi; 0) = 0$ and $g_t(x, \xi; 0) = \delta(x - \xi)$. Its time derivative g_t is an infinite series of delta functions given by Eq. (1.19). The solution to Eqs. (1.75) and (1.76) for $m \neq 0$ and $0 \leq \zeta < 1$ is

$$\tilde{z}(t) = -\frac{1}{m\omega\sqrt{1-\zeta^2}} \int_0^t e^{-\zeta\omega(t-\tau)} \sin[\omega\sqrt{1-\zeta^2}(t-\tau)]\tilde{r}(\tau)\, d\tau, \quad (1.79)$$

where $\omega = \sqrt{k/m}$ and $\zeta = c/2\sqrt{km}$. The displacement $\tilde{z}(t)$ for the critically damped ($\zeta = 1$) and overdamped ($\zeta > 1$) models can be obtained. These expressions and Eq. (1.78) are substituted into the contact condition for a flexible constraint Eq. (1.77b) to yield a Volterra integral equation of the first kind describing the dynamic contact force $\tilde{r}(t)$ between the string and a flexible constraint with nonvanishing mass m:

$$\int_0^t [g(d, d; t-\tau) + \kappa_1(t, \tau)]\tilde{r}(\tau)\, d\tau = -p_1(t), \quad (1.80)$$

where

$$\kappa_1(t,\tau) = \begin{cases} \frac{1}{m\omega\sqrt{1-\zeta^2}} e^{-\zeta\omega(t-\tau)} \sin[\omega\sqrt{1-\zeta^2}(t-\tau)], & 0 \leq \zeta < 1, \\ \frac{1}{m} e^{-\omega(t-\tau)}(t-\tau), & \zeta = 1, \\ \frac{1}{m\omega\sqrt{\zeta^2-1}} e^{-\zeta\omega(t-\tau)} \sinh[\omega\sqrt{\zeta^2-1}(t-\tau)], & \zeta > 1, \end{cases} \quad (1.81)$$

$$p_1(t) = (1-d)[e(t) - e(0)] + \int_0^t\int_0^1 [\tilde{f}(\xi,\tau) - (1-\xi)\ddot{e}(\tau) + 2v\dot{e}(\tau)] \times g(d, \xi; t-\tau)\, d\xi\, d\tau - \dot{e}(0)\int_0^1 (1-\xi) g(d, \xi; t)\, d\xi. \quad (1.82)$$

The discontinuous $g(d, d; t-\tau)$ in the kernel of Eq. (1.80) can produce noise in its numerical solution. Because $p_1(t)$ is continuous, Eq. (1.80) can be differentiated distributionally to eliminate the discontinuity in the kernel. Integration and use of Eq. (1.19) gives

$$\int_0^t g_t(d, d; t-\tau)\tilde{r}(\tau)\, d\tau = \frac{1}{2}\tilde{r}(t) + \sum_{j=1}^{N} \tilde{r}\left(t - \frac{2j}{1-v^2}\right) H\left(t - \frac{2j}{1-v^2}\right) - \frac{1}{2}\sum_{j=1}^{N+1} \tilde{r}\left(t - \frac{2d}{1-v^2} - \frac{2(j-1)}{1-v^2}\right) \times H\left(t - \frac{2d}{1-v^2} - \frac{2(j-1)}{1-v^2}\right)$$

$$-\frac{1}{2}\sum_{j=1}^{N+1}\tilde{r}\left(t+\frac{2d}{1-v^2}-\frac{2j}{1-v^2}\right)$$
$$\times H\left(t+\frac{2d}{1-v^2}-\frac{2j}{1-v^2}\right)\equiv \mathbf{D}[\tilde{r}], \quad (1.83)$$

where $N = \text{Int}[(1-v^2)t_f/2]$ here and **D** is termed as the delay operator. Distributional differentiation of Eq. (1.80) with Eq. (1.83) and $g(d,d;0)=0$ yields the delay-integral equation

$$\int_0^t \kappa(t,\tau)\tilde{r}(\tau)\,d\tau + \mathbf{D}[\tilde{r}] = -p(t), \quad (1.84)$$

where $\kappa(t,\tau) = d\kappa_1(t,\tau)/dt$ and

$$p(t) = dp_1/dt = (1-d)\dot{e}(t) + \int_0^t\int_0^1 [\tilde{f}(\xi,\tau) - (1-\xi)\ddot{e}(\tau) + 2v\dot{e}(\tau)]$$
$$\times g_t(d,\xi;t-\tau)\,d\xi\,d\tau - \dot{e}(0)I_1(t) \quad (1.85)$$

and where $I_1(t) = \int_0^1 (1-\xi)g_t(d,\xi;t)\,d\xi$ is shown to be discontinuous in what follows. The double integral in Eq. (1.85), denoted by $I_2(t)$, is continuous if $\tilde{f}(x,t)$ is piecewise continuous and $e(t)$ is continuously differentiable. Hence $p(t)$ is discontinuous only when $\dot{e}(0) \neq 0$. By Eqs. (1.83) and (1.84), discontinuity of $p(t)$ and nonvanishing initial value $\tilde{r}(0)$ can cause discontinuity in $\tilde{r}(t)$. Here, the initial value is $\tilde{r}(0) = -2p(0) = 0$ by Eqs. (1.84) and (1.85), as $I_1(0) = 1-d$ recalling $g_t(d,\xi;0) = \delta(\xi-d)$. Consequently, $\tilde{r}(t)$ is discontinuous if and only if $\dot{e}(0) \neq 0$, as the discontinuities of the Heaviside functions in $\mathbf{D}[\tilde{r}]$ vanish because $\tilde{r}(0) = 0$.

The Volterra integral equation of the first kind, governing the dynamic contact force $\tilde{r}(t)$ between the string and a rigid constraint, is derived from Eqs. (1.77a) and (1.78):

$$\int_0^t g(d,d;t-\tau)\tilde{r}(\tau)\,d\tau = -p_1(t), \quad (1.86)$$

where $p_1(t)$ is given by Eq. (1.82). Distributional differentiation of Eq. (1.86), using Eq. (1.83), yields the delay equation describing $\tilde{r}(t)$:

$$\mathbf{D}[\tilde{r}] = -p(t), \quad (1.87)$$

where $p(t)$ is given by Eq. (1.85).

For a flexible, massless constraint, Eqs. (1.77b) and (1.78) are substituted into Eq. (1.75) with vanishing m to derive a Volterra integral equation of the second kind governing $\tilde{r}(t)$:

$$\int_0^t [kg(d,d;t-\tau) + cg_t(d,d;t-\tau)]\tilde{r}(\tau)\,d\tau + \tilde{r}(t) + p_2(t) = 0, \quad (1.88)$$

where

$$p_2(t) = \int_0^t\int_0^1 [\tilde{f}(\xi,\tau) - (1-\xi)\ddot{e}(\tau) + 2v\dot{e}(\tau)](k + c\partial/\partial t)$$
$$\times g(d,\xi;t-\tau)\,d\xi\,d\tau - k\dot{e}(0)\int_0^1 (1-\xi)g(d,\xi;t)\,d\xi$$
$$- c\dot{e}(0)I_1(t) + (1-d)[k(e(t)-e(0)) + c\dot{e}(t)]. \qquad (1.89)$$

Use of Eq. (1.83) reduces Eq. (1.88) to the delay-integral equation

$$\int_0^t kg(d,d;t-\tau)\tilde{r}(\tau)\,d\tau + c\mathbf{D}[\tilde{r}] + \tilde{r}(t) + p_2(t) = 0 \qquad (1.90)$$

with the initial value $\tilde{r}(0)=0$ given by Eqs. (1.90) and (1.89). The function $p_2(t)$ is discontinuous if and only if $c\dot{e}(0) \neq 0$. Hence by Eq. (1.90), $\tilde{r}(t)$ is discontinuous if and only if $c\dot{e}(0) \neq 0$. Unlike Eqs. (1.80) and (1.86), distributional differentiation of Eq. (1.90) for nonvanishing $c\dot{e}(0)$ can not be used to eliminate the discontinuity in the kernel because the singular delta functions arise from the discontinuities in $\tilde{r}(t)$ and $p_2(t)$. The delta function terms resulting from the Heaviside functions in $\mathbf{D}[\tilde{r}]$ after differentiation (for instance $\tilde{r}[t-2j/(1-v^2)]\delta[t-2j/(1-v^2)]$) all vanish because $\tilde{r}(0)=0$.

The strategy devised for $c\dot{e}(0) \neq 0$ is to first compute the discontinuities of $\tilde{r}(t)$ from the discontinuities of $I_1(t)$ in $p_2(t)$. By the calculus of generalized functions in Eq. (1.31),

$$I_1(t) = \frac{1+v}{2}\left(\sum_{l_1\le i\le r_1} coef_1(i)\left[1+(1+v)t-d-\frac{2i}{1-v}\right]\right.$$
$$\left.-\sum_{l_4\le i\le r_4} coef_4(i)\left[1+(1+v)t+\frac{1+v}{1-v}d-\frac{2i}{1-v}\right]\right)$$
$$+\frac{1-v}{2}\left(\sum_{l_2\le i\le r_2} coef_2(i)\left[1-(1-v)t-d+\frac{2i}{1+v}\right]\right.$$
$$\left.-\sum_{l_3\le i\le r_3} coef_3(i)\left[1-(1-v)t+\frac{1-v}{1+v}d+\frac{2i}{1+v}\right]\right), \qquad (1.91)$$

where

$$l_1 = (1-v)[(1+v)t-d]/2, \qquad r_1 = (1-v)[1+(1+v)t-d]/2,$$
$$l_2 = (1+v)[(1-v)t+d-1]/2, \qquad r_2 = (1+v)[(1-v)t+d]/2,$$
$$l_3 = -(1-v)d/2 + (1-v^2)t/2 - (1+v)/2,$$
$$r_3 = -(1-v)d/2 + (1-v^2)t/2,$$

$$l_4 = (1 - v^2)t/2 + (1 + v)d/2,$$

$$r_4 = (1 - v)/2 + (1 - v^2)t/2 + (1 + v)d/2.$$

The discontinuities of $I_1(t)$ in $0 \le t \le t_f$ occur at $t_+^{(j)} = 2(j-1)/(1-v^2) + d/(1+v)$ for $j = 1, 2, \ldots, N_+$ and $t_-^{(j)} = 2j/(1-v^2) - d/(1-v)$ for $j = 1, 2, \ldots, N_-$, where $N_+ = 1 + \mathrm{Int}[(1-v^2)t_f/2 - (1-v)d/2]$ and $N_- = \mathrm{Int}[(1-v^2)t_f/2 + (1+v)d/2]$. As shown in Fig. 1.10, $t_+^{(j)}$ and $t_-^{(j)}$ are the instants when propagating wavefronts of the main transmitted wave component in the string interact with the constraint at $x = d$ in $0 \le t \le t_f$. The discontinuity of I_1 at $t_+^{(j)}$, due to the discontinuities in the first and fourth summations in Eq. (1.91), is calculated explicitly to be -1. Likewise, the discontinuity of I_1 at $t_-^{(j)}$ due to the second and third summations in Eq. (1.91) is $+1$. Hence, the discontinuity of $\tilde{r}(t)$ due to the discontinuity of $p_2(t)$ at $t_+^{(j)}$ is obtained from Eq. (1.90):

$$\tilde{r}\left[(t_+^{(j)})^+\right] - \tilde{r}\left[(t_+^{(j)})^-\right] = -\frac{2}{c+2}\left(p_2\left[(t_+^{(j)})^+\right] - p_2\left[(t_+^{(j)})^-\right]\right) = -\frac{2c}{c+2}\dot{e}(0). \tag{1.92}$$

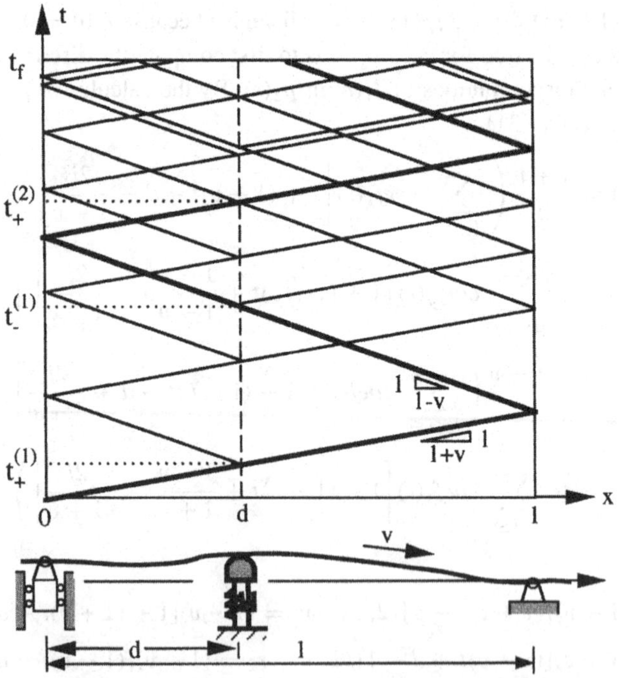

Fig. 1.10. Multiple wave scattering at a flexible constraint in the translating string. Intersections of the characteristic lines with the dashed line $x = d$ are the instants when propagating wavefronts in the string interact with the constraint. The heavy line is the main transmitted wave component.

The discontinuity of $\tilde{r}(t)$ due to the discontinuity of $p_2(t)$ at $t_-^{(j)}$ is similarly obtained:

$$\tilde{r}\left[\left(t_-^{(j)}\right)^+\right] - \tilde{r}\left[\left(t_-^{(j)}\right)^-\right] = \frac{2c}{c+2}\dot{e}(0). \tag{1.93}$$

The discontinuities in Eqs. (1.92) and (1.93) are independent of the transport speed v and vanish when $c\dot{e}(0) = 0$. In particular, the first discontinuity of $\tilde{r}(t)$ at $t_+^{(1)} = d/(1+v)$, given by Eq. (1.92), agrees with that given in Appendix A when a propagating pulse in a stationary string approaches an attached spring and damper.

Distributional differentiation of Eq. (1.90) is then performed, dropping the delta function terms associated with the discontinuities in $\tilde{r}(t)$ and $p_2(t)$ at $t_+^{(j)}$ and $t_-^{(j)}$. The resulting delay-differential equation

$$\begin{aligned} k\mathbf{D}[\tilde{r}] &+ \left(1+\frac{c}{2}\right)\frac{d}{dt}\tilde{r}(t) + c\left[\sum_{j=1}^{N}\frac{d}{dt}\tilde{r}\left(t-\frac{2j}{1-v^2}\right)H\left(t-\frac{2j}{1-v^2}\right)\right.\\ &-\frac{1}{2}\sum_{j=1}^{N+1}\frac{d}{dt}\tilde{r}\left(t-\frac{2d}{1-v^2}-\frac{2(j-1)}{1-v^2}\right)H\left(t-\frac{2d}{1-v^2}-\frac{2(j-1)}{1-v^2}\right)\\ &\left.-\frac{1}{2}\sum_{j=1}^{N+1}\frac{d}{dt}\tilde{r}\left(t+\frac{2d}{1-v^2}-\frac{2j}{1-v^2}\right)H\left(t+\frac{2d}{1-v^2}-\frac{2j}{1-v^2}\right)\right]\\ &+\frac{d}{dt}p_2(t) = 0 \end{aligned} \tag{1.94}$$

is valid at all $t \in [0, t_f]$, except at the $t_+^{(j)}$ and $t_-^{(j)}$ when $c\dot{e}(0) \neq 0$ where discontinuity in $\tilde{r}$ at those instants is supplied by Eqs. (1.92) and (1.93). Here d/dt is the distributional derivative without regard to the discontinuities at $t_+^{(j)}$ and $t_-^{(j)}$ and

$$\frac{dp_2}{dt} = (1-d)[k\dot{e}(t) + c\ddot{e}(t)] - \dot{e}(0)\left(k + c\frac{d}{dt}\right)I_1(t) + \left(k + c\frac{d}{dt}\right)I_2(t), \tag{1.95}$$

where $I_2(t)$ is the continuous double integral in Eq. (1.85). The term $dI_1(t)/dt$ is the distributional derivative of the discontinuous $I_1(t)$, upon ignoring the delta functions at $t_+^{(j)}$ and $t_-^{(j)}$, given by

$$\begin{aligned} \frac{dI_1(t)}{dt} &= \frac{(1+v)^2}{2}\left(\sum_{l_1\le i\le r_1} coef_1(i) - \sum_{l_4\le i\le r_4} coef_4(i)\right)\\ &-\frac{(1-v)^2}{2}\left(\sum_{l_2\le i\le r_2} coef_2(i) - \sum_{l_3\le i\le r_3} coef_3(i)\right). \end{aligned} \tag{1.96}$$

1.3.2.2 Numerical Solution

The delay equation (1.87), the delay-integral equation (1.84), and the delay-differential equation (1.94), along with discontinuity in $\tilde{r}$ supplied by Eqs. (1.92) and (1.93) when $c\dot{e}(0) \neq 0$, are solved numerically for $\tilde{r}(t)$. The temporal domain $[0, t_f]$ is divided into n_p equal subintervals of width $\Delta t = t_f/n_p$.

The double integral $I_2(t)$ in $p(t)$ can be evaluated either by series summation for the Green's function in Eq. (1.14) or by the delta function series in Eq. (1.19) for $g_t(x, \xi; t - \tau)$ using the numerical algorithm developed in Section 1.2.3. The term $dI_2(t)/dt$ in Eq. (1.95) is then evaluated by finite difference. The discontinuous $I_1(t)$ in Eq. (1.85) is computed by Eq. (1.91) to avoid the Gibb's phenomenon that occurs at the discontinuities of $I_1(t)$ when the truncated series in Eq. (1.14) represents the Green's function. The distributional derivative of $I_1(t)$, $dI_1(t)/dt$ in Eq. (1.95), with neglected delta functions at $t_+^{(j)}$ and $t_-^{(j)}$, is computed by Eq. (1.96); the expansion series in Eq. (1.14) is divergent for $dI_1(t)/dt$ and can not be used.

The delay terms in $\mathbf{D}[\tilde{r}]$ are represented by linear interpolation. The integral in Eq. (1.84) is replaced by trapezoidal integration (Linz, 1985) and $d\tilde{r}(t)/dt$ in Eq. (1.94) is replaced by a backward finite difference operator. The derivatives of the delay terms in Eq. (1.94) are evaluated by linear interpolation and then by backward finite differences. With the forcing terms $p(t)$ and dp_2/dt computed, the discretized delay-difference equations for Eqs. (1.84), (1.87), and (1.94) are solved recursively for $\tilde{r}(t)$ from the initial value $\tilde{r}(0) = 0$. For $c\dot{e}(0) \neq 0$, the discontinuities of $\tilde{r}(t)$, given by Eqs. (1.92) and (1.93), are accounted for at $t_+^{(j)}$ and $t_-^{(j)}$ in the recursive solution of the discretized representation of Eq. (1.94). Prediction of a negative contact force $r(t) = r^* + \tilde{r}(t)$ signals contact loss between the string and the constraint.

The string displacement $\tilde{u}(x, t)$ at time t is determined by Eq. (1.78), once $\tilde{r}(t)$ is known. The double integral is calculated by using the Green's function series in Eq. (1.14). The integral $I_3(x) = \int_0^1 (1-\xi) g(x, \xi; t)\, d\xi$ for given t and v can be integrated in closed form using Eqs. (1.14) and (1.15). For example, for $v = 1/2$ and at $t = 1$, $I_3(x)$ equals: $-2x^2/9 + x/3$ for $0 \leq x \leq 1/2$, $-20x^2/9 + 7x/3 - 1/2$ for $1/2 \leq x \leq 5/6$, and $-2x^2/9 + x - 7/9$ for $5/6 \leq x \leq 1$. Because exact integration of $I_3(x)$ requires considerable effort for different t and v, $I_3(x)$ is computed using the expansion series in Eq. (1.14). With the spatial domain divided into 501 points and 200 terms retained in Eq. (1.14) for this example, the maximum relative error of the series solution over the spatial domain is 0.73%. $I_4(x, t) = \int_0^t \tilde{r}(\tau) g(x, d; t-\tau)\, d\tau$ is obtained by integrating $dI_4(x, t)/dt$, which is computed by the generalized function approach similar to Eq. (1.91) with linear interpolation used for $\tilde{r}(t)$. I_4 can also be computed using the expansion series in Eq. (1.14).

1.3.3 Free and Steady Harmonic Forced Response

1.3.3.1 Natural Frequencies and Modes

The natural frequencies and modes of a constrained translating string around its equilibrium are the eigensolutions of Eqs. (1.72), (1.75), and (1.77b) with homogeneous boundary conditions $\tilde{u}(0,t) = \tilde{u}(1,t) = 0$. The external force $\tilde{f}(x,t)$ in Eq. (1.72) is set to zero. Assumption of a separable solution

$$\tilde{u}(x,t) = U(x)e^{\lambda t} = \begin{cases} U_1(x)e^{\lambda t}, & 0 < x < d, \\ U_2(x)e^{\lambda t}, & d < x < 1, \end{cases} \tag{1.97}$$

where $U(x)$ and λ are, in general, complex, and substitution of Eq. (1.97) into Eq. (1.72) and its homogeneous boundary conditions, yields

$$\lambda^2 U_1(x) + 2v\lambda U_1'(x) + (v^2 - 1)U_1''(x) = 0, \quad 0 < x < d, \tag{1.98a}$$

$$\lambda^2 U_2(x) + 2v\lambda U_2'(x) + (v^2 - 1)U_2''(x) = 0, \quad d < x < 1, \tag{1.98b}$$

$$U_1(0) = 0, \qquad U_2(1) = 0. \tag{1.99}$$

Integration of Eq. (1.72) from $x = d^-$ to $x = d^+$ by use of Eqs. (1.97), (1.75), and (1.77b) and continuity of the string displacement at $x = d$ gives

$$U_1(d) = U_2(d), \tag{1.100}$$

$$(v^2 - 1)\left[U_2'(d) - U_1'(d)\right] + [m\lambda^2 + c\lambda + k]U_1(d) = 0. \tag{1.101}$$

Solution to the eigenvalue problem Eqs. (1.98a)–(1.101) leads to the transcendental characteristic equation

$$(m\lambda^2 + c\lambda + k)\sinh\frac{\lambda d}{1 - v^2}\sinh\frac{\lambda(1-d)}{1 - v^2} + \lambda\sinh\frac{\lambda}{1 - v^2} = 0 \tag{1.102}$$

whose roots are the complex eigenvalues $\lambda = \mu + \omega i$, where μ and ω are real and $i = \sqrt{-1}$. The complex roots of Eq. (1.102) appear in complex conjugate pairs; that is, $\lambda_{\pm n} = \mu_n \pm \omega_n i$ $(n = 1, 2, 3, \ldots)$, where the positive ω_n, arranged in ascending order of magnitude, give the sequence of the dimensionless natural frequencies of the system. Temporal variation of the vibration amplitude for mode n is described by $\mu_n = \mathrm{Re}(\lambda_n)$, whose positive and negative values indicate flutter instability and the rate of decay, respectively. For $c = 0$, the eigenvalue is imaginary and Eq. (1.102) reduces to

$$(k - m\omega^2)\sin\frac{\omega d}{1 - v^2}\sin\frac{\omega(1-d)}{1 - v^2} + \omega\sin\frac{\omega}{1 - v^2} = 0. \tag{1.103}$$

The special case of $m = 0$ in Eq. (1.103) returns the characteristic equation for a translating string guided by a single spring (Perkins, 1990). If in addition, $k = 0$ in Eq. (1.103), the positive roots of Eq. (1.103) recover the natural frequencies of the classical moving threadline, $\omega_n = n\pi(1 - v^2)$ ($n = 1, 2, 3, \ldots$) (Archibald and Emslie, 1958).

The solid lines in Fig. 1.11 show the dependence of the first four natural frequencies on the constraint location d, obtained numerically from Eq. (1.102), for the case used in Section 1.3.4 with $k = m = 10, c = 0$, and $v = 0.0631$. The decay constant $\mu = \text{Re}(\lambda) = 0$ because $c = 0$. Two sets of natural frequencies for the classical moving threadlines of lengths d and $1-d$ (i.e., $\omega_n = n\pi(1-v^2)/d$ and $\omega_m = m\pi(1 - v^2)/(1 - d)$ for $n, m = 1, 2, 3, 4$), shown as dashed lines in Fig. 1.11, correspond to the case of a rigid constraint at $x = d$. As expected by Eq. (1.102), the frequencies are symmetric with respect to the center location $d = 1/2$. When $d = 0$ or $d = 1$, the nth natural frequency is that of the classical moving threadline: $n\pi(1 - v^2)$. Because of the presence of mass m each frequency decreases rapidly when the constraint is located slightly away from either of the two supports. Expanded views of the transition bands for the 2nd, 3rd, and 4th modes are shown on the top row of graphs in Fig. 1.11. The higher the mode number, or the larger the mass m relative to the stiffness k, the

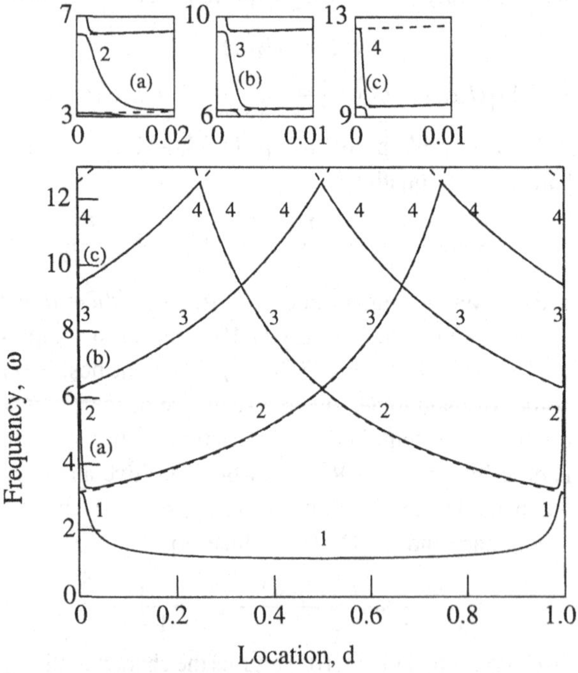

Fig. 1.11. Dependence of the natural frequencies on the location of a rigid (– – –) or flexible (—, $k = m = 10$, $c = 0$) constraint. Expanded views of the transition bands (*a*), (*b*), and (*c*) are shown in the top row of the figure. Mode numbers are labeled.

narrower the transition band. It is also seen that veering of each pair of adjacent frequency loci occurs because of the rather large constraint parameters k and m.

With the addition of a 5% damping ($c = 2 \times 5\% \times \sqrt{km} = 1$) to the constraint in the previous case, the natural frequencies become slightly lower than, but are practically indistinguishable from, the corresponding undamped frequencies shown in Fig. 1.11. The decay constants μ no longer vanish and are symmetric about the center location $d = 1/2$, as shown in Fig. 1.12 for the first four modes. Here $\mu \leq 0$ for the nonzero damping c and subcritical transport speed $v < 1$. It is observed that $\mu = 0$ for $d = 0$ and $d = 1$, as predicted by Eq. (1.103). In addition, $\mu = 0$ for the nth mode when the constraint is located at $d = j/n$ ($j = 1, 2, \ldots, n-1$). This can be shown analytically by substituting $\lambda = \omega i$ into Eq. (1.102) to give

$$(k - m\omega^2)\sin\frac{\omega d}{1 - v^2}\sin\frac{\omega(1-d)}{1-v^2} + \omega\sin\frac{\omega}{1-v^2}$$
$$+ic\omega\sin\frac{\omega d}{1-v^2}\sin\frac{\omega(1-d)}{1-v^2}s = 0. \tag{1.104}$$

Separating the real and imaginary parts in Eq. (1.104), we conclude that either $\sin\omega/(1-v^2) = \sin\omega d/(1-v^2) = 0$ or $\sin\omega/(1-v^2) = \sin\omega(1-d)/$

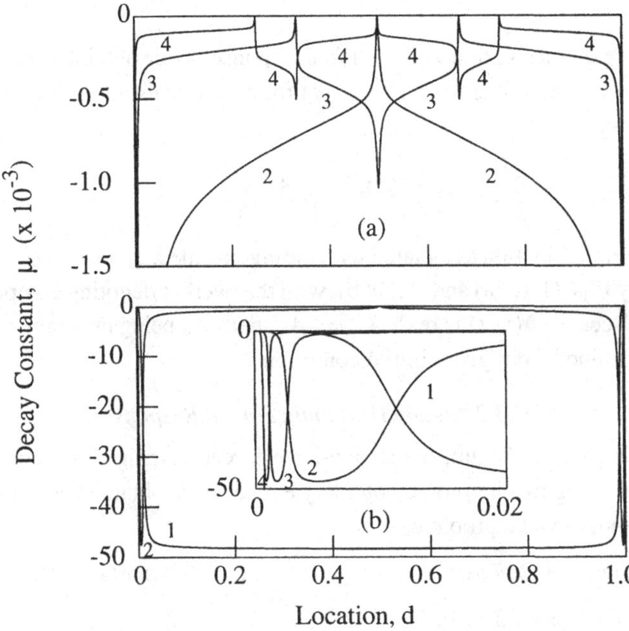

Fig. 1.12. Dependence of the decay constants on the location of a flexible constraint ($k = m = 10$, $c = 1$). (*a*) Mode 2, 3, and 4; (*b*) mode 1 and 2 and the expanded views of the first four modes for d close to the left boundary. Mode numbers are labeled.

$(1 - v^2) = 0$. Hence $\mu = 0$ if and only if $d = j/n$ $(n = 1, 2, 3, \ldots, 0 \leq j \leq n)$. The constraint location coincides with one of the supports for $j = 0$ and $j = n$. The other discrete locations $d = j/n$ for $1 \leq j \leq n - 1$ are the $n - 1$ nodal points of mode n for the classical moving threadline. To provide damping to all of the vibration modes, the constraint location $d \neq j/n$ for all n and $0 < j < n$; hence d must be an irrational number, in agreement with Yang and Mote (1991).

As observed in Fig. 1.12, although large and almost constant damping ($\mu \approx -0.048$) is obtained for the fundamental mode for all d except near one of the supports, the nearly same amount of dissipation can be achieved for the higher modes only at a constraint location in the vicinity of the boundaries. With this approach, the optimal constraint location d can be determined for specific operating speed, constraint parameters, and dominant vibration modes.

The complex eigenfunction $U^{(n)}(x)$ corresponding to the complex eigenvalue λ_n is obtained from Eqs. (1.98a)–(1.101) as follows:

$$U^{(n)}(x) = e^{\frac{\lambda_n}{1-v}x} - e^{-\frac{\lambda_n}{1+v}x}, \quad 0 < x < d, \tag{1.105a}$$

$$U^{(n)}(x) = -e^{-\frac{\lambda_n}{1-v^2}} \frac{\sinh \frac{\lambda_n d}{1-v^2}}{\sinh \frac{\lambda_n (1-d)}{1-v^2}} \left(e^{\frac{\lambda_n}{1-v}x} - e^{\frac{2\lambda_n}{1-v^2}} e^{-\frac{\lambda_n}{1+v}x}\right), \quad d < x < 1. \tag{1.105b}$$

Hence, the general solution $\tilde{u}(x, t)$ of free response can be obtained as the superposition of the separable form Eq. (1.97) for each eigensolution $\{\lambda_n, U^{(n)}(x)\}$ so determined:

$$\tilde{u}(x, t) = \sum_{n=1}^{\infty} \left(A_n U^{(n)}(x) e^{\lambda_n t} + A_{-n} U^{(-n)}(x) e^{\lambda_{-n} t}\right), \tag{1.106}$$

where the eigenfunction associated with the eigenvalue $\lambda_{-n} = \bar{\lambda}_n$ is $U^{(-n)}(x) = \bar{U}^{(n)}(x)$ by Eqs. (1.105a) and (1.105b), with the overbar denoting complex conjugation. Because $\tilde{u}(x, t)$ is real, $A_{-n} = \bar{A}_n$, with A_n being arbitrary constants to be determined from given initial conditions.

1.3.3.2 Steady Harmonic Forced Response

Relative to its equilibrium, the steady-state forced response of a constrained translating string to a harmonic boundary excitation is obtained as the solution to the boundary value problem

$$\tilde{u}_{tt}(x, t) + 2v\tilde{u}_{xt}(x, t) + (v^2 - 1)\tilde{u}_{xx}(x, t) = \tilde{r}(t)\delta(x - d), \tag{1.107}$$

$$\tilde{u}(0, t) = e_0 \sin \Omega t = \mathrm{Im}\{e_0 e^{i\Omega t}\}, \qquad \tilde{u}(1, t) = 0, \quad -\infty < t < \infty, \tag{1.108}$$

$$m\ddot{\tilde{z}}(t) + c\dot{\tilde{z}}(t) + k\tilde{z}(t) = -\tilde{r}(t), \qquad \tilde{z}(t) = \tilde{u}(d, t) \tag{1.109}$$

by assuming harmonic response in the form

$$\tilde{u}(x,t) = \mathrm{Im}\{U(x)e^{i\Omega t}\}, \tag{1.110}$$

where

$$U(x) = \begin{cases} U_1(x), & 0 < x < d, \\ U_2(x), & d < x < 1. \end{cases} \tag{1.111}$$

Substitute Eq. (1.110) into Eq. (1.107), and integrate the resulting expression from $x = d^-$ to $x = d^+$ by use of Eq. (1.109), Eq. (1.111), and continuity of $U(x)$ at $x = d$, giving

$$U_1(d) = U_2(d), \tag{1.112}$$

$$(v^2 - 1)\left[U_2'(d) - U_1'(d)\right] + [-m\Omega^2 + ic\Omega + k]U_1(d) = 0. \tag{1.113}$$

Substitution of Eqs. (1.110) and (1.111) into Eq. (1.107) yields

$$(v^2 - 1)U_1''(x) + 2vi\Omega U_1'(x) - \Omega^2 U_1(x) = 0, \quad 0 < x < d, \tag{1.114a}$$

$$(v^2 - 1)U_2''(x) + 2vi\Omega U_2'(x) - \Omega^2 U_2(x) = 0, \quad d < x < 1. \tag{1.114b}$$

Solutions of Eqs. (1.114a) and (1.114b) are

$$U_1(x) = C_1 e^{\frac{i\Omega}{1-v}x} + C_2 e^{-\frac{1\Omega}{1+v}x}, \tag{1.115a}$$

$$U_2(x) = C_3 e^{\frac{i\Omega}{1-v}x} + C_4 e^{-\frac{i\Omega}{1+v}x}. \tag{1.115b}$$

Substitution of Eqs. (1.115a) and (1.115b) into Eqs. (1.108), (1.112), and (1.113) yields four linear algebraic equations in four unknowns, C_1 to C_4. If the constraint damping c is nonzero, there is a unique solution. For a nondissipative constraint ($c = 0$), the solution is unique if the excitation frequency Ω is not one of the natural frequencies ω_n ($n = 1, 2, 3, \ldots$). In both cases the solution is

$$\begin{aligned} C_1 &= e_0 i\Delta_1/2\Delta, & C_2 &= -e_0 i\Delta_2/2\Delta, \\ C_3 &= e_0\Omega i e^{-\frac{\Omega i}{1-v^2}}/2\Delta, & C_4 &= -e_0\Omega i e^{\frac{\Omega i}{1-v^2}}/2\Delta, \end{aligned} \tag{1.116}$$

where

$$\Delta = (k - m\Omega^2 + ic\Omega)\sin\frac{\Omega(1-d)}{1-v^2}\sin\frac{\Omega d}{1-v^2} + \Omega\sin\frac{\Omega}{1-v^2}, \tag{1.117}$$

$$\Delta_1 = (k - m\Omega^2 + ic\Omega)\sin\frac{\Omega(1-d)}{1-v^2}e^{-\frac{\Omega i d}{1-v^2}} + \Omega e^{-\frac{\Omega i}{1-v^2}}, \tag{1.118}$$

$$\Delta_2 = (k - m\Omega^2 + ic\Omega)\sin\frac{\Omega(1-d)}{1-v^2}e^{\frac{\Omega i d}{1-v^2}} + \Omega e^{\frac{\Omega i}{1-v^2}}. \tag{1.119}$$

If $c = 0$ and Ω is one of the natural frequencies of the system, $\Delta = 0$ by Eq. (1.103) and resonance occurs. Hence the nonresonant steady-state response is obtained in closed form from Eqs. (1.110)–(1.111) and Eqs. (1.115a)–(1.119):

$$u(x,t) = \frac{e_0}{|\Delta|}\left[a_{11}(x)\cos\Omega\left(\frac{vx}{1-v^2}+t\right) + a_{12}(x)\sin\Omega\left(\frac{vx}{1-v^2}+t\right)\right], \quad 0 < x < d, \tag{1.120a}$$

$$u(x,t) = \frac{\Omega e_0}{|\Delta|}\sin\frac{\Omega(1-x)}{1-v^2}\left[a_{21}\cos\Omega\left(\frac{vx}{1-v^2}+t\right) + a_{22}\sin\Omega\left(\frac{vx}{1-v^2}+t\right)\right], \quad d < x < 1, \tag{1.120b}$$

where

$$a_{11}(x) = -c\Omega^2\sin^2\frac{\Omega(1-d)}{1-v^2}\sin\frac{\Omega x}{1-v^2}, \tag{1.121a}$$

$$\begin{aligned} a_{12}(x) = {} & \Omega^2\sin\frac{\Omega}{1-v^2}\sin\frac{\Omega(1-x)}{1-v^2} - [(k-m\Omega^2)^2+\Omega^2c^2]\sin^2\frac{\Omega(1-d)}{1-v^2} \\ & \times \sin\frac{\Omega d}{1-v^2}\sin\frac{\Omega(x-d)}{1-v^2} - \Omega(k-m\Omega^2)\sin\frac{\Omega(1-d)}{1-v^2} \\ & \times\left[\sin\frac{\Omega}{1-v^2}\sin\frac{\Omega(x-d)}{1-v^2} - \sin\frac{\Omega d}{1-v^2}\sin\frac{\Omega(1-x)}{1-v^2}\right], \end{aligned} \tag{1.121b}$$

$$a_{21} = -\Omega c\sin\frac{\Omega(1-d)}{1-v^2}\sin\frac{\Omega d}{1-v^2}, \tag{1.121c}$$

$$a_{22} = (k-m\Omega^2)\sin\frac{\Omega(1-d)}{1-v^2}\sin\frac{\Omega d}{1-v^2} + \Omega\sin\frac{\Omega}{1-v^2}, \tag{1.121d}$$

$$|\Delta| = a_{21}^2 + a_{22}^2. \tag{1.121e}$$

The steady-state contact force $\tilde{r}(t)$ can be subsequently obtained from Eqs. (1.109) and (1.120a)–(1.121e).

1.3.4 Examples and Discussion

The results are presented for a model of a magnetic tape-head system typical of those used in data storage technology. The tape of thickness 28 μm, width 12.7 mm, and density 1.4 g/cm^3, under tension $P = 0.5$ N, is modeled as a translating string of mass per unit length $\rho = 0.49784$ g/m. The tape travels at a speed $V = 2$ m/s ($v = 0.0631$) between two reels separated by L = 200 mm. The dimensionless tape weight $w = \rho g L/P$ is 1.9515×10^{-3}. For a rigid head at $d = 0.25$ or 0.75 with the head penetration $h = 0.015$ (3 mm), the preload is $r^* = 0.08066(0.04033\text{ N})$. For a rigid head at $d = 0.5$ with the same penetration (3 mm), the preload is 0.03705 N. For a flexible head, the preload r^* is chosen to be that for the rigid head with $z_0 = (mw + r^*)/k$ by Eq. (1.70a) for comparison purposes. Two types of boundary disturbances due to the reel eccentricity at the upstream end are used: $e_1(t) = e_0 \sin \Omega t$ and $e_2(t) = e_0(1 - \cos \Omega t)$ with $e_0 = 7.5 \times 10^{-3}$ (1.5 mm) and $\Omega = 4$. Initially the tape is at rest in equilibrium given by Eq. (1.71) and $f(x, t) = 0$. The transient response is calculated for $t_f = 8$.

The solid line in Fig. 1.13(a) shows the dimensional contact force history $R(T) = R^* + \tilde{R}(T)$ between the tape and a rigid head located at $d = 0.25$ under boundary disturbance $e_1(t)$. The preload is $R^* = 0.04033$ N before the disturbance reaches and is reflected by the rigid head at $t_+^{(1)} = d/(1 + v) = 0.2352$ (1.48 ms) when a negative jump in $R(T)$ of $-2\dot{e}_1(0) = -0.06$ (-0.03 N) occurs. This jump is predicted by Eq. (1.87). The jump in $R(T)$ is also -0.03 N at subsequent instants $t_r^{(n)} = d/(1 + v) + 2dn/(1 - v^2) = 0.73716$ (4.65 ms), 1.23916 (7.82 ms), 1.74116 (10.99 ms), etc. ($n = 1, 2, 3, \ldots, 15$) when the reflected wave interacts with the rigid head. The dashed line in Fig. 1.13(a) represents the contact force with $k = m = 10$ and $\zeta = 0.05$. The spring compression at

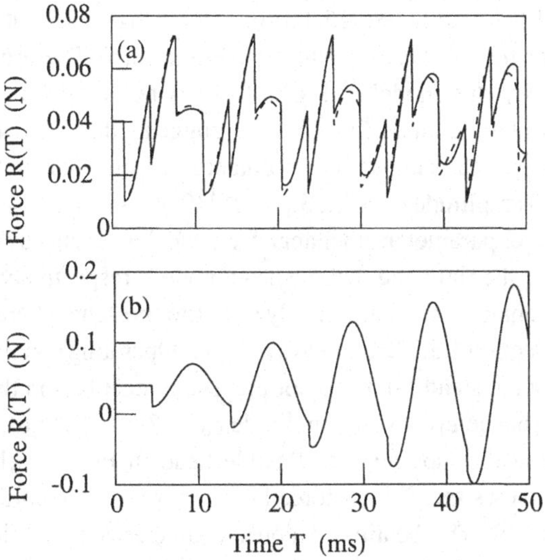

Fig. 1.13. Contact force history under disturbance $e_1(t)$, in which discontinuity exists because $\dot{e}_1(0) \neq 0$: (*a*) rigid (——) and flexible (– – –, $k = m = 10$, $\zeta = 5\%$) head at $d = 0.25$, $n_p = 4{,}000$; (*b*) rigid (——) and flexible (– – –, $k = m = 10$, $\zeta = 5\%$) head at $d = 0.75$, $n_p = 4{,}000$.

equilibrium is $Z_0 = 2.0034$ mm. Unlike the rigid head case, waves are transmitted across the flexible head. However, transmitted waves introduce no additional discontinuity in the contact force because the slope of the tape at a transmitted wavefront is zero (e.g., see Fig. 1.17(b)). Vanishing slope occurs when $m \neq 0$. The contact force history for this relatively stiff head suspension resembles that for a rigid head. For a soft suspension, $k = m = 1$, with the preload R^* and other parameters unaltered, which is not shown here, a contact loss is signaled at $T = 10.98$ ms by a negative R. Also, for a smaller head penetration, say for $h = 0.01$ (2 mm), with other parameters unchanged, the loss of contact occurs at $T = 1.48$ ms when the disturbance reaches the head because of the reduced preload.

The solid and dashed lines in Fig. 1.13(b) show the contact forces for the head positioned at $d = 0.75$ with other parameters unchanged. Again $R(T)$ has negative jumps of -0.03 N at the instants $t_r^{(n)} = d/(1+v) + 2dn/(1-v^2) =$ 0.7055 (4.45 ms), 2.2115 (13.96 ms), 3.7174 (23.46 ms), etc. ($n = 0, 1, 2, 3, 4$), for both rigid and flexible heads. Discontinuity of $R(T)$ vanishes at the instant a transmitted wavefront interacts with the flexible head (e.g., $t_-^{(1)} = 1.2075$ (7.62 ms)) because the slope of the transmitted wavefront is zero. Although the natural frequencies of the coupled tape–flexible head system are identical for symmetric head locations, like $d = 0.25$ and $d = 0.75$, as shown by Eq. (1.102) and Fig. 1.11, the transient responses are different. While contact loss does not occur in Fig. 1.13(a) when $d = 0.25$, loss occurs when $d = 0.75$ at $T = 13.98$ ms for both rigid and flexible heads, as shown in Fig. 1.13(b). Therefore, the contact force predicted by this model in Fig. 1.13(b) is not valid for $T > 13.98$ ms. The initially growing contact force history suggests that a small eccentricity e_0 can lead to tape-head contact loss. Contact loss will occur in this case for a disturbance with amplitude $e_0 > 1.535 \times 10^{-3}$ (0.307 mm). For $e_0 = 2.5 \times 10^{-3}$ (0.5 mm) and other parameters unchanged, contact loss occurs at $T = 42.58$ ms and 42.51 ms for the rigid and flexible head models, respectively. For this case the sinusoidal contact force at the steady state (not presented here), obtained in closed form in Section 1.3.3.2, is always positive. Operating conditions ensuring tape-head contact at steady state do not prevent contact loss in the transient.

For the disturbance $e_1(t)$ with amplitude $e_0 = 2.5 \times 10^{-4}$ (0.05 mm), contact loss between the tape and the flexible head in Fig. 1.13(b) is avoided. Figure 1.14(a) shows the contact force history for a time record of five times that in Fig. 1.13(b). The beating response characterized by a low-frequency modulation of a high-frequency oscillation, which does not occur for the head location $d = 0.25$, is observed for $d = 0.75$. The beating phenomenon arises because the forcing frequency ($\Omega = 4$) is close to the second natural frequency of the system (see Fig. 1.11, $\omega_2 = 4.206$). Also, as expected from Eq. (1.84),

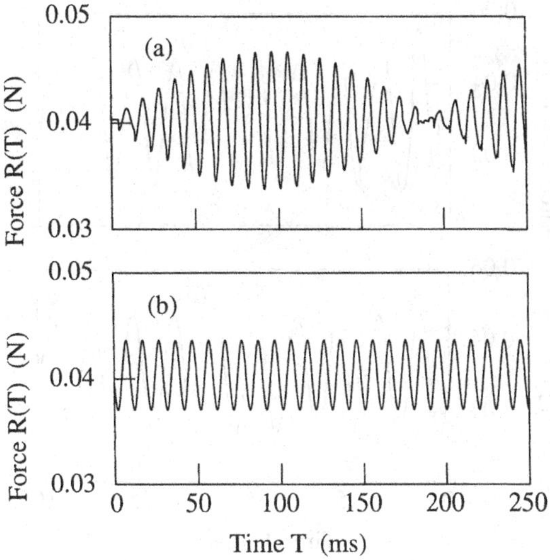

Fig. 1.14. The transient and steady-state contact force for the flexible head case in Fig. 1.13b with $e_0 = 2.5 \times 10^{-4}$ (0.5 mm). (*a*) Transient response; (*b*) closed-form steady-state response.

discontinuity of constant amplitude $2\dot{e}_1(0) = 0.006$ (0.003 N) exists throughout the contact force history. The steady-state contact force under the harmonic disturbance, as determined from the closed-form, steady forced analysis in Section 1.3.3.2, is shown in Fig. 1.14(b). The transient response does not approach to the sinusoidal steady-state response because of absence of damping in the tape model.

Under the same disturbance $e_1(t)$ as in Fig. 1.13, the contact force between the tape and a flexible massless head ($k = c = 10$), which represents an air bearing model, is shown in Fig. 1.15(a) for a final time t_f = 100 ms. The head is positioned at $d = 0.5$, the preload is $R^* = 0.03705$, and $Z_0 = 1.613$ mm. The slope at a transmitted wavefront through the flexible massless head does not vanish, and consequently discontinuity in R occurs at the instant the transmitted wavefront interacts with the head. The head is located at the midspan, so all instants of wave–head interaction are at $t_+^{(j)}$ and $t_-^{(j)}$. The first discontinuity in R at $t_+^{(1)} = 0.4704$ (2.968 ms) equals $-2c\dot{e}_1(0)/(c+2) = -0.05$ (-0.025 N), as predicted by Eq. (1.92). The discontinuity in $R(T)$ at later instants is not that given by Eqs. (1.92) or (1.93). The wave–head interactions that occur at the same time from earlier reflected and transmitted waves, as predicted by the delay terms in Eq. (1.94), sum with the discontinuity due to the main transmitted wave component given by Eqs. (1.92) or (1.93). It is seen that the

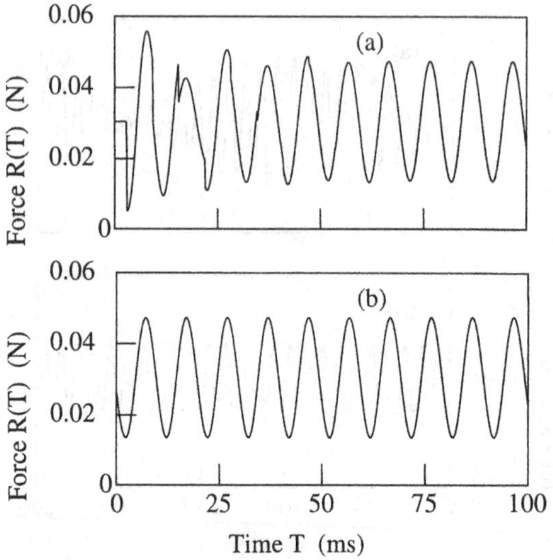

Fig. 1.15. The transient and steady-state contact force for the case of a flexible massless head ($k = c = 10$) at $d = 0.5$. (*a*) Transient response; (*b*) closed-form steady-state response.

discontinuity in $R(T)$ decreases with time and eventually vanishes. After about 60 ms the contact force approaches a sinusoidal steady forced response. The transient is dissipated by a single damper at $x = d$. If we compare the steady-state $R(T)$ in Fig. 1.15(a) with that in Fig. 1.15(b), which is determined by the closed-form, steady forced analysis in Section 1.3.3.2, excellent agreement is observed. Whereas the velocity $\dot{z}(t)$ is continuous by Eq. (1.79) for the case of nonvanishing head mass m, it is discontinuous in the transient for the flexible massless head when $c \neq 0$, as predicted by Eq. (1.75), because $\tilde{r}(t)$ is discontinuous.

Figures 1.16(a–c) show the contact forces under disturbance $e_2(t)$ with all other parameters equal to those in Figs. 1.13 and 1.15(a). The trend and amplitude ranges of the contact force histories in Fig. 1.16 are similar to those in Figs. 1.13 and 1.15(a), except discontinuity in $R(T)$ vanishes in Fig. 1.16 because $\dot{e}_2(0) = 0$. Contact losses occur at $T = 15.35$ ms and $T = 15.42$ ms in Fig. 1.16(b) for the rigid and flexible heads, respectively.

Figures 17(a–e) show the propagation of the disturbance $e_1(t)$ from the left end of the tape for the case in Fig. 1.13(a). The displacements $\tilde{U}(X, T)$ shown are relative to the parabolic static equilibrium $U^*(X)$ given by Eq. (1.71). In the tape modeled as a string, waves (solid lines) are not transmitted through a rigid head and accordingly are confined in the upstream section of the tape

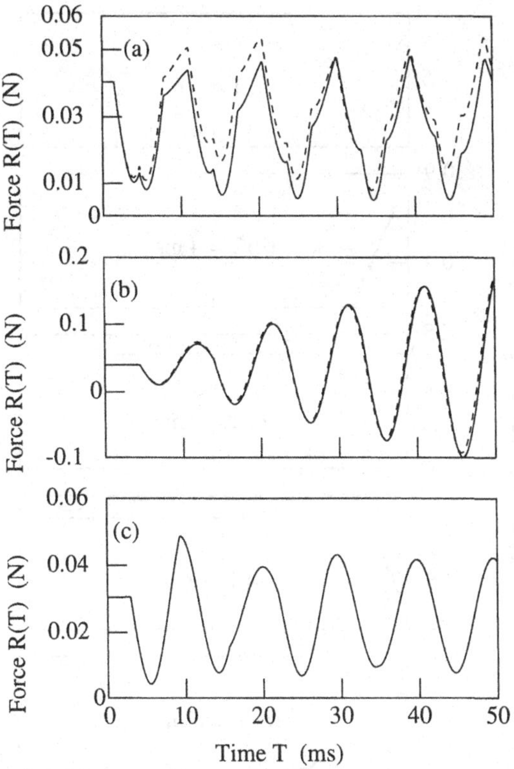

Fig. 1.16. Contact force history under disturbance $e_2(t)$, in which discontinuity is absent because $e_2(0) = \dot{e}_2(0) = 0$: (*a*) rigid (—) and flexible (– – –, $k = m = 10$, $\zeta = 5\%$) head at $d = 0.25$, $n_p = 2{,}000$; (*b*) rigid (—) and flexible (– – –, $k = m = 10$, $\zeta = 5\%$) head at $d = 0.75$, $n_p = 2{,}000$; (*c*) flexible massless head ($k = c = 10$) at $d = 0.5$, $n_p = 3{,}100$.

($X < 50$ mm). Waves are transmitted through the flexible head, however, as shown by the dashed lines. It is seen in Fig. 1.17(b) that the slope of the wavefront at $X = 134.76$ mm transmitted through the flexible head with nonzero mass vanishes.

1.4 Piano String Response under a Hammer Strike

1.4.1 Theoretical Formulation

Consider now a fixed-fixed, piano string of length L, tension P, and linear density ρ struck by a narrow hammer of negligible width with an initial transverse velocity V. The striking point is at a distance D from the left end. The contact force between the string and the narrow hammer is a concentrated force $R(T)$.

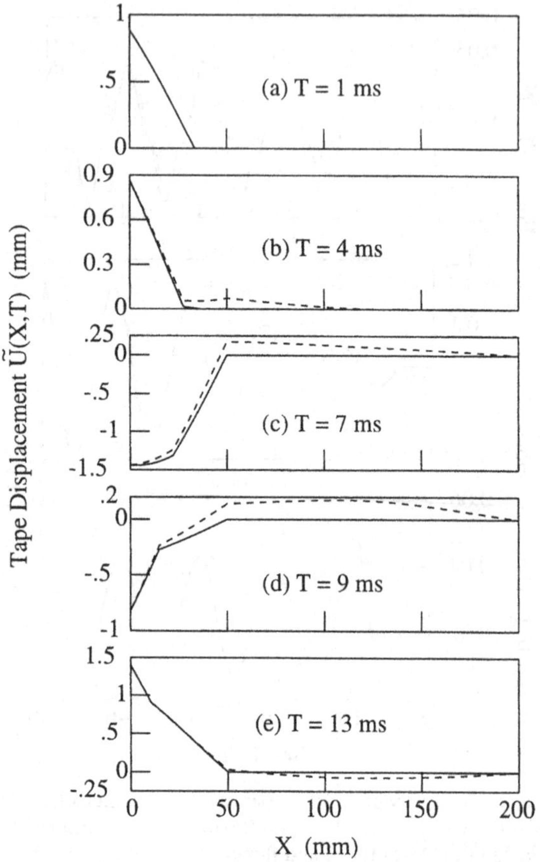

Fig. 1.17. Propagation of boundary disturbance $e_1(t)$ in the tape in contact with a rigid (—) or flexible (- - -, $k = m = 10$, $\zeta = 5\%$) head at $d = 0.25$. The tape displacement $\tilde{U}(X, T)$ is relative to its equilibrium. The spatial domain is divided into 501 points.

The hard, narrow hammer shown in Fig. 1.18(a) is modeled as a concentrated mass M, and the soft, narrow hammer in Fig. 1.18(b) is modeled as a point mass M, backed by either a linear or nonlinear spring of constant K_1 or K_2 and a linear dashpot of damping constant C. The transverse displacements of the string and hammer are denoted by $U(X, T)$ and $Z(T)$, respectively. Displacement $Z(T)$ is measured from the horizontal position of the string for a hard hammer and from the unstretched, fixed, upper end of the spring for a soft hammer. The force–deformation relationship for the nonlinear spring is assumed to follow a power-law dependence $R(T) = K_2[U(D, T) - Z(T)]^{\alpha}$, where $2 < \alpha < 5$, similar to that reported for hammer felts (Boutillon, 1988; Hall, 1992).

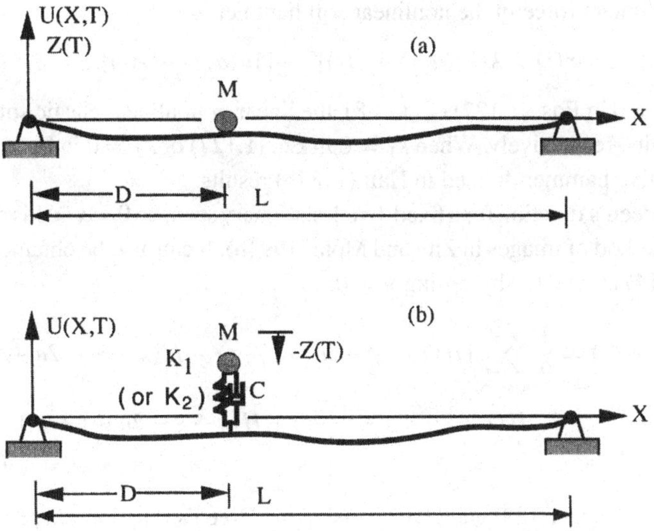

Fig. 1.18. Schematic of the piano string and narrow hammers: (*a*) hard hammer; (*b*) linear or nonlinear soft hammer.

With the dimensionless variables defined below

$$\begin{gathered} x = X/L, \qquad u = U/L, \qquad z = Z/L, \qquad d = D/L, \qquad r = R/P, \\ v = V(\rho/P)^{1/2}, \qquad t = T(P/\rho L^2)^{1/2}, \qquad m = M/\rho L, \\ k_1 = K_1 L/P, \qquad k_2 = K_2 L^{\alpha}/P, \qquad c = C/(P\rho)^{1/2} \end{gathered} \tag{1.122}$$

the wave equation becomes

$$u_{tt}(x,t) - u_{xx}(x,t) = -r(t)\delta(x-a). \tag{1.123}$$

The boundary and initial conditions are

$$\begin{aligned} u(0,t) &= u(1,t) = 0, \quad 0 \le t \le t_f; \\ u(x,0) &= u_t(x,0) = 0, \quad 0 < x < 1, \end{aligned} \tag{1.124}$$

where t_f is an arbitrary final time ($0 < t_f < \infty$). The equation of motion and initial conditions for the hammer mass are

$$m\ddot{z}(t) = r(t); \qquad z(0) = 0, \qquad \dot{z}(0) = -v. \tag{1.125}$$

The kinematic contact condition for a hard hammer is

$$z(t) = u(d,t). \tag{1.126}$$

The contact force in the linear soft hammer is

$$r(t) = k_1[u(d,t) - z(t)] + c[\dot{u}(d,t) - \dot{z}(t)], \tag{1.127}$$

and the contact force in the nonlinear soft hammer is

$$r(t) = k_2[u(d,t) - z(t)]^\alpha + c[\dot{u}(d,t) - \dot{z}(t)]. \tag{1.128}$$

When $c = 0$ in Eqs. (1.127) or (1.128), the linear or nonlinear elastic soft hammer results, respectively. When $k_1 = 0$ in Eq. (1.127) or $k_2 = 0$ in Eq. (1.128), the resistive hammer defined in Hall (1987b) results.

The Green's function for a fixed-fixed, stationary string in $0 < x < 1$ is derived by the method of images in Zhu and Mote (1994b). It can also be obtained from Eqs. (1.14) and (1.15) by setting $v = 0$:

$$\begin{aligned} g(x,\xi;t-\tau) = \frac{1}{2}\sum_{n=-\infty}^{+\infty} [H(x-\xi-2n+t-\tau) - H(x-\xi-2n-t+\tau) \\ - H(x+\xi-2n+t-\tau) + H(x+\xi-2n-t+\tau)]. \end{aligned} \tag{1.129}$$

Solution to Eqs. (1.123) and (1.124), using the Green's function $g(x,\xi;t-\tau)$, yields

$$u(x,t) = -\int_0^t r(\tau) g(x,d;t-\tau)\,d\tau, \tag{1.130}$$

and the solution of Eq. (1.125) is

$$z(t) = \frac{1}{m}\int_0^t (t-\tau) r(\tau)\,d\tau - vt. \tag{1.131}$$

1.4.1.1 Hard Hammer

Substitution of Eqs. (1.130) and (1.131) into the contact condition, Eq. (1.126), leads to the equation for the contact force $r(t)$ between the string and a hard narrow hammer:

$$\int_0^t r(\tau) g(d,d;t-\tau)\,d\tau + \frac{1}{m}\int_0^t (t-\tau) r(\tau)\,d\tau = vt. \tag{1.132}$$

Integration and use of Eq. (1.129) yields the identity similar to Eq. (1.83):

$$\begin{aligned} \int_0^t r(\tau) g_t(d,d;t-\tau)\,d\tau = \frac{1}{2} r(t) + \sum_{j=1}^{N} r(t-2j) H(t-2j) \\ - \frac{1}{2}\sum_{j=1}^{N+1} r[t-2d-2(j-1)] H[t-2d-2(j-1)] \\ - \frac{1}{2}\sum_{j=1}^{N+1} r(t+2d-2j) H(t+2d-2j) \equiv \mathbf{D}[r], \quad N = \mathrm{Int}\left(\frac{t_f}{2}\right), \end{aligned} \tag{1.133}$$

where **D** is the delay operator. Differentiating Eq. (1.132) with the use of the key identity in Eq. (1.133) yields the delay-integral equation describing $r(t)$:

$$\mathbf{D}[r] + \frac{1}{m}\int_0^t r(\tau)\,d\tau = v. \qquad (1.134)$$

The initial value is $r(0) = 2v$ by Eqs. (1.134) and (1.133).

A few remarks may be useful here. First, Eq. (1.134) is exact and without modal truncation. Second, the complex multiple wave scattering is implicitly accounted for by the delay terms in $\mathbf{D}[r]$. The solution for $r(t)$ is determined for any hammer to string mass ratio. Third, because $r(0)$ is normally nonzero, $r(t)$ from Eq. (1.134) is discontinuous due to the presence of the Heaviside functions. These discontinuities are not predicted in a normal mode method. Fourth, linearity of Eq. (1.134) gives the solution $\kappa r(t)$ for an initial hammer velocity κv. Hence the duration of contact, determined by the first zero of $r(t)$ for $t > 0$, is independent of the hammer speed v.

The hammer strikes the string at $x = d$, and a wave propagates in both directions with subsequent wave scattering in the string–hammer system shown in Fig. 1.19. By Eqs. (1.134) and (1.133), discontinuity in $r(t)$ is shown to occur only at the instants, $t_r^{(k)} = 2kd\,(k = 1, 2, \ldots, \text{Int}[t_f/2d])$ and $t_R^{(n)} = 2n(1 - d)(n = 1, 2, \ldots, \text{Int}[t_f/2(1-d)])$, when the reflected waves impinge on the hammer. The discontinuity in $r(t)$ at these instants equals $2v$. The wavefront

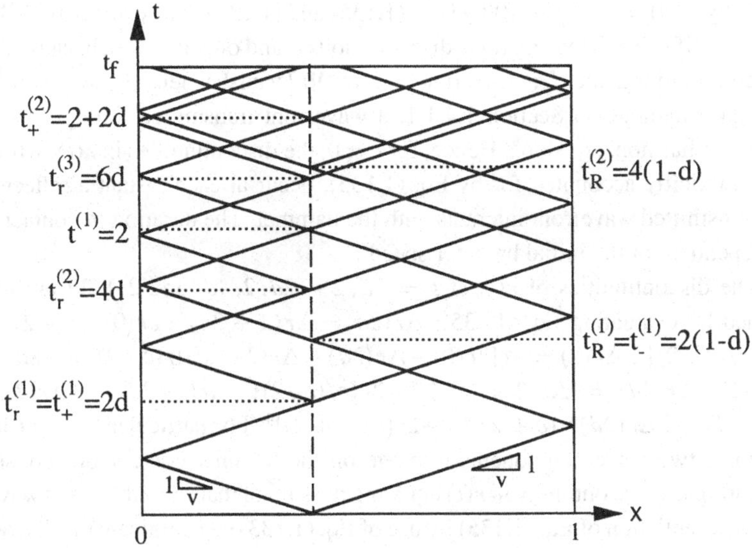

Fig. 1.19. Multiple wave scattering in the string–hammer system. Intersections of the characteristic lines with the dashed line $x = d$ are the instants when propagating wavefronts in the string interact with the hammer.

transmitted through a concentrated mass has zero slope, and therefore it does not cause discontinuity in the contact force when it impinges on the hammer again at $t = 2, 2+2d, 4-2d$, etc. For the hammer location $d = n/(k+n)$, $t_r^{(k)}$ coincides with $t_R^{(n)}$; hence discontinuity of $r(t)$, at $t_r^{(k)} (= t_R^{(n)})$ and its integer multiples, amounts to $4v$.

The first discontinuity of $r(t)$ at $t = 2d$ is obtained by subtracting Eq. (1.134) at $t/2d^-$ from Eq. (1.134) at $t = 2d^+$; evaluating the result gives $\Delta r(2d) = r(0) = 2v$. Similarly, $\Delta r[2(1-d)] = r(0) = 2v$, $\Delta r(2) = \Delta r(2d) + \Delta r[2(1-d)] - 2r(0) = 0$, $\Delta r(4d) = \Delta r(2d) = 2v$, $\Delta r(2+2d) = \Delta r(4d) + r(0) + \Delta r(2) - 2\Delta r(2d) = 0$, $\Delta r[4(1-d)] = \Delta r[2(1-d)] = 2v$, and $\Delta r(4-2d) = -2\Delta r[2(1-d)] + \Delta r[4(1-d)] + \Delta r(2) + r(0) = 0$, etc. When $d = 1/3$, discontinuity in $r(t)$ at $4d (= 2(1-d))$ sums to $4v$ and at $2+2d (= 4(1-d))$ equals $2v$.

1.4.1.2 Linear Soft Hammer

Substitution of Eqs. (1.130) and (1.131) into Eq. (1.127), and use of Eq. (1.133), yields the exact expression

$$r(t) + c\mathbf{D}[r] + \frac{k_1}{m}\int_0^t (t-\tau) r(\tau)\,d\tau + \frac{c}{m}\int_0^t r(\tau)\,d\tau$$
$$+ k_1 \int_0^t r(\tau) g(d, d; t-\tau)\,d\tau = k_1 v t + c v. \qquad (1.135)$$

Initially, $r(0) = 2cv/(c+2)$ by Eqs. (1.135) and (1.133). However, $r(0) \neq 0$ if and only if $c \neq 0$. Hence, $r(t)$ is discontinuous if and only if $c \neq 0$, because the discontinuities in the Heaviside functions in $\mathbf{D}[r]$ vanish when $r(0) = 0$. Unlike the hard hammer in Section 1.4.1.1, a wavefront transmitted through a soft hammer has nonzero slope. Hence, for $c \neq 0$, the discontinuities in $r(t)$, which are implicitly accounted for by Eq. (1.135), occur at each instant a reflected or transmitted wavefront interacts with the hammer. The duration of contact is independent of the initial hammer speed v.

The discontinuities of $r(t)$ at $t = 2d, 2-2d, 2, 4d$, and $2+2d$ are obtained by evaluating Eq. (1.135): $\Delta r(2d) = \Delta r(2-2d) = cr(0)/(c+2) = 2v[c/(c+2)]^2$, $\Delta r(2) = -c[2r(0) - \Delta r(2d) - \Delta r(2-2d)]/(c+2) = -8c^2 v/(c+2)^3$, $\Delta r(4d) = c\Delta r(2d)/(c+2) = 2v[c/(c+2)]^3$, $\Delta r(2+2d) = c[\Delta r(2) + \Delta r(4d) - 2\Delta r(2d)]/(c+2) = -2v[c/(c+2)]^4$. For particular hammer locations, two wave components impinge on the hammer contemporaneously, resulting in a discontinuity in $r(t)$ equaling the sum of that caused by each wave.

Differentiation of Eq. (1.135) by use of Eq. (1.133) can eliminate the discontinuous kernel $g(d, d; t-\tau)$ in Eq. (1.135). However, when $c \neq 0$, singular delta functions arise from differentiating the Heaviside functions in $\mathbf{D}[r]$. For this case one must first account for the discontinuities in $r(t)$ due to the Heaviside functions. The discontinuity in $r(t)$ at $t_+^{(j)} = 2(j-1)+2d$ ($j = 1, 2, \ldots, N+1$) is obtained by evaluating Eq. (1.135) at $(t_+^{(j)})^-$ and $(t_+^{(j)})^+$ and subtracting one from the other:

$$\Delta r\left[t_+^{(j)}\right] = r\left[\left(t_+^{(j)}\right)^+\right] - r\left[\left(t_+^{(j)}\right)^-\right] = cr(0)/(c+2) = 2v[c/(c+2)]^2. \tag{1.136}$$

Likewise, the discontinuity of $r(t)$ at $t_-^{(j)} = 2j - 2d$ ($j = 1, 2, \ldots, N+1$) is

$$\Delta r\left[t_-^{(j)}\right] = r\left[\left(t_-^{(j)}\right)^+\right] - r\left[\left(t_-^{(j)}\right)^-\right] = cr(0)/(c+2) = 2v[c/(c+2)]^2 \tag{1.137}$$

and discontinuity of $r(t)$ at $t^{(j)} = 2j$ ($j = 1, 2, \ldots, N$) is

$$\Delta r[t^{(j)}] = r[(t^{(j)})^+] - r[(t^{(j)})^-] = -2cr(0)/(c+2) = -4v[c/(c+2)]^2. \tag{1.138}$$

Differentiating Eq. (1.135) and dropping the delta functions arising from differentiation of the Heaviside functions in $\mathbf{D}[r]$ gives the delay-integro-differential equation:

$$\left(1 + \frac{c}{2}\right)\frac{d}{dt}r(t) + \frac{k_1}{m}\int_0^t r(\tau)\,d\tau + \frac{c}{m}r(t) + c\mathbf{D}'[r] + k_1\mathbf{D}[r] = k_1 v, \tag{1.139}$$

where operator $\mathbf{D}'$ is defined by

$$\begin{aligned}\mathbf{D}'[r] = &\sum_{j=1}^{N}\frac{d}{dt}r(t-2j)H(t-2j) - \frac{1}{2}\sum_{j=1}^{N+1}\frac{d}{dt}r[t-2d-2(j-1)]\\ &\times H[t-2d-2(j-1)] - \frac{1}{2}\sum_{j=1}^{N+1}\frac{d}{dt}r(t+2d-2j)H(t+2d-2j).\end{aligned} \tag{1.140}$$

Equation (1.139) is valid for all $t \in [0, t_f]$, except $t_+^{(j)}$, $t_-^{(j)}$, and $t^{(j)}$ for $c \neq 0$ where Eq. (1.139) is supplemented by the discarded discontinuities represented by Eqs. (1.136)–(1.138).

1.4.1.3 Nonlinear Soft Hammer

Substitution of Eqs. (1.130) and (1.131) into Eq. (1.128) leads to the nonlinear delay-integral equation governing $r(t)$ between the string and a nonlinear soft hammer:

$$\begin{aligned}r(t) = k_2\Bigg(&-\frac{1}{m}\int_0^t (t-\tau)r(\tau)\,d\tau + vt - \int_0^t r(\tau)g(d, d; t-\tau)\,d\tau\Bigg)^{\alpha}\\ &-\frac{c}{m}\int_0^t r(\tau)\,d\tau + cv - c\mathbf{D}[r].\end{aligned} \tag{1.141}$$

The initial contact force from Eq. (1.141), $r(0) = 2cv/(c+2)$, is equivalent to that for the linear soft hammer. Again, $r(t)$ is discontinuous if and only if $c \neq 0$, and discontinuity of $r(t)$ occurs at each instant a reflected or transmitted wavefront impinges on the hammer. Nonlinearity in the hammer stiffness does not affect discontinuity in $r(t)$. The duration of contact, however, depends on the initial hammer speed v for a nonlinear hammer.

The discontinuous kernel $g(d, d; t-\tau)$ in Eq. (1.141) is treated by the method used for the linear soft hammer. Discontinuities in $r(t)$ at $t_+^{(j)}$ $(j = 1, 2, \ldots, N+1)$, $t_-^{(j)}(j = 1, 2, \ldots, N+1)$, and $t^{(j)}(j = 1, 2, \ldots, N)$ are the same as those in Eqs. (1.136)–(1.138). Differentiating Eq. (1.141) and dropping the delta functions associated with differentiating the Heaviside functions in $\mathbf{D}[r]$ gives

$$\begin{aligned}\left(1+\frac{c}{2}\right)\frac{d}{dt}r(t) = k_2\alpha\Bigg(&-\frac{1}{m}\int_0^t (t-\tau)r(\tau)\,d\tau + vt \\ &- \int_0^t r(\tau)g(d,d;t-\tau)\,d\tau\Bigg)^{\alpha-1} \\ &\times\left(-\frac{1}{m}\int_0^t r(\tau)\,d\tau + v - \mathbf{D}[r]\right) \\ &- \frac{c}{m}r(t) - c\mathbf{D}'[r]. \qquad (1.142)\end{aligned}$$

From Eq. (1.141), we have

$$\begin{aligned}-\frac{1}{m}\int_0^t (t-\tau)r(\tau)\,d\tau + vt - \int_0^t r(\tau)g(d,d;t-\tau)\,d\tau \\ = \left(\frac{1}{k_2}\right)^{\frac{1}{\alpha}}\left(r(t) + \frac{c}{m}\int_0^t r(\tau)\,d\tau - cv + c\mathbf{D}[r]\right)^{\frac{1}{\alpha}}. \qquad (1.143)\end{aligned}$$

Substitution of Eq. (1.143) into Eq. (1.142) yields the nonlinear delay-integro-differential equation describing $r(t)$:

$$\begin{aligned}\left(1+\frac{c}{2}\right)\frac{d}{dt}r(t) = k_2^{\frac{1}{\alpha}}\alpha\left(r(t) + \frac{c}{m}\int_0^t r(\tau)\,d\tau - cv + c\mathbf{D}[r]\right)^{\frac{\alpha-1}{\alpha}} \\ \times\left(-\frac{1}{m}\int_0^t r(\tau)\,d\tau + v - \mathbf{D}[r]\right) - \frac{c}{m}r(t) - c\mathbf{D}'[r]. \\ (1.144)\end{aligned}$$

Equation (1.144) is valid at all $t\varepsilon[0, t_f]$ except $t_+^{(j)}$, $t_-^{(j)}$, and $t^{(j)}$ when $c \neq 0$ at which Eq. (1.144) is supplemented by Eqs. (1.136)–(1.138) describing the discarded terms.

1.4.1.4 Multiple Contacts

The delay-integral equation (1.134), the delay-integro-differential equations (1.139) and (1.144), along with the discontinuity equations (1.136)–(1.138) are discretized in Section 1.4.2 by finite difference. The resulting delay-difference equation for r_i is recursively solved from the initial value $r(0)$. The solution continues until either the contact is lost or $t = t_f$. Contact loss occurs when an r_i is first nonpositive. As evident from experiments (George, 1924, 1925; Ghosh, 1932) and simulations (Hall, 1987a, 1987b), the string can renew and lose contact with the hammer many times subsequent to the initial contact loss.

After the initial contact loss at t_1, the contact force $r(t > t_1)$ is equated to zero, and the string and hammer undergo free response. Determination of renewed contact is predicted from the separation $z(t) - u(d, t)$ for $t > t_1$ from Eqs. (1.130) and (1.131). Renewed contact occurs at t_2 when $z(t_2) - u(d, t_2)$ vanishes. The force of recontact is determined by using the following theorem:

Theorem of Recontact Suppose the string recontacts the hammer for the kth time at t_{2k}. The function $r(t)$ is known for $t \leq t_{2k}$. The force of recontact is governed by Eqs. (1.134) and (1.139) or (1.144) plus Eqs. (1.136)–(1.138) for the hard and soft hammers, respectively. The force of recontact $r(t_{2k})$ is given by the expression for initial contact force with v there replaced by the relative velocity of the point of contact at $t = t_{2k}$. That is, $r(t_{2k}^+) = 2[\dot{u}(d, t_{2k}) - \dot{z}(t_{2k})]$ for the hard hammer, and $r(t_{2k}^+) = 2c[\dot{u}(d, t_{2k}) - \dot{z}(t_{2k})]/(c + 2)$ for the linear and nonlinear soft hammers.

Proof See Appendix B. ■

By the Theorem of recontact, $r(t_2^+)$ is determined from $\dot{u}(d, t) - \dot{z}(t)$ at $t = t_2$:

$$\begin{aligned}\dot{u}(d, t_2) - \dot{z}(t_2) = &-\Bigg(\sum_{j=1}^{N} r(t_2 - 2j)H(t_2 - 2j)\\ &-\frac{1}{2}\sum_{j=1}^{N+1} r[t_2 - 2d - 2(j-1)]H[t_2 - 2d - 2(j-1)]\\ &-\frac{1}{2}\sum_{j=1}^{N+1} r(t_2 + 2d - 2j)H(t_2 + 2d - 2j)\Bigg)\\ &-\frac{1}{m}\int_0^{t_2} r(\tau)\,d\tau + v.\end{aligned} \tag{1.145}$$

Recursive solution to the delay-difference equation continues for $t > t_2$, starting from the new initial value $r(t_2^+)$, until contact loss occurs at t_3 where $r(t_3)$ vanishes. This recontact procedure is repeated until $t = t_f$.

The method described above has a convenient structure and is easy to implement. The treatments of multiple contacts for both the linear and nonlinear hammer models are identical. The contact force during the periods of initial contact and all future recontacts is recursively predicted from one single delay-difference equation for any time interval $[0, t_f]$.

1.4.2 Numerical Solution

The temporal domain $[0, t_f]$ is divided into n_p equal subintervals of width $\Delta t = t_f/n_p$. The integrals in Eqs. (1.134), (1.139), and (1.144) are replaced by trapezoidal integration

$$\int_0^t r(\tau)\, d\tau = \frac{\Delta t}{2}[r_0 + 2r_1 + \cdots + 2r_{i-1} + r_i], \tag{1.146}$$

where $t = i\Delta t$ and $r_i = r(i\Delta t)$. The term $dr(t)/dt$ in Eqs. (1.139) and (1.144) is replaced by the backward finite difference $[r_i - r_{i-1}]/\Delta t$. The derivatives of the delay terms in $\mathbf{D}'[r]$ are replaced by backward differences, when h can be chosen so that it divides $2d$ and 2. Otherwise, the delay terms are approximated by linear interpolations, and the derivatives of the delay terms are approximated by backward finite differences of linear interpolations.

The discretized delay-difference equations for Eqs. (1.134) and (1.139) are linear and not given here. The nonlinear delay-difference equation for Eq. (1.144) is

$$\left[\frac{1}{\Delta t} + c\left(\frac{1}{m} + \frac{1}{2\Delta t}\right)\right] r_i = C_1 + \alpha k_2^{\frac{1}{\alpha}} \left[\left(1 + \frac{c}{2} + \frac{c\Delta t}{2m}\right) r_i + C_2\right]^{\frac{\alpha-1}{\alpha}}$$
$$\times \left[C_3 - \frac{1}{2}\left(1 + \frac{\Delta t}{m}\right) r_i\right], \tag{1.147a}$$

where

$$C_1 = \frac{1}{\Delta t}\left(1 + \frac{c}{2}\right) r_{i-1} - cC_6, \qquad C_2 = cC_4 - cv + cC_5, \tag{1.147b}$$

$$C_3 = -C_4 - C_5 + v, \qquad C_4 = \frac{\Delta t}{2m}[r_0 + 2r_1 + \cdots + 2r_{i-1}], \tag{1.147c}$$

and C_5 is the discretized expression for $\mathbf{D}[r] - r(t)/2$ and C_6 for $\mathbf{D}'[r]$. Therefore C_1 to C_3 in Eq. (1.147a) depend only on the contact force history, r_0 to r_{i-1}. The contact force r_i is determined recursively from known values for earlier initial contact and recontacts. A nonlinear algebraic equation algorithm is required to solve Eq. (1.147a). For a dissipative ($c \neq 0$) soft hammer, the discontinuities in $r(t)$ at $t_+^{(j)}$, $t_-^{(j)}$, and $t^{(j)}$, given by Eqs. (1.136)–(1.138), are added to the computed r_i from Eq. (1.147a) or from the discretized form of Eq. (1.139).

Once the contact force history is known, the response of the string can be computed from Eq. (1.130).

1.4.3 Examples and Discussion

The dimensional contact force between the string and the hard hammer used in Deb (1972) is computed and shown in Fig. 1.20. The parameters used are those from Ghosh (1932) and Deb (1972): $L = 6$ m, $D = 3$ m, $P = 36{,}488.7 \times 10^{-3} \times 9.78$ N, $\rho = 4.95$ g/m, $M = 0.7143\,\rho L$, and $V = 45.5$ cm/s. t_f is chosen to be 2, and $n_p = 2{,}000$. Contact is lost at $T = 0.03057$ s and no further recontact occurs. Comparing $R(T)$ in Fig. 1.20 to that shown in Fig. 2 in Deb (1972), one sees that the normal mode method does not predict the nonzero initial contact force $R(0) = 1.20946$ N nor the jump of 2.4153 N at $T = 0.022346$ s ($t = 1$). This jump has dimensionless amplitude $4v$, as predicted by Eq. (1.134), because $t_r^{(1)} = t_R^{(1)}$ for $d = 1/2$.

The $r(t)$ between the string and hard hammer for the example problem in Fig. 5 in Hall (1987a) is calculated next and shown in Fig. 1.21. The contact force is scaled, $r(t)/2v$, to the definition used in Hall (1987a) for comparison purposes. Here $t_f = 4$, $n_p = 2{,}000$, and $r(0)/2v = 2v/2v = 1$. For Fig. 1.21(a), where $m = 1$ and $d = 0.13$, the initial contact loss is at $t = 0.952$. One recontact occurs at $t = 1.068$ and contact loss occurs at $t = 1.112$. The force of initial recontact is $r(t_2)/2v = 0.3873$. Discontinuity in $r(t)/2v$ of amplitude $2v/2v = 1$ is observed at $t = 2kd = 0.26, 0.52, 0.78$ ($k = 1, 2, 3$). In Fig. 1.21(b), where $m = 4.9$ and $d = 0.13$, contact loss occurs at $t = 2.45$ and

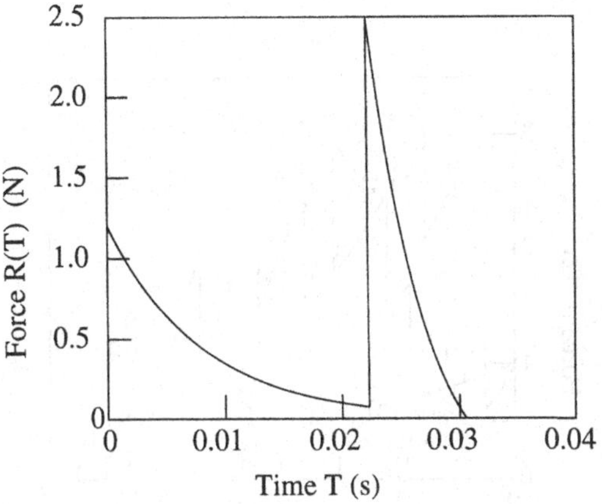

Fig. 1.20. Dimensional contact force $R(T)$ between the string and the hard hammer in Figure 2 in Deb (1972), where $L = 6$ m, $D = 3$ m, $P = 36488.7 \times 10^{-3} \times 9.78$ N, $\rho = 4.95$ g/m, $M = 0.7143\rho$L, and $V = 45.5$ cm/s.

recontact does not occur. Like Fig. 1.21(a), discontinuity in $r(t)/2v$ of amplitude $2v/2v=1$ occurs at $t=2kd=0.26, 0.52, \ldots, 2.34$ ($k=1, 2, \ldots, 9$) and $t=2(1-d)=1.74$. In Fig. 1.21(c), $m=4.9$ and $d=0.45$, and the initial contact is lost at $t=3.26$. Two recontacts occur during [3.302, 3.522] and [3.614, 3.77] with the initial forces of recontact 0.8649 and 0.6798, respectively. Again, discontinuity of amplitude 1 is seen in $r(t)/2v$ at $t=2kd=0.9, 1.8, 2.7$ ($k=1, 2, 3$) and $t=2n(1-d)=1.1, 2.2$ ($n=1, 2$). The discontinuities of $r(t)$ at $t=2, 2+2d=2.9$, and $4-2d=3.1$, when a transmitted wavefront impinges on the hammer, vanish as expected. A discontinuity in the slope of $r(t)/2v$ is noted at $t=2$ in Fig. 1.21(c). The results for the initial contact in all cases are practically indistinguishable from those of Fig. 5 in Hall (1987a) where the traveling wave solution is utilized. The results from both methods for the first recontacts in cases (a) and (c) are also quite close to the extent one can compare with data in the figures. The initial contact force of the second recontact in Fig. 1.21(c) is slightly larger than that in Hall (1987a). As the contact force is predicted from a single relationship in the present approach, the same accuracy is expected for all contacts. In Hall (1987a) each recontact requires individual treatment and solution.

Graphs of $r(t)/2v$ for the linear elastic soft hammers in Fig. 4 in Hall (1987b) are shown in Fig. 1.22. Here $d=0.185$, $m=1.5$, $c=0$, $t_f=2.5$, $n_p=4{,}000$, and k_1 equals 1600/9 and 160/3 for the solid and dashed curves, respectively. These stiffness values correspond to the dimensionless compliances $\sigma=0.06$ and 0.2 in Hall (1987b). The function $r(t)$ is continuous, and $r(0)$, $r(t_2)$, etc. vanish because $c=0$. The steep slopes of $r(t)/2v$ for the large hammer stiffness approach the discontinuous $r(t)$ for a hard hammer as shown

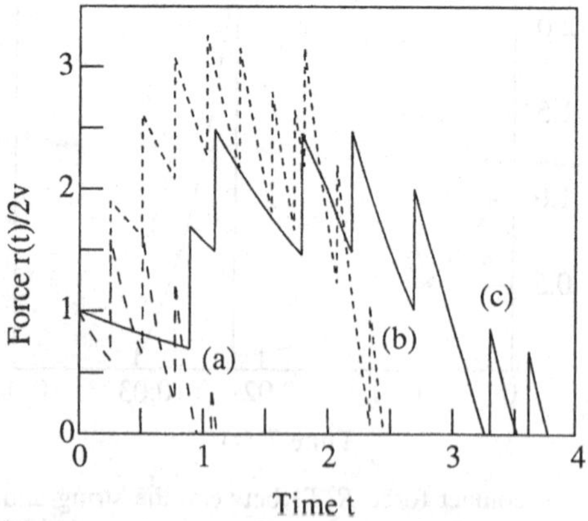

Fig. 1.21. Dimensionless contact forces $r(t)/2v$ between the string and the hard hammers in Fig. 5 in Hall (1987a): (*a*) $m=1, d=0.13$; (*b*) $m=4.9, d=0.13$; (*c*) $m=4.9, d=0.45$.

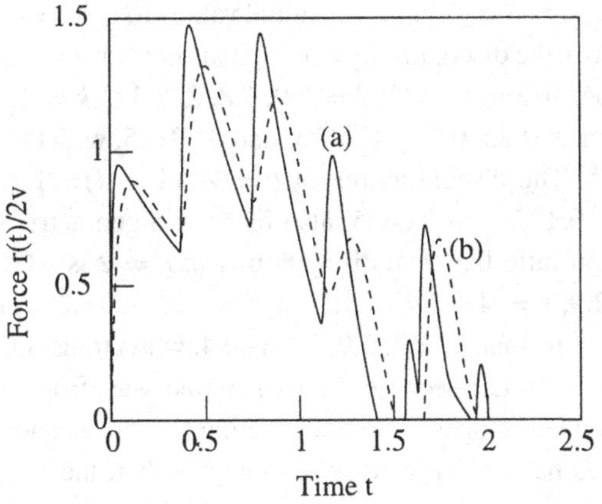

Fig. 1.22. Dimensionless contact forces $r(t)/2v$ between the string and the linear soft hammers in Fig. 4 in Hall (1987b): (*a*) $k_1 = 1600/9$; (*b*) $k_1 = 160/3$.

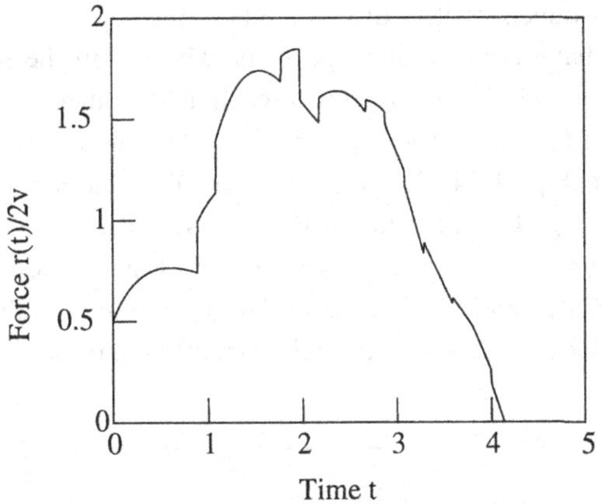

Fig. 1.23. Dimensionless contact force $r(t)/2v$ for a dissipative, linear soft hammer, where $k_1 = 10$, $c = 2$, $m = 4.9$, and $d = 0.45$.

in Fig. 1.21. The differences between the results from the present method and the traveling wave method in Hall (1987b) are indistinguishable for the initial contacts. The force during the first recontact in Fig. 1.22(a) is also very similar to that in Hall (1987b). Some differences exist between the forces from the present method and those in Hall (1987b) during the second recontact in Fig. 1.22(a) and the first recontact in Fig. 1.22(b). The present method is reliable for treating multiple contacts because the unified treatment is independent of the number of contacts.

A graph of $r(t)/2v$ for a dissipative, linear soft hammer is shown in Fig. 1.23. The parameters are those used in Fig. 1.21(c), except the hammer is soft with $k_1 = 10$ and $c = 2$. Here $t_f = 5$ and $n_p = 4{,}000$. The initial contact loss occurs at

$t = 4.134$ and no recontact occurs. The initial value $r(0)/2v = c/(c+2) = 0.5$. Unlike Fig. 1.21(c), the discontinuity in $r(t)/2v$ is not constant. The discontinuities in $r(t)/2v$ occurring at $t = 2kd = 0.9, 1.8, 2.7, 3.6$ ($k = 1, 2, 3, 4$) form a geometric sequence: 0.25, 0.125, 0.0625, and 0.03125, with the common ratio $c/(c+2) = 0.5$. The discontinuities at $t = 2n(1-d) = 1.1, 2.2, 3.3$ ($n = 1, 2, 3$) are 0.25, 0.125, and 0.0625, also forming a geometric sequence with the same common ratio 0.5. The discontinuity at $t = 2$ is -0.25, and that at $t = 2 + 2d = 2.9$, $t = 4 - 2d = 3.1$, or $t = 4$ is -0.0625. The discontinuities occur at those instants $t = 2, 2.9, 3.1$, and 4, when transmitted wavefronts interact with the hammer, because the transmitted wavefronts through a soft hammer have nonzero slopes. These discontinuities are negative because the transmitted waves have negative displacements. When they propagate to the hammer, they affect a release of the string contact from the hammer and hence cause negative discontinuity in the contact force. The negative discontinuity in the contact force has not been discussed previously in the literature. Reflected waves always introduce positive discontinuity in the contact force because the reflected waves have positive displacements. They whip the string upwards against the hammer when the wavefronts reach the hammer.

Graphs of $r(t)$ for dissipative and nondissipative, nonlinear soft hammers are illustrated in Fig. 1.24. The d and m are those used in Fig. 1.22, k_2 equals the k_1 in Fig. 1.22(a), and $v = 0.5$; $c = 5$, $t_f = 2.5$, and $n_p = 2{,}000$ in Figs. 1.24(a,b); $c = 0$, $t_f = 3.5$, and $n_f = 5{,}000$ in Fig. 1.24(c). The exponent $\alpha = 1.2$ in Fig. 1.24(a), and $\alpha = 3$ in Figs. 1.24(b,c). The initial value $r(0) = 2cv/(c+2) = 0.7143$ for the dissipative, nonlinear soft hammers in Figs. 1.24(a,b),

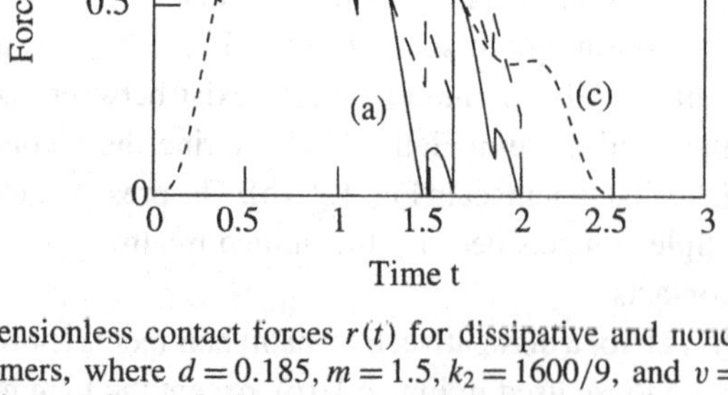

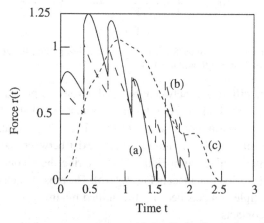

Fig. 1.24. Dimensionless contact forces $r(t)$ for dissipative and nondissipative, nonlinear soft hammers, where $d = 0.185$, $m = 1.5$, $k_2 = 1600/9$, and $v = 0.5$. Other parameters are: (*a*) $\alpha = 1.2$, $c = 5$; (*b*) $\alpha = 3$, $c = 5$; (*c*) $\alpha = 3$, $c = 0$.

whereas $r(0) = 0$ for the nondissipative, nonlinear soft hammer in Fig. 1.24(c). The discontinuities in $r(t)$ at $t = 2kd = 0.37, 0.74, 1.11$ $(k = 1, 2, 3)$ are $(5/7)^2$, $(5/7)^3$, and $(5/7)^4$ in Figs. 1.24(a,b). These discontinuities are independent of the nonlinearity in the hammer stiffness. In Fig. 1.24(a) the initial contact loss occurs at $t_1 = 1.464$, and one recontact occurs at $t_2 = 8d = 1.48$, with $r(t_2^+) = 0.11075$. The discontinuities in $r(t)$, occurring during the period of recontact at $t = 2(1 - d) = 1.63$ and $t = 10d = 1.85$ are $25/49$ and $5r(t_2^+)/7 = 0.0791$, respectively. The recontact loss occurs at $t_3 = 1.9864$, and there is no further contact. In Fig. 1.24(b) the same discontinuities occur at the same instants as those in Fig. 1.24(a). The contact is lost at $t = 2$ when a transmitted wavefront causes a negative jump in the contact force, and there is no recontact. The function $r(t)$ is continuous in Fig. 1.24(c) because $c = 0$. Contact loss occurs at $t = 2.53$ and there is no recontact. It is seen that dissipation in the hammer felt reduces the duration of contact.

In the last example, $r(t)$ for a heavy, hard hammer is shown in Fig. 1.25. The parameters are those of Fig. 1.20, except $D = 2.4$ m $(d = 0.4)$, $M = 100\rho L$ $(m = 100)$, $t_f = 20$, and $n_p = 4{,}000$. This problem can not be solved by the traveling wave method because of the large hammer to string mass ratio, as indicated in Hall (1987a). The initial value $r(0)$ and the discontinuities in $r(t)$ at $t_r^{(k)} = 0.8k (k = 1, 2, \ldots, 18)$ and $t_R^{(n)} = 1.2n$ $(n = 1, 2, \ldots, 12)$ are $2v = 0.0033892$. The discontinuities in $r(t)$ at $t = 2.4l$ $(l = 1, 2, \ldots, 6)$ are $4v = 0.0067784$, because $t_r^{(3l)}$ coincides with $t_R^{(2l)}$. The initial contact loss occurs at $t = 15.15$. One recontact is established at $t = t_r^{(19)} = 15.2$ and is lost at $t_3 = 15.385$. The contact force is nearly constant in this case before the

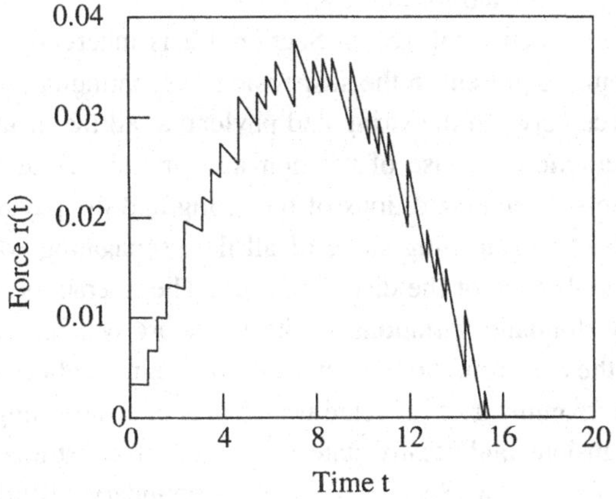

Fig. 1.25. Dimensionless contact force $r(t)$ for a heavy, hard hammer. The parameters are those of Fig. 1.20, except $D = 2.4$ m and $M = 1000\rho L$.

first reflected wave arrives at $t = 0.8$. The maximum contact force at $t = 7.2$ (0.143016 s) equals 0.037702 (12.6619 N).

1.5 Conclusions

A discontinuity associated with multiple wave scattering in the interaction force between a distributed structure and a discrete component occurs at the instants when the propagating wavefronts with nonvanishing slopes collide with the point of attachment of the discrete component. Classical spatial discretization approaches do not normally capture this dynamic behavior. The transient analysis developed exactly accounts for the discontinuities in the interaction force and allows for nonlinearity in the discrete component. It can also handle the problem of multiple contacts between the structure and the component. The main results of three transient problems are summarized below.

1. The Green's function formulation reduces the governing ordinary and partial differential equations for a constrained distributed system to a single delay-integral equation describing the interaction force. The presence of the delay terms accounts for the multiple wave scattering in the constrained distributed structure. The kernel and the dependent variable of the delay-integral equation can be discontinuous. Discontinuity in the kernel of an integral equation can cause numerical instability. A differentiation strategy within the framework of the theory of distributions eliminates the discontinuity in the kernel. The resulting governing equation is supplemented by representation of the discontinuity obtained prior to differentiation at a finite number of instants, such as $t_-^{(j)}$ and $t_+^{(j)}$. The discontinuities at all instants are implicitly accounted for in the analysis.
2. The transport system analyzed in Section 1.2 is inherently time varying. When damping is present in the suspension, discontinuity arises in the interaction force between the string and payload at all the instants of wave–payload interaction because of the nonvanishing slopes of the propagating wavefronts. Initial conditions of the string and the payload satisfying Eq. (1.45) lead to vanishing slope of all the propagating wavefronts and consequently absence of the discontinuities. The interaction force is characterized by dramatic variations in the transport system, as the payload approaches the exit from the span, because of the interaction of the payload with an infinite number of reflected wavefronts of diminishing energy.
3. Both the transient and steady-state responses of constrained translating strings are obtained in Section 1.3. For a boundary disturbance with a nonzero initial rate, all propagating wavefronts will have nonzero slopes

except those that transmit through a flexible constraint with nonzero mass m. The dynamic contact force is hence discontinuous for all modeled constraints other than a single spring element. For a rigid or flexible ($m \neq 0$) constraint, a constant discontinuity $-2\dot{e}(0)$ occurs in the contact force at the instants $t_r^{(n)}$. The contact force under a harmonic disturbance does not approach to the closed-form, harmonic forced response at the steady state. For a flexible massless constraint, discontinuities occur at all instants when propagating wavefronts coincide with the constraint. The first discontinuity in $r(t)$ at $t_+^{(1)}$ is $-2c\dot{e}(0)/(c+2)$. Discontinuities at later instants have decreasing amplitudes when $c \neq 0$. For a harmonic disturbance, the contact force approaches the closed-form, sinusoidal response at the steady state when $c \neq 0$.

4. A continuous forcing integral can be evaluated by using either the series expansion or the generalized function representation for the Green's function. The generalized function approach retains all the modes of the distributed system and gives an exact analysis in the spatial domain. The Gibb's phenomenon is manifested when a discontinuous forcing integral, such as $I_1(t)$ in Section 1.2, is calculated using the truncated Green's function series. For some integrals, such as dI_1/dt in Sections 1.2 and 1.3, their expansion series are divergent. To evaluate them, the generalized function approach must be used. It is easier, however, to use the series expansion to evaluate the continuous double integral, and it gives a virtually identical result to that obtained by the generalized function approach when a sufficient number of terms are retained in the expansion.
5. The application of the transient methodology to the classical problem of a piano string response when struck by a narrow hammer avoids key limitations of both standing and traveling wave methods (Zhu and Mote, 1994b). The standing wave method (Deb, 1972) suffers from a slowly convergent series and fails to predict the discontinuities in the contact force. The traveling method (Hall, 1987a, 1987b) calculates and superposes each individual wave. It cannot be applied to a nonlinear hammer, and suffers difficulty treating the exploding wave population as the number of reflected and transmitted waves to be explicitly calculated grows very substantially with increasing hammer to string mass ratio. The new method solves the string–hammer interaction for both the initial contact and all future recontacts in a unified manner for all hammer models. It can easily solve cases with virtually any large hammer to string mass ratio, and it is especially useful for nonlinear hammers and for multiple contacts where formal solution techniques have not been available.

Acknowledgment

The authors would like to thank the National Science Foundation and the authors' industrial sponsors for their support and interest in this work.

References

Archibald, F. R., and Emslie, A. G., 1958, "The Vibration of a String Having a Uniform Motion Along its Length," *Journal of Applied Mechanics*, Vol. 25, No. 3, pp. 347–48.

Bergman, L. A., and Nicholson, J. W., 1985, "Forced Vibration of a Damped Combined Linear System," *ASME Journal of Vibration, Acoustics, Stress, and Reliability in Design*, Vol. 107, pp. 275–81.

Boutillon, X., 1988, "Model for Piano Hammers: Experimental Determination and Digital Simulation," *Journal of the Acoustical Society of America*, Vol. 83, No. 2, pp. 746–54.

Butkovskiy, A. G., 1983, *Structural Theory of Distributed Systems*, Ellis Horwood, New York.

Deb, K. K., 1972, "Dynamics of the Pianoforte String and Hammer," *Journal of Sound and Vibration*, Vol. 20, No. 1, pp. 1–7.

George, W. H., 1924, "On the Helmholtz Theories for the Struck String – Part II. Experimental," *Philos. Mag.*, Vol. 6, No. 48, pp. 34–43.

George, W. H., 1925, "An Electrical Method for the Study of Impact Applied to the Struck String," *Proc. R. Soc. London Ser. A*, Vol. 108, pp. 284–95.

Ghosh, M., 1932, "Experimental Study of the Duration of Contact of an Elastic Hammer Striking a Damped Pianoforte String," *Indian Journal of Physics*, Vol. 7, pp. 365–82.

Hall, D. E., 1987a, "Piano String Excitation II: General Solution for a Hard Narrow Hammer," *Journal of the Acoustical Society of America*, Vol. 81, No. 2, pp. 535–41.

Hall, D. E., 1987b, "Piano String Excitation III: General Solution for a Soft Narrow Hammer," *Journal of the Acoustical Society of America*, Vol. 81, No. 2, pp. 547–55.

Hall, D. E., 1992, "Piano String Excitation. IV: Nonlinear Modeling," *Journal of the Acoustical Society of America*, Vol. 92, No. 1, pp. 95–105.

Hoskins, R. F., 1979, *Generalised Functions*, Ellis Horwood Limited, Chichester.

Lacey, C. A., and Talke, F. E., 1990, "Tape Dynamics in a High-Speed Helical Recorder," *IEEE Transactions on Magnetics*, Vol. 26, No. 5, pp. 2208–10.

Linz, P., 1985, *Analytical and Numerical Methods for Volterra Equations*, SIAM, Philadelphia.

Meirovitch, L., 1967, *Analytical Methods in Vibrations*, Macmillan Publishing Co., Inc., New York.

Perkins, N. C., 1990, "Linear Dynamics of a Translating String on an Elastic Foundation," *Journal of Vibration and Acoustics*, Vol. 112, No. 5, pp. 2–7.

Smith, C. E., 1964, "Motions of a Stretched String Carrying a Moving Mass Particle," *Journal of Applied Mechanics*, Vol. 31, No. 1, pp. 29–37.

Stahl, K. J., White, J. W., and Deckert, K. L., 1974, "Dynamic Response of Self-Acting Foil Bearings," *IBM J. Res. Develp.*, November, pp. 513–20.

Stakgold, I., 1979, *Green's Functions and Boundary Value Problems*, John Wiley & Sons, Inc., New York.

Sundaram, R., and Benson, R. C., 1990, "Tape Dynamics Following an Impact," *IEEE Transactions on Magnetics*, Vol. 26, No. 5, pp. 2211–13.

Tan, C. A., and Chung, C. H., 1993, "Transfer Function Formulation of Constrained Distributed Parameter Systems, Part I: Theory," *Journal of Applied Mechanics*, Vol. 60, pp. 1004–11.
Ulsoy, A. G., 1984, "Vibration Control in Rotating or Translating Elastic Systems," *Journal of Dynamic Systems, Measurement and Control*, Vol. 106, No. 1, pp. 6–14.
Ulsoy, A. G., and Mote, C. D., Jr., 1982, "Vibration of Wide Band Saw Blades," *Journal of Engineering for Industry*, Vol. 104, No. 1, pp. 71–78.
Wickert, J. A., and Mote, C. D., Jr., 1990, "Classical Vibration Analysis of Axially-Moving Continua," *Journal of Applied Mechanics*, Vol. 57, pp. 738–44.
Wickert, J. A., and Mote, C. D., Jr., 1991, "Traveling Load Response of an Axially-Moving String," *Journal of Sound and Vibration*, Vol. 149, No. 2, pp. 267–84.
Yanagisawa, T., Nakamura, K., and Aiko, H., 1981, "Experimental Study on Force–Time Curve During the Contact Between Hammer and Piano String," *Journal of the Acoustical Society of Japan*, Vol. 37, pp. 627–33.
Yang, B., 1992, "Transfer Functions of Constrained/Combined One-Dimensional Continuous Dynamic Systems," *Journal of Sound and Vibration*, Vol. 156, No. 3, pp. 425–43.
Yang, B., and Mote, C. D., Jr., 1991, "Controllability and Observability of Distributed Gyroscopic Systems," *Journal of Dynamic Systems, Measurement and Control*, Vol. 113, No. 1, pp. 11–17.
Zhu, W. D., and Mote, C. D., Jr., 1994a, "Free and Forced Response of an Axially Moving String Transporting a Damped Linear Oscillator," *Journal of Sound and Vibration*, Vol. 177, No. 5, pp. 591–610.
Zhu, W. D., and Mote, C. D., Jr., 1994b, "Dynamics of the Pianoforte String and Narrow Hammers," *Journal of the Acoustical Society of America*, Vol. 96, No. 4, pp. 1999–2007.
Zhu, W. D., and Mote, C. D., Jr., 1995, "Propagation of Boundary Disturbances in an Axially Moving Strip in Contact with Rigid and Flexible Constraints," *Journal of Applied Mechanics, Vol. 62*, pp. 873–79.

Nomenclature

$a(x), b(x)$	initial displacement and velocity
a	initial pulse location
C or c	damping constant
D or d; $\mathbf{D}$	constraint location; delay operator
$E(t)$ or $e(t)$; e_0	boundary disturbance; disturbance amplitude
$F(X, T)$ or $f(x, t)$	distributed external force
$g(x, \xi; t - \tau)$	Green's function
H or h	penetration displacement
I_i $(i = 1, 2, 3, 4)$	forcing integrals
K or k	stiffness
K_1 or k_1	linear hammer stiffness
K_2 or k_2	nonlinear hammer stiffness
L; l	span length; pulse length
L_m, l_m	lower limits in summation signs
M or m	mass
N, N_+, N_-	number of wave interaction
n_p	number of temporal points

P; $p(t)$, $p_1(t)$, $p_2(t)$	tension; forcing terms
$Q(T)$ or $q(t)$	external force on the payload
$R(T)$ or $r(t)$	interaction force
R_m, r_m	upper limits in summation signs
s	wave speed
T or t; t_f; Δt	time; final time; time step
$U(X, T)$ or $u(x, t)$	displacement
$U^{(n)}(x)$	eigenfunctions
$U(x)$ or $U_1(x)$, $U_2(x)$	separable spatial solution
V or v	transport speed
w or g	gravity constant
X or x	spatial coordinate
$Y(T)$ or $y(t)$	payload displacement
y_0, $\dot{y}_0$	initial payload displacement and velocity
$Z(T)$ or $z(t)$	contact point or hammer displacement

Greek

α	stiffness nonlinearity exponent
$\kappa(t, \tau)$, $\kappa_1(t, \tau)$	kernel
λ, μ	eigenvalue and its real part
ω, ζ	natural frequency and damping ratio
Ω	disturbance frequency
ϕ	sawtooth wave
ρ	mass density
$\Theta(x - a)$	pulse shape function
θ, ψ, η	incident, transmitted, and reflected waves

Superscripts

$-$	complex conjugation
$\hat{}$	transformed variable
$'$	differentiation
$*$	equilibrium
$\tilde{}$	relative to the equilibrium

Appendix A: Discontinuity in the Interaction Force in Constrained Stationary Strings

Consider first a pulse of length l propagating towards a damper attached at the origin $x = 0$ in an infinite, taut string of mass density ρ and tension P, as shown in Fig. 1.1(c) with $k = 0$. The pulse is initially at $x \in [a, a + l]$ with its shape described by $\Theta(x - a)$ $(a \leq x \leq a + l, a > 0)$. Assume $\Theta(0) = \Theta(l) = 0$ to ensure continuous displacement of the string. The incident wave is

$$\theta(x + st) = [H(x + st - a) - H(x + st - a - l)]\Theta(x + st - a), \tag{A.1}$$

where $s = \sqrt{P/\rho}$ is the wave speed in the string. The transmitted and reflected waves from the damper with damping constant C are denoted by $\psi(st + x)$ and $\eta(st - x)$, respectively. The boundary conditions at $x = 0$ are

$$\psi(st) = \theta(st) + \eta(st), \qquad P[\theta'(st) - \eta'(st) - \psi'(st)] = sC\psi'(st), \tag{A.2}$$

where the prime denotes differentiation with respect to the argument. Solution to (A.2) by use of the initial conditions $\psi(0) = \theta(0) = 0$ yields

$$\psi(st) = 2P\theta(st)/(2P + sC), \qquad \eta(st) = -sC\theta(st)/(2P + sC). \tag{A.3}$$

By (A.3) and (A.1), the damping force at $x = 0$ is

$$\begin{aligned} F_d = -Cs\psi'(st) &= -\frac{2CsP}{2P + Cs}\{[\delta(st - a) - \delta(st - a - l)]\Theta(st - a) \\ &\quad + [H(st - a) - H(st - a - l)]\Theta'(st - a)\} \\ &= -\frac{2CsP}{2P + Cs}[H(st - a) - H(st - a - l)]\Theta'(st - a), \end{aligned} \tag{A.4}$$

where the delta function terms vanish because $\Theta(0) = \Theta(l) = 0$. The discontinuities in the damping force at $t = a/s$ and $t = (a + l)/s$, obtained from (A.4), are $-2CsP\Theta'(0)/(2P + Cs)$ and $2CsP\Theta'(l)/(2P + Cs)$, respectively. The discontinuity in the dimensionless damping force F_d/P when the pulse reaches the damper is $-2c/(c + 2)\Theta'(0)$, where $c = C/(P\rho)^{1/2}$.

Likewise, the discontinuities in the spring force, when the pulse reaches and leaves an attached spring, vanish because $\Theta(0) = \Theta(l) = 0$. Hence for the constraint of a spring and a damper shown in Fig. 1.1(c), the discontinuity in the dimensionless constraint force is $-2c/(c + 2)\Theta'(0)$ when the pulse impinges on the constraint, in agreement with the prediction in Eq. (1.92), because the slope at the wavefront $\Theta'(0)$ is equivalent to $\dot{e}(0)$.

For a rigid constraint shown in Fig. 1.1(a), the pulse is reflected when it reaches the constraint. We have $\eta(st) = -\theta(st)$. The interaction force between the string and constraint is

$$\begin{aligned} F_r = -2P\theta'(st) &= -2P\{[\delta(st - a) - \delta(st - a - l)]\Theta(st - a) \\ &\quad + [H(st - a) - H(st - a - l)]\Theta'(st - a))\} \\ &= -2P[H(st - a) - H(st - a - l)]\Theta'(st - a), \end{aligned} \tag{A.5}$$

where the delta function terms vanish because $\Theta(0) = \Theta(l) = 0$. Hence the discontinuities in the dimensionless constraint force F_r/P at $t = a/s$ and $t = (a + l)/s$ are $-2\Theta'(0)$ and $2\Theta'(l)$, respectively. Again the discontinuity in the force when the pulse contacts the constraint agrees with that predicted from Eq. (1.87).

For a flexible constraint represented by a spring–mass–dashpot shown in Fig. 1.1(b), a similar procedure can be adopted to calculate the discontinuity in the interaction force when the pulse reaches the constraint. The result can be alternatively obtained from Eq. (1.84). By Eq. (1.84) the first discontinuity in $\tilde{r}(t)$ at $t_+^{(1)}$ is $-2\dot{e}(0)$. By replacing the $\dot{e}(0)$ there by $\Theta'(0)$, the discontinuity in the dimensionless constraint force when the pulse reaches the constraint in Fig. 1.1(b) is also $-2\Theta'(0)$, the same as that for a rigid constraint.

Appendix B: Proof of the Theorem of Recontact in Section 1.4.1.4

The string and hammer responses for $t \geq t_{2k}$ are expressed by Eqs. (1.130) and (1.131), where $r(t)$ is known for $t < t_{2k}$ and unknown for $t > t_{2k}$. The condition of recontact for $t > t_{2k}$ is given by Eq. (1.126) for a hard hammer and by Eq. (1.127) or (1.128) for a linear or nonlinear soft hammer. The equations describing the force of recontact are those for the initial contact.

For the hard hammer, Equation (1.134) at t_{2k}^+ with $r(t_{2k}^-) = 0$ and Eqs. (1.130),

(1.131), and (1.133) give

$$\begin{aligned}\frac{1}{2}r\left(t_{2k}^{+}\right) = &-\left[\frac{1}{2}r\left(t_{2k}^{-}\right) + \sum_{j=1}^{N} r(t_{2k} - 2j)H(t_{2k} - 2j)\right.\\ &-\frac{1}{2}\sum_{j=1}^{N+1} r[t_{2k} - 2d - 2(j-1)]H[t_{2k} - 2d - 2(j-1)]\\ &\left.-\frac{1}{2}\sum_{j=1}^{N+1} r(t_{2k} + 2d - 2j)H(t_{2k} + 2d - 2j)\right]\\ &-\left(\frac{1}{m}\int_0^{t_{2k}} r(\tau)\,d\tau - v\right) = \dot{u}(d, t_{2k}) - \dot{z}(t_{2k}).\end{aligned} \tag{B.1}$$

Therefore, the force of initial recontact is $r(t_{2k}^{+}) = 2[\dot{u}(d, t_{2k}) - \dot{z}(t_{2k})]$.

For the linear soft hammer, Equation (1.135) at t_{2k}^{+} with $r(t_{2k}^{-}) = 0$ and Eqs. (1.130), (1.131), and (1.133) give

$$\begin{aligned}\left(1 + \frac{c}{2}\right)r\left(t_{2k}^{+}\right) = &-k_1\int_0^{t_{2k}} r(\tau)g(d, d; t-\tau)\,d\tau - \frac{k_1}{m}\int_0^{t_{2k}} (t-\tau)r(\tau)\,d\tau\\ &+k_1 v t_{2k} - c\left[\frac{1}{2}r\left(t_{2k}^{-}\right) + \sum_{j=1}^{N} r(t_{2k} - 2j)H(t_{2k} - 2j)\right.\\ &-\frac{1}{2}\sum_{j=1}^{N+1} r[t_{2k} - 2d - 2(j-1)]H[t_{2k} - 2d - 2(j-1)]\\ &\left.-\frac{1}{2}\sum_{j=1}^{N+1} r(t_{2k} + 2d - 2j)H(t_{2k} + 2d - 2j)\right]\\ &-\frac{c}{m}\int_0^{t_{2k}} r(\tau)\,d\tau + cv = k_1[u(d, t_{2k}) - z(t_{2k})]\\ &+c[\dot{u}(d, t_{2k}) - \dot{z}(t_{2k})] = c[\dot{u}(d, t_{2k}) - \dot{z}(t_{2k})],\end{aligned} \tag{B.2}$$

where the seperation at contact $z(t_{2k}) - u(d, t_{2k}) = 0$ is used. Hence $r(t_{2k}^{+}) = 2c[\dot{u}(d, t_{2k}) - \dot{z}(t_{2k})]/(c + 2)$.

For the nonlinear soft hammer, Equation (1.128) at t_{2k}^{+} with $r(t_{2k}^{-}) = 0$, $z(t_{2k}) = u(d, t_{2k})$, and Eqs. (1.130) and (1.133) yield

$$\begin{aligned}\left(1 + \frac{c}{2}\right)r\left(t_{2k}^{+}\right) = &\, c\dot{u}(d, t_{2k}) - c\left[\frac{1}{2}r\left(t_{2k}^{-}\right) + \sum_{j=1}^{N} r(t_{2k} - 2j)H(t_{2k} - 2j)\right.\\ &-\frac{1}{2}\sum_{j=1}^{N+1} r[t_{2k} - 2d - 2(j-1)]H[t_{2k} - 2d - 2(j-1)]\\ &\left.-\frac{1}{2}\sum_{j=1}^{N+1} r(t_{2k} + 2d - 2j)H(t_{2k} + 2d - 2j)\right]\\ = &\, c[\dot{u}(d, t_{2k}) - \dot{z}(t_{2k})].\end{aligned} \tag{B.3}$$

Hence $r(t_{2k}^{+}) = 2c[\dot{u}(d, t_{2k}) - \dot{z}(t_{2k})]/(c + 2)$.

2

On the Problem of a Distributed Parameter System Carrying a Moving Oscillator

B. YANG, C. A. TAN, and L. A. BERGMAN

Abstract

The problem of an elastic distributed system coupled with a moving oscillator, often referred to as the "moving-oscillator" problem, is studied in this chapter. The problem is formulated using a "relative displacement" model. It is shown that, in the limiting case, the moving-mass problem is recovered. The coupled equations of motion are recast into an integral equation, which is amenable to solution by both iterative and direct numerical procedures. The response of a string with a moving oscillator is studied using the direct numerical method.

2.1 Introduction

2.1.1 Perspective

The prediction of the dynamic response of a distributed elastic system that supports one or more translating elastic subsystems has been a fundamental problem of interest for well over a century. Interest in this problem originates in structural engineering for the design of railroad tracks, railroad bridges, and highway bridges, wherein the accurate calculation of loads is essential for reliable design and accurate life prediction (Stokes, 1883; Ting and Yener, 1983; Tan and Shore, 1968). It has been observed that, as a structure is subjected to moving loads, the dynamic deflection and stresses can be significantly higher than those observed in the static case. Hence, strict design criteria are now required as structural engineers become more aggressive in the use of long, flexible spans in cable-stayed and suspension bridges and compliant bearings in highway bridges to accommodate environmental loads.

Besides remaining a classical mechanics problem, the topic we are addressing here also has application to structures such as parking garages and aircraft carriers, advanced propulsion concepts such as railguns (Johnson, 1993), high-speed precision machining (Chen and Wang, 1994; Katz et al., 1987, 1988), magnetic disk drives (Iwan and Stahl, 1973; Iwan and Moeller, 1976; Shen, 1993), and cable transporting materials (Zhu and Mote, 1994). In machining processes, the interaction of the machine tool and the work-piece generates a traveling force that depends on the tool deflection. Moreover, this moving force is usually feedback-controlled by a computer to minimize the vibration of the cutting tool. In the disk drive problem, the stability of the rotating disk is determined by its complex interaction with a stationary read–write head, which may be modeled as an elastic subsystem.

An important characteristic of the response solution of a distributed parameter structure carrying one or more traveling elastic subsystems is that no steady-state response exists for this problem, and the system stability is determined by the transient response. Hence, classical solution techniques for steady-state response cannot be applied. Moreover, the solution must be obtained by solving a set of coupled partial-ordinary differential equations simultaneously. In general, three types of problems have been considered in the literature. When the inertia of the moving subsystem is small, the constraint force can be treated as a moving load. This is called the *moving-load* problem and occurs in high-speed machining processes. As the coupling stiffness between the distributed structure and the elastic subsystem approaches infinity, we have the *moving-mass* problem. It should be noted that the constraint force that couples the two elastic subsystems remains finite for this problem. The *moving-oscillator* problem is obtained when the coupling stiffness is finite and the inertial effect of the subsystem is not negligible. This problem is found in many realistic engineering problems, for example, a vehicle traveling across a bridge.

While most research has focused on the moving-force and moving-mass problems because of the perception of simpler solutions, the moving-oscillator problem has rarely been addressed in the literature, and to the authors' knowledge few solutions have been obtained. One group of researchers used the finite element method (Lin and Trethewey, 1990), whereas several others employed Fourier series (Inglis, 1934). More recently, the problem was solved by expanding the kernel of the governing integral equation in the eigenfunctions of the continuum (Pesterev and Bergman, 1997). The discussion of the moving-force and moving-mass problems has been properly presented (Sadiku and Leipholz, 1987). As Sadiku and Leipholz pointed out, the analysis of the original "moving-force moving-mass" problem (Timoshenko et al., 1974) is nontrivial, due primarily to one or more singularities in the inertia operator of the system, which depends on both the temporal and spatial variables. Thus, the "moving-force moving-mass" problem is generally approximated to first order by the moving-force problem, which has been well documented (Fryba, 1972).

It should be noted that neither the "moving-force" solution given by Fryba nor the "moving-force moving-mass" solution given by Sadiku and Leipholz can adequately account for the complex and important dynamic effects caused by the compliance of the moving elastic subsystem (i.e., the oscillator). To date, the only rigorous and general solution for the moving-oscillator problem to appear in the literature is that of Pesterev and Bergman, 1997. In this paper, we present a formulation for the transient response of a one-dimensional elastic structure carrying a single degree-of-freedom undamped oscillator and propose several additional methods for its solution.

2.1.2 Literature Survey

There are a number of early works on the dynamic response of a structure under the influence of a moving-force system (Mathews, 1958; Kenny, 1954; Fryba, 1957), in which the response of infinite beams on elastic foundations and under moving loads was studied. The results of Mathews and Kenny were extended to a Timoshenko beam, and the effects of shear deformation on the beam response and critical speeds were examined (Chonan, 1976, 1978; Achenbach and Sun, 1965). The dynamic response to a moving load of a tensioned Timoshenko beam on a Winkler foundation was examined (Prasad, 1981), as was the response of a finite Mindlin beam under moving loads (Mackertich, 1990). The response of an infinite string or beam under moving loads was obtained by a wave-number approach (Pierucci, 1993). Here, relations for the wave numbers as a function of the traveling speed of the load were established, and critical Mach numbers were identified. The response of a Timoshenko beam on an elastic foundation and subject to a moving step load was analyzed by the Fourier transform method (Felszeghy, 1995a, 1995b). Besides string and beam problems, the response of other elastic structures under moving loads include infinite plates (Achenbach et al., 1967; Yen and Chou, 1970; Chonan, 1976; Reismann, 1964; Alder and Reismann, 1974), rotating shafts (Katz et al., 1988; Zu et al., 1994), rotating cylindrical shells (Huang and Hsu, 1990; Chonan, 1977), and rotating plates for disk drive and circular saw applications (Iwan and Stahl, 1973; Iwan and Moeller, 1976; Shen, 1993; Mote, 1970, 1977; Yu and Mote, 1987).

The moving-mass problem, involving one-dimensional distributed elastic structures, has been examined (Sadiku and Leipholz, 1987; Saigal, 1986; Stanisic and Hardin, 1969; Stanisic et al., 1974; Stanisic, 1985; Hutton and Counts, 1974; Wilson and Wilson, 1984; Nelson and Conover, 1971; Benedetti, 1974). Among these papers, Sadiku and Leipholz provide the most comprehensive discussion. A Green's function approach is proposed for the "moving-force moving-mass" problem, and the solution by successive approximations is shown to converge. Other papers employ an eigenfunction expansion method to determine the response solution (Stanisic et al., 1974; Stanisic, 1985). The response of plates under moving masses has also been considered (Saigal et al., 1987; Agrawal and Saigal, 1986), as has the response of an infinite beam to a moving vibrating mass (Duffy, 1990).

As discussed earlier, the moving-oscillator problem has rarely been discussed in the literature. A finite element solution for the response of a finite beam carrying a damped oscillator has been presented (Lin and Trethewey, 1990), as have several Fourier series solutions (Inglis, 1934). Good agreement between the finite element analysis and some experimental data was obtained. The finite element method has also been applied to the moving-load problem (Hino et al., 1984, 1985; Yoshimura et al., 1986; Filho, 1978).

Other pertinent studies related to the three types of problems addressed earlier include the response of beams under the combined effect of an axial force and a moving load (Kerr, 1972; Chonan, 1975) and the random vibration of elastic structures under moving loads (Fryba, 1976, 1977, 1986; Bolotin, 1965, 1982; Crandall and Mark, 1963; Knowles, 1968, 1970). It should be noted that, because of the nature of the coupled problem, most analytical approaches use the Green's function and integral formulation ideas (Wickert and Mote, 1991; Smith, 1964; Rodeman et al., 1976; Ting et al., 1974). Also, a solution procedure, based on the method of undetermined coefficients, to evaluate the response of a simply supported beam subjected to a constant-velocity moving two-degree-of-freedom system with a pulsating force was examined (Licari and Wilson, 1962).

2.1.3 Research Objectives

The objectives of this paper are twofold. First, the formulation of the problem of a simple oscillator traversing a one-dimensional elastic continuum is presented in Section 2.2, with careful attention to the relation between this problem and the moving-mass problem. It is shown that, as the stiffness of the spring approaches infinity, we recover the moving-mass problem discussed in Sadiku and Leipholz. Moreover, the present formulation leads to a much simpler solution procedure that does not explicitly include the inertial convective term, though its effect has been carefully included in the formulation. The second objective is to present an exact integral formulation for the response solution of the coupled dynamic system. This is given in Section 2.3. The resulting integral equations require only integration in the time variables, and the solution procedures are shown to be more straightforward than previously suggested methods for the moving-force and moving-mass problems. Two methods are introduced to solve the integral equations: an iterative scheme based upon successive approximations and a scheme based upon numerical integration. The latter provides a novel and direct approach that avoids tedious recursive iterations. The proposed solution technique is applied to a taut, uniform string traversed by a simple, undamped oscillator. Results are presented in the context of resonance of the underlying structural subsystems.

2.2 Problem Formulation

2.2.1 Equation of Motion

Consider a one-dimensional continuum carrying a moving oscillator of stiffness k and inertia m, shown in Fig. 2.1, where c is the constant speed of the oscillator, L is the length of the continuum, and $f(x,t)$ is the external force acting on the distributed system. The displacement $w(x,t)$ of the distributed parameter system is governed by the partial differential equation

$$\rho w_{tt}(x,t) + \mathbf{K}w(x,t) = f(x,t) + q_c(t)\delta(x-ct), \qquad x \in \Omega \tag{2.1a}$$

with the homogeneous boundary condition

$$\mathbf{B}w(x,t) = 0, \qquad x \in \partial\Omega \tag{2.1b}$$

and the initial conditions

$$w(x,0) = a(x), \qquad w_t(x,0) = b(x), \qquad x \in \Omega, \tag{2.1c}$$

where $w_t = \partial w/\partial t$, the domain $\Omega = (0, L)$, the boundary includes the two end points $x = 0$ and $x = L$, ρ is the mass density of the distributed system, $q_c(t)$ is the constraint force applied to the distributed system by the moving oscillator

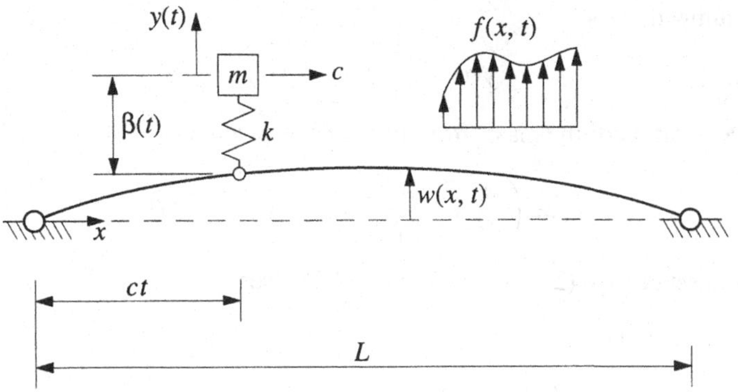

Fig. 2.1. Distributed parameter system with a moving oscillator.

at their contact point $x = ct$, $\mathbf{B}$ is the boundary operator, and $a(x)$ and $b(x)$ are the initial displacement and initial velocity of the continuum, respectively.

The positive definite spatial operator, $\mathbf{K}$, describes the stiffness or elastic restoring force of the distributed system. For example, for a taut uniform string under tension p, $\mathbf{K} = -p\partial^2/\partial x^2$; and for a uniform Euler–Bernoulli beam with bending stiffness EI, $\mathbf{K} = EI\partial^4/\partial x^4$.

The constraint force is given by

$$q_c(t) = k\beta(t), \tag{2.2}$$

where $\beta(t)$ is the relative displacement between the distributed system and the oscillator (refer to Fig. 2.1),

$$\beta(t) = y(t) - w(ct, t). \tag{2.3}$$

Here, $y(t)$ is the displacement of the lumped mass, which is along the same line as the displacement of the continuum at the contact point $x = ct$. Both $y(t)$ and $w(x, t)$ are the absolute displacements measured in a fixed frame. The motion of the lumped mass satisfies

$$m\ddot{y}(t) = -q_c(t) - mg = -k\beta(t) - mg. \tag{2.4}$$

By Eq. (2.4), Eq. (2.1a) can be expressed as

$$\rho w_{tt}(x, t) + \mathbf{K}w(x, t) = f(x, t) + k\beta(t)\delta(x - ct), \qquad x \in \Omega. \tag{2.5}$$

2.2.2 *Asymptotic Behavior: The Moving Mass Problem*

A continuum carrying a moving mass is a special case of the above coupled system model when the spring coefficient k of the oscillator goes to infinity. To show this, consider the relation between the absolute and relative displacements of the lumped mass

$$y(t) = w(ct, t) + \beta(t). \tag{2.6}$$

The acceleration of the mass, from Eq. (2.6), is given by

$$\ddot{y}(t) = \left(\frac{\partial}{\partial t} + c\frac{\partial}{\partial x}\right)^2 w(x, t)|_{x=ct} + \ddot{\beta}(t). \tag{2.7}$$

Substitution of Eqs. (2.4) and (2.7) into (2.5) gives

$$\begin{aligned} &\rho w_{tt}(x, t) + \mathbf{K}w(x, t) \\ &\quad = f(x, t) - m\left[\left(\frac{\partial}{\partial t} + c\frac{\partial}{\partial x}\right)^2 w(x, t)|_{x=ct} + \ddot{\beta}(t) + g\right]\delta(x - ct) \end{aligned} \tag{2.8}$$

for $x \in \Omega$, where g is the acceleration of gravity. When the spring coefficient k is allowed to become infinite, the relative displacement $\beta(t)$ is zero, implying that $\ddot{\beta}(t)$ will be zero for any t. Thus, Eq. (2.8) becomes

$$\rho w_{tt}(x,t) + \mathbf{K}w(x,t) = f(x,t) - m\left[\left(\frac{\partial}{\partial t} + c\frac{\partial}{\partial x}\right)^2 w(x,t) + g\right]_{x=ct} \times \delta(x - ct), \qquad x \in \Omega, \tag{2.9}$$

which is identical to that used in the previous studies (Sadiku and Leipholz, 1987). The constraint force in this limiting case is

$$q_c(t) = k\beta(t) = -m\left[\left(\frac{\partial}{\partial t} + c\frac{\partial}{\partial x}\right)^2 w(x,t)\right]_{x=ct} - mg. \tag{2.10}$$

Clearly, as $k \to \infty$, $\beta(t) \to 0$; but the product $k\beta(t)$ must be determinate and finite.

Equations (2.5) and (2.8) are equivalent insofar as they both describe the response of the coupled system. The latter, however, has been shown to be advantageous for demonstrating the relationship between the moving-mass and moving-oscillator problems. The former, on the other hand, is simpler for developing solution procedures because its right-hand side terms do not explicitly involve spatial and temporal terms like $(\partial/\partial t + c\partial/\partial x)^2$.

2.3 General Solution Procedures

As shown in earlier sections, the fundamental problem is to solve Eqs. (2.4) and (2.5), subject to the boundary and initial conditions given in Eq. (2.1). Note that the Dirac delta function $\delta(x - ct)$ in Eq. (2.5) indicates that the oscillator is initially at the left end of the continuum, $x = 0$, and travels to the right end, $x = L$, at time $t = L/c$. Hence, $w(x,t)$ in the interval $0 < t \leq L/c$ is to be determined. In the following analysis, the coupled system is first decomposed into its component distributed and lumped subsystems, and the Green's function for each is determined. The subsystems are then coupled through the use of the Green's functions, leading to an integral equation for the response of the coupled system.

2.3.1 Response of the Distributed System without the Moving Oscillator

The displacement $z(x,t)$ of the distributed system without the moving oscillator is described by the equation

$$\rho z_{tt}(x,t) + \mathbf{K}z(x,t) = f(x,t), \qquad x \in \Omega$$

subject to the boundary and initial conditions

$$\begin{aligned} \mathbf{B}z(x,t) &= 0, & x &\in \partial\Omega, \\ z(x,0) &= a(x), & z_t(x,0) &= b(x). \end{aligned} \tag{2.11}$$

The Green's function formula for such a system is

$$\begin{aligned} z(x,t) = &\int_\Omega \rho\left(\frac{\partial}{\partial t}G(x,\xi,t)a(\xi) + G(x,\xi,t)b(\xi)\right)d\xi \\ &+ \int_0^t\int_\Omega G(x,\xi,t-\tau)f(\xi,\tau)\,d\xi\,d\tau. \end{aligned} \tag{2.12}$$

The distributed system space–time Green's function is expressed by the eigenfunction expansion

$$G(x,\xi,t) = \sum_{j=1}^{\infty} \frac{1}{\omega_j}\phi_j(x)\phi_j(\xi)\sin\omega_j t, \tag{2.13}$$

where ω_j and $\phi_j(x)$, $j = 1, 2, \ldots$, are the solutions of the eigenvalue problem

$$\mathbf{K}\phi_j = \omega_j^2\rho\phi_j \tag{2.14}$$

with the eigenfunctions ϕ_j satisfying the orthonormality condition

$$\int_\Omega \rho\phi_j(x)\phi_l(x)\,dx = \begin{cases} 1, & j = l, \\ 0, & j \neq l. \end{cases} \tag{2.15}$$

2.3.2 Response of the Moving Oscillator

By Eqs. (2.3) and (2.4), the equation of motion for the moving oscillator is

$$\ddot{y}(t) + \mu^2 y(t) = \mu^2 w(ct,t) - g, \qquad \mu = \sqrt{k/m}, \tag{2.16}$$

where μ is the natural frequency of the oscillator. Physically, $w(ct,t)$ is the unknown motion of the foundation (or support) of the oscillator. The response of the oscillator is

$$y(t) = y(0)\cos\mu t + \frac{1}{\mu}\dot{y}(0)\sin\mu t + \int_0^t h(t-\tau)w(c\tau,\tau)\,d\tau - \frac{g}{\mu^2}(1-\cos\mu t), \tag{2.17}$$

where $y(0)$ an $\dot{y}(0)$ are the initial displacement and velocity of the mass, respectively, and $h(t)$ is μ^2 multiplied by the influence or Green's function of the oscillator due to the foundation motion, which is given by

$$h(t) = \mu\sin\mu t. \tag{2.18}$$

2.3.3 Coupling of the Distributed and Lumped Subsystems

Application of Eq. (2.12) to (2.5) leads to

$$w(x,t) = z(x,t) + k\int_0^t G(x, c\tau, t-\tau)(y(\tau) - w(c\tau,\tau))\,d\tau$$

or, alternately,

$$w(x,t) = w_{eI}(x,t) + w_c(x,t) + w_l(x,t) \tag{2.19}$$

where $z(x,t)$ is given in Eq. (2.12),

$$w_{eI}(x,t) = z(x,t) + k\int_0^t G(x, c\tau, t-\tau)$$
$$\times\left(y(0)\cos\mu\tau + \frac{1}{\mu}\dot{y}(0)\sin\mu\tau - \frac{g}{\mu^2}(1-\cos\mu t)\right)d\tau,$$

$$w_c(x,t) = -k\int_0^t G(x, c\tau, t-\tau)w(c\tau,\tau)\,d\tau,$$

and

$$w_l(x,t) = k\int_0^t \left\{G(x, c\tau, t-\tau)\int_0^\tau h(\tau-\eta)w(c\eta,\eta)\,d\eta\right\}d\tau. \tag{2.20}$$

Note that, for given initial conditions and external disturbances, $w_{eI}(x,t)$ is known. The $w_l(x,t)$ in the last of Eqs. (2.20) involves double integration. Interchanging the order of the double integration yields

$$w_l(x,t) = k\int_0^t \left\{w(c\eta,\eta)\int_\eta^t G(x, c\tau, t-\tau)h(\tau-\eta)\,d\tau\right\}d\eta$$
$$= k\int_0^t R(x,t,\eta)w(c\eta,\eta)\,d\eta, \tag{2.21}$$

where

$$R(x,t,\eta) = \int_\eta^t G(x, c\tau, t-\tau)h(\tau-\eta)\,d\tau = \mu\sum_{j=1}^{\infty}\frac{1}{\omega_j}\phi_j(x)r_j(t,\eta) \tag{2.22}$$

with

$$r_j(t,\eta) = \int_\eta^t \phi_j(c\tau)\sin\omega_j(t-\tau)\sin\mu(\tau-\eta)\,d\tau. \tag{2.23}$$

It follows that the response of the coupled system is governed by the integral equation

$$w(x,t) = w_{eI}(x,t) + \int_0^t J(x,t,\eta)w(c\eta,\eta)\,d\eta, \tag{2.24}$$

where the integral kernel is given by

$$J(x,t,\eta) = kR(x,t,\eta) - kG(x,c\eta,t-\eta). \tag{2.25}$$

By eigenfunction expansions, the kernel $J(x,t,\eta)$ is of the form

$$J(x,t,\eta) = k\sum_{j=1}^{\infty} \frac{1}{\omega_j}\phi_j(x)\{\mu r_j(t,\eta) - \phi_j(c\eta)\sin\omega_j(t-\eta)\}. \tag{2.26}$$

Since the eigensolutions of the uncoupled distributed parameter system can be accurately determined by many well-developed techniques, the integral kernel can be precisely and systematically evaluated.

The boundary–initial value problem for the coupled system is defined by the differential equations (2.4) and (2.5), which are subject to arbitrary boundary and initial conditions. Through use of the Green's function formulas (2.12) and (2.17), and interchange of the order of the double integration (2.21), the original problem is transformed into Eq. (2.24), which requires only a single integration with respect to time. Solution methods based on integral equation formulations have many advantages, particularly with respect to accuracy, convergence, and efficiency. Accordingly, Eq. (2.24) provides a foundation for the development of systematic and accurate methods for the moving-oscillator problem.

2.4 Solution Procedures

In this section, two methods are proposed for the solution of the integral equation (2.24): successive approximations and a new numerical integration formulation.

2.4.1 Solution by Successive Approximations

Equation (2.24) can be written as

$$w = w_{eI} + \mathbf{T}[w], \tag{2.27}$$

where the integral operator $\mathbf{T}$ is defined by

$$\mathbf{T}[u] = \int_0^t J(x,t,\eta)u(c\eta,\eta)\,d\eta, \qquad u \in \mathbf{H}, \tag{2.28}$$

and $\mathbf{H}$ is a Hilbert space. The proposed method considers a sequence of functions, $w_0(x,t)$, $w_1(x,t)$, ..., that are generated by the following successive integrations:

$$\begin{aligned} w_0(x,t) &= w_{eI}(x,t), \\ w_j &= w_{eI} + \mathbf{T}[w_{j-1}], \qquad j = 1, 2, \ldots. \end{aligned} \tag{2.29}$$

If the operator $\mathbf{T}$ represents a contraction mapping, namely $\|\mathbf{T}\| \leq \theta < 1$, the sequence $\{w_j\}$ converges to the solution of the integral equation

$$w(x,t) = \lim_{j\to\infty} w_j(x,t). \tag{2.30}$$

Equation (2.27) has been shown to be a contraction mapping for the case of the "moving-mass" problem (Sadiku and Leipholz, 1987). It is hoped that a formal proof for convergence of the "moving-oscillator" problem can be shown at a future time.

2.4.2 Solution by Numerical Integration

This method requires the discretization of the integral Eq. (2.24) over the space–time continuum. Define a solution region $\mathbf{S}$ by

$$\mathbf{S} = \{(x,t) \mid (0 \leq x \leq L, 0 \leq t \leq L/c)\}, \tag{2.31}$$

where L is the length of the distributed system and c is the (constant) traveling speed of the oscillator. Here, the upper bound of the variable t is the time needed for the oscillator to travel from the left end ($x = 0$) to the right end ($x = L$) of the continuum. In Fig. 2.2, the region $\mathbf{S}$ is divided into an $N \times N$ equally spaced mesh, where

$$x_j = j\Delta x, \qquad t_j = j\Delta t, \qquad j = 1, 2, \ldots, N \tag{2.32}$$

and

$$\Delta x = \frac{L}{N}, \qquad \Delta t = \frac{1}{c}\Delta x = \frac{L}{cN}. \tag{2.33}$$

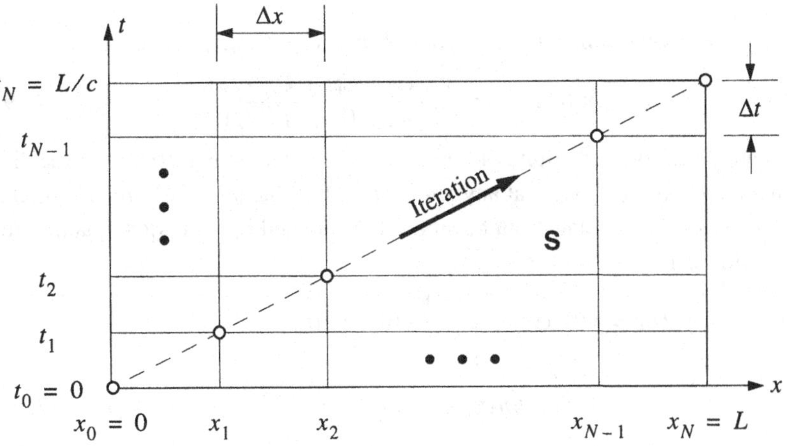

Fig. 2.2. Numerical integration scheme.

The response of the distributed system at times $t = (t_1, t_2, \ldots, t_N)$ is evaluated. Consider the response of the continuum at time t_m, $1 \le m \le N$,

$$w(x, t_m) = w_{eI}(x, t_m) + \int_0^{t_m} J(x, t_m, \eta) w(c\eta, \eta)\, d\eta. \tag{2.34}$$

Assume that the mesh of the region **S** is sufficiently dense. Then, the integral in Eq. (2.34) can be approximated by the numerical quadrature

$$w(x, t_m) = w_{eI}(x, t_m) + \sum_{i=0}^{m} d_i J(x, t_m, t_i) w(ct_i, t_i), \qquad 0 \le x \le ct, \tag{2.35}$$

where the d_i are the constants of a chosen quadrature formula. For the trapezoidal rule, for instance,

$$d_0 = \frac{1}{2}\Delta t, \qquad d_1 = d_2 = \cdots = d_{m-1} = \Delta t, \qquad d_m = \frac{1}{2}\Delta t. \tag{2.36}$$

It is clear from Eq. (2.35) that evaluation of $w(x, t_m)$ for any x requires only knowledge of $w(ct_i, t_i)$, $i = 0, 1, \ldots, m - 1$. This is demonstrated in Fig. 2.2, where circles on the diagonal of the rectangle mark the points $w(ct_i, t_i)$.

Suppose that $w(ct_0, t_0), w(ct_1, t_1), \ldots, w(ct_{m-1}, t_{m-1})$ have been determined. Then Eq. (2.35) can be rewritten as

$$w(x, t_m) = w_{eI}(x, t_m) + U_m(x) + d_m J(x, t_m, t_m) w(ct_m, t_m), \qquad 0 \le x \le ct, \tag{2.37}$$

where

$$U_m(x) = \sum_{i=0}^{m-1} d_i J(x, t_m, t_i) w(ct_i, t_i), \tag{2.38}$$

Replacing x by ct_m in Eq. (2.37) and solving for $w(ct_m, t_m)$ gives

$$w(ct_m, t_m) = \frac{w_{eI}(ct_m, t_m) + U_m(ct_m)}{1 - d_m J(ct_m, t_m, t_m)}. \tag{2.39}$$

Equation (2.39) is substituted into Eq. (2.37) to obtain the response of the distributed parameter system at any $x \in [0, L]$. Once the $w(ct_i, t_i)$ are known, the response of the oscillator can be similarly evaluated by numerical quadrature; e.g., Eq. (2.17),

$$y(t_m) = y(0)\cos\mu t_m + \frac{1}{\mu}\dot{y}(0)\sin\mu t_m + \sum_{i=0}^{m} d_i h(t_m - t_i) w(ct_i, t_i) - \frac{g}{\mu^2}(1 - \cos\mu t_m). \tag{2.40}$$

This suggests an iterative procedure for the solution of the "moving-oscillator" problem. To evaluate the response of the coupled system at $t = t_n$, $1 \leq n \leq N$:

(i) Set $m = 0$ and $w(ct_0, t_0) = a(0)$, where the function $a(x)$ is the given initial displacement of the distributed system (See Eq. (2.1)).
(ii) Set $m = m + 1$; calculate $w_{el}(ct_m, t_m)$ and $U_m(ct_m)$ by Eq. (2.20) and Eq. (2.38), respectively; determine $w(ct_m, t_m)$ by Eq. (2.39).
(iii) If $m < n$, return to step (ii); otherwise, go to step (iv).
(iv) For a given x, calculate $w_{el}(x, t_n)$ and $U_n(x)$; determine $w(x, t_n)$ and $y(t_n)$ by Eq. (2.37) and Eq. (2.40), respectively.
(v) Repeat until the solution is known everywhere in **S**.

2.5 Case Study: The Taut Uniform String

The uniform taut string, fixed at $x = 0$ and $x = L$, was chosen for the case study to be reported herein. There are sound reasons for this choice, among them relative simplicity, which will prove useful for demonstrating the method, and the fact that some components of the solution can be evaluated analytically, leading to a reasonably concise presentation of the problem.

2.5.1 Equation of Motion and Nondimensionalization

The response of the taut string is given by the solution to Eq. (2.1), where the stiffness operator **K** is $-p(\partial^2/\partial x^2)$. Let

$$\bar{x} = \frac{x}{L}, \qquad \bar{t} = \Omega t, \qquad \bar{w}(\bar{x}, \bar{t}) = \frac{w(x,t)}{L}, \qquad \bar{y}(\bar{t}) = \frac{y(t)}{L},$$

$$\bar{\delta}() = L\delta(), \qquad \bar{c} = \frac{c}{L\Omega}, \qquad \Omega^2 = \frac{p}{\rho L^2}, \qquad \bar{\omega}_j = \frac{\omega_j}{\Omega}, \qquad \bar{\mu} = \frac{\mu}{\Omega}.$$

Following Sadiku and Leipholz (1987), the external force is assumed to be zero,

$$f(x, t) = 0.$$

Then, letting

$$\bar{k} = \frac{kL}{p}, \qquad \bar{m} = \frac{m}{\rho L}, \qquad \bar{W} = \frac{mg}{p}, \qquad \bar{g} = \frac{\bar{W}}{\bar{m}} = \frac{\rho L g}{p}$$

and dropping the overbars, we arrive at the nondimensional form of the equation of motion,

$$w_{tt}(x,t) = w_{xx}(x,t) + k(y(t) - w(ct,t))\delta(x - ct), \qquad x \in [0, 1] \qquad (2.41a)$$

subject to the boundary conditions

$$w(0,t)=w(1,t)=0 \tag{2.41b}$$

and assumed quiescent initial conditions

$$w(x,0)=w_t(x,0)=y(0)=\dot{y}(0)=0. \tag{2.41c}$$

Note that the effect of the weight of the moving mass is not an external force to the continuum, but its influence is through the constraint $k\beta(t)$.

2.5.2 The Moving-force Solution

The moving-force problem, defined in Eq. (2.11), becomes trivial, since initial conditions as well as the distributed force are defined to be zero. However, for comparison with the moving-oscillator solution, we seek a solution to the moving-force problem predicted upon a force equal to the weight of the oscillator moving at constant speed, or

$$\tilde{z}_{tt}(x,t)=\tilde{z}_{xx}(x,t)-W\delta(x-ct), \qquad x\in[0,1], \tag{2.42}$$

subject to the same boundary an initial conditions shown above. The system eigenvalues and normalized eigenfunctions are given by $\omega_j=j\pi$ and $\phi_j(x)=\sqrt{2}\sin j\pi x$, respectively, $j=1,2,\ldots$. The space–time Green's function is

$$G(x,\xi,t-\tau)=\sum_{j=1}^{\infty}\frac{2}{j\pi}\sin j\pi x\sin j\pi\xi\sin j\pi(t-\tau), \tag{2.43}$$

and the moving-force solution, from Eq. (2.12), can be expressed as

$$\tilde{z}(x,t)=\frac{2W}{c^2-1}\sum_{j=1}^{\infty}\left(\frac{1}{j\pi}\right)^2\sin j\pi x(c\sin j\pi t-\sin j\pi ct). \tag{2.44}$$

2.5.3 The Moving-force Moving-oscillator Solution

The solution can be expressed, as shown in Eq. (2.19), as

$$w(x,t)=w_{eI}(x,t)+w_c(x,t)+w_I(x,t),$$

where now

$$w_{eI}(x,t)=-k\int_0^t G(x,c\tau,t-\tau)\frac{g}{\mu^2}(1-\cos\mu\tau)\,d\tau,$$

$w_c(x,t)$ is as defined in the second of Eqs. (2.20) with the space–time Green's function given by Eq. (2.43), and $w_I(x,t)$ is as defined in Eqs. (2.21) and

(2.22), with $r_j(t, \eta)$ given by Eq. (2.23). However, this can now be integrated, giving

$$\begin{aligned} r_j(t,\eta) &= \sqrt{2}\int_\eta^t \sin j\pi c\tau \sin j\pi(t-\tau)\sin\mu(\tau-\eta)\,d\tau \\ &= -\frac{\sqrt{2}}{2}\left[\frac{\mu}{j^2\pi^2(1-c)^2-\mu^2}\cos j\pi(t-(1-c)\eta)\right. \\ &\quad - \frac{\mu}{j^2\pi^2(1+c)^2-\mu^2}\cos j\pi(t-(1+c)\eta) \\ &\quad + \frac{j\pi}{(\mu+j\pi c)^2-j^2\pi^2}\cos(\mu(t-\eta)+j\pi ct) \\ &\quad \left. - \frac{j\pi}{(\mu-j\pi c)^2-j^2\pi^2}\cos(\mu(t-\eta)-j\pi ct)\right]. \end{aligned} \tag{2.45}$$

The function $J(x, t, \eta)$ in Eq. (2.26) can thus be written in the form

$$J(x,t,\eta)=\sqrt{2}k\sum_{j=1}^{\infty}\sin j\pi x[\mu r_j(t,\eta)-\sin j\pi c\eta\sin j\pi(t-\eta)], \tag{2.46}$$

where $r_j(t, \eta)$ is given in Eq. (2.45), and the solution can then be written in the form of Eq. (2.24), repeated below,

$$w(x,t)=w_{eI}(x,t)+\int_0^t J(x,t,\eta)w(c\eta,\eta)\,d\eta.$$

2.5.4 Numerical Solution by the Method of Successive Approximations

Recognizing that $w_0(x, t) = w_{eI}(x, t)$ for the problem as defined, a solution to the moving-force moving-oscillator problem is given by

$$w_1(x,t)=w_{eI}(x,t)+\mathbf{T}[w_0(x,t)], \tag{2.47}$$

$$w_2(x,t)=w_{eI}(x,t)+\mathbf{T}[w_1(x,t)], \tag{2.48}$$

and so forth, where, from Eq. (2.24), $\mathbf{T}[w_n(x, t)]$ is given by

$$\mathbf{T}[w_n(x,t)]=\int_0^t J(x,t,\eta)w_{n-1}(c\eta,\eta)\,d\eta. \tag{2.49}$$

Numerical experiments have shown that convergence for the string is slow,

Table 2.1. *System physical parameters, dimensional.*

Figures	m (kg)	L (m)	ρ (kg/m)	k (N/m)	P(N)	c (m/s)
3 a,b,c	1,500	10	8,000	100,000	20,000,000	30
4 a,b,c	1,500	25	8,000	150,000	50,000,000	20
5 a,b	1,500	25	8,000	600,000	50,000,000	20

Table 2.2. *System physical parameters, normalized.*

Figures	c	k	g	μ	W
3 a,b,c	0.6	0.05	0.0392	1.633	0.000736
4 a,b,c	0.253	0.075	0.0392	3.162	0.000294
5 a,b	0.253	0.3	0.0392	6.325	0.000294

requiring more than one iteration. Thus, the direct numerical procedure has been selected to investigate the example problems, as shown below.

2.5.5 Solution by the Direct Numerical Procedure

The analysis of Section 2.4.2 has been programmed into MATLAB®, utilizing the Green's function of Eq. (2.43) given in the form of an infinite series. Numerical simulations show that there is no significant difference between results obtained using 20 terms and 40 terms in the series. Hence, all results presented in this section were obtained using 20 terms.

The response solutions of the oscillator and the string depend on the normalized parameters c, k, g, and μ. It should be noted that these parameters are not all independent. Consider a practical problem of a vehicle traveling on a bridge, in which the vehicle and the bridge are modeled by the moving oscillator and the string, respectively. Based on typical values of the physical parameters, values of normalized parameters used in the simulations are summarized in Tables 2.1 and 2.2. In Table 2.1, L may be interpreted as the length between successive bridge supports.

Figures 2.3a and 2.3b compare the moving-force solution to the moving-oscillator solution. The negative values are plotted since the deflection of the string is downward due to the weight. Note that the "wavefronts" of the two solutions are quite different. The difference in the two solutions is shown in Fig. 2.3c. It is seen that the difference is of the same order of magnitude as

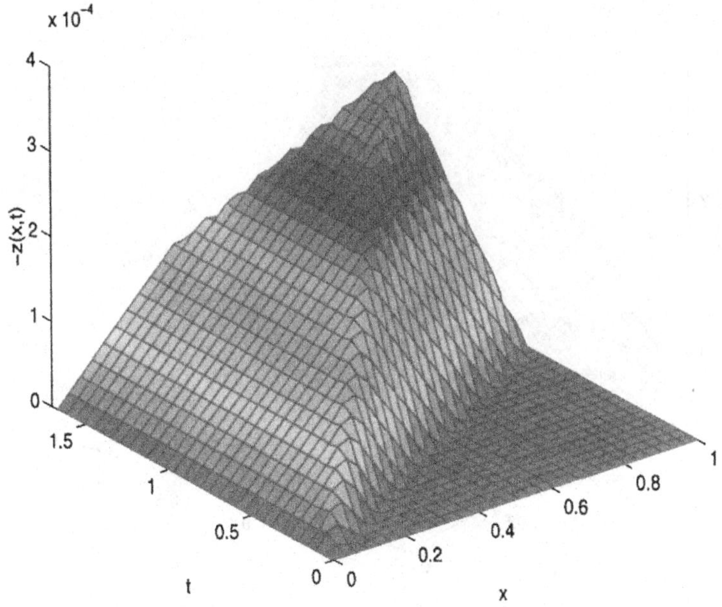

(a)

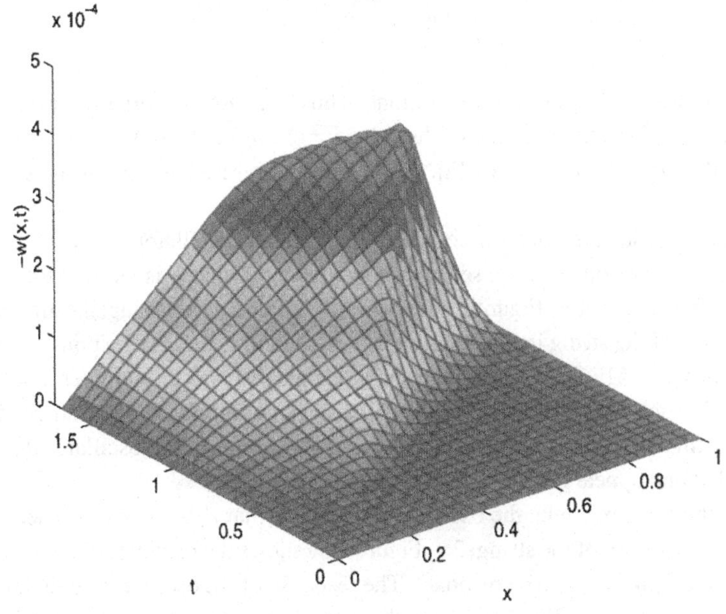

(b)

Fig. 2.3. (*a*) Moving-force solution. (*b*) Moving-oscillator solution. (*c*) Difference of moving-force and -oscillator solutions.

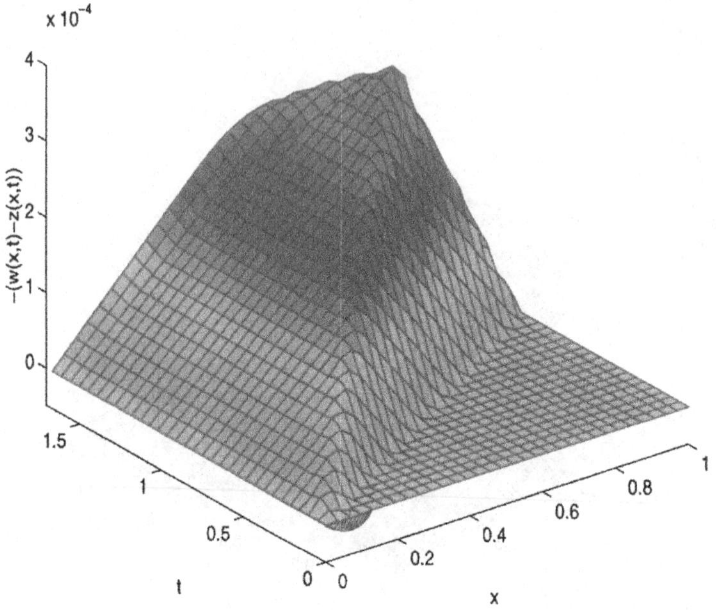

(c)

Fig. 2.3. (*cont.*)

those of the moving-oscillator solutions. Thus, the moving-force solution can give a very misleading picture of the behavior of this coupled dynamic system. From the Green's function of Eq. (2.43), the natural frequencies of the string are $j\pi$, $j = 1, 2, \ldots$.

It is interesting to determine whether a moving oscillator can excite the continuum to resonance. We select the parameters so that μ is close to the first natural frequency of π. Figure 2.4a shows the response of the string; the growing amplitude of the string indicates a resonance phenomenon. The response of the string at $x = 0.613$ is also plotted to show that the resonance resembles a flutter type of instability. It should be noted that the moving-force solution cannot predict this resonance. Figure 2.4c shows the response of the oscillator that is bounded and appears not to grow with time.

For the final example, the oscillator is tuned so that μ is close to the second natural frequency of the string, 2π. Figure 2.5a shows the response of the string, which is again shown to be resonant. The response of the oscillator, again fairly constant, is shown in Fig. 2.5b. Note that the latter plot would be smoother had a finer mesh been employed for the computation.

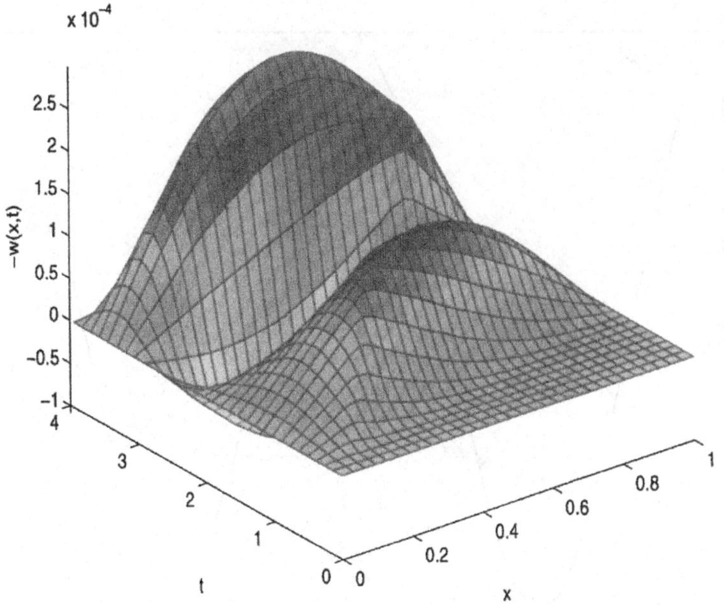

(*a*)

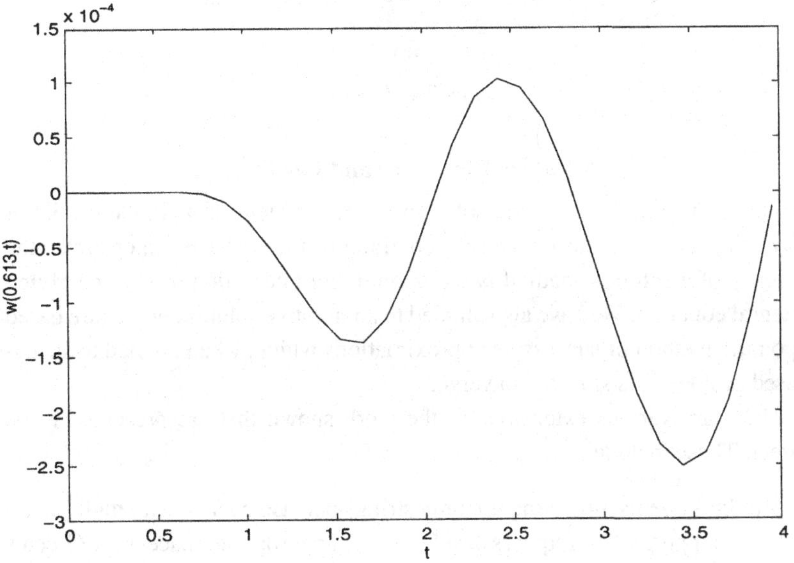

(*b*)

Fig. 2.4. (*a*) String response, moving-oscillator solution. (*b*) String response at $x = 0.613$. (*c*) Oscillator response.

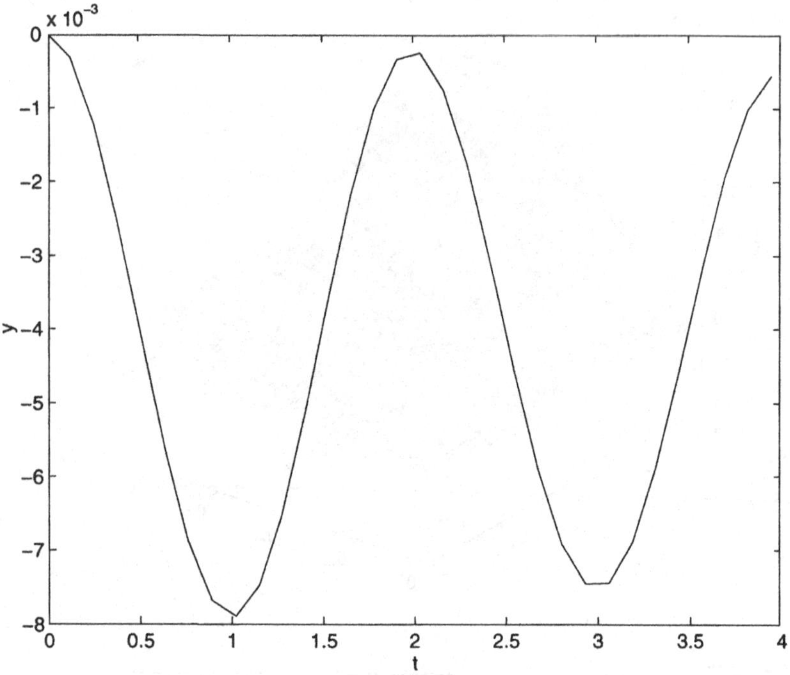

(c)

Fig. 2.4. (*cont.*)

2.6 Future Directions and Possibilities

We have examined herein the solution to the problem of a simple oscillator traversing a one-dimensional elastic continuum, and we have demonstrated the efficacy of a solution method based upon direct integration of the underlying integral equation. We have also alluded to an iterative solution procedure based upon the method of successive approximations which, when applied to string-based problems, is slow to converge.

There are various extensions to the work shown that are presumed to be direct. These include:

- Similar systems, involving not only strings but also beams, with multiple interior supports. This requires development of appropriate space–time Green's functions.
- Similar systems, wherein the elastic continuum is two dimensional, such as a plate or shell. Again, appropriate space–time Green's functions must be developed.

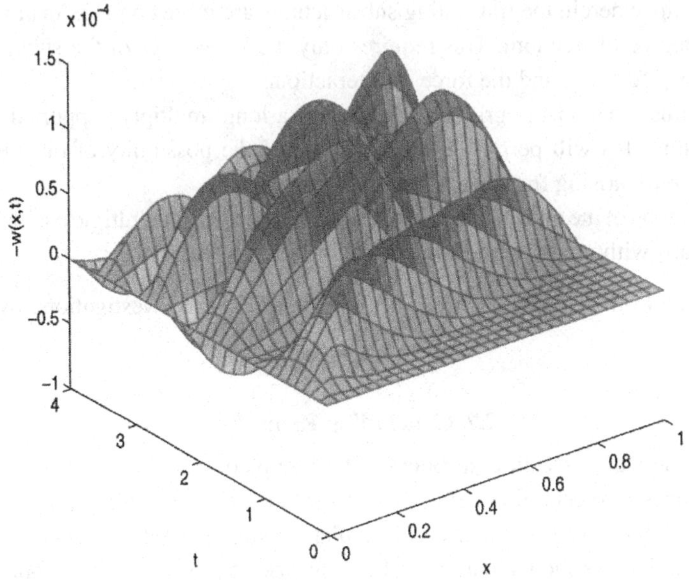

(*a*)

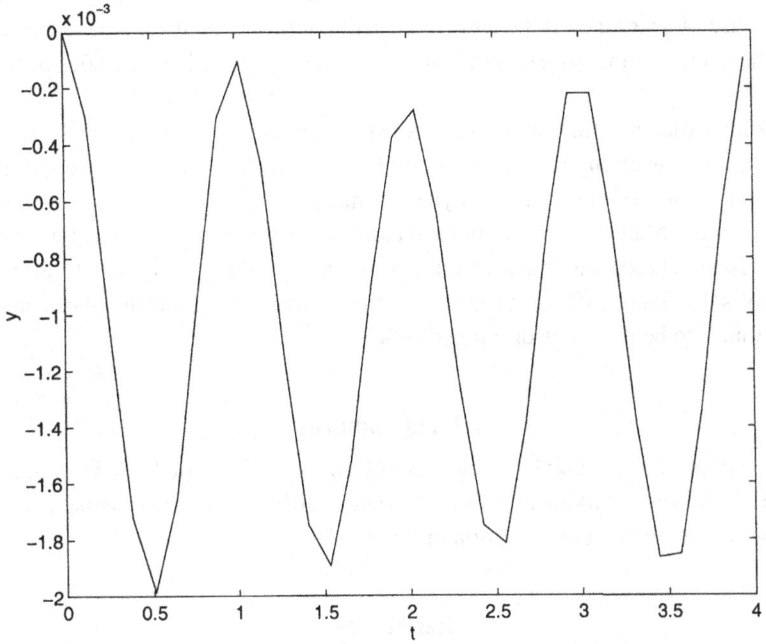

(*b*)

Fig. 2.5. (*a*) String response. (*b*) Oscillator response.

- Systems wherein the translating substructures are more complex than a single degree of freedom. This requires only the knowledge of the subsystem Green's function and the force of interaction.
- Systems wherein the primary structure is a long, multiply supported continuum. This will permit the investigation of the possibility of divergence instability arising from subsystem motion.
- Variations of the above, taking into account the effects of multiple oscillators, perhaps with random times of arrival.

These, and other topics, will be the subjects of future investigations by the authors.

2.7 Concluding Remarks

The problem of an elastic distributed system coupled with an oscillator translating at constant speed, often referred to as the "moving-oscillator" problem, has been examined. The problem was formulated using a "relative displacement" model, and it was shown that, in the limiting case, the moving-mass problem is recovered. The coupled equations of motion were recast into an integral equation, and both iterative and direct numerical procedures were presented. The response of a string carrying a moving oscillator was studied, and the response of the continuum was seen to be sensitive to the dynamic characteristics of the oscillator.

With structures such as bridges being constructed in new and innovative ways, often involving lightweight materials, long spans, and flexible supports, the importance of modeling the dynamic interaction between subsystems (i.e., between the bridge and the vehicles it carries) is likely to become increasingly important. The development of efficient semi-analytical procedures to effectively solve these difficult problems is the motivation behind this paper and continues to be the focus of our research.

Acknowledgment

The authors wish to thank Mr. Shenger (Ted) Ying of Wayne State University and Mr. Steven Wojtkiewicz of the University of Illinois for their assistance in completing the numerical portion of this work.

References

Achenbach, J. D., Keshava, S. P., and Hermann, G., 1967, "Moving Load on a Plate Resting on an Elastic Half Space," *ASME Journal of Applied Mechanics*, Vol. 34, No. 4, pp. 910–14.

Achenbach, J. D., and Sun, C. T., 1965, "Moving Load on a Flexibly Supported Timoshenko Beam," *International Journal of Solids and Structures*, Vol. 1, pp. 353–70.

Agrawal, O. P., and Saigal, S., 1986, "Dynamic Responses of Orthotropic Plates Under Moving Masses," in *Recent Trends in Aeroelasticity, Structures, and Structural Dynamics*, pp. 313–33, ed. P. Hajela, University of Florida Press, Gainesville.

Alder, A. A., and Reismann, H., 1974, "Moving Loads on an Elastic Plate Strip," *ASME Journal of Applied Mechanics*, Vol. 41, pp. 713–18.

Benedetti, G. A., 1974, "Dynamic Stability of a Beam Loaded by a Sequence of Moving Mass Particles," *ASME Journal of Applied Mechanics*, Vol. 41, Dec., pp. 1069–71.

Bolotin, V. V., 1965, *Statistical Methods in Engineering Mechanics*, Strojizdat, Moscow.

Bolotin, V. V., 1982, *Application of the Theory of Probability and of the Theory of Reliability in the Analysis of Structures*, Strojizdat, Moscow.

Chen, C. H., and Wang, K. W., 1994, "An Integrated Approach Toward the Modeling and Dynamic Analysis of High-Speed Spindles, Part II: Dynamics Under Moving End Load," *ASME Journal of Vibration and Acoustics*, Vol. 116, pp. 514–22.

Chonan, S., 1975, "The Elastically Supported Timoshenko Beam Subjected to an Axial Force and a Moving Load," *International Journal of Mechanical Science*, Vol. 17, pp. 573–81.

Chonan, S., 1976, "Critical Velocity of a Load Moving on a Beam Supported by an Elastic Stratum," *Bulletin of JSME*, Vol. 19, pp. 604–9.

Chonan, S., 1976, "Moving Load on a Pre-Stressed Plate Resting on a Fluid Half-Space," *Ingenieur-Archiv*, Vol. 45, pp. 171–78.

Chonan, S., 1977, "Response of Fluid-Filled Cylindrical Shell to a Moving Load," *Journal of Sound and Vibration*, Vol. 55, pp. 419–30.

Chonan, S., 1978, "Moving Harmonic Load on an Elastically Supported Timoshenko Beam," *ZAMM*, Vol. 58, pp. 9–15.

Crandall, S. J., and Mark, W. D., 1963, *Random Vibration in Mechanical Systems*, Academic Press, New York.

Duffy, D. G., 1990, "The Response of an Infinite Railroad Track to a Moving Vibrating Mass," *ASME Journal of Applied Mechanics*, Vol. 57, pp. 66–73.

Felszeghy, S. F., 1996a, "The Timoshenko Beam on an Elastic Foundation and Subject to a Moving Step Load, Part I: Steady-State Response," *ASME Journal of Vibration and Acoustics*, Vol. 118, pp. 277–84.

Felszeghy, S. F., 1996b, "The Timoshenko Beam on an Elastic Foundation and Subject to a Moving Step Load, Part II: Transient Response," *ASME Journal of Vibration and Acoustics*, Vol. 118, pp. 285–91.

Filho, F. V., 1978, "Finite Element Analysis of Structures Under Moving Loads," *Shock and Vibration Digest*, Vol. 10, pp. 27–35.

Fryba, L., 1957, "Infinite Beam on an Elastic Foundation Subjected to a Moving Load," *Aplikace Matematiky* (in Czech), Vol. 2, No. 2, pp. 105–32.

Fryba, L., 1972, *Vibration of Solids and Structures Under Moving Loads*, Noordhoff International Publishing, Groningen, Prague.

Fryba, L., 1976, "Non-Stationary Response of a Beam to a Moving Random Force," *Journal of Sound and Vibration*, Vol. 46, No. 3, pp. 323–38.

Fryba, L., 1977, "Stationary Respone of a Beam to a Moving Continuous Random Load," *Acta Technica CSAV*, Vol. 22, No. 4, pp. 444–79.

Fryba, L., 1986, "Random Vibration of a Beam on Elastic Foundation Under Moving Force," in *Random Vibration-Status and Recent Developments*, pp. 127–47, ed. I. Elishakoff and R. H. Lyon, Elsevier Science Publishers, Dordrecht.

Hino, J., Yoshimura, T., and Ananthanarayana, N., 1985, "Vibration Analysis of Nonlinear Beams Subjected to a Moving Load Using the Finite Element Method," *Journal of Sound and Vibration*, Vol. 100. pp. 477–91.

Hino, J., Yoshimura, T., and Konishi, K., 1984, "A Finite Element Method for Prediction of the Vibration of a Bridge Subjected to a Moving Vehicle Load," *Journal of Sound and Vibration*, Vol. 96, pp. 45–53.

Huang, S. C., and Hsu, B. S., 1990, "Resonant Phenomena of a Rotating Cylindrical Shell Subjected to a Harmonic Moving Load," *Journal of Sound and Vibration*, Vol. 136, No. 2, pp. 215–28.

Hutton, D. V., and Counts, J., 1974, "Deflection of a Beam Carrying a Moving Mass," *ASME Journal of Applied Mechanics*, Vol. 41, Sept., pp. 803–4.

Inglis, C. E., 1934, A Mathematical Treatise on Vibrations in Railway Bridges, Cambridge University Press, Cambridge.

Iwan, W. D., and Moeller, T. L., 1976, "The Stability of Spinning Elastic Disk With a Transverse Load System," *ASME Journal of Applied Mechanics*, Vol. 43, pp. 485–90.

Iwan, W. D., and Stahl, K. J., 1973, "The Response of an Elastic Disk With a Moving Mass System," *ASME Journal of Applied Mechanics*, Vol. 40, pp. 445–51.

Johnson, E. A., 1993, "Firing-Induced Vibration of an Electromagnetic Railgun Rail," M. S. Thesis, Department of Aeronautical and Astronautical Engineering, University of Illinois at Urbana-Champaign.

Katz, R., Lee, C. W., Ulsoy, A. G., and Scott, R. A., 1987, "Dynamic Stability and Response of a Beam Subject to a Deflection Dependent Moving Load," *ASME Journal of Vibration, Acoustics, Stress and Reliability in Design*, Vol. 109, pp. 361–65.

Katz, R., Lee, C. W., Ulsoy, A. G., and Scott, R. A., 1988, "The Dynamic Response of a Rotating Shaft Subject to a Moving Load," *Journal of Sound and Vibration*, Vol. 122, No. 1, pp. 131–48.

Kenny, J. T., 1954, "Steady-State Vibrations of Beam on Elastic Foundation for Moving Load," *ASME Journal of Applied Mechanics*, Vol. 21, No. 4, pp. 359–64; discussion Vol. 22, No. 3, 1955, p. 436.

Kerr, A. D., 1972, "The Continuously Supported Rail Subjected to an Axial Force and a Moving Load," *International Journal of Mechanical Science*, Vol. 14, pp. 71–78.

Knowles, J. K., 1968, "On the Dynamic Response of a Beam to a Randomly Moving Load," *ASME Journal of Applied Mechanics*, Vol. 35, No. 1, pp. 1–6.

Knowles, J. K., 1970, "A Note on the Response of a Beam to a Randomly Moving Force," *ASME Journal of Applied Mechanics*, Vol. 37, No. 4, pp. 1192–94.

Licari, J. S., and Wilson, E. N., 1962, "Dynamic Response of a Beam Subjected to a Moving Force System," *Proceedings of the 4th U.S. National Congress of Applied Mechanics*, Vol. 1, pp. 419–25.

Lin, Y.-H., and Trethewey, M. W., 1990, "Finite Element Analysis of Elastic Beams Subjected to Moving Dynamic Loads," *Journal of Sound and Vibration*, Vol. 136, No. 2, pp. 323–42.

Mackertich, S., 1990, "Moving Load on a Timoshenko Beam," *Journal of the Acoustical Society of America*, Vol. 88, pp. 1175–78.

Mathews, P. M., 1958, "Vibrations of a Beam on Elastic Foundation," *ZAMM*, Vol. 38, pp. 105–15.

Mote, C. D., Jr., 1970, "Stability of Circular Plates Subjected to Moving Loads," *Journal of Franklin Institute*, Vol. 290, No. 4, Oct., pp. 329–44.

Mote, C. D., Jr., 1977, "Moving-Load Stability of a Circular Plate on a Floating Central Collar," *Journal of the Acoustical Society of America*, Vol. 61, No. 2, pp. 439–47.

Nelson, H. D., and Conover, R. A., 1971, "Dynamic Stability of a Beam Carrying Moving Masses," *ASME Journal of Applied Mechanics*, Vol. 38, pp. 1003–6.

Pesterev, A. V., and Bergman, L. A., 1997, "Response of Elastic Continuum Carrying Moving Linear Oscillator," *ASCE Journal of Engineering Mechanics*, Vol. 123, No. 8, pp. 878–84.

Pierucci, M., 1993, "The Flexural Wave-Number Response of a String and a Beam Subjected to Moving Harmonic Force," *Journal of the Acoustical Society of America*, Vol. 93, No. 4, pp. 1908–17.

Prasad, B., 1981, "On The Response of a Timoshenko Beam Under Initial Stress to a Moving Load," *International Journal of Engineering Science*, Vol. 19, pp. 615–28.

Reismann, H. 1964, "Response of a Pre-Stressed Elastic Plate Strip to a Moving Pressure Load," *Journal of Franklin Institute*, Vol. 278, No. 1, pp. 8–19.

Rodeman, R., Longcope, D. B., and Shampine, L. F., 1976, "Response of a String to an Accelerating Mass," *ASME Journal of Applied Mechanics*, Vol. 43, pp. 675–80.

Sadiku, S., and Leipholz, H. H. E., 1987, "On the Dynamics of Elastic Systems With Moving Concentrated Masses," *Ingenieur-Archiv*, Vol. 57, pp. 223–42.

Saigal, S., 1986, "Dynamic Behavior of Beam Structures Carrying Moving Masses," *ASME Journal of Applied Mechanics*, Vol. 53, pp. 222–24.

Saigal, S., Agrawal, O. P., and Stanisic, M. M., 1987, "Influence of Moving Masses on Rectangular Plate Dynamics," *Ingenieur-Archiv*, Vol. 57, pp. 187–96.

Shen. I. Y., 1993, "Response of a Stationary, Damped, Circular Plate Under a Rotating Sliding Bearing System," *ASME Journal of Vibration and Acoustics*, Vol. 115, pp. 65–69.

Smith, C. E., 1964, "Motions of s Stretched String Carrying a Moving Mass Particle," *ASME Journal of Applied Mechanics*, Vol. 31, No. 1, pp. 29–37.

Stanisic, M. M., 1985, "On a New Theory of the Dynamic Behavior of the Structures Carrying Moving Masses," *Ingenieur-Archiv*, Vol. 55, pp. 176–85.

Stanisic, M. M., Euler, J. A., and Montgomery, S. T., 1974, "On a Theory Concerning the Dynamical Behavior of Structures Carrying Moving Masses," *Ingenieur-Archiv*, Vol. 43, pp. 295–305.

Stanisic, M. M., and Hardin, J. C., 1969, "On Response of Beams to an Arbitrary Number of Moving Masses," *Journal of Franklin Institute*, Vol. 287, pp. 115–23.

Stokes, G. G., 1883, "Discussions of a Differential Equation Relating to the Breaking of Railway Bridges," *Mathematical and Physical Papers*, Vol. 2, pp. 178–220.

Tan, C. P., and Shore, S., 1968, "Response of Horizontally Curved Bridge to Moving Load," *ASCE Journal of the Structural Division*, Vol. 94, pp. 2135–51.

Timoshenko, S., Young, D. H., and Weaver, W., 1974, *Vibration Problems in Engineering*, pp. 448–53, 4th ed., Wiley, New York.

Ting, E. C., Genin, J., and Ginsberg, J. H., 1974, "A General Algorithm for Moving Mass Problems," *Journal of Sound and Vibration*, Vol. 33, No. 1, pp. 49–58.

Ting, E. C., and Yener, M., 1983, "Vehicle–Structure Interactions in Bridge Dynamics," *Shock and Vibration Digest*, Vol. 15, No. 2, pp. 3–9.

Wickert, J. A., and Mote, C. D., Jr., 1991, "Traveling Load Response of an Axially Moving String," *Journal of Sound and Vibration*, Vol. 149, No. 2, pp. 267–84.

Wilson, J. F., and Wilson, D. M., 1984, "Responses of Continuous, Inertialess Beams to Traversing Mass – A Generalization of Stokes' Problem," *International Journal of Mechanical Sciences*, Vol. 26, No. 2, pp. 105–12.

Yen, D. H. Y., and Chou, C. C., 1970, "Response of a Plate Supported by a Fluid Half Space to a Moving Pressure," *ASME Journal of Applied Mechanics*, Vol. 37, pp. 1050–54.

Yoshimura, T., Hino, J., and Ananthanarayana, N., 1986, "Vibration Analysis of Nonlinear Beams Subjected to Moving Loads by Using the Galerkin Method," *Journal of Sound and Vibration*, Vol. 104, pp. 179–86.

Yu, R. C., and Mote, C. D., Jr., 1987, "Vibration and Parametric Excitation in Asymmetric Circular Plates Under Moving Loads," *Journal of Sound and Vibration*, Vol. 119, No. 3, pp. 409–27.

Zhu, W. D., and Mote, C. D., Jr., 1994, "Free and Forced Response of an Axially Moving String Transporting a Damped Linear Oscillator," *Journal of Sound and Vibration*, Vol. 177, No. 5, pp. 591–610.

Zu, J. W.-Z., and Han, R. P. S., 1994, "Dynamic Response of a Spinning Timoshenko Beam With General Boundary Conditions and Subjected to a Moving Load," *ASME Journal of Applied Mechanics*, Vol. 61, pp. 152–60.

Nomenclature

a, b	initial conditions
$\mathbf{B}$	boundary operator
c	constant translational velocity
f	distributed load
g	acceleration of gravity
G	Green's function of the string
h	Green's function of the oscillator
k	oscillator stiffness coefficient
$\mathbf{K}$	stiffness operator
L	length of the string
p	string tension
q_c	force of constraint
t	independent temporal coordinate
w	transverse displacement of the string
W	weight of the translating inertia
x	independent spatial coordinate
y	absolute displacement of the oscillator
z	string displacement under a moving force

Greek

β	relative displacement of the oscillator
δ	Dirac delta function
μ	oscillator natural frequency
ρ	mass per unit length of the continuum
ϕ	eigenfunction of the continuum
ω	circular frequency
Ω	nondimensional frequency parameter
$\boldsymbol{\Omega}$	domain of the continuum

3

Nonlinear Normal Modes and Wave Transmission in a Class of Periodic Continuous Systems

VALERY N. PILIPCHUK and ALEXANDER F. VAKAKIS

Abstract

We present two methodologies for studying periodic oscillations and wave transmission in periodic continuous systems with strongly nonlinear supporting or coupling stiffnesses. The first methodology is based on a nonsmooth transformation of the spatial variable and eliminates the singularities (generalized functions) from the governing partial differential equation of motion. The resulting smooth partial differential equation is then analyzed asymptotically. This method is used to study localized nonlinear normal modes (NNMs) of an infinite linear string supported by a periodic array of nonlinear stiffnesses. A second methodology is developed to study primary pulse transmission in a periodic system composed of linear layers coupled by means of strongly nonlinear stiffnesses. A piecewise transformation of the time variable is introduced, and the scattering of the primary pulse at the nonlinear stiffnesses is reduced to solving a set of strongly nonlinear first-order ordinary differential equations. Approximate analytical and exact numerical solutions of this set are presented, and the methodology is employed to study primary pulse transmission in a system with clearance nonlinearities.

3.1 Introduction

Flexible periodic structures such as truss aerospace systems, periodically supported or stiffened shells and plates, and turbine cyclic assemblies are very common in engineering applications. Periodic structures are composed of a number of identical (or near identical) coupled substructures. The majority of works in the literature dealing with periodic structures are based on the assumption of linearity. Linear periodic systems were analyzed utilizing the concepts of "propagation constant" and "characteristic receptance" (Mead, 1975a, 1997b, 1986), by employing modal analysis and substructure synthesis (Engels and

Meirovitch, 1978), or by applying techniques from structural wave propagation (Delph et al., 1979, 1980; Cetinkaya et al., 1995). Anderson (1958) investigated spatial localization of motions ("Anderson localization") in randomly disordered linear chains of particles. Additional works (Kissel, 1987; Wei and Pierre, 1988) focused on Anderson localization in discrete and continuous linear periodic systems with structural disorders.

The analysis of nonlinear periodic systems requires the application of special analytical techniques. In Kosevich and Kovalev (1975), Vedenova and Manevich (1981), Peyrard and Kruskal (1984), Rosenau (1987), and Vakakis et al. (1994) continuum approximations are utilized to study nonlinear periodic systems of infinite spatial extent; in this approach, small variations in the amplitudes of adjacent subsystems are assumed, and the infinite set of the governing ordinary differential equations is replaced by a single nonlinear partial differential equation. Nonlinear standing waves in structural periodic assemblies with and without cyclic symmetry were studied in Vakakis and Cetinkaya (1993), Vakakis et al. (1993), and Aubrecht and Vakakis (1995) using asymptotic and numerical techniques. These studies proved that in contrast to linear theory spatial motion confinement in nonlinear periodic structures occurs even in the absence of structural disorder. In a recent work by Vakakis and King (1995) wave propagation and attenuation in a one-dimensional nonlinear periodic system of infinite spatial extent was studied. Nonlinear standing waves in the periodic system were analyzed by forming infinite sets of homogeneous and inhomogeneous difference equations and solving them using computer algebra. The analysis of that work is computationally involved, and its extension to multicoupled systems poses a challenging task.

In additional works, unidirectional wave propagation in linear repetitive structures was studied (Mead, 1975a, 1975b, 1986). It was shown that linear periodic structures possess propagation and attenuation zones (PZs and AZs) in the frequency domain; waves with frequencies inside PZs are transmitted unattenuated through the periodic system, whereas waves in AZs spatially attenuate and are near-field solutions. In Vakakis and King (1995) the notions of PZs and AZs are extended to nonlinear periodic systems. Von Flotow and coworkers (1986) analyzed transient (nonharmonic) wave propagation in periodic structural components using concepts from network theory and employing transfer function formulations.

Waves in systems with discrete or distributed nonlinearities were also studied in the literature. In the work by Nayfeh and Mook (1984) analytical techniques for analyzing dispersive or nondispersive nonlinear waves were given. The scattering of traveling waves by nonlinear stiffnesses was examined in Cekirge and Varley (1973) and in Nayfeh et al. (1993) by means of an analysis based on the method of characteristics. Employing a similar approach, Achenbach and coworkers (1982, 1991) studied the scattering of incident waves by cracks modeled as nonlinear elasticities and the scattering of ultrasonic waves by adhesive nonlinear layers. Hirose (1991) analyzed the nonlinear scattering of elastic waves by three-dimensional inclusions possessing interface stiffnesses with elastoplastic characteristics by formulating and solving boundary integral equations. Naik and Crews (1989) used finite elements to model the dynamics of joints with clearance nonlinearities. Finally, in an additional class of works analytical and numerical methods for studying traveling and standing waves in nonlinear elastic media governed by nonlinear partial differential equations were developed (Engelbrecht et al., 1990; Wegner and Norwood, 1993; Jeffrey and Engelbrecht, 1994; Kleinman et al., 1993).

In this work nonlinear oscillations and wave transmission in two nonlinear periodic systems are analyzed. Both systems consist of finite linear continua supported or coupled by nonlinear elastic stiffnesses. The first system consists of an infinite linear string supported by a periodic array of strongly nonlinear stiffnesses. A new technique is developed for analyzing localized nonlinear normal modes (NNMs) of this periodic system, based on a nonsmooth spatial transformation that permits the elimination of singularities (generalized functions) from the governing equation of motion. In essence, the described nonsmooth transformations lead to a transformed set of governing equations which is amenable to regular or singular perturbation analysis. The second system considered in this work is composed of an infinite number of linear rods coupled by means of strongly nonlinear stiffnesses. Only primary pulse transmission and reflection at each nonlinear element is considered, and a piecewise smooth transformation of the temporal variable is introduced. This permits the reduction of the problem to an infinite set of first-order, strongly nonlinear, ordinary differential equations. A subset of these equations is solved both analytically and numerically. This method enables the study of primary pulse transmission in systems with strong nonlinearities such as clearances or impacts. An alternative continuum approximation methodology is also outlined that reduces the problem of primary pulse transmission to solving a single nonlinear partial differential equation.

3.2 Localized Nonlinear Normal Modes of a String Resting on Nonlinear Elastic Stiffnesses

3.2.1 Nonsmooth Transformations–General Methodology

The first continuous system we consider consists of a linear string of infinite spatial extent, supported by means of an infinite set of periodically spaced nonlinear stiffnesses, and forced by distributed external excitation. This system was studied extensively in Pilipchuk and Vakakis (1995); the following analysis closely follows that work.

The partial differential equation governing small transverse oscillations of the string is of the following form:

$$\rho\frac{\partial^2 u}{\partial t^2} - T\frac{\partial^2 u}{\partial y^2} + 2f(u)\sum_{k=-\infty}^{\infty}\delta\left(\frac{y}{\varepsilon} - 1 - 2k\right) = q\left(\frac{y}{\varepsilon}, y, t\right), \quad -\infty < y < \infty, \tag{3.1}$$

where $u = u((y/\varepsilon)y, t)$ denotes the transverse displacement, ρ and T the (uniform) density and tension of the string, $f(u)$ the nonlinear characteristics of the stiffnesses, and $q((y/\varepsilon), y, t)$ the external distributed load. Imposing the condition $0 < \varepsilon \ll 1$, the distance between adjacent stiffnesses is assumed to be of $O(\varepsilon) \ll 1$, and the transverse displacement and the external excitation are assumed to possess slow- as well as fast-scale spatial dependencies. Time-periodic solutions of system (3.1) will be constructed by employing nonsmooth transformations of the fast-scale spatial variable (y/ε). These transformations will lead to the elimination of the singular terms from the equation of motion and will allow the use of analytical techniques from the theory of nonlinear oscillations to analyze localized and nonlocalized vibrations of the string.

Before proceeding with the analysis, a brief synopsis of the methodology of nonsmooth transformations will be given; more details of the method can be found in the works by Pilipchuk and coworkers (Pilipchuk, 1985, 1988; Pilipchuk et al., 1997; Vakakis et al., 1996). Considering a generic real variable z, the following nonsmooth transformation is introduced:

$$z \to (\tau(z), e(z)), \qquad \tau(z) = (2/\pi)\sin^{-1}(\sin(\pi z/2)), \\ e(z) = \tau'(z), \qquad z \in R, \tag{3.2}$$

where derivatives and equalities should be understood in the generalized sense of the theory of distributions (Richtmyer, 1985). By construction, the nonsmooth variables $\tau(z)$ and $e(z)$ are periodic in z (cf. Figure 3.1), and $e(z)$ satisfies the relation $e^2(z) = 1$, $z \in R$. As discussed in Pilipchuk (1985, 1988) and Vakakis et al. (1996), a general periodic function $g = g(z)$ with normalized period $\Pi = 4$ can be expressed in terms of the nonsmooth variables $\tau(z)$ and $e(z)$ as follows:

$$g(z) = g(\tau, e) = X(\tau) + eY(\tau), \tag{3.3a}$$

where

$$X(\tau) = (1/2)[g(\tau) + g(2-\tau)], \qquad Y(\tau) = (1/2)[g(\tau) - g(2-\tau)], \\ g(\tau, e) = g(\tau(z), e(z)). \tag{3.3b}$$

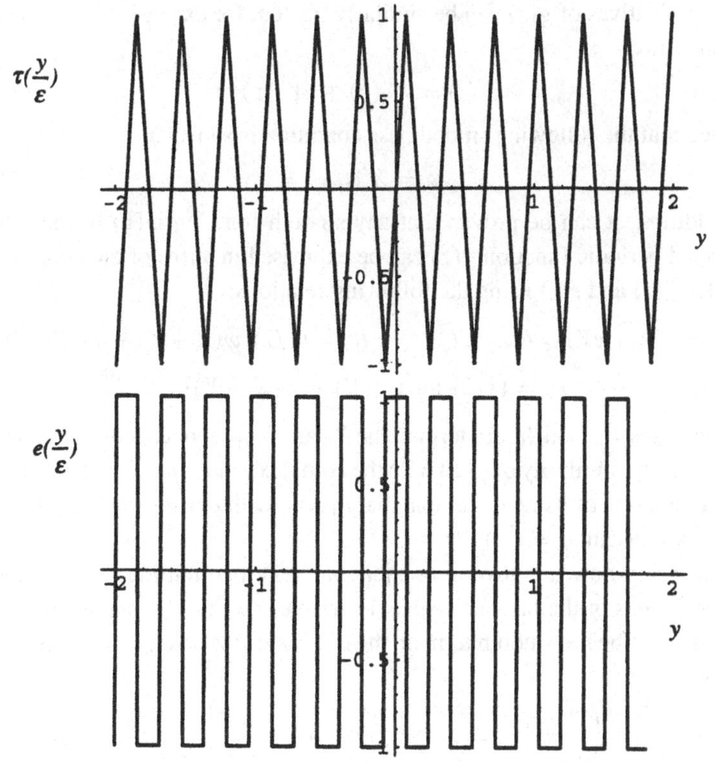

Fig. 3.1. The nonsmooth variables $\tau(y/\varepsilon)$ and $e(y/\varepsilon)$ for $\varepsilon = 0.08$.

The derivative of $g(z)$ with respect to z can be similarly expressed in terms of the nonsmooth variables by means of the relation

$$g'(z) = Y'(\tau) + eX'(\tau) + e'Y(\tau), \tag{3.4a}$$

where primes denote differentiation with respect to the argument. The last term in (3.4a) is singular since it represents an infinite sequence of Dirac functions:

$$e'(z) = 2 \sum_{k=-\infty}^{\infty} [\delta(z + 1 - 4k) - \delta(z - 1 - 4k)]. \tag{3.4b}$$

The derivative (3.4a) is free of the singularities (3.4b) if the function $g(z)$ is continuous at the set of points $z \in N = \{z/\tau(z) = \pm 1\}$ (cf. Figure 3.1). It follows that the elimination of the singularities in (3.4a) requires the imposition of the following additional smoothness conditions:

$$Y|_{z \in N} = Y|_{\tau=\pm 1} = 0. \tag{3.4c}$$

If the function $g(z)$ possesses discontinuities at points $z \in N$, the last term in (3.4a) must be included, and the generalized derivative e' is expressed by (3.4b).

Higher derivatives of $g(z)$ can be similarly treated; for example, the following relation holds:

$$g''(z) = X''(\tau) + eY''(\tau) \tag{3.5a}$$

provided that the following smoothness condition is satisfied:

$$X'|_{z\in N} = X'|_{\tau=\pm 1} = 0. \tag{3.5b}$$

In addition, it can be proven that any smooth function $w(g)$ of the aforementioned periodic function $g(z)$ can be expressed in terms of the nonsmooth variables $\tau(z)$ and $e(z)$ using the following relations:

$$w(g) = w(X+eY) = R_w + eI_w, \qquad R_w = (1/2)[w(X+Y) + w(X-Y)],$$
$$I_w = (1/2)[w(X+Y) - w(X-Y)]. \tag{3.6}$$

The quantities R_w and I_w are termed the R- and I-components of the function $w(g)$. Note the similarity of (3.6) with the complex representation of a complex variable in terms of its real and imaginary parts, with e being analogous to the imaginary constant $j = (-1)^{1/2}$.

Considering now the governing equation (3.1), one introduces nonsmooth transformations of the fast-scale spatial variable, $\tau = \tau(y/\varepsilon)$ and $e = e(y/\varepsilon)$, and expresses the nonsmooth term on the left-hand side using relation (3.4b) as follows:

$$2f(u)\sum_{k=-\infty}^{\infty}\delta\left(\frac{y}{\varepsilon} - 1 - 2k\right) = -f(u)\mathrm{sgn}(\tau)\tau'', \qquad \tau = \tau(y/\varepsilon). \tag{3.7}$$

Assuming that the transverse displacement and the external excitation are periodic functions of (y/ε), one can also express these variables in terms of the nonsmooth variables, using the relations (cf. equations (3.3))

$$\begin{aligned} u &= U(\tau, y, t) + V(\tau, y, t)\tau', \\ q &= Q(\tau, y, t) + P(\tau, y, t)\tau'. \end{aligned} \tag{3.8}$$

These expressions are analogous to the general formula (3.3) for the decomposition of a periodic function in terms of nonsmooth variables. Substituting (3.7) and (3.8) into (3.1) one obtains the following alternative expression for the governing partial differential equation:

$$\begin{aligned} &\rho\left(\frac{\partial^2 U}{\partial t^2} + \frac{\partial^2 V}{\partial t^2}\tau'\right) - T\left[\frac{\partial^2 U}{\partial y^2} + \frac{2}{\varepsilon}\frac{\partial^2 V}{\partial y\partial\tau} + \frac{1}{\varepsilon^2}\frac{\partial^2 U}{\partial\tau^2}\right. \\ &\quad \left. + \left(\frac{\partial^2 V}{\partial y^2} + \frac{2}{\varepsilon}\frac{\partial^2 U}{\partial y\partial\tau} + \frac{1}{\varepsilon^2}\frac{\partial^2 V}{\partial\tau^2}\right)\tau' + \left(\frac{1}{\varepsilon}\frac{\partial V}{\partial y} + \frac{1}{\varepsilon^2}\frac{\partial U}{\partial\tau}\right)\tau''\right] \\ &= \mathrm{sgn}(\tau)\tau''(R_f + I_f\tau') + Q(\tau, y, t) + P(\tau, y, t)\tau', \end{aligned} \tag{3.9}$$

where R_f and I_f are the R- and I-components, respectively, of the function $f(u)$. Note that in (3.9) the dependent variables U and V are functions of the nonsmooth variable $\tau \in [-1, 1]$ and the original variables y and t. Since by construction $\tau = \tau(y/\varepsilon)$ is a periodic function of the fast-scale spatial variable (y/ε), the periodicity of U and V with respect to (y/ε) is guaranteed.

Note that the relation

$$\tau'\tau'' \equiv ee' = 0 \tag{3.10}$$

holds. A proof of this statement based on the theory of distributions is given in Pilipchuk and Vakakis (1995). In addition, since τ and e are independent variables, the following equivalence relation is satisfied:

$$X(\tau) + eY(\tau) = 0 \Leftrightarrow X(\tau) = 0 \quad \text{and} \quad Y(\tau) = 0. \tag{3.11}$$

Taking into consideration these remarks, Equation (3.9) is decomposed into the two following relations:

$$\rho\frac{\partial^2 U}{\partial t^2} - T\left[\frac{\partial^2 U}{\partial y^2} + \frac{2}{\varepsilon}\frac{\partial^2 V}{\partial y \partial \tau} + \frac{1}{\varepsilon^2}\frac{\partial^2 U}{\partial \tau^2}\right] = Q(\tau, y, t), \tag{3.12a}$$

$$\rho\frac{\partial^2 V}{\partial t^2} - T\left[\frac{\partial^2 V}{\partial y^2} + \frac{2}{\varepsilon}\frac{\partial^2 U}{\partial y \partial \tau} + \frac{1}{\varepsilon^2}\frac{\partial^2 V}{\partial \tau^2}\right] = P(\tau, y, t) \tag{3.12b}$$

for τ in the range $-1 \le \tau \le 1$. The equations of motion (3.12a,b) are complemented by the following smoothness boundary conditions that guarantee the elimination of all singularities (i.e., generalized functions) from the original equation (3.9) and the partial derivatives of U and V with respect to τ:

$$-\left(\frac{1}{\varepsilon}\frac{\partial V}{\partial y} + \frac{1}{\varepsilon^2}\frac{\partial U}{\partial \tau}\right)\bigg|_{\tau=\pm 1} = \pm\frac{1}{T}R_f, \tag{3.12c}$$

$$\frac{\partial V}{\partial y}\bigg|_{\tau=\pm 1} = V|_{\tau=\pm 1} = 0. \tag{3.12d}$$

Rearranging the set (3.12) one derives the following final set of equations governing the motion of the nonlinearly supported string with respect to the fast-scale variable (y/ε):

$$\frac{\partial^2 U}{\partial \tau^2} = -2\varepsilon\frac{\partial^2 V}{\partial y \partial \tau} + \varepsilon^2\left[\frac{\rho}{T}\frac{\partial^2 U}{\partial t^2} - \frac{\partial^2 U}{\partial y^2} - \frac{Q(\tau, y, t)}{T}\right], \tag{3.13a}$$

$$\frac{\partial^2 V}{\partial \tau^2} = -2\varepsilon\frac{\partial^2 U}{\partial y \partial \tau} + \varepsilon^2\left[\frac{\rho}{T}\frac{\partial^2 V}{\partial t^2} - \frac{\partial^2 V}{\partial y^2} - \frac{P(\tau, y, t)}{T}\right], \tag{3.13b}$$

$$V|_{\tau=\pm 1} = 0, \tag{3.13c}$$

$$-\frac{\partial U}{\partial \tau}\bigg|_{\tau=\pm 1} = \pm\frac{\varepsilon^2}{T}R_f = \pm\frac{\varepsilon^2}{2T}[f(U+V) + f(U-V)]. \tag{3.13d}$$

Comparing the set of Equations (3.13) to the original Equation (3.1) the advantage of using the aforementioned nonsmooth transformations becomes clear; in contrast to (3.1), the set (3.13) does not contain terms with generalized functions, and perturbation methods based on analytic functions can be applied to study its solutions. *Hence, the use of the nonsmooth transformations of one of the independent variables resulted in the smoothening of the governing equations of motion.* In addition, it will be shown that the solutions of the set (3.13) can be obtained by constructing successive approximations based on *regular* perturbation series, a feature that considerably simplifies the analysis.

3.2.2 Perturbation Analysis for the Unforced Case

The perturbation analysis of the set (3.13) will be carried only for the unforced case corresponding to $Q(\tau, y, t) = P(\tau, y, t) = 0$; however, a similar analysis can be performed when forcing exists. The solutions are assumed in the form

$$U(\tau, y, t) = \sum_{k=0}^{\infty} \varepsilon^k U_k(\tau, y, t), \qquad V(\tau, y, t) = \sum_{k=0}^{\infty} \varepsilon^k V_k(\tau, y, t), \quad (3.14)$$

where all variables other than $\varepsilon \ll 1$ are assumed to be of $O(1)$. Substituting (3.14) into (3.13) and matching coefficients of respective powers of ε, one solves an infinite sequence of problems at various orders of approximation. The accuracy of the perturbation solution depends on the number of terms retained in the series (3.14).

At $O(\varepsilon^0)$ one computes the following leading-order approximations:

$$\begin{aligned} \frac{\partial^2 U_0}{\partial \tau^2} &= 0 \Rightarrow U_0(\tau, y, t) = A_0(y, t)\tau + B_0(y, t), \\ \frac{\partial^2 V_0}{\partial \tau^2} &= 0 \Rightarrow V_0(\tau, y, t) = C_0(y, t)\tau + D_0(y, t). \end{aligned} \quad (3.15a)$$

Taking into account the leading-order smoothness conditions,

$$\left.\frac{\partial U_0}{\partial \tau}\right|_{\tau=\pm 1} = 0 \quad \text{and} \quad V_0|_{\tau=\pm 1} = 0, \quad (3.15b)$$

one expresses the $O(1)$ solutions as follows:

$$U_0(\tau, y, t) = B_0(y, t), \qquad V_0(\tau, y, t) = 0. \quad (3.15c)$$

At $O(\varepsilon)$ one obtains the corresponding set of equations and smoothness conditions governing the approximations $U_1(\tau, y, t)$, $V_1(\tau, y, t)$, possessing the following trivial solutions:

$$\begin{gathered} \frac{\partial^2 U_1}{\partial \tau^2} = -2\frac{\partial^2 V_0}{\partial y \partial \tau}, \qquad \frac{\partial^2 V_1}{\partial \tau^2} = -2\frac{\partial^2 U_0}{\partial y \partial \tau}, \qquad \left.\frac{\partial U_1}{\partial \tau}\right|_{\tau=\pm 1} = 0, \\ V_1|_{\tau=\pm 1} = 0 \Rightarrow U_1(\tau, y, t) = 0, \quad V_1(\tau, y, t) = 0. \end{gathered} \quad (3.16)$$

Considering terms of $O(\varepsilon^2)$ one derives the set of equations

$$\frac{\partial^2 U_2}{\partial \tau^2} = -2\frac{\partial^2 V_1}{\partial y \partial \tau} + \frac{\rho}{T}\frac{\partial^2 U_0}{\partial t^2} - \frac{\partial^2 U_0}{\partial y^2}, \tag{3.17a}$$

$$\frac{\partial^2 V_2}{\partial \tau^2} = -2\frac{\partial^2 U_1}{\partial y \partial \tau} + \frac{\rho}{T}\frac{\partial^2 V_0}{\partial t^2} - \frac{\partial^2 V_0}{\partial y^2} \tag{3.17b}$$

complemented by the $O(\varepsilon^2)$ smoothness conditions

$$V_2|_{\tau=\pm 1} = 0, \qquad \left.\frac{\partial U_2}{\partial \tau}\right|_{\tau=\pm 1} = \pm\frac{1}{2T}[f(U_0+V_0) + f(U_0-V_0)]. \tag{3.17c}$$

Taking into account the relations (3.15)–(3.16), we can express the solution of the set (3.17) as

$$U_2(\tau, y, t) = -\frac{f(B_0(y,t))}{T}\frac{\tau^2}{2} + B_2(y,t), \qquad V_2(\tau, y, t) = 0, \tag{3.18}$$

where the $O(1)$ approximation $B_0(y, t)$ is governed by the following nonlinear partial differential equation in y and t:

$$\rho\frac{\partial^2 B_0(y,t)}{\partial t^2} - T\frac{\partial^2 B_0(y,t)}{\partial y^2} + f(B_0(y,t)) = 0, \quad -\infty < y < \infty. \tag{3.19}$$

This is the modulation equation governing the spatial and temporal dependence of the $O(1)$ amplitude of the motion. Higher-order approximations were studied in Vakakis and Pilipchuk (1995).

In summary, the solutions of the system (3.13), which correspond to periodic solutions in the fast-scale spatial variable (y/ε) of the nonlinearly supported string, are computed as

$$U(\tau, y, t) = U_0(\tau, y, t) + \varepsilon U_1(\tau, y, t) + \varepsilon^2 U_2(\tau, y, t) + \cdots,$$
$$V(\tau, y, t) = V_0(\tau, y, t) + \varepsilon V_1(\tau, y, t) + \varepsilon^2 V_2(\tau, y, t) + \cdots,$$

where the various terms in the series are computed by (3.15)–(3.19).

The general methodology will now be applied to compute time-periodic solutions of a string supported by an infinite periodic array of nonlinear stiffnesses with hardening cubic nonlinearities. For the system with cubic hardening nonlinearities the approximation $B_0(y, t)$ is governed by the partial differential equation

$$\frac{\partial^2 B_0(y,t)}{\partial t^2} - \mu\lambda\frac{\partial^2 B_0(y,t)}{\partial y^2} + B_0(y,t) + \mu\alpha B_0^3(y,t) = 0, \quad \lambda > 0,$$
$$\alpha > 0, \quad -\infty < y < \infty, \tag{3.20}$$

where the notation $T/\rho = \mu\lambda$ was used, and the additional small parameter $0 < \mu$ was introduced to scale the spatial partial derivative and the nonlinear term. Normalized coefficients were used to describe the nonlinear stiffness of the supports.

Clearly, Equation (3.20) admits different classes of solutions, and in this work only time-periodic, standing wave solutions (*breathers* (Flytzanis et al., 1985; Weinstein, 1985; Eleonsky, 1991)) will be considered; these are motions during which *all* material points of the string execute synchronous oscillations, reaching their extreme values and passing through their equilibrium positions at the same instants of time. In essence, these motions are nonlinear normal modes (NNMs) of the system under consideration. Only the subclass of NNMs with spatially localized envelopes will be examined herein; however, similar results can be derived for waves with envelopes possessing spatial periodicity in y.

Standing wave solutions with localized amplitude or slope envelopes for homogeneous and nonhomogeneous equations of the general form (3.20) were studied in previous works by Vakakis and coworkers (King and Vakakis, 1994; Vakakis et al., 1994). In these studies it was shown that for $\mu, \lambda, \alpha > 0$ Equation (3.20) can only possess nonlinear standing wave solutions with localized slopes and amplitudes that reach constant limits as $y \to \pm\infty$. This type of standing wave solution can be computed by either applying the concept of harmonic balance or by employing an asymptotic methodology based on a nonlinear separation of variables (King and Vakakis, 1994). Following this later work, the asymptotic, localized standing wave solution of Equation (3.20) can be expressed as

$$B_0(y,t) = \left[a_1^{(0)}(y) + \mu a_1^{(1)}(y)\right]u_0(t) + \mu a_3^{(1)}(y)u_0^3(t) + O\left(\mu u_0^5(t), \mu^2\right), \tag{3.21}$$

where

$$a_1^{(0)}(y) = [(\beta-\kappa)/\beta]^{1/2}\tanh[(\beta-\kappa/2]^{1/2}y, \tag{3.22a}$$

$$a_3^{(1)}(y) = \frac{-\lambda[(\beta-\kappa)/\beta]^{1/2}}{6u_0^{*2}}\{(\beta-\kappa)\,\mathrm{sech}^2[(\beta-\kappa)/2]^{1/2}$$
$$\times\, y\tanh[(\beta-\kappa)/2]^{1/2}y + \kappa\tanh[(\beta-\kappa)/2]^{1/2}y\}, \tag{3.22b}$$

$$\beta = (3\alpha/4\lambda)u_0^{*2}, \tag{3.22c}$$

and κ is determined through the relation

$$[(\beta-\kappa)/\beta]^{1/2}\tanh[(\beta-\kappa)/2]^{1/2}y_0 = 1. \tag{3.23}$$

In the expression above, y_0 defines an arbitrary position ("reference point") of the string (King and Vakakis, 1994), and it is required that $y_0 \neq 0$. The coefficient $a_1^{(1)}(y)$ in (3.21) is computed by the analytical expressions listed in the appendix, and the quantity $u_0(t)$ represents the time-periodic oscillation of the "reference point." It is evaluated asymptotically by solving the following nonlinear ordinary differential equation:

$$\frac{d^2u_0}{dt^2} + (1-\mu\lambda\kappa)u_0 + \mu\alpha\left[1+(\mu\alpha/\lambda)a_3^{(1)''}(y_0)\right]u_0^3 + O\left(\mu u_0^5, \mu^2\right) = 0,$$

$$u_0(0) = u_0^*, \qquad \frac{du_0}{dt}(0) = 0. \tag{3.24}$$

The approximate solution can be computed using a standard perturbation method as follows:

$$u_0(t) \cong u_0^* \cos(\omega t), \qquad \omega^2 \cong (1-\mu\lambda\kappa) + \frac{3}{4}\mu\alpha\left[1+(\mu\alpha/\lambda)a_3^{(1)''}(y_0)\right]u_0^{*2}. \tag{3.25}$$

The solution (3.21) is not spatially localized, since the envelope of the wave reaches the nonzero limits ± 1 as $y \to \pm\infty$. Indeed, solution (3.21) describes a wave with spatially extended envelope and spatially localized slope of the envelope.

As a numerical application of the derived theoretical results, consider a system with parameters

$$[(\beta-\kappa)/\beta]^{1/2} = 1, \qquad (\beta-\kappa)/2 = 3\alpha u_0^{*2}/8\lambda,$$

$$\varepsilon = 0.08, \qquad \mu = 0.7, \qquad \lambda = 3, \qquad \alpha = .7, \qquad u_0^* = 0.4. \tag{3.26}$$

Employing the previous results, we can approximate the transverse displacement of the string during the localized standing wave as

$$u\left(\frac{y}{\varepsilon}, y, t\right) = B_0(y,t) + O(\varepsilon^2) = a_1^{(0)}(y)u_0(t) + O(\mu u_0(t), \mu^2) + O(\varepsilon^2), \tag{3.27}$$

whereas the corresponding slope of the transverse displacement is computed as

$$\begin{aligned}
&\frac{\partial u}{dy}\left(\frac{y}{\varepsilon}, y, t\right) \\
&= \frac{\partial B_0}{\partial y}(y,t) + \varepsilon\frac{\partial U_2}{\partial \tau}\left(\frac{y}{\varepsilon}, y, t\right)e\left(\frac{y}{\varepsilon}\right) + O(\varepsilon^2) \\
&= \frac{\partial B_0}{\partial y}(y,t) - \left[\frac{\varepsilon B_0(y,t)}{\mu\lambda} + \frac{\varepsilon\alpha B_0^3(y,t)}{\lambda}\right]\tau\left(\frac{y}{\varepsilon}\right)e\left(\frac{y}{\varepsilon}\right) + O(\varepsilon^2) \\
&= a_1^{(0)'}(y)u_0(t) - \left[\frac{\varepsilon a_1^{(0)}(y)u_0(t)}{\mu\lambda} + \frac{\varepsilon\alpha a_1^{(0)^3}(y)u_0^3(t)}{\lambda}\right]\tau\left(\frac{y}{\varepsilon}\right)e\left(\frac{y}{\varepsilon}\right) \\
&\quad + O(\mu u_0(t), \mu^2, \varepsilon u_0(t)) + O(\varepsilon^2).
\end{aligned} \tag{3.28}$$

Taking into account the numerical values for the parameters (3.26) we conclude that $u_0(t) = O(0.1)$. Moreover, for the sake of simplicity only the leading-order approximation for $B_0(y, t)$ was considered. Examining (3.27) and (3.28) it is noted that correct to $O(\varepsilon^2)$ there are no discreteness effects in the expression of the transverse displacement. However, *there exist $O(\varepsilon)$ discreteness effects in the expression of the slope*, represented by the term containing the product of nonsmooth variables $\tau(y/\varepsilon)e(y/\varepsilon)$.

3.2.3 Concluding Remarks

Nonsmooth transformations of an independent variable were introduced to eliminate singularities from the governing equation of motion of a string supported by periodically spaced nonlinear stiffnessess. The transformed smooth set of equations of motion was analyzed using regular perturbation series, resulting in a considerable simplification of the analysis; the modulation relations obtained were analyzed using standard asymptotic techniques from the theory of smooth nonlinear dynamical systems. This methodology permits the computation of discreteness effects in the motion of the string, a feature not feasible in techniques based on continuum approximations. Indeed, for the specific system with cubic nonlinearities studied here, $O(\varepsilon)$ discreteness effects in the spatial distribution of the slope of the transverse motion were computed and graphically depicted.

In the next section we analyze primary pulse transmission in a periodic system with strongly nonlinear coupling elements. A different methodology which models exactly the transmission and reflection of the transient waves at the nonlinear elements, is followed.

3.3 Wave Transmission in a Strongly Nonlinear Periodic Continuous System

3.3.1 Problem Formulation

We consider a periodic system composed of a semi-infinite array of linear elastic layers jointed by means of coupling elements possessing stiffness and/or damping nonlinearities. The following analysis follows closely that of Pilipchuk et al. (1996).

The equation of motion governing longitudinal waves in the ith layer is

$$\frac{\partial^2 u_i}{\partial t^2} - \frac{\partial^2 u_i}{\partial x^2} = 0, \tag{3.29}$$

where $u_i = u_i(t, x)$ denotes the longitudinal displacement, i numbers the layers, and all variables are normalized so that $0 \leq x \leq 1$. Furthermore, it is assumed that at $t = 0$ the ith layer is at rest and $u_i = 0$, $\partial u_i/\partial t = 0$ for $t = 0$. We denote by $F_i(t)$ the force acting on the left boundary of the ith layer. Clearly, when $i = 1$ we have that $F_1(t) \equiv F(t)$, that is, the force is equal to the externally applied impulsive load. For $i \neq 1$ the force $F_i(t)$ is equal to the force exerted by the nonlinear coupling element and is expressed in symbolic form as

$$F_i(t) = K(z_{i-1}, \partial z_{i-1}/\partial t), \tag{3.30}$$

where $K(z, \partial z/\partial t)$ is the characteristic and z_{i-1} the relative displacement of the $(i-1)$-th coupling element (cf. Figure 3.1). It follows that Equation (3.29) is complemented by the following set of boundary conditions:

$$\left.\frac{\partial u_i}{\partial x}\right|_{x=0} = F_i(t), \qquad \left.\frac{\partial u_i}{\partial x}\right|_{x=1} = K(z_i, \partial z_i/\partial t) \equiv K(t), \tag{3.31}$$

where the relative displacement is given by

$$z_i(t) = u_{i+1}(t, 0) - u_i(t, 1). \tag{3.32}$$

Similarly, the motion of the $(i+1)$-th layer is governed by

$$\frac{\partial^2 u_{i+1}}{\partial t^2} - \frac{\partial^2 u_{i+1}}{\partial x^2} = 0 \tag{3.33}$$

and

$$\left.\frac{\partial u_{i+1}}{\partial t}\right|_{x=0} = K(t) \equiv F_{i+1}(t-1), \tag{3.34}$$

where only the boundary condition at the left boundary is given. It is assumed that at $t = 1$ the $(i+1)$-th layer is at rest, that is,

$$u_{i+1} \equiv 0, \qquad \frac{\partial u_{i+1}}{\partial t} \equiv 0 \quad \text{for } t = 1.$$

Consider now waves propagating in the ith layer. Since the layer is a nondispersive medium we can use D'Alembert's representation for the solution as follows:

$$\frac{\partial u_i}{\partial x} = F_i(t-x) \quad \text{for } t < 1 \quad \text{and} \quad \frac{\partial u_i}{\partial x} = F_i(t-x) + \Phi_i(t+x) \quad \text{for } 1 < t < 2, \tag{3.35}$$

where Φ_i denotes the profile of the reflected wave and is defined using the second of relations (3.31),

$$F_i(t-1) + \Phi_i(t+1) = K(t) \Rightarrow \Phi_i(\xi) = -F_i(\xi - 2) + K(\xi - 1). \tag{3.36}$$

Substituting (3.36) into (3.35) one obtains the relation

$$\frac{\partial u_i}{\partial x} = F_i(t - x) - F_i(t + x - 2) + K(t + x - 1) \quad \text{for} \quad 1 < t < 2. \tag{3.37}$$

The motion in the $(i + 1)$-th layer then satisfies the following relation:

$$\frac{\partial u_{i+1}}{\partial x} = K(t - x) \quad \text{for } 1 < t < 2. \tag{3.38}$$

Integrating equations (3.9) and (3.11) we obtain

$$u_i(t, x) = u_i(t, 0) - \int_t^{t-x} F_i(\xi)\, d\xi - \int_t^{t+x} F_i(\xi - 2)\, d\xi$$
$$+ \int_t^{t+x} K(\xi - 1)\, d\xi \quad \text{for } 1 < t < 2 \tag{3.39}$$

and

$$u_{i+1}(t, x) = u_{i+1}(t, 0) - \int_t^{t-x} K(\xi)\, d\xi \quad \text{for } 1 < t < 2. \tag{3.40}$$

Taking into account the impulsive character of the propagating wave, we impose the conditions $u_i(t, 0) = const$ and $u_{i+1}(t, 1) = 0$ when $1 < t < 2$. Setting $x = 1$ in (3.39) and (3.40) and taking into account these conditions, we can compute the quantities $u_{i+1}(t, 0)$ and $u_i(t, 1)$, and by means of (3.32) $z_i(t)$. Hence, we can compute the relative displacement of the coupling element connecting the ith and $(i + 1)$-th layers as follows:

$$z_i(t) = u_{i+1}(t, 0) - u_i(t, 1) = \int_t^{t-1} K(\xi)\, d\xi$$
$$- \left[const - \int_t^{t-1} F_i(\xi)\, d\xi - \int_t^{t+1} F_i(\xi - 2)\, d\xi \right.$$
$$\left. + \int_t^{t+1} K(\xi - 1) d\xi \right] \quad \text{for } 1 < t < 2. \tag{3.41}$$

Differentiating the above expression with respect to t we obtain the relation

$$\dot{z}_i = 2K(t - 1) - 2K(t) + 2F_i(t - 1) - F_i(t) - F_i(t - 2) \quad \text{for } 1 < t < 2. \tag{3.42}$$

Finally, taking into account that $2K(t-1) - F_i(t) - F_i(t-2) = 0$ for $1 < t < 2$, we obtain the following initial value problem (IVP) governing $z_i(t)$:

$$\frac{1}{2}\dot{z}_1 + K(z_i, \dot{z}_i) = F_i(t); \quad z_i|_{t=0} = 0, \tag{3.43}$$

where the time shift $(t - 1) \to t$ was imposed to set the origin of time at $t = 1$. Recalling at this point that $K(z_i, \dot{z}_i) = F_{i+1}(t)$ provides the profile of the traveling stress wave in the $(i+1)$-th layer, we can rewrite relation (3.43) as

$$F_{i+1}(t) = F_i(t) - \frac{1}{2}\dot{z}_i. \tag{3.44}$$

The above expression describes the transformation of the profile of the stress wave as it passes from the ith to the $(i+1)$-th layer.

Equations (3.43) and (3.44) solve iteratively the problem of primary pulse transmission through the repetitive medium of Figure 3.3. In the initial step one sets $i = 1$ and $F_i(t) = F_1(t)$ in (3.43) and solves the first-order differential equation for $z_1(t)$. Then, Equation (3.44) provides the profile of the stress pulse in the second layer $F_2(t)$. The iterative process can then be continued for $i = 2, 3, \ldots$ to study the transmission of the primary pulse through subsequent layers of the system. Note that due to the previous shift in time, the time of arrival of the primary pulse at an arbitrary layer is always set equal to zero.

Assume a coupling characteristic of the form

$$K(z, \dot{z}) = A_1 z + L(z) + B_1\dot{z} + M(\dot{z}), \tag{3.45}$$

where A_1 and B_1 are the coefficients of the linear terms and $L(z)$ and $M(\dot{z})$ denote the stiffness and damping nonlinearities of the coupling elements, respectively. Taking into account (3.44), relation (3.43) is expressed as

$$\lambda z_i + \dot{z}_i + \phi(z_i) + \psi(\dot{z}_i) = \bar{F}_i(t); \qquad z_i|_{t=0} = 0, \tag{3.46a}$$

where

$$\begin{aligned} \lambda &= 2A_1/(1+2B_1), & \bar{F}_i(t) &= 2F_i(t)/(1+2B_1), \\ \phi(z) &= 2L(z)/(1+2B_1), & \psi(\dot{z}) &= 2M(\dot{z})/(1+2B_1). \end{aligned} \tag{3.46b}$$

In terms of the new notation, Equation (3.16) is written as

$$\bar{F}_{i+1}(t) = \bar{F}_i(t) - \frac{1}{(1+2B_1)}\dot{z}_i. \tag{3.46c}$$

In the following sections analytical and numerical solutions of the IVPs (3.46) will be developed in order to study the primary pulse transmission in the repetitive system for various coupling element configurations.

3.3.2 Analytical Approximations

The solution of the IVPs (3.46) will be obtained for an applied impulsive force $F_1(t)$ (Pilipchuk et al., 1996). Since the only nonzero segment of the applied force is for $0 < t \le \varepsilon$, we introduce a piecewise continuous transformation of the time variable as follows:

$$s = \begin{cases} t/\varepsilon, & \text{for } t \le \varepsilon, \\ t - \varepsilon + 1, & \text{for } t > \varepsilon, \end{cases} \tag{3.47a}$$

where s denotes the new time variable. The inverse transformation is given by

$$t = \frac{s' - 1}{1/\varepsilon - 1}\varepsilon s + \frac{s' - 1/\varepsilon}{1 - 1/\varepsilon}(s + \varepsilon - 1), \tag{3.47b}$$

where primes denote the derivative with respect to t. The reason for introducing this transformation will become apparent from the following analysis. At this point we make the following remarks.

(i) Almost everywhere we have that

$$s'^2 = \frac{1}{\varepsilon^2}\frac{s' - 1}{1/\varepsilon - 1} + \frac{s' - 1/\varepsilon}{1 - 1/\varepsilon}, \tag{3.48}$$

that is, s'^2 is the linear form of s'. It follows that the right part of (3.48) is an element of an algebra.

(ii) Any function $z(t)$ can be represented as an element of this algebra using the relation

$$z(t) = U(s) + V(s)s'. \tag{3.49}$$

Substituting (3.19b) into (3.21) we obtain the representations

$$z_-(s) \equiv z(\varepsilon s) = U + V/\varepsilon \text{ for } t \le \varepsilon \quad \text{and} \quad z_+(s) \equiv z(s + \varepsilon - 1)$$
$$= U + V \quad \text{for } t > \varepsilon. \tag{3.50}$$

Alternatively, $z(t)$ can be expressed in terms of $z_-(s)$ and $z_+(s)$ as follows:

$$z = \frac{\varepsilon}{1 - \varepsilon}(z_+/\varepsilon - z_-) + \frac{\varepsilon}{1 - \varepsilon}(z_- - z_+)s'. \tag{3.51}$$

(iii) The time derivative of $z(t)$ is also an element of the algebra, that is, $\dot{z}(t)$ can be expressed in a form similar to (3.21) if the condition

$$(z_- - z_+)|_{s=1} = 0 \tag{3.52}$$

is imposed.

Taking into account the remarks (i)–(iii) and substituting (3.23) into (3.18) we obtain the following two subproblems describing the deformation of the ith coupling element for $t \le \varepsilon$ and $t > \varepsilon$, respectively:

$$\frac{1}{\varepsilon}z'_{i-} + \lambda z_{i-} + \psi\left(\frac{1}{\varepsilon}z'_{i-}\right) + \phi(z_{i-}) = \bar{F}_i^- \quad \text{and}$$
$$z'_{i+} + \lambda z_{i+} + \psi(z'_{i+}) + \phi(z_{i+}) = \bar{F}_i^+, \tag{3.53}$$

where the forcing terms on the right-hand sides of the above equations are defined according to the transformation formulas (3.50). Transforming the initial condition and taking into account (3.52), we obtain the following set of conditions complementing (3.53):

$$z_{i-}(0) = 0 \quad \text{and} \quad z_{i+}(1) = z_{i-}(1). \tag{3.54}$$

The transformed forcing terms in Equation (3.53) are computed using the previous properties of the piecewise transformation, in accordance with (3.50). The relation (3.46c) determining the dispersion of the profile of the stress wave from layer (i) to layer $(i+1)$ is similarly transformed as follows:

$$\bar{F}_{i+1}^{-} = \bar{F}_{i}^{-} - \frac{1}{\varepsilon(1+2B_1)} z'_{i-} \quad \text{and} \quad \bar{F}_{i+1}^{+} = \bar{F}_{i}^{+} - \frac{1}{(1+2B_1)} z'_{i+}. \tag{3.55}$$

Hence, the piecewise transformation of the independent variable enables us to study the wave transmission through the system in two distinct time intervals: variables with $(-)$ sign refer to dynamic response in the time period $0 < t \leq \varepsilon$, whereas those with $(+)$ sign refer to dynamic response in the period $t > \varepsilon$. *Implicit in this scheme are the previously mentioned shifts of the origin of the time axis which assures that the stress wave arrives at each layer at time $t = 0$.* To study the wave transmission through the system we solve Equations (3.53) for $i = 1, 2, 3, \ldots$ with boundary conditions (3.54) taking into account that for the first layer the force is prescribed as

$$F_1^{-} \equiv F(\varepsilon s) \quad \text{and } F_1^{+} = 0. \tag{3.56}$$

The profile of the stress wave in subsequent layers $i = 2, 3, \ldots$ is then computed by (3.27).

In the next section we study the primary wave transmission in systems with nonanalytic coupling characteristics by performing direct numerical integrations of the governing scattering equations.

3.3.3 Computational Results

We now study numerically primary pulse transmission in periodic systems with clearance nonlinearities. Such nonanalytic nonlinearities are of considerable practical importance since they appear often in practical engineering applications. Due to the complexity of the problem an alternative way for studying the primary pulse transmission is followed, by directly (numerically) integrating the governing scattering Equations (3.53)–(3.55). The primary transmitted pulse in the ith layer is computed in terms of the components $\bar{F}_i^{-}(s)$ and $\bar{F}_i^{+}(s)$ using the following inverse transform expression:

$$\begin{aligned} \bar{F}_i(t) = {} & \frac{\varepsilon}{1-\varepsilon}\left(\bar{F}_i^{+}(s(t))/\varepsilon - \bar{F}_i^{-}(s(t))\right) \\ & + \frac{\varepsilon}{1-\varepsilon}\left(\bar{F}_i^{-}(s(t)) - \bar{F}_i^{+}(s(t))\right)s'(t), \qquad i = 1, 2, \ldots. \end{aligned} \tag{3.57}$$

The coupling elements of the system are assumed to be undamped and to possess clearance stiffness nonlinearities. In terms of the notation used in

(3.53)–(3.55) we have that

$$\phi(z)=\begin{cases}-\lambda z, & \text{for } |z|\le z_0\\ -\lambda z_0 \operatorname{sgn}(z), & \text{for } |z|>z_0\end{cases},\quad \psi(\dot{z})=0,\quad B_1=0,\quad \bar{F}_i(t)=2F_i(t). \tag{3.58}$$

Note that clearances introduce strong (essential) nonlinearities into the system, which due to their piecewise differentiability are difficult to study analytically. In the numerical simulations the impulsive force was assumed to be of the form

$$F(t)=\begin{cases}Pt^2(t-\varepsilon)^2\sin(t/\varepsilon), & \text{for } 0\le t\le \varepsilon,\\ 0, & \text{for } \varepsilon < t\end{cases} \tag{3.59}$$

with $P = 5.2 \times 10^7$ and $\varepsilon = 0.1$.

Numerical integrations of Equations (3.53)–(3.55) were carried out for various values of the clearance parameters λ and z_0 and the corresponding primary transmitted pulses (3.57) are depicted in Figures 3.2 and 3.3. Figure 3.2 depicts the primary pulse in the first fifteen layers of the system with $\lambda = 10$ and $z_0 = 0.0, 0.1, 0.2$. By the definition of the clearance nonlinearity (3.58) $z_0 = 0.0$ corresponds to the *linear* system, whereas an increase in z_0 corresponds to increasingly stronger clearance nonlinearities. From the numerical results we note that *for* $z_0 = 0.1$ *no primary pulse is transmitted above the eighth layer, whereas for* $z_0 = 0.2$ *no primary wave transmission takes place above the fourth layer*. In physical terms, the transmitted pulse diminishes in amplitude after the initial dispersion in the leading layers of the system; when the amplitude of the primary wave becomes smaller than the clearance between layers no further transmission is possible. Clearly, *secondary* pulses generated from the complicated pattern of transmissions and reflections at the nonlinear joints of the system can still be transmitted through the system but only after a time delay. *Hence, clearance nonlinearities restrict primary pulse transmission through the layered system; as a result, energy entrapment in the leading layers may occur and energy can only be transmitted to higher-order layers through secondary pulses, with time delays.* This interesting result is of engineering importance.

Primary pulse entrapment is also observed in the numerical simulations presented in Figure 3.3 ($\lambda = 100$ and $z_0 = 0.0, 0.03, 0.06, 0.08, 0.1, 0.2, 0.3, 0.4, 0.5$). We note that for the same level of clearance, the primary pulse entrapment is enhanced with increased stiffness characteristic λ. For given values of λ and z_0 it would be of interest to analytically find the maximum number of layers of primary pulse propagation. This appears to be a demanding mathematical

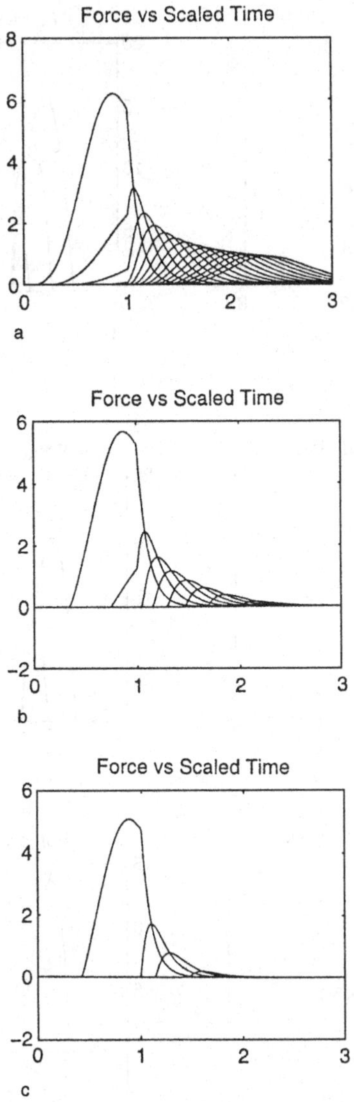

Fig. 3.2. Transmitted primary stress waves in the leading 15 layers of the periodic system with clearance nonlinearities with $\lambda = 10$ and (*a*) $z_0 = 0.0$, (*b*) $z_0 = 0.1$, and (*c*) $z_0 = 0.2$. Note primary pulse entrapment due to the clearances.

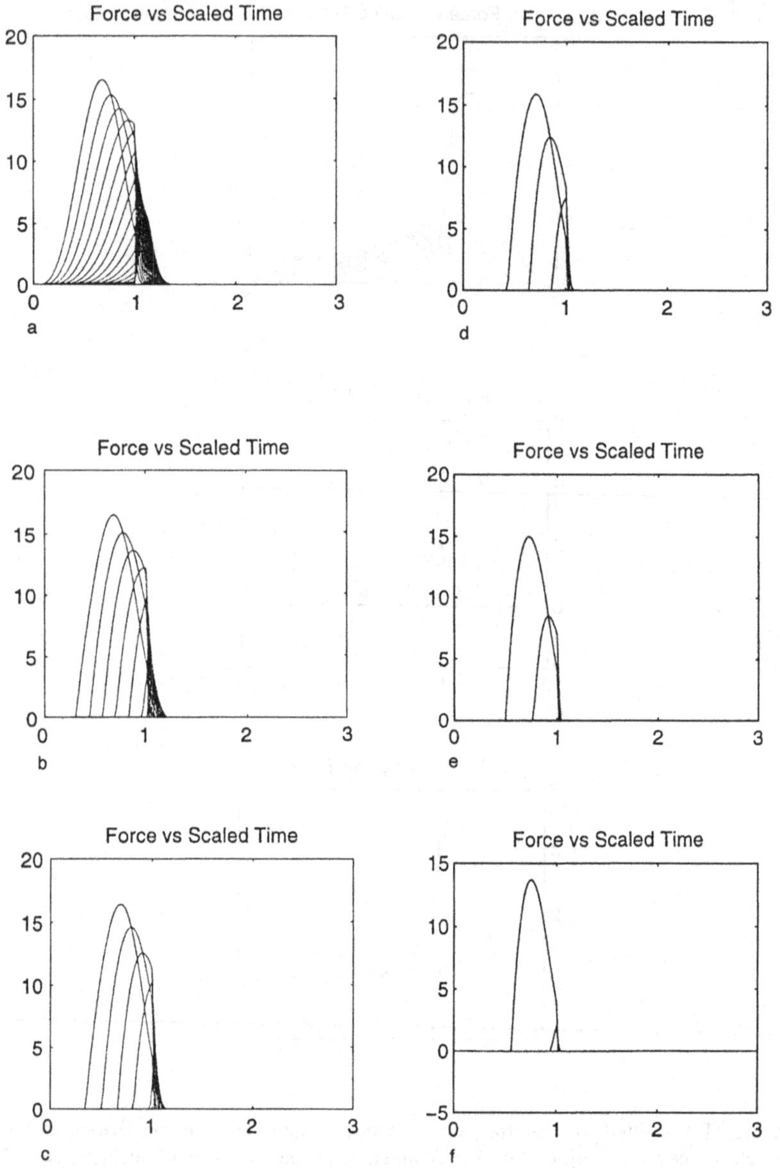

Fig. 3.3. Transmitted primary stress waves in the leading 15 layers of the periodic system with clearance nonlinearities with $\lambda = 100$ and (*a*) $z_0 = 0.0$, (*b*) $z_0 = 0.06$, (*c*) $z_0 = 0.1$, (*d*) $z_0 = 0.2$, (*e*) $z_0 = 0.3$, and (*f*) $z_0 = 0.4$.

problem, and a formulation that can contribute to its solution is outlined in the following section.

3.3.4 Continuum Approximation

We now provide an alternative formulation of the primary pulse transmission problem by performing a continuum approximation of the scattering equations. Recall at this point the original relations governing the transmission and reflection of the primary pulse at the $(i+1)$-th coupling element:

$$\frac{1}{2}\dot{z}_{i+1}+K(z_{i+1},\dot{z}_{i+1}) = F_{i+1}(t); \qquad z_{i+1}|_{t=0}=0, \qquad K(z_1,\dot{z}_i)=F_{i+1}(t). \tag{3.60}$$

Suppose that $\dot{z}_i(t)$ does not vary very much between subsequent coupling elements and impose the following continuum approximation:

$$z_i(t)\equiv z(i,t)\Rightarrow z_{i+1}(t)\equiv z(i+1,t)=z(i,t)+\frac{\partial z(i,t)}{\partial i}+\frac{1}{2}\frac{\partial^2 z(i,t)}{\partial i^2}+\cdots. \tag{3.61}$$

Similarly, the quantity $K(z_{i+1},\dot{z}_{i+1})$ is approximated as follows:

$$\begin{aligned} K(z_{i+1},\dot{z}_{i+1}) &= K\left[z(i,t)+\frac{\partial z(i,t)}{\partial t}+\cdots,\ \frac{\partial z(i,t)}{\partial t}+\frac{\partial^2 z(i,t)}{\partial t\partial i}+\cdots\right] \\ &= K\left[z(i,t),\frac{\partial z(i,t)}{\partial i}\right]+\frac{\partial K}{\partial z}\frac{\partial z(i,t)}{\partial i}+\frac{\partial K}{\partial(\partial z/\partial t)}\frac{\partial^2 z(i,t)}{\partial t\partial i}+\cdots. \end{aligned} \tag{3.62}$$

Expressing the force $F_{i+t}(t)$ using the second of relations (3.60) and imposing the previous continuum approximations we obtain the following approximate partial differential equation governing the evolution of the continuous variable $z(i,t)$:

$$\frac{\partial z(i,t)}{\partial t}+2\frac{\partial K}{\partial z}\frac{\partial z(i,t)}{\partial i}+\left[1+2\frac{\partial K}{\partial(\partial z/\partial t)}\right]\frac{\partial^2 z(i,t)}{\partial t\partial i}+\cdots=0. \tag{3.63a}$$

This equation must be solved in the domain $0\le t\le 1$, $1\le i<\infty$, with the nonlinear boundary condition

$$\frac{1}{2}\frac{\partial z(1,t)}{\partial t}+K\left[z(1,t),\frac{\partial z(1,t)}{\partial t}\right]=F(t) \tag{3.63b}$$

and the initial condition

$$z(i,0)=0, \tag{3.63c}$$

where $F(t)$ is the applied external force at the first layer. The solution of the nonlinear problem (3.66) provides a continuum approximation to the deformations of the coupling elements that can then be used to compute the dispersed shape of the stress wave through the second of relations (3.60). It must be noted that only leading-order discretization effects are taken into account in (3.60), and more accurate continuum approximations can be constructed by including higher-order partial derivatives of the continuous variables in the governing equation and the boundary condition.

Although no exact solution to the nonlinear problem (3.66) appears to exist, one can compute the solution of the *linearized* problem corresponding to $\partial K/\partial z = A_1$, $\partial K/\partial(\partial z/\partial t) = B_1$ by employing an integral transform analysis (Pilipchuk et al., 1996). However, the formulation can be of use for studying the problem of primary pulse transmission in the system with clearance nonlinearities discussed in Section 3.3.3. The interested reader is referred to Pilipchuk et al. (1996).

3.3.5 Concluding Remarks

In this section we provided analytical and numerical results for primary pulse transmission in a strongly nonlinear periodic system. Under the assumption of sufficiently small duration of the external impulsive excitation, a piecewise differentiable coordinate transformation was introduced which enabled us to compute the primary transmitted pulse in two distinct time phases. The scattering of the primary pulse at nonlinear coupling elements was found to be governed by a strongly nonlinear ordinary differential equation of the first order. This equation can be solved by analytical as well as numerical techniques.

For a system with clearance nonlinearities, it was found that the transmitted primary pulse propagates to only a finite number of layers. As a result, transmission of energy into further layers can occur only through secondary pulses or can not occur at all. Hence, depending on the strength of the clearance nonlinearities there is a critical number of leading layers above which there is a time-delayed secondary transmission of the impulsive energy, or there is no transmitted energy. This result shows that clearance nonlinearities in a periodic layered system can lead to energy entrapment in the leading layers. Since clearances are a common form of strong nonlinearities occurring often in engineering applications, this result might be of significant practical importance. An alternative continuum approximation methodology was also presented. The solution of the resulting continuum problem represents a challenging mathematical task.

Secondary pulse transmissions can be studied similarly, by solving analytically or numerically the corresponding secondary scattering relations. It must be pointed out, however, that since the principle of linear superposition does not apply in the nonlinear systems under consideration, special care should be taken to formulate the appropriate ordinary differential equations governing the scattering of the secondary waves.

Acknowledgments

The authors would like to thank Dr. Michael El-Raheb, Central Research, Dow Chemical Company, Midland, Michigan, for discussions concerning the problem analyzed in Section 3.3, and Mr. M. A. F. Azeez, Department of Mechanical & Industrial Engineering, University of Illinois, for his help with the numerical simulations of Section 3.3. Special thanks are also due to Mrs. Tammy Smith, Department of Mechanical Engineering, University of Illinois, for her help with the typing of the manuscript.

References

Achenbach, J. D., and Norris, A. N., 1982, Loss of specular reflection due to nonlinear crack-face interaction. *Journal of Nondestructive Evaluation*, 3, pp. 229–39.

Achenbach, J. D., and Parikh, O. K., 1991, Ultrasonic analysis of nonlinear response and strength of adhensive bonds. *Journal of Adhesion Science and Technology*, 5, pp. 601–18.

Anderson, P. W., 1958, Absence of diffusion in certain random lattices. *Physical Review*, 109, pp. 1492–505.

Aubrecht, J., and Vakakis, A. F., 1996, Localized and nonlocalized nonlinear normal modes in a multi-span beam with geometric nonlinearities. *J. Vib. Acoust.* 118, (4), pp. 533–42.

Baker, G. A., and Graves-Morris, P., 1981, *Pade' Approximants*, Addison-Wesley Publication Company, New York.

Cekirge, H. M., and Varley, E., 1973, Large amplitude waves in bounded media I. Reflection and transmission of large amplitude shockless pulses at an interface. *Philos. Trans. R. Soc. London*, A273, pp. 261–313.

Cetinkaya, C., Vakakis, A. F., El-Raheb, M., 1995, Axisymmetric elastic wave propagation in weakly coupled layered media of infinite radial extent. *J. Sound Vib.*, 182 (2), pp. 283–302.

Delph, T. J., Herrmann, G., and Kaul, R. K., 1979, Harmonic wave propagation in a periodically layered infinite elastic body: plane strain, analytical results. *J. Appl. Mech.*, 46, pp. 113–19.

Delph, T. J., Herrmann, G., and Kaul, R. K., 1980, Harmonic wave propagation in a periodically layered infinite elastic body: plane strain, numerical results. *J. Appl. Mech.*, 47, pp. 531–37.

Eleonsky, V., 1991, Problems of existence of nontopological solitons (breathers) for nonlinear Klein–Gordon equations. *Proceedings of a NATO Advanced Research Workshop on Asymptotics Beyond All Orders*, La Jolla, California.

Engelbrecht, J., Peipman, T., and Valdek, U., 1990, Nonlinear effects in acoustics of solids. In *Frontiers of Nonlinear Acoustics: Proceedings of 12th ISNA*, eds. M. F. Hamilton and D. T. Blackstock. Elsevier Science Publishers, London.

Engels, R. C., and Meirovitch, L., 1978, Response of periodic structures by modal analysis. *J. Sound Vib.*, 56 (4), pp. 481–93.

Flytzanis, N., Pnevmatikos, S., and Remoissenet, M., 1985, Kink, breather, and asymmetric envelope or dark solitons in nonlinear chains. *J. Phys. C: Solid State Physics*, 18, pp. 4603–20.

Hirose, S., 1991, Boundary integral equation method for transient analysis of 3-D cavities and inclusions. *Engineering Analysis with Boundary Elements*, 8, pp. 146–54.

Jeffrey, A., and Engelbrecht, J., 1994, *Nonlinear Waves in Solids*, CISM Courses and Lectures No. 341, Springer-Verlag, New York and Berlin.

King, M. E., and Vakakis, A. F., 1994, A method for studying waves with spatially localized envelopes in a class of nonlinear partial differential equations. *Wave Motion*, 19, pp. 391–405.

Kissel, G. J., 1987, *Localization in Disordered Periodic Structures*, Ph.D. Thesis, Massachusetts Institute of Technology, Cambridge, Massachusetts.

Kleinman, R., Angell, T., Colton, D., Santosa, F., and Stakgold, I. (eds.), 1993, *Mathematical and Numerical Aspects of Wave Propagation*, SIAM Publication, Philadelphia.

Kosevich, A. M., and Kovalev, A. S., 1975, Self-localization of vibrations in a one-dimensional anharmonic chain. *Soviet Physics-JETP* 40(5), pp. 891–96.

Maslov, V. P., and Omelianov, G. A., 1981, Asymptotic soliton-like solutions of equations with small dispersion. *Advances in Mathematical Sciences–Usp. Mat. Nauk* 36(3), pp. 63–126.

Mead, D. J., 1975a, Wave propagation and natural modes in periodic systems: I. Mono-coupled systems. *J. Sound Vib.*, 40 (1), pp. 1–18.

Mead, D. J., 1975b, Wave propagation and natural modes in periodic systems: II. Multi-coupled systems with and without damping. *J. Sound Vib.*, 40 (1), pp. 19–39.

Mead, D. J., 1986, A new method of analyzing wave propagation in periodic structures; applications to periodic Timoshenko beams and stiffened plates. *J. Sound Vib.*, 104 (1), pp. 9–27.

Naik, R. A., and Crews, J. H., 1989, Stress analysis method for clearance-fit joints with bearing-bypass loads, *AIAA paper 89-1230-CP*.

Nayfeh, A. H., and Mook, D., 1984, *Nonlinear Oscillations*, J. Wiley & Sons, New York.

Nayfeh, A. H., Vakakis, A. F., and Nayfeh, T. A., 1993, A method for analyzing the interaction of nondispersive structural waves and nonlinear joints. *Journal of the Acoustical Society of America*, 93, pp. 849–56.

Peyrard, M, and Kruskal, M. D., 1984, Kink dynamics in the highly discrete Sine–Gordon system. *Physica D*, 14, pp. 88–102.

Pilipchuk, V. N., 1985, The calculation of strongly nonlinear systems close to vibration-impact systems. PMM, 49 (5), pp. 572–78.

Pilipchuk, V. N., 1988, A transformation for vibrating systems based on a non-smooth periodic pair of functions. *Doklady AN Ukr. SSR, Ser. A*, 4, pp. 37–40 (in Russian).

Pilipchuk, V. N., and Vakakis, A. F., 1997, Study of the oscillations of a nonlinearly supported string using nonsmooth transformations. *J. Vib. Acoust.* (in press).

Pilipchuk, V. N., Azeez, M. A. F., and Vakakis, A. F., 1996, Primary pulse transmission in a strongly nonlinear periodic systems. *Nonlinear Dynamics*, 11(1), pp. 61–81.

Pilipchuk, V. N., Vakakis, A. F., and Azeez, M. A. F., 1997, Study of a class of subharmonic motions using a nonsmooth temporal transformation (NSTT). *Physica D*, 100, pp. 145–64.

Richtmyer, R. D., 1985, *Principles of Advanced Mathematical Physics*, Vol. I., Springer-Verlag, New York.

Rosenau, P., 1987, Quasi-continuous spatial motion of a mass-spring chain, *Physica D*, 27, pp. 224–34.

Vakakis, A. F., 1994, Exponentially small splittings of manifolds in a rapidly forced Duffing system. *J. Sound Vib.*, 170 (1), 119–29.
Vakakis, A. F., and Cetinkaya, C., 1993, Mode localization in a class of multi-degree-of-freedom nonlinear systems with cyclic symmetry. *SIAM J. Appl. Math.*, 53, pp. 265–82.
Vakakis, A. F., and King, M. E., 1995, Nonlinear wave transmission in a monocoupled elastic periodic system. *J. Acoust. Soc. Am.*, 98(3), pp. 1534–46.
Vakakis, A. F., Nayfeh, T., and King, M., 1993, A multiple-scales analysis of nonlinear localized modes in a cyclic periodic system. *J. Appl. Mech.*, 60 (2), pp. 388–97.
Vakakis, A. F., King, M. E., and Pearlstein, A. J., 1994, Forced localization in a periodic chain of nonlinear oscillators. *Int. J. Non-linear Mech.*, 29 (3), pp. 429–47.
Vakakis, A. F., Manevitch, L. I., Mikhlin, Yu. I., Pilipchuk, V. N., and Zevin, A. A., 1996, *Normal Modes and Localization in Nonlinear Systems*, Wiley Interscience, New York.
Vedenova, E. G., and Manevich, L. I., 1981, Periodic and localized waves in vibro-impact systems of regular configuration, *Mashinovedenie*, 4, pp. 21–32.
Von Flotow, A., 1986, Disturbance propagation in structural networks. *Journal of Sound and Vibration*, 106, pp. 433–50.
Wegner, J. L., and Norwood, F. R. (eds.), 1993, Nonlinear waves in solids. *Applied Mechanics Reviews (special issue)*, 46, (12), Part 1.
Wei, S.-T., and Pierre, C., 1988, Localization phenomena in mistuned assemblies with cyclic symmetry: I: Free vibrations, II: Forced vibrations. *J. Vib. Acoust. Rel. Design*, 110, pp. 429–48.
Weinstein, A., 1985, Periodic nonlinear waves on a half-line. *Commun. Math. Phys.*, 99, pp. 385–88.

Nomenclature

$0 < \varepsilon \ll 1$	perturbation parameter
$\tau(z)$	nonsmooth variable
$0 < \mu$	perturbation parameter
$\Omega(s)$	attenuation exponent in the transformed domain
$B_0(y, t)$	leading order approximation of amplitude modulation
$e(z)$	nonsmooth variable
$f(u)$	nonlinear stiffness characteristic
$F_i(t)$	force acting on the left boundary of the ith layer
I_f	function of the nonsmooth variables
$K(z, \partial z/\partial t)$	nonlinear stiffness characteristic
$q(\frac{y}{\varepsilon}, y, t)$	external distributed excitation of the string
R_f	function of the nonsmooth variables
s	piecewise smooth time variable
$S = s - 1$	piecewise smooth time variable
$u = u(\frac{y}{\varepsilon}y, t)$	tranverse displacement of the string
$u_0(t)$	oscillation of the reference point
$u_i(t, x)$	longitudinal displacement of the ith layer
U	function of the nonsmooth variables
V	function of the nonsmooth variables
y_0	position of the reference point
z_{i-1}	relative displacement of the $(i-1)$-th coupling stiffness

Appendix : The Analytic Expression for $a_1^{(1)}(y)$ (King and Vakakis, 1994)

The coefficient $a_1^{(1)}(y)$ in expression (3.24) is computed by the following analytic expression:

$$a_1^{(1)}(y) = -\left[\int_0^x V(z)\psi^{(2)}(z)\,dz\right]\psi^{(1)}(x) + \left[\int_0^x V(z)\psi^{(1)}(z)\,dz\right]\psi^{(2)}(x), \quad \text{(A.1)}$$

where

$$x = (\beta - \kappa)^{1/2} y, \qquad \psi^{(1)} = \text{sech}^2(2^{-1/2}x),$$

$$\psi^{(2)}(x) = \text{sech}^2(2^{-1/2}x)\left[\frac{3x}{8} + \frac{2^{1/2}}{4}\sinh(2^{1/2}x) + \frac{2^{1/2}}{32}\sinh(2^{3/2}x)\right], \quad \text{(A.2)}$$

$$V(z) = Z_1 \frac{d^2 a_3^{(1)}((\beta-\kappa)^{-1/2}z)}{dz^2} + Z_2 a_3^{(1)}((\beta-\kappa)^{1/2}z)$$
$$+ Z_3 a_1^{(0)2}((\beta-\kappa)^{-1/2}z) a_3^{(1)}((\beta-\kappa)^{-1/2}z)$$
$$+ Z_4 a_1^{(0)}((\beta-\kappa)^{-1/2}z).$$

The coefficients Z_1, Z_2, Z_3, and Z_4 in the expression for $V(z)$ are defined as follows:

$$Z_1 = -\frac{3}{4}u_0^{*2}, \qquad Z_2 = \frac{3}{2\lambda}\left(\hat{K}_1 u_0^{*2} + \hat{K}_2 u_0^{*4}\right) + 3\beta u_0^{*2} - \frac{9}{4}\kappa u_0^{*2}, \qquad Z_3 = 3\beta u_0^{*2},$$

$$Z_4 = \lambda\,\text{sech}(2^{-1/2}x_0)\big[(2^{7/2}33\beta - 2^{9/2}21\beta^2 - 2^{5/2}81\kappa + 2^{9/2}51\beta\kappa - 2^{11/2}5\kappa^2)$$
$$\times u_0^{*2} x_0 \cosh(2^{-1/2}x_0) + 2^{9/2}(5\beta + 3\kappa)x_0\cosh(2^{-1/2}x_0)Z_2$$
$$+ (-720\beta + 544\beta^2 + 744\kappa - 1184\beta\kappa + 640\kappa^2)u_0^{*2}\sinh(2^{-1/2}x_0)$$
$$+ 32(\beta-\kappa)Z_2\sinh(2^{-1/2}x_0) + (54\beta + 36\beta^2 - 27\kappa - 156\beta\kappa + 120\kappa^2)$$
$$\times u_0^{*2}\sinh(2^{-1/2}3x_0) + (36\beta - 28\kappa)Z_2\sinh(2^{-1/2}3x_0)$$
$$+ (6\beta + 4\beta^2 - 3\kappa + 4\beta\kappa - 8\kappa^2)u_0^{*2}\sinh(2^{-1/2}5x_0)$$
$$+ 4(\beta+\kappa)Z_2\sinh(2^{-1/2}5x_0)\big]\left[96u_0^{*2}\left[\sinh(2^{3/2}x_0) - 2^{3/2}x_0\right]\right]^{-1}, \quad \text{(A.3)}$$

and the various variables in (A.3) are evaluated by

$$x_0 = (\beta-\kappa)^{1/2}y_0, \qquad \hat{K}_1 = \frac{\lambda\int_{-\infty}^{\infty} a_1^{(0)'^2}(s)\,ds}{\int_{-\infty}^{\infty} a_1^{(0)^2}(s)\,ds},$$

$$\hat{K}_2 = \frac{\int_{-\infty}^{\infty}\left[2a_1^{(0)}(s)a_3^{(1)}(s) + \frac{\alpha}{2}a_1^{(0)^4}(s)\right]ds}{\int_{-\infty}^{\infty} a_1^{(0)^2}(s)\,ds}.$$

The expression for Z_4 was derived using *Mathematica*. The integrals in (A.1) were evaluated also symbolically using *Mathematica* but their expressions are too lengthy to be reproduced here.

4

Dynamics and Control of Articulated Anisotropic Timoshenko Beams

A. V. BALAKRISHNAN

Abstract

This chapter illustrates the use of continuum models in control design for stabilizing flexible structures. A 6-degree-of-freedom anisotropic Timoshenko beam with discrete nodes where lumped masses or actuators are located provides a sufficiently rich model to be of interest for mathematical theory as well as practical application. We develop concepts and tools to help answer engineering questions without having to resort to ad hoc heuristic ("physical") arguments or faith. In this sense the paper is more mathematically oriented than engineering papers and vice versa at the same time. For instance we make precise time-domain solutions using the theory of semigroups of operators rather than formal "inverse Laplace transforms." We show that the modes arise as eigenvalues of the generator of the semigroup, which are then related to the eigenvalues of the stiffness operator. With the feedback control, the modes are no longer orthogonal and the question naturally arises as to whether there is still a modal expansion. Here we prove that the eigenfunctions yield a biorthogonal Riesz basis and indicate the corresponding expansion. We prove mathematically that the number of eigenvalues is nonfinite, based on the theory of zeros of entire functions. We make precise the notion of asymptotic modes and indicate how to calculate them. Although limited by space, we do consider the root locus problem and show for instance that the damping at first increases as the control gain increases but starts to decrease at a critical value and goes to zero as the gain increases without bound. The undamped oscillatory modes remain oscillatory and the rigid-body modes go over into deadbeat modes.

The Timoshenko model dynamics are translated into a canonical wave equation in a Hilbert space. The solution is shown to require the use of an "energy" norm, which is no more than the total energy: potential plus kinetic. We show that, under an appropriate extension of the notion of controllability, rate feedback with a collocated sensor can stabilize the structure in the sense that all modes are damped and the energy decays to zero. An example (nonnumeric) is worked out in some detail illustrating the concepts and theory developed.

4.1 Introduction

The purpose of this chapter is to illustrate the use of continuum models in control design for flexible structures: to provide the tools necessary to address relevant engineering issues. This is admittedly hazardous on two counts: On the one hand the complicated three-dimensional (3-D) geometry of realistic structures makes it almost impossible to use continuum models; while on the other hand the mathematics of continuum models is, for the most part, of mathematical interest only, and reduction to engineering practice is seldom undertaken. The alternative, universally the rule now, is to stay with the finite-dimensional "finite element" models. However, the latter has the drawback that any control design is limited to specific numerical values of system parameters, and the dimensions, can be prohibitively large.

The six-degree-of-freedom (6-DOF) anisotropic 1-D Timoshenko beam model is a convenient compromise from both sides. As shown in Noor and Anderson (1979), Noor and Russell (1986), Wang (1994), and Balakrishnan (1992) it is excellent for modeling lattice trusses. Moreover the mathematical theory strikes a good balance between the trivial and the nontractable.

The purpose of the control is to enhance the stability of the system, and the main interest in the theory centers on the modes and the damping attainable – the eigenvalue problem. A purely formal Laplace transform analysis can yield an entire function whose zeros in the complex plane are the eigenvalues. But the main difficulty is in determining the time-domain solution and the nature of the stability. Here is where it becomes necessary to use the theory of semigroups of operators and associated techniques from abstract (functional) analysis. In particular, the system with control is no longer self-adjoint – the mode shapes (eigenfunctions) are no longer orthogonal – and the problem of modal expansion of the solution must draw substantially on non-self-adjoint operator theory. A standard reference for the latter is Gohberg-Krein (1969).

Although some background in Hilbert space and linear operator theory is assumed (and almost all that is needed is covered in Achieser-Glassman (1966), Riesz-Nagy (1955), and Balakrishnan (1981)) every effort has been made to make the treatment self-contained.

We begin in Section 4.2 by translating the basic beam equations into a canonical abstract "vibration" or "wave" equation in a Hilbert space. The main feature of the choice of state is the inclusion of the displacements at the discrete nodes (the "boundary" points) in it – as pioneered by the author for the Bernoulli beam model of SCOLE in Balakrishnan (1991a). Not unlike the FEM version, the "boundary" conditions for the elastic equations are so chosen (in particular making the stiffness operator self-adjoint and nonnegative definite) as to yield the correct form for the potential energy.

Section 4.3 deals with spectral analysis. We show that the undamped structure modes are the zeros of an entire function and that the mode shapes yield an orthogonal basis for the space. We characterize the rigid-body modes showing that they span a six-dimensional space. We develop the Green's function for the eigenvalue problem. We show the relation of the square root of the stiffness operator to the potential energy of the beam.

Section 4.4 treats the time domain solutions of the structure dynamics, including "weak" and "strong" solutions. We introduce the notion of the "energy" inner product and show how the theory of semigroups of operators applies and the relation of the resolvent to the familiar Laplace transform.

In Section 4.5 we show that, under an appropriate generalization of the controllability condition to infinite dimensions, rate feedback using a collocated sensor can stabilize the structure in the sense that all modes are damped and the elastic energy decays to zero. We show that the damping coefficient goes to zero as the mode number increases without bound.

Section 6 is devoted to calculating the asymptotic modes of the structure with rate feedback. We construct the 6×6 matrix that defines the mode shapes, the modes being the zeros of the determinant. We show that the latter is an entire function of exponential type. We show that there are deadbeat modes, equal in number exactly to the (dimension) number of rigid-body modes. The oscillatory modes of the undamped structure remain oscillatory regardless of the control gain. Because of space limitation we do not go into detail on the root locus problem. Making precise the notion of "asymptotic modes," we show that the asymptotic modes are the "clamped" modes where all nodes are clamped – no displacement is allowed.

Section 7 deals with modal expansion. Since with feedback control the modes are no longer orthogonal we have to use a "biorthogonal" system. We show that the eigenfunctions do provide a Riesz basis and develop a "modal" expansion.

Finally, in Section 8, we include a nonnumerical example, albeit simple, to illustrate the concepts and theory developed in the paper.

4.2 The Anisotropic Timoshenko Beam: Dynamics and State Space Formulation

We begin by describing the dynamics of a 6-DOF 1-D Timoshenko beam articulated with lumped masses and/or control actuators at a finite number of "nodes" distributed along the beam, including the end points where the lumped masses may also be "offset." This is a natural extension of the familiar 1-DOF 1-D Timoshenko beam models found in standard texts (e.g., Timoshenko et al., 1974; Meirovich, 1967).

The model does not include the inherent damping since we are concerned only with the damping attainable with control – we refer to this as the "undamped" structure.

Next we choose an appropriate Hilbert space as the state space and define the mass-inertia operator, the stiffness operator, and the control operator (extending the notions familiar in finite dimensions). In particular, the definition of the stiffness operator is based on the (elastic) potential energy of the structure. With these definitions the partial differential equations translate into a "vibration" or "wave" equation in a Hilbert space which needs to be interpreted appropriately but provides the canonical model for the rest of the paper.

The anisotropic Timoshenko beam model adopted here appears to have been introduced by Noor and Anderson (1979) and refined in Noor and Russell (1986) to model the lattice truss structures used for deployment in space. Later Wang (1994) showed how such models could be derived starting from the general elastic solid equations using homogenization theory. Details of deriving the beam equations and the elastic and mass-inertia parameters therein illustrated by a specific example can be found in Balakrishnan (1992), which includes a comparison with FEM for calculating the step response. To minimize complexity we only consider the case of a single flexible beam. An example of a multibeam structure modeled as interconnected Timoshenko beams is given in Balakrishnan (1991b).

Let (x_1, x_2, x_3) denote the coordinates of a rectangular coordinate system, and let the beam axis be the x_1 axis. We shall use "s" to denote the position along the beam: $0 \leq s \leq L$, where L is the length of the beam. We consider the "uniform" case where the beam properties do not depend on s. Let s_i denote discrete points (referred to as "nodes") along the axis

$$s_i < s_{i+1}, \qquad i = 2, \ldots, m-1;$$
$$s_1 = 0; \qquad s_m = L.$$

Between nodes, that is to say for $s_i < s < s_{i+1}$, we have the basic anisotropic Timoshenko beam equations governing the beam displacements, $u(\cdot)$, $v(\cdot)$, $w(\cdot)$:

$u(\cdot)$ axial displacement (or x_1 component),

$v(\cdot)$ bending displacement in the x_1–x_2 plane (or x_2 component),

$w(\cdot)$ bending displacement in the x_1–x_3 plane (or x_3 component), and the torsion angles $\phi_1(\cdot), \phi_2(\cdot), \phi_3(\cdot)$:

$\phi_1(\cdot)$ rotation angle about the x_1 axis,
$\phi_2(\cdot)$ rotation angle about the x_2 axis,
$\phi_3(\cdot)$ rotation angle about the x_3 axis

are given by

$$
\begin{aligned}
m_{11}\ddot{u} - c_{11}u'' - c_{14}v'' - c_{15}w'' - c_{15}\phi_2' + c_{14}\phi_3' = 0,\\
m_{22}\ddot{v} - c_{44}v'' - c_{14}u'' - c_{45}\phi_2' - c_{45}w'' = 0,\\
m_{33}\ddot{w} - c_{55}w'' - c_{15}u'' - c_{45}\phi_2' - c_{45}v'' = 0,\\
m_{44}\ddot{\phi}_1 - c_{66}\phi_1'' - c_{36}\phi_2'' - c_{26}\phi_3'' = 0,\\
m_{55}\ddot{\phi}_2 + m_{56}\ddot{\phi}_3 + c_{15}u' + c_{55}w' - c_{36}\phi_1'' + c_{55}\phi_2 - c_{33}\phi_2'' - c_{23}\phi_3'' - c_{45}\phi_3 = 0,\\
m_{66}\ddot{\phi}_3 + m_{56}\ddot{\phi}_2 - c_{14}u' + c_{44}v' - c_{26}\phi_1'' - c_{23}\phi_2'' + c_{44}\phi_3 - c_{22}\phi_3'' - c_{45}\phi_2 = 0,
\end{aligned} \tag{4.2.1}
$$

where the superdots denote time derivatives and the primes the space derivatives (with respect to s). The matrices

$$
C_1 = \begin{vmatrix} c_{11} & c_{14} & c_{15} \\ c_{14} & c_{44} & c_{45} \\ c_{15} & c_{45} & c_{55} \end{vmatrix}, \qquad C_3 = \begin{vmatrix} c_{66} & c_{36} & c_{26} \\ c_{36} & c_{33} & c_{23} \\ c_{26} & c_{23} & c_{22} \end{vmatrix}
$$

are both strictly positive definite. We shall also use the notation

$$
C_2 = \begin{vmatrix} 0 & -c_{15} & c_{14} \\ 0 & -c_{45} & c_{44} \\ 0 & -c_{55} & c_{45} \end{vmatrix}, \qquad C_4 = \begin{vmatrix} 0 & 0 & 0 \\ 0 & c_{55} & -c_{45} \\ 0 & -c_{45} & c_{44} \end{vmatrix}.
$$

With f denoting the 6×1 (column) vector

$$
f = \begin{vmatrix} u \\ v \\ w \\ \phi_1 \\ \phi_2 \\ \phi_3 \end{vmatrix}, \tag{4.2.2}
$$

these equations can be conveniently rewritten in the vector form

$$
M_0\ddot{f} - A_2 f'' + A_1 f' + A_0 f = 0, \quad s_i < s < s_{i+1}, \tag{4.2.3}
$$

where M_0 is the mass/inertia matrix

$$M_0 = \begin{vmatrix} m_{11} & 0 & 0 & 0 & 0 & 0 \\ 0 & m_{22} & 0 & 0 & 0 & 0 \\ 0 & 0 & m_{33} & 0 & 0 & 0 \\ 0 & 0 & 0 & m_{44} & 0 & 0 \\ 0 & 0 & 0 & 0 & m_{55} & m_{56} \\ 0 & 0 & 0 & 0 & m_{56} & m_{66} \end{vmatrix},$$

where M is also required to be strictly positive definite;

$$A_2 = \begin{vmatrix} C_1 & 0 \\ 0 & C_3 \end{vmatrix},$$

$$A_1 = \begin{vmatrix} 0 & C_2 \\ -C_2* & 0 \end{vmatrix},$$

$$A_0 = \begin{vmatrix} 0 & 0 \\ 0 & C_4 \end{vmatrix};$$

and where f stands for

$$f(t, s), \qquad 0 < t, \qquad 0 < s < L.$$

Here the m_{ij} are the "mass-inertia" coefficients and $\{c_{ij}\}$ the "flexibility" coefficients. They are constrained by the positive definite requirements of C_1 and C_3; in particular, $c_{ii} > 0$, $m_{ii} > 0$ for all i.

The nodes $\{s_i\}$ are points where the slope – spatial derivative $f'(\cdot)$ – is discontinuous because of controllers or lumped masses located there. The displacements at the nodes thus have to be included as part of the definition of the state. Let $L_2(0, L)^6$ denote the L_2-space of 6×1 vector functions $f(\cdot)$ and let $\mathcal{H}$ denote the Hilbert Space

$$\mathcal{H} = L_2(0, L)^6 \times (R^6)^m,$$

where we use the notation $(R^6)^m$ rather than R^{6m} to indicate 6×1 vectors replicated m times. For elements x in $\mathcal{H}$ we shall use the notation

$$x = \begin{vmatrix} f \\ b \end{vmatrix},$$

where $f \in L_2(0, L)^6$ and $b \in (R^6)^m$. To avoid possible confusion let us note explicitly that the inner product in $\mathcal{H}$ is given by

$$[x, y] = [f, g]_{L_2(0,L)^6} + \sum_1^m [b_i, c_i]_{R^6}, \tag{4.2.4}$$

where

$$x = \begin{vmatrix} f \\ b \end{vmatrix}, \qquad b = \begin{vmatrix} b_1 \\ b_2 \\ \vdots \\ b_m \end{vmatrix},$$

$$y = \begin{vmatrix} g \\ c \end{vmatrix}, \qquad c = \begin{vmatrix} c_1 \\ c_2 \\ \vdots \\ c_m \end{vmatrix}.$$

(To avoid excessive notation we may often delete the signature under the inner products if they are clear from the context.) The elements of $\mathcal{H}$ will be our "states." We begin with the stiffness operator since the mass operator will depend on the control masses. The stiffness operator is defined as the "differential" operator A with domains, denoted $\mathcal{D}(A)$, in $\mathcal{H}$ given by

$$\mathcal{D}(A) = \left[x = \begin{vmatrix} f \\ b \end{vmatrix} \quad \text{where } f_i(s) = f(s), \quad s_i < s < s_{i+1}, \quad i = 1, \ldots, m-1, \right.$$

$$f_i(\cdot), \quad f_i'(\cdot), \quad \text{and } f_i''(\cdot) \in L_2(s_i, s_{i+1})^6,$$

$$\left. \lim_{s \uparrow s_{i+1}} f_i(s) = \lim_{s \downarrow s_{i+1}} f_{i+1}(s), \quad i = 1, \ldots, m-1 \right],$$

$$b = \begin{vmatrix} f(s_1) \\ \vdots \\ f(s_m) \end{vmatrix}.$$

In other words the functions $f(\cdot)$ in $\mathcal{D}(A)$ are "piecewise smooth"; they are continuous in $0 \le s \le L$, but the derivative can have a jump discontinuity at each node s_i. We refer to b as the "boundary value." The operator A is defined by

$$Ax = y, \qquad x = \begin{vmatrix} f \\ b \end{vmatrix}, \qquad y = \begin{vmatrix} g \\ c \end{vmatrix}, \tag{4.2.5}$$

where

$$g(s) = -A_2 f''(s) + A_1 f'(s) + A_0 f(s), \quad s_i < s < s_{i+1}, \quad i = 1, \ldots, m-1, \tag{4.2.6}$$

and c is defined by

$$c = \begin{vmatrix} -L_1 f(0+) - A_2 f'(0+) \\ -A_2(f'(s_2+) - f'(s_2-)) \\ \vdots \\ -A_2(f'(s_{m-1}+) - f'(s_{m-1}-)) \\ L_1 f(L-) + A_2 f'(L-) \end{vmatrix}, \tag{4.2.7}$$

where

$$L_1 = \begin{vmatrix} 0 & -C_2 \\ 0 & 0 \end{vmatrix}.$$

We shall find it convenient to use the notation

$$c = A_b f.$$

This definition is made so that we get the right expression for the potential energy:

$$\frac{[Ax, x]}{2} = \text{Potential Energy}, \tag{4.2.8}$$

where the potential energy of the structure is defined by

$$\int_0^L \left[C_1 \begin{vmatrix} u' \\ v' - \phi_3 \\ w' + \phi_2 \end{vmatrix}, \begin{vmatrix} u' \\ v' - \phi_3 \\ w' + \phi_2 \end{vmatrix} \right] ds + \int_0^L \left[C_3 \begin{vmatrix} \phi_1' \\ \phi_2' \\ \phi_3' \end{vmatrix}, \begin{vmatrix} \phi_1' \\ \phi_2' \\ \phi_3' \end{vmatrix} \right] ds. \tag{4.2.9}$$

Let us verify (4.2.8). Integration by parts yields

$$\int_{s_i}^{s_i+1} [A_2 f''(s), f(s)]\, ds = [A_2 f'(s_{i+1}-), f(s_{i+1}-)] - [A_2 f'(s_i+), f(s_i+)]$$
$$- \int_{s_i}^{s_i+1} [A_2 f'(s), f'(s)]\, ds.$$

Hence

$$-\int_0^L [A_2 f''(s), f(s)]\, ds = -\sum_{i=1}^{m-1} [A_2 f'(s_{i+1}-), f(s_{i+1}-)]$$
$$+ \sum_{i=1}^{m-1} [A_2 f'(s_i+), f(s_i+)]$$
$$+ \int_0^L [A_2 f'(s), f(s)]\, ds.$$

But

$$-\sum_{i=1}^{m-1}[A_2 f'(s_{i+1}-), f(s_{i+1}-)] = \sum_{i=2}^{m}[A_2 f'(s_i-), f(s_i-)].$$

And using

$$f(s_i-) = f(s_i+) = f(s_i)$$

(x being in $\mathcal{D}(A)$), we have that

$$\begin{aligned}
&-\sum_{i=0}^{m-1}[A_2 f'(s_{i+1}-), f(s_{i+1}-)] + \sum_{i=1}^{m-1}[A_2 f'(s_i+), f(s_i+)] \\
&\quad = [A_2 f'(L-), f(L-)] \\
&\qquad + \sum_{2}^{m-1}[A_2(f'(s_i+) - f'(s_i-)), f(s_i)][A_2 f'(0+), f(0+)].
\end{aligned}$$

Also, noticing that

$$A_1 = L_1^* - L_1,$$

we can calculate that

$$\begin{aligned}
\int_0^L [A_1 f'(s), f(s)]\, ds &= -[L_1 f(L-), f(L-)] + [L_1 f(0+), f(0+)] \\
&\quad + \int_0^L [L_1 f(s), f'(s)]\, ds + \int_0^L \left[L_1^* f'(s), f(s)\right] ds.
\end{aligned}$$

But

$$\begin{aligned}
[A_b f, b] &= [(L_1 f(L-) + A_2 f'(L-)), f(L-)] \\
&\quad + \sum_{2}^{m-1}[A_2(f'(s_i-) - f'(s_i+)), f(s_i)] \\
&\quad - [L_1 f(0+) + A_2 f'(0+), f(0+)].
\end{aligned}$$

Hence

$$\begin{aligned}
[Ax, x] &= [A_b f, b] + \int_0^L [(A_0 f + A_1 f' - A_2 f''), f(s)]\, ds \\
&= \int_0^L [A_2 f'(s), f'(s)]\, ds + \int_0^L [L_1 f(s), f'(s)]\, ds \\
&\quad + \int_0^L \left[L_1^* f'(s), f(s)\right] ds + \int_0^L [A_0 f(s), f(s)]\, ds.
\end{aligned}$$

Hence

$$[Ax, x] = \int_0^L \left[H \begin{vmatrix} f' \\ f \end{vmatrix}, \begin{vmatrix} f' \\ f \end{vmatrix}\right] ds, \tag{4.2.10}$$

where

$$H = \begin{vmatrix} C_1 & 0 & 0 & -C_2 \\ 0 & C_3 & 0 & 0 \\ 0 & 0 & & \\ -C_{2*} & 0 & & A_0 \end{vmatrix},$$

from which (4.2.9) readily follows, noting that

$$C_2 = -C_1 D_3,$$

$$D_3 = \begin{vmatrix} 0 & 0 & 0 \\ 0 & 0 & -1 \\ 0 & 1 & 0 \end{vmatrix},$$

$$C_4 = D_3^* C_1 D_3,$$

or, equivalently,

$$A_0 = L_1^* A_2^{-1} L_1.$$

This technique of defining the "boundary conditions" on the "differential" operator appropriately to make it self-adjoint and nonnegative definite should be compared with the FEM method where (in the Hamiltonian) the potential energy is specified and the stiffness matrix derived therefrom.

4.2.1 Remark

By our definition, a node is a point where there is a controller or lumped mass. There may be neither at the end points $s = 0$ and $s = L$, and if so, they are not included in the nodes, and additional conditions need to be imposed. Typically, either

$$f(0) = 0 \quad \text{(clamped at zero)} \tag{4.2.11}$$

or

$$L_1 f(0) + A_2 f'(0) = 0 \quad \text{(free at zero)} \tag{4.2.12}$$

and similarly at $s = L$.

4.2.2 Mass/Inertia Operator

To obtain the mass/inertia operator we have to specify the control mass/inertia and end masses, which are possibly offset (modeling antennas). Let us begin with the interior controllers, force (reaction jets, proof-mass actuators), and moment (cmgs). Then we have for $i \neq 1$ or m

$$M_{b,i}\ddot{f}(t, s_i) + A_2(f'(t, s_i-) - f'(t, s_i+)) + U_i(t) = 0, \qquad (4.2.13)$$

where

$$M_{b,i} = \begin{vmatrix} m_i & 0 & 0 & 0 & 0 & 0 \\ 0 & m_i & 0 & 0 & 0 & 0 \\ 0 & 0 & m_i & 0 & 0 & 0 \\ 0 & 0 & 0 & & & \\ 0 & 0 & 0 & & I_i & \\ 0 & 0 & 0 & & & \end{vmatrix}$$

and U_i is the 6×1 vector of force and moment controls at $s = s_i$; m_i is the control mass and I_i the corresponding moment of inertia (matrix). For the ends, allowing for offset end masses m_0 at $s = 0$ and m_L at $s = s_L$ and I_0 and I_c the moments of inertia of antennas and controller, each about the center of gravity, and r_0 and r_L the 3×1 position vectors of the centers of gravity respectively of the end masses we have at $s = 0$:

$$M_{b,0}\ddot{f}(t, 0+) - L_1 f(t, 0+) - A_2 f'(t, 0+) + U_0(t) = 0, \qquad (4.2.14)$$

where

$$r_0 = \begin{vmatrix} 0 \\ r_{0,2} \\ r_{0,3} \end{vmatrix}$$

and

$$M_{b,0} = \left|\begin{array}{ccc|ccc} m_0 + m_1 & 0 & 0 & 0 & 0 & 0 \\ 0 & m_0 + m_1 & 0 & m_0 r_{0,2} & 0 & 0 \\ 0 & 0 & m_0 + m_1 & m_0 r_{0,3} & 0 & 0 \\ \hline 0 & m_0 r_{0,2} & m_0 r_{0,3} & & & \\ 0 & 0 & 0 & & \hat{I}_0 & \\ 0 & 0 & 0 & & & \end{array}\right|,$$

$$\hat{I}_0 = I_0 + I_c + (m_0 + m_1) \begin{vmatrix} r_{0,2}^2 + r_{0,2}^2 & 0 & 0 \\ 0 & r_{0,2}^2 & -r_{0,2} r_{0,3} \\ 0 & -r_{0,2} r_{0,3} & r_{0,3}^2 \end{vmatrix},$$

and where m_1 is the controller mass; and finally U_0 is the control vector (forces and moments).

Similarly for the end $s = L$, we have

$$M_{b,L}\ddot{f}(t, L-) + L_1 f(t, L-) + A_2 f'(t, L-) + U_L(t) = 0, \qquad (4.2.15)$$

where

$$r_L = \begin{vmatrix} L \\ r_{L,2} \\ r_{L,3} \end{vmatrix},$$

$$M_{b,L} = \left|\begin{array}{ccc|ccc} m_L + m_m & 0 & 0 & 0 & 0 & 0 \\ 0 & m_L + m_m & 0 & m_L r_{L,2} & 0 & 0 \\ 0 & 0 & m_L + m_m & m_L r_{L,3} & 0 & 0 \\ \hline 0 & m_L r_{L,2} & m_L r_{L,3} & & & \\ 0 & 0 & 0 & & \hat{I}_L & \\ 0 & 0 & 0 & & & \end{array}\right|,$$

$$\hat{I}_L = I_L + I_c + (m_L + m_m) \begin{vmatrix} r_{L,2}^2 + r_{L,2}^2 & 0 & 0 \\ 0 & r_{L,2}^2 & -r_{L,2} r_{L,3} \\ 0 & -r_{L,2} r_{L,3} & r_{L,3}^2 \end{vmatrix},$$

and where m_m is the controller mass and U_L is the control vector of forces and moments. Again, $(m_L + m_m)$ is the total moving mass.

Let M_b denote the composite matrix of all the control and end masses/inertia:

$$M_b = \begin{vmatrix} M_{b,0} & & & & \\ & M_{b,2} & & & \\ & & \ddots & & \\ & & & M_{b,m-1} & \\ & & & & M_{b,L} \end{vmatrix}.$$

The mass-inertia operator, denoted M, is then defined by

$$Mx = y, \qquad x \begin{vmatrix} f \\ b \end{vmatrix}, \qquad y = \begin{vmatrix} M_0 f \\ M_b b \end{vmatrix} \qquad (4.2.16)$$

and is a linear bounded self-adjoint nonnegative definite opeator $\mathcal{H}$ onto $\mathcal{H}$ with a linear bound inverse

$$M^{-1}x = \begin{vmatrix} M_0^{-1} b \\ M_b^{-1} b \end{vmatrix}.$$

We note that

$$M_b^{-1}b = \begin{vmatrix} M_{b,0}^{-1}b_1 \\ \vdots \\ M_{b,i}^{-1}b_i \\ \vdots \\ M_{b,L}^{-1}b_m \end{vmatrix}.$$

Also,

$$\sqrt{M}x = \begin{vmatrix} \sqrt{M_0}\ f \\ \sqrt{M_b}\ b \end{vmatrix},$$

$$\sqrt{M_b}b = \begin{vmatrix} \sqrt{M_{b,0}}\ b \\ \vdots \\ \sqrt{M_{b,i}}\ b_i \\ \vdots \\ \sqrt{M_{b,L}}\ b_m \end{vmatrix}.$$

4.2.3 Control Operator

The control operator maps the control inputs into $\mathcal{H}$. We denote it by B. We note that it is possible that not all nodes may have controllers. Let the number of controllers – force, moment or proof-mass – be denoted m_c,

$$m_c < 6^m.$$

Then we can regard any control u as an $m_c \times 1$ column vector. We define

$$B_u u = b,$$

where

$$b = \begin{vmatrix} B_1 u \\ \vdots \\ B_m u \end{vmatrix},$$

where each B_i is a $6 \times m_c$ matrix such that $B_i^* B_i$ is nonsingular and further

$$\left[B_i^* b,\ B_j^* b\right] = 0, \qquad i \neq j$$

or

$$B_i B_j^* = 0, \qquad i \neq j.$$

In particular it follows that

$$B_u B_u^* b = \begin{vmatrix} B_1 B_1^* b_1 \\ \vdots \\ B_m B_m^* b_m \end{vmatrix}.$$

We shall use the notation

$$B_i B_i^* = D_i,$$

where the D_i, which are 6×6 each, may be taken to be diagonal with entries 1 or 0. Finally,

$$Bu = \begin{vmatrix} 0 \\ B_u u \end{vmatrix},$$

$$BB^* \begin{vmatrix} f \\ b \end{vmatrix} = \begin{vmatrix} 0 \\ B_u B_u^* b \end{vmatrix}.$$

Note that if

$$B^* x = 0, \qquad x = \begin{vmatrix} f \\ b \end{vmatrix}$$

we must have

$$B_u^* b = 0$$

or b must be such that

$$B_i^* b_i = 0, \qquad i = 1, \ldots, m,$$

or

$$D_i b_i = 0, \qquad i = 1, \ldots, m.$$

Another equally useful representation for BB^* can be obtained in the following way. Let $\{e_i\}$, $i = 1, \ldots, m_c$, denote the unit coordinate vectors in $\mathcal{R}^{m_c}$. Then let

$$B_{(i)} = Be_i.$$

Then

$$\left[B_{(i)}, B_{(j)}\right] = 0, \qquad i \neq j$$

and

$$BB^* = \sum_1^{m_c} B_{(i)} B_{(i)}^*,$$

where the $B_{(i)}$ are orthogonal, and each $B_{(i)}$ is nonzero.

With these definitions we can assemble the canonical "state space" version of the dynamic equations (4.2.1), (4.2.13), (4.2.14), and (4.2.15):

$$M\ddot{x}(t) + Ax(t) + Bu(t) = 0, \quad t > 0, \tag{4.2.17}$$

where, formally, we have taken

$$\ddot{x}(t) = \begin{vmatrix} \ddot{f}(t,\cdot) \\ \ddot{b}(t,\cdot) \end{vmatrix}.$$

We recognize (4.2.17) as "an abstract wave equation in a Hilbert space"; the precise relationship between the original space–time dynamic equations and this abstract version will be clarified later, in stages.

4.3 Spectral Analysis: Modes/Modal Expansion

Before we proceed to "solve" (4.2.17) we need to examine the spectrum of the stiffness operator A. The eigenvalues of A (with respect to the mass operator M) are the modes of the undamped structure. A crucial result is that the modes are the zeros of an entire function – the determinant of the $m \times m$ "condensed dynamic stiffness matrix." The corresponding "mode shapes," the eigenfunctions, yield an M-orthogonal basis for the Hilbert space. We also characterize the "rigid-body" modes, the eigenfunctions corresponding to the zero eigenvalue. We also obtain the "Green's function" for the eigenvalue problem. Finally, we define the square root of the stiffness operator and indicate its relation to the potential energy.

4.3.1 Elementary Properties of the Stiffness Operator

Let us begin cataloging the elementary yet crucial properties of the stiffness operator A. We have already seen that it is self-adjoint,

$$[Ax, y] = [x, Ay] \quad \text{for } x, y \text{ in } \mathcal{D}(A),$$

and nonnegative definite,

$$[Ax, x] \geq 0 \quad \text{for } x \in \mathcal{D}(A).$$

The next property is that the domain of A is dense in $\mathcal{H}$. In other words, while not every element x in $\mathcal{H}$ is in the domain of A as the definition of A clearly shows, we can find elements in the domain of A that approximate x as closely as needed. More precisely, we can find a sequence $\{x_n\}$ in $\mathcal{D}(A)$ such that

$$\|x - x_n\| \to 0.$$

This is a feature of "differential" operators; but since in our case we also have "boundary values" to contend with, we shall present a formal argument. Thus let

$$x = \begin{vmatrix} f \\ b \end{vmatrix}.$$

Let

$$f_i(s) = f(s), \qquad s_i \le s \le s_{i+1}.$$

It should be noted that b is arbitrary and is not necessarily the vector of "boundary" values of $f(\cdot)$ at $s = s_i$; the latter need not of course be even defined. But for each i, we can find a sequence of functions $\{f_{i,n}(\cdot)\}$ such that $f_{i,n}(s)$ is "infinitely" differentiable in $s_i < s < s_{i+1}$, and

$$\int_{s_i}^{s_{i+1}} \| f_i(s) - f_{i,n}(s)\|^2 \, ds \to 0 \quad \text{as } n \to \infty$$

and

$$f_{i,n}(s_i) = b_i, \qquad f_{i,n}(s_{i+1}) = b_{i+1}.$$

Defining

$$f_n(s) = f_{i,n}(s), \qquad s_i < s < s_{i+1}, \qquad f_n(s_i) = f_{i,n}(s),$$

we see that $f_n(\cdot)$ is continuous and

$$x_n = \begin{vmatrix} f_n \\ b \end{vmatrix}$$

belongs to the domain of A, and thus

$$\|x - x_n\|^2 \to 0 \quad \text{as } n \to \infty$$

as required.

4.3.2 Eigenvalues and Eigenfunctions

If we proceed formally and take the Laplace transform of (4.2.17), setting

$$\phi = \int_0^{\infty} e^{-\lambda t} x(t)\, dt, \qquad \operatorname{Re} \lambda > 0,$$

we obtain for fixed λ:

$$\lambda^2 M\phi + A\phi - \psi = 0, \tag{4.3.1}$$

where ψ is an element of $\mathcal{H}$. The solution of this equation plays an important role in the theory. We now formulate this more precisely as: Find ϕ in $\mathcal{D}(A)$ such that

$$\lambda^2 M\phi + A\phi = \psi, \qquad \psi \in \mathcal{H} \tag{4.3.2}$$

for given λ. We see that for (4.3.2) to have a unique solution the "homogeneous" equation

$$\lambda^2 M\phi + A\phi = 0 \tag{4.3.3}$$

can only have the zero solution. Rewriting this equation as

$$A\phi = -\lambda^2 M\phi$$

we may consider more generally

$$A\phi = \gamma M\phi, \qquad \phi \neq 0. \tag{4.3.4}$$

Here γ is called an eigenvalue of A with respect to the mass matrix M, and ϕ is called a corresponding eigenfunction. From

$$0 \leq [A\phi, \phi] = \gamma[M\phi, \phi]$$

it follows that

$$\gamma \geq 0.$$

If γ_1, γ_2 are two distinct eigenvalues with corresponding eigenvectors ϕ_1, ϕ_2 respectively, we have

$$[A\phi_1, \phi_2] = \gamma_1[M\phi_1, \phi_2] = [\phi_1, A\phi_2] = \gamma_2[M\phi_1, \phi_2]$$

and hence we must have

$$[M\phi_1, \phi_2] = 0;$$

in other words, eigenfunctions corresponding to distinct eigenvalues must be M-orthogonal.

We shall show now that the set of eigenvalues is nonfinite, is countable, and can be taken as

$$\{\omega_k^2\}, \qquad \omega_k \leq \omega_{k+1}, \qquad k = 0, 1, 2, \ldots,$$

where

$$\omega_k \to \infty \quad \text{as } k \to \infty$$

and

$$A\phi_k = \omega_k^2 M\phi_k.$$

First let us consider the eigenvalue zero.

4.3.3 Rigid-Body Modes

Because we are considering the case where both $s = 0$ and $s = L$ are nodes, we shall show that zero is an eigenvalue. An eigenvector corresponding to the zero eigenvalue is called a rigid-body mode. This is because if

$$A\phi = 0$$

we have

$$[A\phi, \phi] = 0$$

or, from (4.2.9), the associated potential energy is zero. Let

$$\phi = \begin{vmatrix} f \\ b \end{vmatrix},$$

where b is the boundary vector, ϕ being in $\mathcal{D}(A)$; if the potential energy is zero, we have from (4.2.9) that $f(\cdot)$ must be of the form

$$f(s) = \begin{vmatrix} u(s) = u(0) \\ v(s) = v(0) + s\phi_3(0) \\ w(s) = w(0) - s\phi_2(0) \\ \phi_1(s) = \phi_1(0) \\ \phi_2(s) = \phi_2(0) \\ \phi_3(s) = \phi_3(0) \end{vmatrix}. \tag{4.3.5}$$

The derivative $f'(s)$ is continuous in $0 \leq s \leq L$. Note that the "free–free" boundary conditions (4.2.12) are satisfied at both ends, and $f(\cdot)$ is the same regardless of the number and location of the interior nodes. Also, the dimension of the eigenfunction space is six. In our notation we shall set

$$\omega_0 = 0.$$

4.3.4 Nonzero Eigenvalues

Let us now consider the general case in the form

$$\lambda^2 M\phi + A\phi = 0.$$

Let

$$\phi = \begin{vmatrix} f \\ b \end{vmatrix}.$$

Then

$$\lambda^2 M_0 f + g = 0, \tag{4.3.6}$$

where $g(\cdot)$ is defined by (4.2.6), and

$$\lambda^2 M_b b + A_b f = 0. \tag{4.3.7}$$

To solve (4.3.3), let

$$\mathcal{A}(\lambda) = \begin{vmatrix} 0 & I_6 \\ A_2^{-1}(A_0 + \lambda^2 M_0) & A_2^{-1}A_1 \end{vmatrix}. \tag{4.3.8}$$

Then

$$f(s) = e^{\mathcal{A}(\lambda)(s-s_i)} \begin{vmatrix} f(s_i) \\ f'(s_i+) \end{vmatrix}, \qquad s_i \leq s \leq s_{i+1}, \qquad i = 1, 2, \ldots, m-1. \tag{4.3.9}$$

Let

$$\mathfrak{a} = \begin{vmatrix} f(s_1) \\ \delta f'(s_1) \\ \delta f'(s_2) \\ \vdots \\ \delta f'(s_{m-1}) \end{vmatrix},$$

where

$$\delta f'(s_i) = f'(s_i+) - f'(s_i-),$$

and set

$$f'(0-) = 0.$$

Then

$$\begin{vmatrix} f(s) \\ f'(s) \end{vmatrix} = e^{\mathcal{A}(\lambda)s} \begin{vmatrix} f(0) \\ 0 \end{vmatrix} + \sum_{j=1}^{i} e^{\mathcal{A}(\lambda)(s-s_j)} \begin{vmatrix} 0 \\ \delta f'(s_j) \end{vmatrix},$$

$$s_i < s < s_{i+1}, \qquad i = 1, \ldots, m-1. \tag{4.3.10}$$

Hence we can write

$$b = L(\lambda)\mathfrak{a},$$

where $L(\lambda)$ is a $6m \times 6m$ block lower-triangular matrix:

$$L(\lambda) = \{\ell_{i,j}(\lambda)\},$$

$$\ell_{i,j}(\lambda) = 0, \qquad j > i,$$

$$\ell_{11}(\lambda) = I_6.$$

For $i > 1$

$$\begin{aligned} \ell_{i,1}(\lambda) &= P_{11}(\lambda, s_i), \\ \ell_{i,2}(\lambda) &= P_{12}(\lambda, s_i), \\ \ell_{i,3}(\lambda) &= P_{12}(\lambda, s_i - s_2), \\ &\vdots \\ \ell_{i,i-1}(\lambda) &= P_{12}(\lambda, s_i - s_{i-2}), \\ \ell_{i,i}(\lambda) &= P_{12}(\lambda, s_i - s_{i-1}), \end{aligned}$$

where

$$e^{\mathcal{A}(\lambda)s} = \begin{vmatrix} P_{11}(\lambda, s) & P_{12}(\lambda, s) \\ P_{21}(\lambda, s) & P_{22}(\lambda, s) \end{vmatrix},$$

$$P_{21}(\lambda, s) = \frac{\partial}{\partial s} P_{11}(\lambda, s), \qquad P_{22}(\lambda, s) = \frac{\partial}{\partial s} P_{12}(\lambda, s).$$

Note that

$$\det L(\lambda) = \det P_{12}(\lambda, s_2) \cdots \det P_{12}(\lambda, s_i - s_{i-1}) \cdots \det P_{12}(\lambda, L - s_{m-1}). \tag{4.3.11}$$

Also, we can express $A_b f$ in terms of $\mathfrak{a}$, using (4.3.10). Thus we can write

$$A_b f = K(\lambda)\mathfrak{a},$$

where $K(\lambda)$ is also a $6m \times 6m$ (block lower-triangular) matrix, and hence (4.3.3) yields

$$(\lambda^2 M_b L(\lambda) + K(\lambda))\mathfrak{a} = 0. \tag{4.3.12}$$

Hence the eigenvalues are the roots of

$$\det(\lambda^2 M_b L(\lambda) + K(\lambda)) = 0. \tag{4.3.13}$$

The left side defines an entire function of the complex number λ and can have at most a countable number of zeros, which must grow without bound in magnitude, if nonfinite in number. The corresponding eigenfunction is determined by (4.3.10). Also, (4.3.12) shows that the dimension of each eigenfunction space is finite, not more than $6m$.

We can now deduce that the set of eigenvalues that is countable is not finite. For if it were finite, the number of eigenfunctions would be finite. But the eigenfunctions must be complete in the M-inner product. If not, there must be a function ψ that is M-orthogonal to all the eigenfunctions. The class of such functions form a Hilbert space and A must map a dense domain of this space into itself. However, since A is self-adjoint, we can now recall the fundamental result (see, for example Riesz-Nagy, 1955) that it must have at least one eigenvalue, which leads to a contradiction. But since the dimension of $\mathcal{H}_2$ is nonfinite, we see that the set of eigenvalues cannot be finite. In particular, we see that the eigenfunctions are complete and M-orthogonal. We can therefore make them an M-orthonormal basis for $\mathcal{H}$. For each eigenvalue ω_k^2, let P_k denote the projection (operator) corresponding to the eigenfunction space:

$$A(P_k\phi) = \omega_k^2(P_k\phi).$$

Since the dimension of each eigenfunction space is finite it is convenient to continue to use $\{\phi_k\}$ to denote the M-orthonormalized basis in each eigenfunction space, "counting each eigenfunction as many times as the dimension," as is customary. Then we can write

$$x = \sum_0^\infty \alpha_k \phi_k, \qquad (4.3.14)$$

where

$$\alpha_k = [x, M\phi_k]$$

and

$$\sum_0^\infty |\alpha_k|^2 = [Mx, x].$$

Hence

$$M^{-1}x = \sum_0^\infty [x, \phi_k]\, \phi_k$$

or

$$x = \sum_0^\infty [x, \phi_k]\, M\phi_k. \qquad (4.3.14a)$$

Using (4.3.14) and (4.3.14a) we have

$$[x, x] = \sum_0^\infty [x, \phi_k][M\phi_k, x];$$

conversely

$$x = \sum_1^\infty \alpha_k \phi_k \in \mathcal{H}$$

if and only if

$$\sum_1^\infty |\alpha_k|^2 < \infty$$

and

$$[x, Mx] = \sum_1^\infty |\alpha_k|^2.$$

Since

$$A\phi_k = \omega_k^2 M\phi_k$$

we have that

$$A\sum_1^N [x, M\phi_k]\,\phi_k = \sum_1^N [x, M\phi_k]\,\omega_k^2 M\phi_k;$$

it follows that

$$x \in \mathcal{D}(A)$$

if and only if

$$\sum_1^\infty |[x, M\phi_k]|^2 \omega_k^4 < \infty.$$

Also, for $x,\ y$ in $\mathcal{H}$,

$$[x, My] = \sum_0^\infty [x, M\phi_k][M\phi_k, y], \tag{4.3.15}$$

$$[x, y] = \sum_0^\infty [x, \phi_k][M\phi_k, y]. \tag{4.3.15a}$$

What we are exploiting here is the fact that the sequences $\{\phi_k\}$, $\{M\phi_k\}$ are "biorthogonal" and complete. We shall return to this concept later in Section 4.6. We refer to (4.3.14) as a "modal" expansion. Expanding on (4.3.14), let

$$x = \begin{vmatrix} f \\ b \end{vmatrix}, \qquad \phi_k = \begin{vmatrix} f_k \\ b_k \end{vmatrix}.$$

Then (4.3.14) yields

$$b = \sum_0^\infty [x, M\phi_k]\, b_k.$$

Taking

$$x = \begin{vmatrix} 0 \\ b \end{vmatrix},$$

we have

$$b = \sum_0^\infty [b,\ M_b b_k]\, b_k.$$

Hence

$$\sum |[b,\ M_b b_k]|^2 < \infty,$$

and hence it follows that

$$\sum_0^\infty \|M_b b_k\|^2 < \infty$$

and therefore that

$$\sum_0^\infty \|b_k\|^2 < \infty\,.$$

In particular, therefore,

$$b_k \to 0 \quad \text{as } k \to \infty\,.$$

Returning now to (4.3.10) we have

$$f(s) = P_{11}(\lambda, s)\, f(0) + \sum_{j=1}^{i} P_{12}(\lambda,\ s - s_j)\, \delta f'(s_j), \qquad s_i < s < s_{i+1}, \tag{4.3.16}$$

which we can express as a linear transformation:

$$f = \mathcal{L}(\lambda)\mathfrak{a}, \tag{4.3.17}$$

where $\mathcal{L}(\lambda)$ maps $(R^6)^m$ into $L_2(0, L)^6$. Let

$$f_1 = \mathcal{L}(\lambda)\mathfrak{a}_1, \qquad \mathfrak{a}_1 \in (R^6)^m,$$
$$f_2 = \mathcal{L}(\lambda)\mathfrak{a}_2, \qquad \mathfrak{a}_2 \in (R^6)^m.$$

Then

$$\psi_1 = \begin{vmatrix} \mathcal{L}(\lambda)\mathfrak{a}_1 \\ L(\lambda)\mathfrak{a}_1 \end{vmatrix}, \qquad \psi_2 = \begin{vmatrix} \mathcal{L}(\lambda)\mathfrak{a}_2 \\ L(\lambda)\mathfrak{a}_2 \end{vmatrix}$$

are elements in $\mathcal{D}(A)$ and

$$\begin{aligned} [A\psi_1, \psi_2] &= [A_b f_1, b_2] - \lambda^2 [M_0 f_1,\ f_2] \\ &= [K(\lambda)\, \mathfrak{a}_1,\ L(\lambda)\mathfrak{a}_2] - \lambda^2 [M_0 f_1,\ f_2] \\ &= [\psi_1, A\psi_2] \\ &= [L(\lambda)\mathfrak{a}_1,\ K(\lambda)\mathfrak{a}_2] - \bar{\lambda}^2 [M_0 f_1,\ f_2]. \end{aligned}$$

Hence for λ such that λ^2 is real we have that

$$L(\lambda)^* K(\lambda)$$

is self-adjoint and is nonnegative definite if λ is real, since

$$0 \leq [A\psi_1, \psi_1].$$

Also, $L(\lambda)$ is an entire function of the complex variable λ, and

$$\det|L(\lambda)| = 0$$

or, equivalently,

$$\det P_{12}(\lambda;\ s_i - s_{i-1}) = 0$$

for some $i \geq 2$, from (4.3.11) for at most a countable set of λ. Omitting this set (which are recognized as "clamped" modes of the structure, where every node is clamped so that the displacement is zero), we have

$$\mathfrak{a} = L(\lambda)^{-1} b$$

and

$$(\lambda^2 M_b L(\lambda) + K(\lambda))\mathfrak{a} = (\lambda^2 M_b + K(\lambda)L(\lambda)^{-1})\, b.$$

Hence we can express (4.3.12) as

$$(\lambda^2 M_b + K(\lambda)L(\lambda)^{-1})\, b = 0.$$

Following Wittrick-Williams (1971), we shall call

$$\lambda^2 M_b + K(\lambda)L(\lambda)^{-1}$$

the "condensed dynamic stiffness matrix." Let

$$T(\lambda) = K(\lambda)L(\lambda)^{-1}.$$

Then we have

$$\begin{aligned}
[A\phi_1, \phi_2] &= [A_b f_1, b_2] - \lambda^2 [M_0 f_1,\ f_2] \\
&= [K(\lambda)L(\lambda)^{-1} b_1,\ b_2] - \lambda^2 [M_0 f_1,\ f_2] \\
&= [\phi_1, A\phi_2] \\
&= [b_1,\ K(\lambda)L(\lambda)^{-1} b_2] - \lambda^{-2} [M_0 f_1,\ f_2].
\end{aligned}$$

It follows that for λ such that λ^2 is real, $T(\lambda)$ is self-adjoint; and if λ is real, also nonnegative definite. Also, if

$$K(\lambda)\mathfrak{a} = 0, \qquad \mathfrak{a} \neq 0,$$

taking

$$f = \mathcal{L}(\lambda)\mathfrak{a}, \qquad b = L(\lambda)\mathfrak{a}, \qquad \phi = \begin{vmatrix} f \\ b \end{vmatrix},$$

we have

$$A_b f = 0$$

and hence

$$\begin{aligned}[A\phi, \phi] &= [A_b f, b] - \lambda^2 [M_0 f, f] \\ &= -\lambda^2 [M_0 f, f],\end{aligned}$$

which is impossible unless λ is purely imaginary. Zero is an eigenvalue of $K(0)$. By the "zeros of the dynamic stiffness matrix" we mean the roots of

$$\det(\lambda^2 M_b + T(\lambda)) = 0,$$

which are of course the same as those of

$$\det(\lambda^2 M_b L(\lambda) + K(\lambda)) = 0.$$

The poles of the condensed dynamic stiffness matrix are the zeros of

$$\det L(\lambda),$$

or the "clamped" modes of the structure. The dimension of the eigenfunction space for $\lambda \neq 0$ is equal to 1 if the condensed dynamic stiffness matrix has distinct eigenvalues.

4.3.5 Green's Function

We are now ready to solve the nonhomogeneous equation

$$(\lambda^2 M + A)\,\phi = \psi. \tag{4.3.18}$$

In fact for λ such that

$$\det(\lambda^2 M_b + T(\lambda)) \neq 0$$

or

$$\lambda^2 \neq -\omega_k^2,$$

we can calculate the solution in two ways. First we can obtain a "modal" solution using the modal expansion (4.3.14). For this purpose we note first that

$$(\lambda^2 M + A)^{-1} M \phi_k = \frac{\phi_k}{\lambda^2 + \omega_k^2}$$

since

$$(\lambda^2 M + A)\,\phi_k = \left(\lambda^2 + \omega_k^2\right) M\phi_k.$$

Hence we obtain

$$(\lambda^2 M + A)^{-1} x = \sum_0^\infty \frac{[x, \phi_k]\,\phi_k}{\lambda^2 + \omega_k^2}$$

and

$$[(\lambda^2 M + A)^{-1} x,\ y] = \sum_0^\infty \frac{[x, \phi_k][\phi_k, y]}{\lambda^2 + \omega_k^2}\,.$$

For λ^2 positive we see that

$$(\lambda^2 M + A)^{-1}$$

is self-adjoint and nonnegative definite, and thus

$$[(\lambda^2 M + A)^{-1} \phi_k,\ M\phi_k] = \frac{1}{\lambda^2 + \omega_k^2}$$

so that

$$\sum_0^\infty [(\lambda^2 M + A)^{-1} \phi_k,\ M\phi_k] = \sum_0^\infty \frac{1}{\lambda^2 + \omega_k^2}\,. \tag{4.3.19}$$

We shall show that the right side is finite. For this purpose we need to derive the Green's function for the eigenvalue problem (4.3.18).

For this purpose let

$$\phi = \begin{vmatrix} f \\ b \end{vmatrix}, \qquad \psi = \begin{vmatrix} g \\ c \end{vmatrix}.$$

Then (4.3.18) becomes

$$\lambda^2 M_0 f + h = g,$$

where $h(\cdot)$ is defined by (4.2.6), and

$$\lambda^2 M_b b + A_b f = c. \tag{4.3.20}$$

With $\mathcal{A}(\lambda)$, as before we see that we can solve (4.2.38) as

$$\begin{vmatrix} f(s) \\ f'(s) \end{vmatrix} = e^{\mathcal{A}(\lambda)s} \begin{vmatrix} f(0) \\ 0 \end{vmatrix} + \sum_{j=1}^{i} e^{\mathcal{A}(\lambda)(s-s_j)} \begin{vmatrix} 0 \\ \delta f'(s_j) \end{vmatrix} + \int_0^s e^{\mathcal{A}(\lambda)(s-\sigma)} \begin{vmatrix} g(\sigma) \\ 0 \end{vmatrix} d\sigma. \tag{4.3.21}$$

Or, with $\mathfrak{a}$, $\mathcal{L}(\lambda)$ as before, we have

$$f = \mathcal{L}(\lambda)\mathfrak{a} + \mathcal{N}(\lambda)\, g,$$

where the operator $\mathcal{N}(\lambda)$ mapping $L_2[0, L]^6$ into itself is defined by

$$h = \mathcal{N}(\lambda)g; \qquad h(s) = \int_0^s P_{11}(\lambda; s - \sigma)\, g(\sigma)\, d\sigma, \qquad 0 < s < L.$$

Also, we have

$$b = L(\lambda)\mathfrak{a} + N(\lambda)g,$$

where

$$N(\lambda)g = \begin{vmatrix} h(0) \\ \vdots \\ h(L) \end{vmatrix}$$

and

$$A_b f = K(\lambda)\mathfrak{a} + M(\lambda)g,$$

where

$$M(\lambda)g = A_b(\mathcal{N}(\lambda)g).$$

Hence (4.3.20) becomes

$$(\lambda^2 M_b L(\lambda) + K(\lambda))\mathfrak{a} = c - N(\lambda)g - M(\lambda)g$$

and thus for

$$\lambda^2 + \omega_k^2 \neq 0,$$

we have

$$\mathfrak{a} = (\lambda^2 M_b L(\lambda) + K(\lambda))^{-1}(c - N(\lambda)g - M(\lambda)g).$$

Hence finally

$$f = \mathcal{L}(\lambda)\mathfrak{a}(\lambda) + \mathcal{N}(\lambda)g, \tag{4.3.22}$$

$$b = L(\lambda)\mathfrak{a}(\lambda) = (\lambda^2 M_b + T(\lambda))^{-1}(c - N(\lambda)g - M(\lambda)g), \tag{4.3.23}$$

where

$$\mathfrak{a}(\lambda) = (\lambda^2 M_b L(\lambda) + K(\lambda))^{-1}(c - N(\lambda)g - M(\lambda)g).$$

From (4.3.22) we see that we can write

$$f(s) = \int_0^L G(\lambda, s, \sigma)\, g(\sigma)\, d\sigma, \qquad 0 < s < L,$$

$$b = (\lambda^2 M_b + T(\lambda))^{-1}c + \int_0^L G_b(\lambda, \sigma)\, g(\sigma)\, d\sigma = N(\lambda)f,$$

where the kernel $G(\lambda, s, \sigma)$ is continuous in $0 \leq s, \sigma \leq L$ and $G_b(\lambda, \sigma)$ is continuous in $0 \leq \sigma \leq L$. It follows from this that

$$(\lambda^2 M + A)^{-1}$$

is nuclear (see Balakrishnan (1981) for the definition) and in particular (4.3.19) is finite; hence it follows also that

$$\sum_1^\infty \frac{1}{\omega_k^2} < \infty.$$

Of particular interest to us is the solution of (4.3.18) when ψ is of the form

$$\psi = Bu$$

and Equations (4.3.22), (4.3.23) simplify to

$$f = \mathcal{L}(\lambda)(\lambda^2 M_b L(\lambda) + K(\lambda))^{-1} B_u u, \tag{4.3.24}$$

$$b = (\lambda^2 M_b + T(\lambda))^{-1} B_u u. \tag{4.3.25}$$

Hence

$$B^*(\lambda^2 M + A)^{-1} B = B_u^*(\lambda^2 M_b + T(\lambda))^{-1} B_u. \tag{4.3.26}$$

4.3.6 The Square Root of the Stiffness Operator

Finally we note that we can define $\sqrt{A}$ as a self-adjoint nonnegative operator. This is treated in standard texts (e.g., Riesz-Nagy (1955)). It is known that

$$D(\sqrt{A}) \supset D(A).$$

Unfortunately we cannot use the expansion (4.3.14), since we cannot evaluate

$$\sqrt{A}\phi_k \quad \text{or} \quad \sqrt{A}M\phi_k.$$

However, we do know that

$$\begin{aligned}[\sqrt{A}\phi_k, \sqrt{A}\phi_j] &= [A\phi_k, \phi_j] \\ &= 0, \qquad k \neq j \\ &= \omega_k^2, \qquad k = j.\end{aligned}$$

Using this we have

$$[\sqrt{A}x, \sqrt{A}x] = \sum_1^\infty \omega_k^2 |[x, M\phi_k]|^2, \tag{4.3.27}$$

which if $x \in \mathcal{D}(A)$ is

$$= [Ax, x] = 2 \text{ (Potential Energy)}.$$

Hence we see that if

$$x \in \mathcal{D}(\sqrt{A})$$

then we can define

$$(\text{Potential Energy}) = \frac{\|\sqrt{A}x\|^2}{2}.$$

Thus we can extend the definition of potential energy for all x in $\mathcal{D}(\sqrt{A})$, which is larger than $\mathcal{D}(A)$.

Because the stiffness operator A has nothing to do with the mass operator M, we may consider the case

$$M = \text{Identity}$$

in which case, our eigenfunctions become

$$A\tilde{\phi}_k = \tilde{\omega}_k^2 \tilde{\phi}_k.$$

Then in terms of these eigenfunctions we can define

$$\mathcal{D}(\sqrt{A}) = \left[x \,\middle|\, \sum_1^\infty \tilde{\omega}_k^2 |[x, \tilde{\phi}_k]|^2 < \infty \right] \tag{4.3.28}$$

and

$$\sqrt{A}x = \sum_1^\infty \tilde{\omega}_k [x, \tilde{\phi}_k] \tilde{\phi}_k,$$

$$[Ax, x] = \sum_1^\infty \tilde{\omega}_k^2 |[x, \tilde{\phi}_k]|^2,$$

which is then also the domain on which the potential energy can be defined. However, we need a characterization similar to that for A, which in particular does not invoke eigenfunction. For more on this see Balakrishnan (1990).

4.4 Time-domain Analysis

We have tools enough to consider the "time domain" solution of the dynamic equations: (4.2.14) and in turn (4.2.1). Unlike the finite-dimensional case, we can have more than one kind of solution depending on the interpretation of the equation (4.2.14) and the properties demanded of the solution. For the most satisfactory form of solution, we need to introduce the notion of the "energy norm" space in which the norm is determined in terms of the total system energy (kinetic plus potential) and the theory of semigroups of operators applies.

We begin with the weaker notion first.

4.4.1 Weak Solution/Modal Solution

With respect to (4.2.14) we have to specify first how the time derivative therein is to be defined and whether the solution $x(t),\ 0 < t$, which will depend only on the initial conditions at $t = 0$ and the input $u(\cdot)$, is required to be in the domain of A. Whatever the definition, we must have that for every ϕ in $\mathcal{D}(A)$:

$$[M\ddot{x}(t),\ \phi] + [Ax(t), \phi] + [Bu(t), \phi] = 0.$$

We can rewrite this as

$$\frac{d^2}{dt^2}[x(t),\ M\phi] + [x(t),\ A\phi] + [Bu(t), \phi] = 0. \tag{4.4.1}$$

As for the input $u(\cdot)$, we assume that

$$\int_0^T |u(t)|^2\, dt\ <\ \infty, \quad \text{for every } T, \qquad 0 < T < \infty.$$

That is, $u(\cdot) \in L_2[0,\ T]^{m_c}$ for every $T < \infty$. Note that in (4.4.1) we only need the "weak derivative" (see Balakrishnan (1981) for definition) and further we have circumvented the requirement that $x(t) \in \mathcal{D}(A)$. By a "weak solution" of (4.2.14) we mean a function $x(t),\ t \geq 0$, such that for every ϕ in $\mathcal{D}(A)$,

$$[x(t),\ M\phi]$$

and its derivative are absolutely continuous in $t > 0$, and the function must satisfy (4.4.1) and the initial conditions

$$[x(t),\ M\phi] \to [x_1,\ M\phi] \quad \text{as } t \to 0+,$$

$$\frac{d}{dt}[x(t),\ M\phi] \to [x_2,\ M\phi] \quad \text{as } t \to 0+,$$

where x_1, x_2 are given elements in $\mathcal{H}$.

Even weaker (and easiest to construct) is the notion of a "modal solution" where we only require that (4.4.1) hold for every eigenfunction ϕ_k. In this case (4.4.1) becomes

$$\frac{d^2}{dt^2}[x(t),\ M\phi_k] + \omega_k^2[x(t),\ M\phi_k] + [u(t),\ B^*\phi_k] = 0. \tag{4.4.2}$$

Let

$$a_k(t) = [x(t),\ M\phi_k],$$

$$u_k(t) = [u(t),\ B^*\phi_k].$$

Then for each k, we have the ordinary differential equation

$$\ddot{a}_k(t) + \omega_k^2 a_k(t) + u_k(t) = 0, \qquad \text{a.e. } t \geq 0 \tag{4.4.3}$$

with the initial conditions

$$a_k(0) = [x_1, M\phi_k],$$
$$\dot{a}_k(0) = [x_2, M\phi_k],$$

which we can solve, to yield, for $\omega_k \neq 0$:

$$a_k(t) = \frac{\dot{a}_k(0)\sin\omega_k t}{\omega_k^2} + a_k(0)\cos\omega_k t + \int_0^t \frac{1}{\omega_k}\sin\omega_k(t-s)\, u_k(s)\, ds \tag{4.4.4}$$

and for $\omega_k = 0$:

$$a_k(t) = a_k(0) + t\dot{a}_k(0) + \int_0^t (t-s)u_k(s)\, ds. \tag{4.4.5}$$

It is easy to verify that

$$\sum_0^\infty |a_k(t)|^2 < \infty, \qquad 0 < t < \infty.$$

Hence we can define

$$x(t) = \sum_0^\infty a_k(t)\,\phi_k\,. \tag{4.4.6}$$

Thus defined, we do have

$$a_k(t) = [x(t), M\phi_k]$$

and $x(\cdot)$ satisfies (4.4.2). Hence we have a "modal" solution. However $x(\cdot)$ need not qualify as a weak solution in general and additional restrictions will need to be placed on the initial conditions – on x_1, x_2. Thus for

$$u(t) = 0, \qquad t > 0$$

we would need to require that

$$x_1 \in \mathcal{D}(A)$$

so that

$$\sum_1^\infty a_k(0)^2 \omega_k^4 < \infty\,.$$

In this case

$$\sum_1^\infty |\dot{a}_k(t)|^2 + \sum_1^\infty |\ddot{a}_k(t)|^2 < \infty$$

and hence for ϕ in $\mathcal{D}(A)$

$$\begin{aligned}\frac{d^2}{dt^2}[x(t), M\phi] &= \sum_0^\infty \ddot{a}_k(t)[\phi_k, M\phi] \\ &= \sum_0^\infty a_k(t)\left[(-\omega_k^2)M\phi_k,\ \phi\right] \\ &= -\sum_0^\infty a_k(t)[\phi_k, A\phi] \\ &= -[x(t),\ A\phi]\end{aligned}$$

as required.

A better technique, avoiding these special considerations, is to go over to "state-space" solution.

4.4.2 State-space Solution: Need for Energy Norm

To proceed further with (4.2.14), let us cast it in "state-space" form. Thus we let

$$Y(t) = \begin{vmatrix} x(t) \\ \dot{x}(t) \end{vmatrix}$$

and, formally, (4.2.14) goes over into

$$\dot{Y}(t) = \mathcal{A}Y(t) + \mathcal{B}u(t), \tag{4.4.7}$$

where

$$Y(t) \in \mathcal{H} \times \mathcal{H},$$

$$\mathcal{A} = \begin{vmatrix} 0 & I \\ -M^{-1}A & 0 \end{vmatrix},$$

$$\mathcal{B} = \begin{vmatrix} 0 \\ -M^{-1}B \end{vmatrix},$$

and

$$\mathcal{D}(\mathcal{A}) = \left[Y = \begin{vmatrix} y_1 \\ y_2 \end{vmatrix},\ \begin{matrix} y_1 \in \mathcal{D}(A) \\ y_2 \in \mathcal{H} \end{matrix}\right].$$

We interpret (4.4.7) in the weak sense, satisfying also the initial condition

$$\|Y(t) - Y\| \to 0 \quad \text{as } t \to 0+.$$

In addition we require that for each t, $Y(t)$ be continuous with respect to Y:

$$\|Y(t)\| \to 0 \quad \text{as } \|Y\| \to 0.$$

It is shown in Balakrishnan-Triggiani (1993) that this is impossible unless we change the space $\mathcal{H} \times \mathcal{H}$, that is, change the product-norm, which is physically meaningless, to the "energy" norm. The total energy

$$\begin{aligned} &= \text{Potential (Elastic) Energy} + \text{Kinetic Energy} \\ &= \frac{[Ay_1,\ y_1]}{2} + \frac{[My_2,\ y_2]}{2}, \end{aligned}$$

where

$$Y = \begin{vmatrix} y_1 \\ y_2 \end{vmatrix}, \qquad y_1 \in \mathcal{D}(A).$$

As we have seen, we can extend the definition of potential energy to $\mathcal{D}(\sqrt{A})$. Also the total energy vanishes for

$$Y = \begin{vmatrix} y_1 \\ 0 \end{vmatrix}, \quad y_1 \in \text{nullspace of } A.$$

Let $\mathcal{H}_1$ denote the M-orthogonal complement of the null space of A:

$$\mathcal{H}_1 = [x \mid [Mx, \phi] = 0, \quad \phi \in \text{nullspace of } A].$$

Then

$$\phi_k \in \mathcal{H}_1 \quad \text{for every } k, \qquad \omega_k \neq 0.$$

In other words we consider $\mathcal{H}$ under the M-inner product,

$$[x, y]_M = [Mx, y]$$

which is equivalent to the original inner product since M has a bounded inverse. We now define the energy space $\mathcal{H}_E$ by

$$\mathcal{H}_E = (\mathcal{D}(\sqrt{A}) \cap \mathcal{H}_1) \times \mathcal{H}$$

with inner product defined by

$$[Y, Z]_E = [\sqrt{A}y_1,\ \sqrt{A}z_1] + [My_2, z_2],$$

where

$$Y = \begin{vmatrix} y_1 \\ y_2 \end{vmatrix}, \qquad Z = \begin{vmatrix} z_1 \\ z_2 \end{vmatrix}.$$

Note that

$$[Y, Y]_E = 2[\text{Total Energy}]$$

and hence the name "energy norm" space is used. We shall show now that $\mathcal{H}_E$ is actually complete. Let

$$Y_n = \begin{vmatrix} x_n \\ z_n \end{vmatrix}$$

be a Cauchy sequence in $\mathcal{H}_E$. Then

$$\|Y_n - Y_m\|_E^2 = \|\sqrt{A}(x_n - x_m)\|^2 + \|\sqrt{M}(z_n - z_m)\|^2.$$

Since $\sqrt{M}$ has a bounded inverse,

$$z_n \to z \quad \text{in } \mathcal{H}.$$

Now

$$x_n \in \mathcal{H}_1,$$

and $\sqrt{A}$ restricted to $\mathcal{H}_1$ has a bounded inverse. Hence

$$x_n = (\sqrt{A})^{-1}\sqrt{A}x_n$$

converges, and denoting the limit by x, we have

$$x = (\sqrt{A})^{-1}(\lim \sqrt{A}x_n)$$

and

$$x \in \mathcal{D}(\sqrt{A}) \cap \mathcal{H}_1.$$

Hence $\mathcal{H}_E$ is complete.

Let $\phi_{0,k}$, $k = 1, \ldots, 6$, denote an M-orthonormal basis for the null space of A:

$$[M\phi_{0,k}, \phi_{0,j}] = \delta_j^k.$$

Define the projection operator (self-adjoint in the M-inner product) by

$$\mathcal{P}_0 x = \sum_1^6 [x, M\phi_{0,k}]\,\phi_{0,k}$$

and let

$$\mathcal{P}_1 = I - \mathcal{P}_0$$

so that

$$[\mathcal{P}_1 x, M\phi] = 0 \quad \text{if } \mathcal{P}_0\phi = \phi.$$

Then

$$\mathcal{P}_1\mathcal{H} = \mathcal{H}_1.$$

The operator $\mathcal{A}$ is defined then as follows:

$$\mathcal{D}(\mathcal{A}) = \left[y = \begin{vmatrix} x_1 \\ x_2 \end{vmatrix}, \begin{array}{l} x_1 \in \mathcal{D}(A) \cap \mathcal{H}_1 \\ x_2 \in \mathcal{D}(\sqrt{A}) \end{array} \right], \tag{4.4.8}$$

$$\mathcal{A} = \begin{vmatrix} 0 & \mathcal{P}_1 \\ -M^{-1}A & 0 \end{vmatrix},$$

$$\mathcal{A}Y = \begin{vmatrix} \mathcal{P}_1 x_2 \\ -M^{-1}Ax_1 \end{vmatrix};$$

thus defined $\mathcal{A}$ is closed and has a dense domain.

Let $\mathcal{A}^*$ denote the adjoint. Then

$$\mathcal{A}^* = \begin{vmatrix} 0 & -\mathcal{P}_1 \\ M^{-1}A & 0 \end{vmatrix} = -\mathcal{A}.$$

To prove this, let

$$Y = \begin{vmatrix} y_1 \\ y_2 \end{vmatrix}, \qquad Z = \begin{vmatrix} z_1 \\ z_2 \end{vmatrix}, \qquad Y, Z \in \mathcal{D}(\mathcal{A}).$$

Then

$$\begin{aligned} [\mathcal{A}Y, Z]_{\mathrm{E}} &= [\sqrt{A}y_2, \ \sqrt{A}z_1] - [Ay_1, z_2] \\ &= [My_2, \ M^{-1}Az_1] - [\sqrt{A}y_1, \sqrt{A}z_2] \\ &= [Y, \ \mathcal{A}^* z]_{\mathrm{E}} \\ &= -[Y, \mathcal{A}z]_{\mathrm{E}}. \end{aligned}$$

In what follows we shall omit the subscript E in inner products and norms where elements of $\mathcal{H}_{\mathrm{E}}$ are involved. In particular, for Y in the domain of $\mathcal{A}$:

$$[\mathcal{A}Y, Y] + [Y, \mathcal{A}Y] = \mathrm{Re}\,[\mathcal{A}Y, Y] = 0.$$

Hence (see Balakrishnan (1981)), $\mathcal{A}$ generates a strongly continuous semigroup $S(t), t \geq 0$; actually it is a group:

$$S(t)^* = (S(t))^{-1} = S(-t).$$

For Y in the domain of $\mathcal{A}$, $S(t)Y$ is also in the domain of $\mathcal{A}$, and

$$\frac{d}{dt} S(t)Y = \mathcal{A}S(t)Y.$$

Hence the equation

$$\dot{Y}(t) = \mathcal{A}Y(t)$$

with the derivative interpreted in the strong sense,

$$\left\| \dot{Y}(t) - \frac{Y(t+\Delta) - Y(t)}{\Delta} \right\| \to 0 \quad \text{as } \Delta \to 0,$$

has the unique solution

$$Y(t) = S(t)Y(0)$$

such that

$$\|Y(t) - Y(0)\| \to 0.$$

Note that for Y in $\mathcal{D}(\mathcal{A})$, we can exploit strong differentiability to obtain

$$\frac{d}{dt}[S(t)Y, S(t)Y] = [\mathcal{A}S(t)Y, S(t)Y] + [S(t)Y, \mathcal{A}S(t)Y] = 0. \tag{4.4.9}$$

Therefore

$$\|S(t)Y\|^2 = \|Y\|^2.$$

The domain of $\mathcal{A}$ being dense, we have that

$$\|S(t)Y\| = \|Y\|, \qquad Y \in \mathcal{H}_E.$$

In particular, the energy stays constant in time.

More generally, for any $Y(0)$ in $\mathcal{H}_E$,

$$\dot{Y}(t) = \mathcal{A}Y(t) + \mathcal{B}u(t)$$

interpreted in the weak sense,

$$\frac{d}{dt}[Y(t), Y] = [Y(t), \mathcal{A}^*Y] + [\mathcal{B}u(t), Y],$$

for every Y in the domain of $\mathcal{A}^*$, has the unique solution (see Balakrishnan (1981))

$$Y(t) = S(t)Y(0) + \int_0^t S(t-\sigma)\mathcal{B}u(\sigma)\, d\sigma, \qquad t \geq 0$$

for

$$\int_0^t \|u(\sigma)\|^2 \, d\sigma < \infty \qquad \text{for every } t > 0.$$

Remark

We can finally relate (4.2.14) to (4.2.1). Requiring the initial condition vector Y to be in $\mathcal{D}(\mathcal{A})$ means in particular that we can define the necessary partial derivatives in (4.2.1) and thus have a solution in the "ordinary" or pointwise sense.

4.4.3 Spectral Properties of $\mathcal{A}$

The spectral properties of $\mathcal{A}$ are readily deduced from the spectral properties of A. Thus the equation

$$\lambda X - \mathcal{A}X = Y, \qquad X = \begin{vmatrix} x_1 \\ x_2 \end{vmatrix}, \qquad Y = \begin{vmatrix} y_1 \\ y_2 \end{vmatrix}$$

yields for X:

$$\lambda x_1 - \mathcal{P}_1 x_2 = y_1, \qquad \lambda x_2 + M^{-1} A x_1 = y_2.$$

Or

$$\left.\begin{aligned} \lambda^2 M x_1 + A x_1 &= M(y_2 + \lambda y_1 - \lambda \mathcal{P}_0 x_2) \\ \mathcal{P}_1 x_2 &= \lambda x_1 - y_1 \end{aligned}\right\}. \tag{4.4.10}$$

The eigenvalues of $\mathcal{A}$ are thus given by

$$\left.\begin{aligned} \lambda^2 M x_1 + A x_1 &= \lambda \mathcal{P}_0 x_2 \\ \mathcal{P}_1 x_2 &= \lambda x_1 \end{aligned}\right\}. \tag{4.4.11}$$

Zero is an eigenvalue of $\mathcal{A}$, since zero is an eigenvalue of A. In fact

$$\mathcal{A}\begin{vmatrix} 0 \\ \phi \end{vmatrix} = \begin{vmatrix} \mathcal{P}_1\phi \\ 0 \end{vmatrix} = \begin{vmatrix} 0 \\ 0 \end{vmatrix}, \qquad \text{if } A\phi = 0.$$

Thus the dimension of the null space of $\mathcal{A}$ is 6.

For nonzero λ, we note that

$$\Phi_k = \begin{vmatrix} \phi_k \\ i\omega_k\phi_k \end{vmatrix}$$

is an eigenvector with eigenvalue $(i\omega_k)$. For,

$$[M\phi_k, \mathcal{P}_0 x] = 0, \qquad \text{or } \phi_k \in \mathcal{D}(A) \cap \mathcal{H}_1,$$

we have that

$$\mathcal{P}_0\phi_k = 0, \qquad -\omega_k^2 M\phi_k + A\phi_k = 0,$$

and hence (4.4.11) is satisfied. The dimension of the eigenfunction space is equal to the dimension of the eigenfunction space of A corresponding to ω_k^2. We shall for simplicity take this dimension to be equal to one. Hence the eigenvalues of $\mathcal{A}$ may be enumerated as

$$\omega_0 = 0,$$

$$\{\pm i|\omega_k|\}, \qquad \omega_k \neq 0, \qquad k > 0, \qquad \omega_{k+1} > \omega_k,$$

$$\mathcal{A}\Phi_k = i|\omega_k|\Phi_k, \qquad \omega_k \neq 0,$$

$$\mathcal{A}\bar{\Phi}_k = -i|\omega_k|\bar{\Phi}_k, \qquad \omega_k \neq 0, \quad \text{since } \phi_k = \bar{\phi}_k,$$

where

$$\Phi_k = \begin{vmatrix} \phi_k \\ i|\omega_k|\phi_k \end{vmatrix}, \qquad \Phi_{0,i}, \qquad i = 1, \ldots, 6 = \begin{vmatrix} 0 \\ \phi_{0,i} \end{vmatrix},$$

and where

$$\bar{\Phi}_{0,i} = \Phi_{0,i}, \qquad [M\phi_{0,i}, \phi_{0,j}] = \delta^i_j.$$

Note that

$$\begin{aligned} \mathcal{A}^*\Phi_k &= -i\omega_k\Phi_k, \\ \mathcal{A}^*\bar{\Phi}_k &= i|\omega_k|\bar{\Phi}_k, \\ \mathcal{A}^*\Phi_{0,i} &= 0. \end{aligned}$$

These eigenfunctions are orthogonal:

$$\begin{aligned} \left[\Phi_k, \bar{\Phi}_j\right] &= 0, \\ \left[\Phi_k, \Phi_j\right] &= 0, \qquad k \neq j, \\ [\Phi_k, \Phi_k] &= 2\omega_k^2 \, [M\phi_k, \phi_k]. \end{aligned}$$

To orthonormalize the Φ_k, we need only to take

$$[M\phi_k, \phi_k] = \frac{1}{2\omega_k^2}, \tag{4.4.12}$$

which we shall assume in what follows.

Let us show that the $\{\Phi_k, \bar{\Phi}_k\}$ are complete. Suppose for some Ψ in $\mathcal{H}_{\mathrm{E}}$

$$[\Phi_k, \psi] = [\bar{\Phi}_k, \psi] = 0 \quad \text{for every } k.$$

Writing

$$\Psi = \begin{vmatrix} \psi_1 \\ \psi_2 \end{vmatrix}$$

we have

$$\begin{aligned} 0 &= [\Phi_k, \Psi] = \omega_k^2[M\phi_k, \psi_1] + i\omega_k[M\phi_k, \psi_2], \\ 0 &= \left[\bar{\Phi}_k, \Psi\right] = \omega_k^2[M\phi_k, \psi_1] - i\omega_k[M\phi_k, \psi_2], \\ 0 &= [\Phi_{0,k}, \Psi] = 0, \end{aligned}$$

from which it follows that

$$[M\phi_k, \psi_2] = 0 \quad \text{for every } k$$

and from the completeness of $\{\phi_k\}$ we have that

$$\psi_2 = 0.$$

We also have that

$$[M\phi_k, \psi_1] = 0, \qquad \omega_k \neq 0,$$

and by definition ψ_1 is in $\mathcal{H}_1$. Hence it follows that

$$\psi_1 = 0.$$

As a consequence we have the modal expansion:

$$Y = \sum_1^\infty [Y, \Phi_k]\Phi_k + \sum_1^\infty [Y, \bar{\Phi}_k]\bar{\Phi}_k + \sum_1^6 [Y, \Phi_{0,k}]\Phi_{0,k}. \tag{4.4.13}$$

Since

$$S(t)\Phi_k = e^{i\omega_k t}\Phi_k, \qquad S(t)\bar{\Phi}_k = e^{-i\omega_k t}\bar{\Phi}_k, \qquad S(t)\Phi_{0,k} = e^{i\omega_k t}\Phi_{0,k}$$

we have also

$$S(t)Y = \sum_1^\infty [Y, \Phi_k]e^{i\omega_k t}\Phi_k + \sum_1^\infty \left[Y, \bar{\Phi}_k\right] e^{-i\omega_k t}\bar{\Phi}_k + \sum_1^6 [Y, \Phi_{0,k}]\Phi_{0,k}. \tag{4.4.14}$$

4.4.4 Resolvent of $\mathcal{A}$

For

$$\lambda \neq \pm i|\omega_k|, \qquad \lambda \neq 0,$$

we shall now show that

$$\lambda I - \mathcal{A}$$

has a bounded inverse. To calculate this inverse we go back to (4.4.8), where we let

$$\mathcal{P}_0 x_2 = z.$$

Then

$$\lambda^2 M x_1 + A x_1 = M y_2 + \lambda M y_1 - \lambda M z.$$

For $\lambda \neq \pm i|\omega_k|$, $\lambda \neq 0$, we can define

$$w = (\lambda^2 M + A)^{-1}(M y_2 + \lambda M y_1).$$

Then

$$\lambda^2 M(x_1 - w) + A(x_1 - w) = -\lambda M z.$$

Therefore

$$\lambda^2 M(x_1 - \mathcal{P}_1 w) + A(x_1 - \mathcal{P}_1 w) = \lambda^2 M \mathcal{P}_0 w - \lambda M z = 0$$

if we let

$$z = \lambda \mathcal{P}_0 w,$$

or

$$x_2 = \lambda \mathcal{P}_1 w - y_1 + \lambda \mathcal{P}_0 w = \lambda w - y_1.$$

Hence we obtain

$$\begin{vmatrix} x_1 \\ x_2 \end{vmatrix} = \begin{vmatrix} \mathcal{P}_1 w \\ \lambda w - y_1 \end{vmatrix} = (\lambda I - \mathcal{A})^{-1} \begin{vmatrix} y_1 \\ y_2 \end{vmatrix}, \tag{4.4.15}$$

where

$$w = (\lambda^2 M + A)^{-1}(M y_2 + \lambda y_1)$$

and

$$w \in \mathcal{D}(A).$$

We use the notation

$$R(\lambda, \mathcal{A}) = (\lambda I - \mathcal{A})^{-1}$$

and refer to the left side as the "resolvent of $\mathcal{A}$." We have also the modal representation

$$R(\lambda, \mathcal{A})Y = \sum_1^{\infty} \frac{[Y, \Phi_k]}{\lambda - i\omega_k} \Phi_k + \sum_1^{\infty} \frac{[Y, \bar{\Phi}_k]}{\lambda + i\omega_k} \bar{\Phi}_k + \sum_1^{6} \frac{[Y, \Phi_{0,k}]}{\lambda} \Phi_{0,k}. \tag{4.4.16}$$

It is indeed the Laplace transform:

$$R(\lambda, \mathcal{A})Y = \int_0^{\infty} e^{-\lambda t} S(t) Y \, dt, \qquad \operatorname{Re} \lambda > 0, \tag{4.4.17}$$

but of course it is defined and analytic in λ except for poles at the eigenvalues as follows from (4.4.13) and (4.4.14). Note in particular that

$$\mathcal{B}^* \mathcal{R}(\lambda, \mathcal{A}) Y = -B^*(\lambda w - y_1),$$

and hence it follows that

$$\mathcal{B}^* \mathcal{R}(\lambda, \mathcal{A}) \mathcal{B} = \lambda B^* (\lambda^2 M + A)^{-1} B = \lambda B_u^* (\lambda^2 M_b + T(\lambda))^{-1} B_u. \tag{4.4.18}$$

4.5 Controllability and Stabilizability

In this section we show that under a controllability condition, rate feedback using a collocated sensor can stabilize the system. All modes will decay even though the damping coefficient will decrease with mode frequency. Whatever the initial conditions, the elastic energy will eventually dissipate to zero. (This is known as "strong" stability.) In our model we neglcct any inherent damping in the structure. However, the controller is robust in the sense that it will not destabilize any mode – it will only increase the damping. In other words, we always have stability enhancement.

The main results are known – see Balakrishnan (1981) and the references therein. The presentation here tries as far as possible to be self-contained, using the specific features of the problem at hand rather than merely quoting general results.

In reference to our dynamic equation:

$$\dot{Y}(t) = \mathcal{A}Y(t) + \mathcal{B}u(t) \tag{4.5.1}$$

we do not include a damping operator; $\mathcal{A}$ is not "stable." In fact

$$\|S(t)Y\|_{\mathrm{E}} = \|Y\|_{\mathrm{E}}.$$

The elastic energy is not dissipated in the absence of control. In finite dimensional theory, the system is "exponentially" stabilizable, if (A, B) is "exactly" controllable – by that we mean given any Y_1, Y_2 we can find a control $u(\cdot)$ such that

$$Y_2 = S(t)Y_1 + \int_0^t S(t-\sigma)\mathcal{B}u(\sigma)\,d\sigma$$

for some $t \geq 0$. This is impossible in our case because (the range of) $\mathcal{B}$ is finite dimensional. (See Balakrishnan (1981), for a proof.) The next best thing we can do is to require just "controllability." Thus we say that (A, B) is "controllable" if

$$\bigcup_{t\geq 0} (\text{range of } S(t)\mathcal{B})$$

is dense in $\mathcal{H}_{\mathrm{E}}$. This is equivalent to saying that

$$\bigcup_{t\geq 0}\left(\int_0^t S(t-\sigma)\mathcal{B}u(\sigma)\,d\sigma;\ u(\cdot) \in L_2(0,t)\right)$$

is dense in $\mathcal{H}_{\mathrm{E}}$. In other words we require that the states "reachable" from the zero state are dense in $\mathcal{H}_{\mathrm{E}}$.

Theorem 4.5.1 $\mathcal{A} \sim \mathcal{B}$ is controllable in $\mathcal{H}_{\mathrm{E}}$ if and only if

$$B^*\phi \neq 0 \tag{4.5.2}$$

for any eigenfunction ϕ, defined by

$$A\phi = \omega^2 M\phi, \qquad \phi \neq 0.$$

Equivalently,

$$\mathcal{B}^*\Phi \neq 0 \tag{4.5.3}$$

for any mode Φ:

$$\mathcal{A}\Phi = i\omega\Phi, \qquad \Phi \neq 0.$$

Proof It is convenient to use a modal expansion for any Y as

$$Y = \sum_1^\infty P_k Y + \sum_1^\infty P_{-k} Y + P_0 Y,$$

where P_k, P_{-k}, and P_0 are the Projectors onto the eigenfunction space corresponding to the eigenvalues $i\omega_k$, $-i\omega_k$, and onto the null space of $\mathcal{A}$, respectively. Then

$$S(t)Y = \sum_1^\infty e^{i\omega_k t} P_k Y + \sum_1^\infty e^{-i\omega_k t} P_{-k} Y + P_0 Y. \tag{4.5.4}$$

If the set

$$\bigcup_{t \geq 0} S(t)\mathcal{B}u$$

where u which ranges over all of $\mathcal{R}^{m_c}$ is *not* dense in $\mathcal{H}_E$, we can find a nonzero element Y in $\mathcal{H}_E$ such that

$$[S(t)\mathcal{B}u, Y] = 0, \qquad t \geq 0, \qquad u \in \mathcal{R}^{m_c}.$$

Hence using (4.5.4) we must have

$$\sum_1^\infty e^{i\omega_k t}\mathcal{B}[u, \mathcal{B}^* P_k Y] + \sum_1^\infty e^{-i\omega_k t}[u, \mathcal{B}^* P_{-k} Y] + [u, \mathcal{B}^* P_0 Y] = 0, \quad t \geq 0.$$

But the left side is an almost periodic function in t and can vanish identically if and only if for every k

$$0 = [u, \mathcal{B}^* P_k y] = [u, \mathcal{B}^* P_{-k} Y] = [u, \mathcal{B}^* P_0 Y].$$

Since Y is not zero, there must be at least one k such that

$$P_k Y \neq 0$$

or

$$P_k Y = \Phi_k,$$

where

$$\mathcal{A}\Phi_k = i\omega_k\Phi_k, \qquad \Phi_k \neq 0, \qquad [u,\, \mathcal{B}^*\Phi_k] \doteq 0.$$

But since u is arbitrary, we must have

$$\mathcal{B}^*\Phi_k = 0,$$

which is a contradiction. Since Φ_k must be of the form

$$\Phi_k = \begin{vmatrix} \phi_k \\ i\omega_k\phi_k \end{vmatrix}$$

it follows that

$$B^*\phi_k = 0,$$

which is again impossible by assumption.

Next, suppose $\mathcal{A} \sim \mathcal{B}$ is controllable. Suppose

$$B^*\phi = 0$$

for some mode ϕ and

$$A\phi = \omega^2 M\phi, \qquad \phi \neq 0.$$

Then for

$$\Phi = \begin{vmatrix} \phi \\ i\omega\phi \end{vmatrix},$$

$$[S(t)\mathcal{B}u,\, \Phi] = e^{i\omega t}[\mathcal{B}u,\, \Phi] = e^{i\omega t}(i\omega)[u,\, B^*\phi] = 0.$$

Hence Φ is orthogonal to

$$S(t)\mathcal{B}u$$

for every t and $u \in \mathcal{R}^{m_c}$. Hence ($\mathcal{A} \sim \mathcal{B}$) is not controllable – contradicting the hypothesis.

We shall show that ($\mathcal{A} \sim \mathcal{B}$) is controllable for our system in Section 4.4 where we study the eigenvalue problem.

Corollary Suppose ($\mathcal{A} \sim \mathcal{B}$) is controllable. Then the number of controls (the dimension of the control space) must be at least 6. More generally, the dimension of the control space must be at least equal to the largest eigenfunction-space dimension.

Proof Let $\phi_{0,i}$, $i = 1, \ldots, 6$, be a basis for the null space of A. Suppose $B^*\phi_{0,i}$ are linearly dependent. Then

$$\sum_1^6 a_k B^*\phi_{0,k} = 0, \quad \text{not all } a_k = 0,$$

or

$$B^* \left(\sum_1^6 a_k \phi_{0,k} \right) = 0.$$

But

$$\sum_1^6 a_k \phi_{0,k}$$

is a nonzero eigenfunction function of A corresponding to the eigenvalue zero and controllability is thus violated.

Next we shall prove the fundamental relationship of controllability to stabilizabilty.

Theorem 4.5.2 Suppose $(\mathcal{A} \sim \mathcal{B})$ is controllable. Then the feedback control

$$u(t) = -\alpha \mathcal{B}^* Y(t), \qquad \alpha > 0 \tag{4.5.5}$$

is such that

$$\|Y(t)\| \to 0 \quad \text{as } t \to \infty$$

for every $Y(0)$.

Proof The closed-loop system dynamic is now

$$\left. \begin{aligned} \dot{Y}(t) = (\mathcal{A} - \alpha \mathcal{B}\mathcal{B}^*) Y(t), \\ Y(0) \quad \text{given.} \end{aligned} \right\} \tag{4.5.6}$$

Let

$$Y(t) = \begin{vmatrix} x_1(t) \\ x_2(t) \end{vmatrix}.$$

Then we have

$$\dot{x}_1(t) = \mathcal{P}_1 x_2(t), \tag{4.5.7}$$

$$\dot{x}_2(t) = -M^{-1} A x_1(t) - \alpha M^{-1} B B^* x_2(t). \tag{4.5.8}$$

To relate this to (4.2.17), let the initial conditions for the latter be given as

$$\begin{vmatrix} x(0) \\ \dot{x}(0) \end{vmatrix}.$$

For (4.3.6) let

$$Y(0) = \begin{vmatrix} x_1(0) \\ x_2(0) \end{vmatrix},$$

where

$$x_1(0) = \mathcal{P}_1 x(0),$$
$$x_2(0) = \dot{x}(0).$$

Define

$$x(t) = x_1(t) + \mathcal{P}_0 x(0) + \int_0^t \mathcal{P}_0 x_2(\sigma)\, d\sigma.$$

Then

$$\dot{x}(t) = \dot{x}_1(t) + \mathcal{P}_0 x_2(t)$$

and (4.5.7) and (4.5.8) yield

$$\begin{aligned} \dot{x}(t) &= x_2(t), \\ \dot{x}_2(t) &= \mathcal{P}_1 \dot{x}_2(t) + \mathcal{P}_0 \dot{x}_2(t) \\ &= \ddot{x}(t) \\ &= -M^{-1}Ax(t) - \alpha M^{-1}BB^*\dot{x}(t) \end{aligned}$$

or

$$M\ddot{x}(t) + Ax(t) + \alpha BB^*\dot{x}(t) = 0 \qquad (4.5.9)$$

so that the control $u(t)$ is now

$$u(t) = \alpha B^*\dot{x}(t) = -\alpha\mathcal{B}^*Y(t). \qquad (4.5.10)$$

In other words we have rate feedback using a collocated sensor – a feedback principle for stabilization that is age-old, but still requires proof in our infinite-dimensional context.

First we note that for Y in $\mathcal{D}(\mathcal{A})$:

$$\begin{aligned} \operatorname{Re}[(\mathcal{A} - \alpha\mathcal{B}\mathcal{B}^*)Y, Y] &= -\alpha\|\mathcal{B}^*Y\|^2 \\ &= \operatorname{Re}[(\mathcal{A}^* - \alpha\mathcal{B}\mathcal{B}^*)Y, Y]. \end{aligned}$$

Hence (see Balakrishnan (1981))

$$(\mathcal{A} - \alpha\mathcal{B}\mathcal{B}^*)$$

generates a dissipative, strongly continuous semigroup. Denoting the latter by

$$S_\alpha(t), \qquad t \ge 0$$

we note that

$$\|S_\alpha(t)\| \le 1.$$

4.5.1 Eigenvalues

Let us consider next the eigenvalues

$$(\mathcal{A} - \alpha \mathcal{B}\mathcal{B}^*)Y = \lambda Y \tag{4.5.11}$$

or, then,

$$(\operatorname{Re} \lambda)[Y, Y] = -\alpha \|\mathcal{B}^* Y\|^2.$$

Here

$$\|\mathcal{B}^* Y\| \quad \text{cannot be zero}$$

for if

$$\mathcal{B}^* Y = 0,$$

then

$$(A - \alpha \mathcal{B}\mathcal{B}^*)Y = \mathcal{A}Y = \lambda Y$$

and we violate the controllability condition. Hence it follows that

$$\operatorname{Re} \lambda = \frac{-\alpha \|\mathcal{B}^* Y\|^2}{[Y, Y]} < 0.$$

Hence every eigenvalue has a strictly negative real part. Rewriting (4.5.11) as

$$(\lambda I - \mathcal{A} + \alpha \mathcal{B}\mathcal{B}^*)Y = 0$$

and multiplying on the left by $\mathcal{R}(\lambda, \mathcal{A})$, we have

$$Y + \alpha \mathcal{R}(\lambda, \mathcal{A})\, \mathcal{B}\mathcal{B}^* Y = 0.$$

Therefore

$$\mathcal{B}^* Y + \alpha \mathcal{B}^* \mathcal{R}(\lambda, \mathcal{A})\, \mathcal{B}\mathcal{B}^* Y = 0$$

or

$$(I + \alpha \mathcal{B}^* \mathcal{R}(\lambda, \mathcal{A})\mathcal{B})\mathcal{B}^* Y = 0,$$

where inside the parentheses on the left side is an $m_c \times m_c$ matrix. Hence the eigenvalues are the roots of

$$D(\lambda, \alpha) = 0, \tag{4.5.12}$$

where

$$D(\lambda, \alpha) = \det(I + \alpha \mathcal{B}^* \mathcal{R}(\lambda, A)\mathcal{B}). \tag{4.5.13}$$

But using (4.4.18), we have

$$D(\lambda, \alpha) = \det\left[I + \alpha \lambda B_u^* (\lambda^2 M_b + T(\lambda))^{-1} B_u\right]. \tag{4.5.14}$$

Let $\{\lambda_k\}$ denote the eigenvalues, where we know that λ_k must have the form

$$\lambda_k = -|\sigma_k| + i\nu_k, \qquad \sigma_k, \nu_k \text{ real.}$$

The corresponding eigenfunctions (denote them Y_k) are then given by

$$Y_k = \mathcal{R}(\lambda_k, \mathcal{A})\, \mathcal{B}u(\lambda_k), \tag{4.5.15}$$

where

$$(I + \alpha\mathcal{M}(\lambda_k))u(\lambda_k) = 0, \qquad \|u(\lambda_k)\| = 1$$

and where

$$\mathcal{M}(\lambda) = \mathcal{B}^*\mathcal{R}(\lambda, \mathcal{A})\mathcal{B}.$$

We note that the dimension of the eigenfunction space is the dimension of the eigenvector space corresponding to the eigenvalue zero of the $m_c \times m_c$ matrix

$$I + \alpha\mathcal{M}(\lambda_k))$$

and is thus less than m_c. It is equal to one, if we assume that the matrix has distinct eigenvalues, which we shall, for simplicity, in what follows.

At this point we leave open whether the sequence $\{\lambda_k\}$ is finite or not. (We shall eventually see that it is not.)

The eigenfunctions $\{Y_k\}$ it must be noted are *not* orthogonal. We shall normalize them so that

$$\|Y_k\| = 1.$$

Lemma 4.5.1

$$\sum_1^\infty(-\sigma_k) \leq \alpha \text{ Tr}\, \mathcal{B}\mathcal{B}^* = \alpha \text{ Tr}\, B_u^* B_u. \tag{4.5.16}$$

Proof We follow essentially Gohberg-Krein (1969, p. 101). Let us orthogonalize $\{Y_k\}$ following the Gram–Schmidt procedure. Then

$$Z_k = Y_k - \sum_{j=1}^{k-1} a_{kj} Y_j, \qquad [Z_k Z_j] = 0, \qquad k \neq j.$$

Recall that the $\{Y_k\}$, $k = 1, \ldots, n$, cannot be linearly dependent for any n, since the $\{\lambda_k\}$ are distinct. Hence

$$\|Z_k\| \neq 0.$$

Now

$$(\mathcal{A} - \alpha\mathcal{B}\mathcal{B}^*)Z_k = \lambda_k Y_k - \sum_{j=1}^{k-1} a_{kj}\lambda_j Y_j.$$

Hence

$$[(\mathcal{A} - \alpha\mathcal{B}\mathcal{B}^*)Z_k,\ Z_k] = \lambda_k[Y_k,\ Z_k] = \lambda_k[Z_k,\ Z_k].$$

Therefore

$$\operatorname{Re}\lambda_k = \frac{-\alpha[\mathcal{B}\mathcal{B}^* Z_k,\ Z_k]}{[Z_k,\ Z_k]}.$$

Now $\mathcal{B}\mathcal{B}^*$ is nuclear, since $\mathcal{B}$ is finite dimensional. Because

$$\left\{\frac{Z_k}{\sqrt{[Z_k,\ Z_k]}}\right\}$$

is now an orthonormal sequence, we have

$$\sum_1^\infty \frac{[\mathcal{B}\mathcal{B}^* Z_k,\ Z_k]}{[Z_k, Z_k]} \leq \operatorname{Tr}\mathcal{B}\mathcal{B}^*.$$

We note that

$$\operatorname{Tr}\mathcal{B}\mathcal{B}^* = \operatorname{Tr}\mathcal{B}^*\mathcal{B} = \operatorname{Tr} B_u^* B_u.$$

Hence Equation (4.3.9) follows. We have an obvious corollary:

Corollary If the sequence $\{\lambda_k\}$ is not finite, then

$$|\sigma_k| \to 0 \quad \text{as } k \to \infty,$$
$$|\nu_k| \to \infty \quad \text{as } k \to \infty.$$

Proof The first part of the statement follows from (4.3.16) and the second part from the fact that $\det(I + \alpha\mathcal{M}(\lambda))$ is an analytic function for $\operatorname{Re}\lambda$ negative, so that no subsequence of the sequence $\{\lambda_k\}$ can have a finite limit point. Hence

$$|\lambda_k| \to \infty \quad \text{as } k \to \infty.$$

This will also follow from the fact that the resolvent

$$\mathcal{R}(\lambda,\ \mathcal{A} - \alpha\mathcal{B}\mathcal{B}^*), \quad \lambda \neq \lambda_k$$

is compact, as we shall see presently. Nevertheless, we still have to prove that the sequence $\{\lambda_k\}$ is *not* finite, which we shall do in Section 4.7.

Now

$$S_\alpha(t)Y_k = e^{-|\sigma_k|t}e^{i\mu_k t}Y_k, \qquad t \geq 0,$$

and hence each mode is damped with damping coefficient σ_k. However, the number of modes is not finite, and the damping coefficient eventually goes to zero, so that we cannot guarantee a finite gain margin. In particular, the fact that each mode decays is not enough to prove (4.5.6).

For this purpose we can invoke a general result due to Benchimol (1978). We are assuming that $\mathcal{A} \sim \mathcal{B}$ is controllable and we have seen that the resolvent of $\mathcal{A}$ is compact and that

$$\mathcal{A} + \mathcal{A}^* = 0.$$

Hence by a theorem of Benchimol (1978) it follows that the semigroup $S_\alpha(\cdot)$ is "strongly stable":

$$\|S_\alpha(t)Y\| \to 0 \quad \text{as } t \to \infty.$$

Since

$$Y(t) = S_\alpha(t)Y(0),$$

the theorem is proved. We shall give an independent and self-contained proof in Section 4.7.

Corollary The solution of Equation (4.5.9) is such that the total energy (elastic plus kinetic),

$$\|\sqrt{A}x(t)\|^2 + [M\dot{x}(t),\ \dot{x}(t)],$$

is monotonically nonincreasing as t increases and decays to zero as t increases without bound. Moreover,

$$\mathcal{P}_0(x(t)) \to \mathcal{P}_0(x(0) - \dot{x}(0)) \quad \text{as } t \to \infty. \tag{4.5.17}$$

Proof We have only to note that

$$Y(t) = \begin{vmatrix} \mathcal{P}_1 x(t) \\ \dot{x}(t) \end{vmatrix}$$

and

$$\|Y(t)\|^2 = \|\sqrt{A}x(t)\|^2 + [M\dot{x}(t),\ \dot{x}(t)].$$

Thus the elastic energy decays to zero. The rigid-body component is given by

$$\mathcal{P}_0 x(t) = \int_0^t \mathcal{P}_0 \dot{x}(\sigma)\, d\sigma + \mathcal{P}_0 x(0),$$

where we so far only know that

$$\|\mathcal{P}_0 \dot{x}(t)\| \to 0 \quad \text{as } t \to \infty.$$

If the initial state is such that

$$\begin{vmatrix} \mathcal{P}_1 x(0) \\ \dot{x}(0) \end{vmatrix} = Y_k$$

is an eigenfunction, then

$$Y(t) = e^{\lambda_k t} Y_k$$

so that

$$\mathcal{P}_0 \dot{x}(t) = \lambda_k e^{\lambda_k t} \mathcal{P}_0 \dot{x}(0)$$

and hence

$$\begin{aligned} \mathcal{P}_0(x(t)) &= \mathcal{P}_0(x(0)) + e^{\lambda_k t} \mathcal{P}_0(\dot{x}(0)) - \mathcal{P}_0(\dot{x}(0)) \\ &\to \mathcal{P}_0(x(0)) - \mathcal{P}_0(\dot{x}(0)) \quad \text{as } t \to \infty. \end{aligned}$$

We shall show in Section 4.7 that this holds generally, using the modal expansion.

4.5.2 Resolvent

Let us see how the resolvent of $(\mathcal{A} - \alpha \mathcal{B}\mathcal{B}^*)$ can be expressed in terms of the resolvent of $\mathcal{A}$. We have

$$\mathcal{R}(\lambda, \ \mathcal{A} - \alpha \mathcal{B}\mathcal{B}^*)X = Y$$

or

$$(\lambda I - \mathcal{A} + \alpha \mathcal{B}\mathcal{B}^*)Y = X.$$

For $\lambda \neq \lambda_k$ and $\lambda \neq i\omega_k, 0$, we can multiply on the left by $\mathcal{R}(\lambda, \mathcal{A})$ and obtain

$$Y + \alpha \mathcal{R}(\lambda, \mathcal{A})\, \mathcal{B}\mathcal{B}^* Y = \mathcal{R}(\lambda, \mathcal{A})X.$$

Hence

$$\begin{aligned} \mathcal{B}^* Y + \alpha \mathcal{B}^* \mathcal{R}(\lambda, \mathcal{A})\, \mathcal{B}\mathcal{B}^* Y &= \mathcal{B}^* \mathcal{R}(\lambda, \mathcal{A})X, \\ (I + \alpha \mathcal{M}(\lambda))\, \mathcal{B}^* Y &= \mathcal{B}^* \mathcal{R}(\lambda, \mathcal{A})X, \end{aligned}$$

and since the matrix on the left side is nonsingular,

$$\mathcal{B}^* Y = (I + \alpha \mathcal{M}(\lambda))^{-1} \mathcal{B}^* R(\lambda, A)X. \tag{4.5.18}$$

Hence it follows that

$$\mathcal{R}(\lambda, \ \mathcal{A} - \alpha \mathcal{B}\mathcal{B}^*) = \mathcal{R}(\lambda, \mathcal{A}) - \alpha \mathcal{R}(\lambda, \mathcal{A})\, \mathcal{B}(I + \alpha \mathcal{M}(\lambda))^{-1} \mathcal{B}^* \mathcal{R}(\lambda, \mathcal{A}). \tag{4.5.19}$$

It follows in particular that the resolvent of $(\mathcal{A} - \alpha \mathcal{B}\mathcal{B}^*)$ has all the properties of $\mathcal{R}(\lambda, \mathcal{A})$ such as being compact, Hilbert–Schmidt, etc., since it is a perturbation of the latter by a finite-dimensional operator.

4.6 Asymptotic Modes

In this section we examine in more detail the modes, both open-loop system (the undamped structure) and closed-loop (with rate feedback). In particular, we obtain asymptotic estimates.

As we have seen in Section 4.2, the open-loop mode frequencies are the roots of

$$\det[-\omega^2 M_b + T(i\omega)] = 0, \tag{4.6.1}$$

where $T(\lambda)$ is self-adjoint, and $T(i\omega)$ is a function of ω^2. The mode shape is determined by

$$\mathcal{L}(i\omega_k)\, L(i\omega_k)^{-1} b(i\omega_k), \tag{4.6.2}$$

where

$$\left(-\omega_k^2 M_b + T(i\omega_k)\right) b(i\omega_k) = 0. \tag{4.6.3}$$

Our first step is to show how the dimension of the matrix that determines the eigenvalues can be reduced. In (4.6.1) this dimension is $6m \times 6m$. We shall show that it can be reduced to 6×6, regardless of how large m is.

We shall actually consider the closed-loop system, with the corresponding eigenvalue problem

$$(\mathcal{A} - \alpha \mathcal{B}\mathcal{B}^*)Y = \lambda Y,$$

and proceed in a slightly different way than before. Let

$$Y = \begin{vmatrix} y_1 \\ y_2 \end{vmatrix}.$$

Then we have

$$\lambda y_1 = \mathcal{P}_1 y_2, \qquad \lambda \neq 0,$$
$$\lambda y_2 + M^{-1} A y_2 + \alpha M^{-1} B B^* y_2 = 0.$$

Hence

$$\lambda y_2 + \frac{1}{\lambda} M^{-1} A y_2 + \alpha M^{-1} B B^* y_2 = 0$$

and so we need to find y_2 satisfying

$$\lambda^2 M y_2 + A y_2 + \alpha B B^* y_2 = 0 \tag{4.6.4}$$

and then

$$y_1 = \frac{\mathcal{P}_1 y_2}{\lambda}. \tag{4.6.5}$$

We shall call y_2 the mode shape even though it is not purged of rigid-body modes, as is (4.6.5). The advantage in going to (4.6.4) is that we get the undamped mode frequencies by setting $\alpha = 0$, which we cannot do with (4.5.13).

To proceed with (4.6.4), let

$$y_2 = \begin{vmatrix} f \\ b \end{vmatrix}.$$

We have

$$-A^2 f''(s) + A_1 f'(s) + A_0 f(s) + \lambda^2 M_0 f = 0, \qquad s_i < s < s_{i+1}, \tag{4.6.6}$$

$$\lambda^2 M_b b + \alpha\lambda B_u B_u^* b + A_b f = 0. \tag{4.6.7}$$

As we have seen in Section 4.2,

$$B_u B_u^* b = Db,$$

where D is the "diagonal" in the sense that

$$Db = \begin{vmatrix} D_1 b_1 \\ \vdots \\ D_m b_m \end{vmatrix},$$

where D_i, is a 6×6, self-adjoint, nonnegative definite, diagonal and

$$\begin{vmatrix} f(s) \\ f'(s) \end{vmatrix} = e^{\mathcal{A}(\lambda)(s-s_i)} \begin{vmatrix} f(s_i) \\ f'(s_i+) \end{vmatrix}, \qquad s_i < s < s_{i+1}, \tag{4.6.8}$$

where

$$\mathcal{A}(\lambda) = \begin{vmatrix} 0 & I \\ A_2^{-1}(A_0 + \lambda^2 M_0) & A_2^{-1} A_1 \end{vmatrix}.$$

The boundary conditions (4.2.7) relate $f'(s_i+)$ to $f(s_i)$ and $f'(s_i-)$:

$$f'(0+) = A_2^{-1}(-L_1 + \alpha\lambda D_1 + \lambda^2 M_{b,0}) f(0),$$

$$f'(s_i+) = f'(s_i-) + A_2^{-1}(\alpha\lambda D_i + \lambda^2 M_{b,i}) f(s_i),$$

$$2 \le i \le m-1,$$

and for $i = m$:

$$f'(L) = -A_2^{-1}(L_1 + \lambda\alpha D_m + \lambda^2 M_{b,L}) f(L). \tag{4.6.9}$$

Hence it follows that we can calculate $f(L)$ and $f'(L)$ in terms of $f(0)$ and then invoke (4.6.9) to obtain

$$h(\lambda, \alpha) f(0) = 0,$$

where

$$h(\lambda;\alpha) = \begin{vmatrix} A_2^{-1}(L_1 + \alpha\lambda D_m + \lambda^2 M_{b,L}) & I \end{vmatrix}$$

$$\cdot\, e^{\mathcal{A}(\lambda)(L-s_{m-1})} \begin{vmatrix} I & 0 \\ A_2^{-1}(\alpha\lambda D_{m-1} + \lambda^2 M_{b,m-1}) & I \end{vmatrix}$$

$$\cdots\, e^{\mathcal{A}(\lambda)(s_{i+1}-s_i)} \begin{vmatrix} I & 0 \\ A_2^{-1}(\alpha\lambda D_i + \lambda^2 M_{b,i}) & I \end{vmatrix}$$

$$\cdots\, e^{\mathcal{A}(\lambda)s_2} \begin{vmatrix} I \\ A_2^{-1}(-L_1 + \alpha\lambda D_1 + \lambda^2 M_{b,0}) \end{vmatrix} \tag{4.6.10}$$

and where I is the 6×6 identity matrix.

Let us use the notation

$$\Delta_i = s_{i+1} - s_i, \quad i = 1, \ldots, m-1,$$

so that

$$\Delta_1 = s_2 - s_1 = s_2, \qquad \Delta_{m-1} = L - s_{m-1}.$$

Let

$$t_i(\lambda,\ \alpha) = A_2^{-1}(\lambda^2 M_{b,i} + \alpha\lambda D_i), \qquad 1 \le i \le m, \tag{4.6.11}$$

$$T_i(\lambda,\ \alpha) = \begin{vmatrix} I & 0 \\ t_i(\lambda,\ \alpha) & I \end{vmatrix}, \qquad i = 2, 3, \ldots, m-1, \tag{4.6.12}$$

$$T_1(\lambda;\alpha) = \begin{vmatrix} A_2^{-1}L_1 + t_m(\lambda,\ \alpha) & I \end{vmatrix},$$

$$T_m(\lambda;\ \alpha) = \begin{vmatrix} I \\ -A_2^{-1}L_1 + t_1(\lambda,\ \alpha) \end{vmatrix}.$$

Then we can write $h(\lambda,\ \alpha)$ as

$$h(\lambda,\ \alpha) = T_m(\lambda;\ \alpha)\, Q(\lambda,\ \alpha) T_1(\lambda;\ \alpha), \tag{4.6.13}$$

where

$$Q(\lambda,\ \alpha) = e^{\mathcal{A}(\lambda)\Delta_{m-1}} T_{m-1}(\lambda;\ \alpha)\ \cdots\ e^{\mathcal{A}(\lambda)\Delta_2} T_2(\lambda,\ \alpha)\ \cdot\ e^{\mathcal{A}(\lambda)\Delta_1}.$$

We note that the $t_i(\lambda;\ \alpha)$ as well as $T_i(\lambda;\ \alpha)$ are polynomials in λ whereas

$$e^{\mathcal{A}(\lambda)\Delta_i}$$

involve transcendental functions of λ.

The coefficients of the terms of the highest degree in λ as well as α are contained in the term

$$A_2^{-1}(\alpha\lambda D_m + \lambda^2 M_{b,L}) P_{12}(\lambda;\, \Delta_{m-1}) \cdot A_2^{-1}(\alpha\lambda D_{m-1} + \lambda^2 M_{b,m-1}) \cdots \cdot P_{12}(\lambda;\, \Delta_2)\, A_2^{-1}(\alpha\lambda D_2 + \lambda^2 M_{b,2}) \cdot P_{12}(\lambda;\, \Delta_1)\, A_2^{-1}(\alpha\lambda D_1 + \lambda^2 M_{b,0}). \tag{4.6.14}$$

In particular, the term containing highest powers of λ that occurs is

$$\lambda^{2m} A_2^{-1} M_{b,L} P_{12}(\lambda;\, \Delta_{m-1})\, A_2^{-1} M_{b,m-1} \cdots P_{12}(\lambda;\, \Delta_1)\, A_2^{-1} M_{b,0}. \tag{4.6.15}$$

The term containing the highest powers of α that occurs is

$$\alpha^m \left(\lambda^m A_2^{-1} D_m P_{12}(\lambda;\, \Delta_{m-1}) A_2^{-1} D_{m-1} \cdots P_{12}(\lambda;\, \Delta_1) A_2^{-1} D_1\right). \tag{4.6.16}$$

From (4.13) we have that

$$\|h(\lambda;\alpha)\| \le \|T_1(\lambda;\alpha)\| \cdots \|T_m(\lambda;\alpha)\| \cdot \left\|e^{\mathcal{A}(\lambda)\Delta_1}\right\| \cdots \left\|e^{\mathcal{A}(\lambda)\Delta_{m-1}}\right\|, \tag{4.6.17}$$

where $\|\cdot\|$ denotes the matrix norm.

The corresponding mode shape function $f(\cdot)$ (corresponding to y_2) is given by

$$f(s) = |I \quad 0| e^{\mathcal{A}(\lambda)(s-s_i)} T_i(\lambda;\alpha) \cdot e^{\mathcal{A}(\lambda)\Delta_{i-1}} T_{i-1}(\lambda;\alpha) \cdots e^{\mathcal{A}\lambda\Delta_i} \begin{vmatrix} I \\ A_2^{-1}(-L_1 + \lambda^2 M_{b,0} + \lambda\alpha D) \end{vmatrix} f(0), \quad s_i \le s \le s_{i+1}. \tag{4.6.18}$$

Let

$$d(\lambda;\alpha) = \det h(\lambda;\alpha).$$

Then $d(\lambda;\alpha)$ is an entire function of λ and the eigenvalues $\{\lambda_k\}$ of $(\mathcal{A} - \alpha\mathcal{B}\mathcal{B}^*)$ are the nonzero roots of

$$d(\lambda;\alpha) = 0.$$

Behavior at $\lambda = 0$ Let us first consider

$$\lambda = 0.$$

We have

$$h(0,\alpha) = h(0,0) = \left|A_2^{-1} L_1 \quad I\right| e^{\mathcal{A}(0)L} \begin{vmatrix} I \\ A_2^{-1} L_1 \end{vmatrix} = 0.$$

Moreover, we have the power series expansion about zero:

$$h(\lambda; \alpha) = \lambda h'(0, \alpha) + \frac{\lambda^2}{2} h''(0, \alpha) + \text{terms of higher order in } \lambda$$

and correspondingly

$$d(\lambda; \alpha) = \lambda^6 \alpha^6 d_6(\alpha) + \text{terms of higher order in } \lambda, \qquad (4.6.19)$$

where

$$d_6(0) \neq 0.$$

Thus $d(\lambda; \alpha)$ for nonzero α has a zero of order 6 at $\lambda = 0$, whereas $d(\lambda, 0)$ has a zero of order 12 at $\lambda = 0$.

Relation of $d(\lambda; \alpha)$ to $D(\lambda; \alpha)$ Let us examine next the relation of $d(\lambda; \alpha)$ to $D(\lambda, \alpha)$, where the latter is defined in (4.5.13). Now

$$\begin{aligned} \mathcal{M}(\lambda) &= \mathcal{B}^* \mathcal{R}(\lambda, \mathcal{A}) \mathcal{B} \\ &= \sum_1^\infty \frac{\mathcal{B}^* P_k \mathcal{B}}{\lambda - i\omega_k} + \sum_1^\infty \frac{\mathcal{B}^* P_{-k} \mathcal{B}}{\lambda + i\omega_k} + \frac{\mathcal{B}^* P_0 \mathcal{B}}{\lambda} \end{aligned} \qquad (4.6.20)$$

with P_k as in Section 4.3 (see also (4.4.16)). Hence as λ goes to zero

$$\begin{aligned} D(\lambda, \alpha) &\sim \det\left(I + \frac{\alpha \mathcal{B}^* P_0 \mathcal{B}}{\lambda} \right) \\ &\sim \left(\frac{\alpha}{\lambda}\right)^{m_c} \det\left[\frac{\lambda}{\alpha} + \mathcal{B}^* P_0 \mathcal{B}\right]. \end{aligned}$$

Now the range space of $\mathcal{B}^* P_0 \mathcal{B}$ is of dimension 6 and hence the null space is of dimension $m_c - 6$ (which is nonnegative by virtue of the controllability assumption!). Hence

$$\det\left[\frac{\lambda}{\alpha} + \mathcal{B}^* P_0 \mathcal{B}\right] = \left(\frac{\lambda}{\alpha}\right)^{m_c - 6} \cdot \text{(nonzero constant)}.$$

It follows that

$$D(\lambda; \alpha) \sim \left(\frac{\alpha}{\lambda}\right)^6 \cdot \text{(nonzero constant)}, \quad \text{as } \lambda \to 0.$$

In a similar way we see from (4.6.20) that the nonzero poles of $D(\lambda, \alpha)$ are the zeros of $d(\lambda, 0)$ to the same order. The dimension of the eigenfunction space of $\mathcal{A}$ for $\lambda \neq 0$ may be taken to be unity, since we see from (4.6.18) that the dimension is equal to the dimension of the eigenvector space of $h(\lambda, 0)$ corresponding to the zero eigenvalue, and if $h(\lambda, 0)$ has distinct eigenvalues, the dimension is equal to one. We can make a similar statement for the eigenfunction spaces corresponding to the eigenvalues with nonzero imaginary parts of $\mathcal{A} - \alpha\mathcal{B}\mathcal{B}^*$. Hence it follows that

$$d(\lambda; 0)\, D(\lambda; \alpha)$$

is an entire function with zeros coinciding with that of $d(\lambda; \alpha)$. Thus we know that we must have

$$d(\lambda; \alpha) = e^{q(\lambda)} d(\lambda; 0)\, D(\lambda; \alpha), \tag{4.6.21}$$

where $q(\lambda)$ is an entire function.

Order of $d(\lambda; \alpha)$ Next we shall show that the order of the entire function $d(\lambda; \alpha)$ is less than or equal to one. Let

$$m(r; \alpha) = \max_{|\lambda|=r} |d(\lambda, \alpha)|.$$

Now, since the determinant of a matrix is the product of the eigenvalues, we have

$$\begin{aligned} |d(\lambda; \alpha)| &\le (\text{spectral radius of } h(\lambda; \alpha))^6 \\ &\le \|h(\lambda; \alpha)\|^6. \end{aligned}$$

From (4.6.17) we see that

$$\log\|h(\lambda; \alpha)\| \le \sum_{i=1}^{m} \log\|T_i(\lambda; \alpha)\| + \sum_{i=1}^{m-1} \log\left\|e^{\mathcal{A}(\lambda)\Delta_i}\right\|.$$

Since the $T_i(\lambda; \alpha)$ are polynomials in λ, we need only to consider the order of

$$\|e^{\mathcal{A}(\lambda)\Delta}\|.$$

For this purpose, we proceed to evaluate the eigenvalues of $\mathcal{A}(\lambda)$.

An eigenvector of $\mathcal{A}(\lambda)$, corresponding to the eigenvalue $\gamma(\lambda)$, must be of the form

$$\begin{vmatrix} y(\lambda) \\ \gamma(\lambda) y(\lambda) \end{vmatrix},$$

where

$$\gamma(\lambda)^2 y(\lambda) = A_2^{-1}(\lambda^2 M_0 + A_0) y(\lambda) + \gamma(\lambda) A_2^{-1} A_1 y(\lambda). \tag{4.6.22}$$

Hence

$$\gamma(\lambda)^2 A_2\, y(\lambda) = \lambda^2 M_0 y(\lambda) + A_0 y(\lambda) + \gamma(\lambda) A_1 y(\lambda). \tag{4.6.23}$$

Let

$$a_2(\lambda) = [A_2 y(\lambda), y(\lambda)], \qquad a_1(\lambda) = [A_1 y(\lambda), y(\lambda)],$$
$$m(\lambda) = \lambda^2 m_0(\lambda) + [A_0 y(\lambda), y(\lambda)],$$
$$m_0(\lambda) = [M_0 y(\lambda), y(\lambda)].$$

Then solving the quadratic equation

$$\gamma(\lambda)^2 a_2(\lambda) = m(\lambda) + \gamma(\lambda) a_1(\lambda)$$

we have

$$\gamma(\lambda) = \frac{1}{2a_2(\lambda)}\left(+a_1(\lambda) \pm \sqrt{a_1(\lambda)^2 + 4m(\lambda)a_2(\lambda)}\right). \qquad (4.6.24)$$

From (4.6.22) we have

$$\left(\frac{\gamma(\lambda)}{\lambda}\right)^2 y(\lambda) = \left(A_2^{-1} M_0 + \frac{1}{\lambda^2} A_0 + \left(\frac{\gamma(\lambda)}{\lambda}\right)\frac{1}{\lambda} A_1\right) y(\lambda).$$

Since

$$a_2(\lambda) \geq (\text{smallest eigenvalue of } A_2)\|y(\lambda)\|^2$$

and

$$|a_1(\lambda)| \quad \text{is bounded,}$$

it follows from (4.6.24) that

$$\left\|\frac{\gamma(\lambda)}{\lambda}\right\| \quad \text{is bounded.}$$

Therefore

$$\left\|\left(A_2^{-1} M_0 + \frac{1}{\lambda^2} A_0 + \frac{\gamma(\lambda)}{\lambda}\frac{1}{\lambda} A_1\right) - A_2^{-1} M_0\right\| = O\left(\frac{1}{|\lambda|}\right) \quad \text{as } \lambda \to \infty.$$

Thus if we "normalize" the eigenvectors so that

$$\|y(\lambda)\| = 1,$$

then every sequence $\{y(\lambda_n)\}$ has a subsequence that converges to one of the eigenvectors of

$$A_2^{-1} M_0$$

as $|\lambda_n| \to \infty$. We assume (for simplicity) that the eigenvalues of this matrix are distinct, and they are of course strictly positive. Let μ_i, $i = 1, \ldots, 6$, denote these eigenvalues and $\mathbf{e}_i$ the corresponding eigenvector of unit norm. Then for $|\lambda|$ large enough we can arrange so that the eigenvalues of $\mathcal{A}(\lambda)$ are

$$\gamma_i^+(\lambda), \gamma_i^-(\lambda), \qquad i = 1, \ldots, 6,$$

$$\left| \left(\frac{\gamma_i^{\pm}(\lambda)}{\lambda} \right)^2 - \mu_i \right| \to 0 \quad \text{as } |\lambda| \to \infty$$

with corresponding eigenvectors

$$\left| \begin{matrix} y_i(\lambda) \\ \gamma_i^+(\lambda) y_i(\lambda) \end{matrix} \right|, \qquad \left| \begin{matrix} y_i(\lambda) \\ \gamma_i^-(\lambda) y_i(\lambda) \end{matrix} \right|,$$

$$\|y_i(\lambda)\| = 1, \qquad i = 1, \ldots, 6, \qquad y_i(\lambda) \to \mathbf{e}_i.$$

In particular

$$[A_1 y_i(\lambda),\ y_i(\lambda)] \to [A_1 \mathbf{e}_i,\ \mathbf{e}_i].$$

But the $\mathbf{e}_i$ are real valued and

$$A_1 = -L_1 + L_1^*,$$

so we have that

$$[A_1 y_i(\lambda),\ y_i(\lambda)] \to 0.$$

As a result, in (4.6.24)

$$\gamma^{\pm}(\lambda) - \frac{\pm 1}{2a_2(\lambda)} \sqrt{a_1(\lambda)^2 + 4m(\lambda)a_2(\lambda)} \to 0 \quad \text{as } |\lambda| \to \infty$$

and hence

$$\gamma_i^{\pm}(\lambda) - \pm\lambda\sqrt{\mu_i} \to 0 \quad \text{as } |\lambda| \to \infty. \tag{4.6.25}$$

Thus we have

$$\left| \frac{e^{\gamma_i^{\pm}(\lambda)\Delta}}{e^{\pm\lambda\sqrt{\mu_i}\Delta}} \right| \to 1 \quad \text{as } |\lambda| \to \infty.$$

Hence

$$\text{spectral radius of } e^{\mathcal{A}(\lambda)\Delta} = \max_k \left| e^{\pm\lambda\sqrt{\mu_k}\Delta} \right|,$$

for $|\lambda|$ sufficiently large. Again, since $y_i(\lambda) \to \mathbf{e}_i$ we can find a constant M such that

$$\left\| e^{\mathcal{A}(\lambda)\Delta} \right\| \le M \max_k \left| e^{\pm\lambda\sqrt{\mu_k}\Delta} \right|, \qquad 0 < \Delta, \tag{4.6.26}$$

for all $|\lambda|$ sufficiently large. Hence it follows that

$$\max_{|\lambda|=r} \left\| e^{\mathcal{A}(\lambda)\Delta} \right\| \le e^{r\Delta \max\sqrt{\mu_k}}$$

and thus

$$\max_{|\lambda|=r} \sum_{i=1}^{m-1} \log \|e^{\mathcal{A}(\lambda)\Delta_i}\| \sim r\left(\max \sqrt{\mu_k}\right) L.$$

Therefore we have that $d(\lambda; \alpha)$ is of order less than or equal to one (or is of "exponential type" in the terminology of Levin (1980)).

Since

$$|D(\lambda, \alpha)| \to 1 \quad \text{as } |\lambda| \to \infty,$$

we note that

$$d(\lambda, 0)\, D(\lambda, \alpha)$$

is also of exponential type and hence (Levin, 1980, p. 24) we can sharpen (4.6.21) to

$$d(\lambda, \alpha) = e^{p_1(\lambda)} d(\lambda, 0) D(\lambda, \alpha), \tag{4.6.27}$$

where $p_1(\lambda)$ is a polynomial of degree one at most.

4.6.1 Deadbeat Modes

An eigenvalue that is real is often referred to as a "deadbeat" mode. They occur in closed loops only if there are rigid-body modes (zero eigenvalues in the open loop). In fact we have:

Theorem 4.6.1 For each $\alpha > 0$, the number of deadbeat modes is equal to the number of (linearly independent) rigid-body modes (= dimension of the null space of A).

Proof Let Π_k denote the M-orthogonal projection operator projecting $\mathcal{H}$ into the eigenfunction space corresponding to the eigenvalue ω_k^2 of A. Then we have

$$H(\lambda, \alpha) = I + \frac{\alpha B^* \Pi_0 B}{\lambda} + \alpha \sum_1^{\infty} \frac{\lambda B^* \Pi_k B}{\lambda^2 + \omega_k^2}. \tag{4.6.28}$$

We are only interested in $\lambda < 0$. Hence we can write

$$|\lambda| H(\lambda, \alpha) = |\lambda| - \alpha B^* \Pi_0 B - \alpha \sum_1^{\infty} \frac{\lambda^2}{\lambda^2 + \omega_k^2} B^* \Pi_k B, \tag{4.6.29}$$

where we note

$$\sum_1^{\infty} \frac{\lambda^2}{\lambda^2 + \omega_k^2}\, B^* \Pi_k B \;\to 0 \quad \text{as } |\lambda| \to 0.$$

Hence

$$|\lambda| H(\lambda, \alpha) \to -\alpha B^* \Pi_0 B \quad \text{as } |\lambda| \to 0.$$

Let

$$\gamma_k(\lambda, \alpha), \qquad k = 1, \ldots, m_c,$$

denote the eigenvalues of $H(\lambda, \alpha)$. Then

$$|\lambda| \gamma_k(\lambda, \alpha) \to (-\alpha) \cdot \text{eigenvalues of } B^* \Pi_0 B, \quad \text{as } |\lambda| \to 0.$$

Now $B^* \Pi_0 B$ has exactly six nonzero eigenvalues. Because B is one-to-one, the range space of $B^* \Pi_0 B$ is the same,

$$\{B^* \phi\}, \qquad \phi \, \varepsilon \text{ null space of } A,$$

and the latter, as we have seen, has dimension 6, by controllability. Hence

$$|\lambda| \gamma_k(\lambda, \alpha) \to -|\gamma_k|, \qquad k = 1, \ldots, 6, \qquad \text{as } |\lambda| \to 0.$$

Therefore

$$\gamma_k(\lambda, \alpha) \to -\infty, \qquad k = 1, \ldots, 6, \qquad \text{as } |\lambda| \to 0.$$

However,

$$H(\lambda, \alpha) \to I \quad \text{as } |\lambda| \to \infty$$

and hence

$$\gamma_k(\lambda, \alpha) > 0 \quad \text{for } |\lambda| > \lambda_0.$$

Thus it follows that

$$\gamma_k(\lambda, \alpha) = 0 \qquad \text{for some } \lambda, \qquad -|\lambda_0| < \lambda < 0.$$

Hence it follows that

$$D(\lambda_k, \alpha) = 0, \qquad \lambda_k < 0, \quad k = 1, \ldots, 6,$$

and we have exactly six deadbeat modes.

Remark For $\alpha = 0$, the eigenvalues are the zeros of $d(\lambda, 0)$. We want to consider now the limiting case $\alpha = \infty$. For this purpose we consider

$$\left(\frac{I}{\alpha} + \mathcal{M}(\lambda)\right)$$

and note that the zeros of

$$\det\left(\frac{I}{\alpha} + \mathcal{M}(\lambda)\right) \tag{4.6.30}$$

are the same as those of $D(\lambda, \alpha)$. However, the form (4.6.30) allows us to consider the case for large α, or α = infinity. The matrix

$$\frac{I}{\alpha} + \mathcal{M}(\lambda) \to \mathcal{M}(\lambda) \quad \text{as } \alpha \to \infty \tag{4.6.31}$$

for each $\lambda \neq i\omega_k$, and hence (4.6.30) implies

$$m(\lambda) = \det \mathcal{M}(\lambda).$$

Therefore we define the eigenvalues corresponding to $\alpha = +\infty$ as the roots of

$$m(\lambda) = 0. \tag{4.6.32}$$

We shall show now that the roots are purely imaginary. Suppose

$$\mathcal{M}(\lambda)u = 0.$$

Then

$$[\mathcal{R}(\lambda)\mathcal{B}u,\ \mathcal{B}u] = 0.$$

Let

$$\mathcal{R}(\lambda)\mathcal{B}u = Y.$$

Then

$$[\mathcal{R}(\lambda)\mathcal{B}u,\ \mathcal{B}u] = [Y,\ \lambda I - \mathcal{A}Y] = \bar{\lambda}[Y, Y] - [Y, \mathcal{A}Y].$$

Hence

$$\text{Re}[\mathcal{A}(\lambda)\mathcal{B}u,\ \mathcal{B}u] = \text{Re}\big(\bar{\lambda}\big)\,[Y, Y]$$

($\mathcal{B}^*\mathcal{R}(\lambda)\mathcal{B}$ is a "positive-real" matrix) and hence

$$\text{Re}(\bar{\lambda}) = 0$$

or λ is pure imaginary.

Let us examine the eigenvalue further. We have

$$\lambda B_u^*(\lambda^2 M_b + T(\lambda))^{-1} B_u u = 0.$$

We assume that no zero of $m(\lambda)$ is a zero of $L(\lambda)$. In that case we can write

$$\lambda B_u^* L(\lambda)(\lambda^2 M_b L(\lambda) + K(\lambda))^{-1} B_u u = 0.$$

Since

$$T(\lambda)^* = T(\lambda),$$

we have

$$\begin{aligned}(L(\lambda)(\lambda^2 M_b L(\lambda) + K(\lambda))^{-1})^* &= (\lambda^2 M_b L(\lambda)^* + K(\lambda)^*)^{-1} L(\lambda)^* \\ &= L(\lambda)(\lambda^2 M_b L(\lambda) + K(\lambda))^{-1}.\end{aligned}$$

Hence

$$\lambda B_u^*(\lambda^2 M_b L(\lambda)^* + K(\lambda)^*)^{-1} L(\lambda)^* B_u u = 0.$$

In the special case where

$$m_c = 6m, \qquad B_u = \text{identity},$$

we have that the eigenvalues corresponding to $\alpha = \infty$ are the "clamped" modes. This is not true in general, as in fact the example in Section 4.8 shows.

The mode "shape" associated with these eigenvalues is given by

$$\phi = \left| \frac{\mathcal{L}(\lambda)\mathfrak{a}(\lambda)}{L(\lambda)\mathfrak{a}(\lambda)} \right|,$$

where

$$\mathfrak{a}(\lambda) = (\lambda^2 M_b + K(\lambda))^{-1} B_u u.$$

Then letting

$$f = \mathcal{L}(\lambda)\mathfrak{a}(\lambda)$$

we see that

$$\lambda^2 M_0 f + g = 0, \tag{4.6.33}$$

where g is defined by (4.2.6), and

$$B_u^* L(\lambda)\mathfrak{a}(\lambda) = 0. \tag{4.6.34}$$

Thus these are modes in which the control nodes are clamped. We shall characterize these modes more precisely below.

4.6.2 Root Locus

From Theorem 4.6.1, it follows that oscillatory modes of the undamped structure remain oscillatory for *all* values of α, however large. The behavior of the set of eigenvalues is such that as α increases from zero they migrate from the imaginary axis to the left half-plane and then back to the imaginary axis. It is possible to define the eigenvalues each as a function of α and show that the real part decreases first and then at a critical value of α starts to increase as α increases, going to zero as α increases to infinity. We can also show that the critical value increases as the mode number increases. The loci describe differential arcs in the complex plane. Thus let

$$\lambda_k(0) = i\omega_k,$$

where $i\omega_k$ is a zero of $d(\lambda, 0)$. Then we define $\lambda_k(\alpha)$ using the derivatives at $\alpha = 0$. Thus

$$\left.\frac{d\lambda_k}{d\alpha}\right|_{\alpha=0} = \frac{-d_\alpha(i\omega_k, 0)}{d_\lambda(i\omega_k, 0)}, \tag{4.6.35}$$

where the subscripts denote partial derivatives, and we calculate similarly higher order derivatives using the identity

$$d(\lambda_k(\alpha), \alpha) = 0.$$

We can show that it is real and negative, and in particular leading to an approximation for σ_k, the real part, via the Newton formula:

$$\sigma_k \sim -\alpha \frac{d_\alpha(i\omega_k, 0)}{d_\lambda(i\omega_k, 0)} \quad \text{for small } \alpha. \tag{4.6.36}$$

Owing to space limitations we must stop here and refer to the example in Section 4.8 for more details.

4.6.3 Asymptotic Modes

The modes are the roots of the equation

$$d(\lambda; \alpha) = 0.$$

Our interest is not in evaluating the roots – which in a given case will be a problem in numerical analysis – but rather in their asymptotic behavior as the mode number increases without bound.

Let $\{\lambda_k\}$ denote a sequence of modes, where we note that

$$|\lambda_k| \to \infty \quad \text{as } k \to \infty.$$

We shall say that the sequence $\{\tilde{\lambda}_k\}$ is asymptotically equivalent if the sequence

$$|\tilde{\lambda}_k - \lambda_k|$$

is bounded. In our case we shall show that it actually goes to zero. We call $\{\tilde{\lambda}_k\}$ "asymptotic" modes. Note that the "percent error"

$$\frac{|\tilde{\lambda}_k - \lambda_k|}{\lambda_k} \to 0 \quad \text{as } k \to \infty.$$

Since all mode determination is approximate only, this is clearly the best we can do.

The zeros of $d(\lambda, \alpha)$ for each $\alpha \geq 0$ are confined to the strip

$$-|\sigma| \leq \mathrm{Re}\lambda \leq 0, \qquad \sigma = \sup|\sigma_i| < \infty \tag{4.6.37}$$

and from (4.6.26) we have the important result that in this strip

$$\left\| e^{A(\lambda)\Delta_i} \right\|, \qquad i = 1, \ldots, m-1,$$

is bounded. Now we can express $h(\lambda; \alpha)$ as

$$h(\lambda, \alpha) = \sum_0^{2m} \lambda^k h_k(\lambda; \alpha), \tag{4.6.38}$$

where the coefficient matrices $h_k(\cdot, \cdot)$ are bounded in the strip (4.6.37). From Eq. (4.6.14) the coefficient of λ^{2m} is given by

$$h_{2m}(\lambda; \alpha) = A_2^{-1} M_{b,L} P_{12}(\lambda, \Delta_{m-1}) \cdots P_{12}(\lambda, \Delta_1) A_2^{-1} M_{b,0} \tag{4.6.39}$$

and does not depend on α. The zeros of $d(\lambda, \alpha)$ are those of

$$\det\left(\frac{h(\lambda, \alpha)}{\lambda^{2m}}\right)$$

and hence are "asymptotically" those of

$$\det h_{2m}(\lambda; \alpha)$$

as $|\lambda| \to \infty$. Since A_2 and $M_{b,i}$ are nonsingular we have that

$$\det h_{2m}(\lambda; \alpha) = \prod_1^{m-1} \det P_{12}(\lambda; \Delta_i). \tag{4.6.40}$$

Next we shall show that:

Lemma 4.6.1

$$\det P_{12}(\lambda; \Delta) = \prod_{k=1}^{6} \left(\frac{\sinh \Delta\gamma_k(\lambda)}{\Delta\gamma_k(\lambda)}\right), \tag{4.6.41}$$

where

$$|\gamma_k(\lambda) - \lambda\sqrt{\mu_k}| \to 0, \qquad |\lambda| \to \infty.$$

Proof Using

$$\begin{vmatrix} y_k(\lambda) \\ \gamma_k(\lambda) y_k(\lambda) \end{vmatrix} - \begin{vmatrix} y_k(\lambda) \\ -\gamma_k(\lambda) y_k(\lambda) \end{vmatrix} = \begin{vmatrix} 0 \\ 2\gamma_k(\lambda) y_k(\lambda) \end{vmatrix}$$

we have

$$\begin{aligned} P_{12}(\lambda; \Delta) y_k(\lambda) &= |I \quad 0| e^{A(\lambda)\Delta} \begin{vmatrix} 0 \\ y_k(\lambda) \end{vmatrix} \\ &= \frac{\sinh \gamma_k(\lambda)\Delta}{\gamma_k(\lambda)\Delta} y_k(\lambda). \end{aligned}$$

Hence

$$\det P_{12}(\lambda;\,\Delta) = \prod_{k=1}^{6} \frac{\sinh \gamma_k(\lambda)\Delta}{\gamma_k(\lambda)\Delta},$$

where by (4.6.25),

$$\left|\gamma_k(\lambda) - \lambda\sqrt{\mu_k}\right| \to 0, \qquad \text{as } |\lambda| \to \infty$$

as required.

Next, let

$$\mathcal{A}_\infty(\lambda) = \begin{vmatrix} 0 & I \\ \lambda^2 A_2^{-1} M_0 & 0 \end{vmatrix}$$

and let

$$A_s = \sqrt{A_2^{-1} M_0},$$

where the eigenvalues of A_s are

$$\sqrt{\mu_k}, \qquad k = 1, \ldots, 6.$$

Then

$$e^{\mathcal{A}_\infty(\lambda)\Delta} = \begin{vmatrix} \cosh(\lambda A_s \Delta) & (\lambda A_s \Delta)^{-1} \sinh(\lambda A_s \Delta) \\ (\lambda A_s \Delta) \sinh(\lambda A_s \Delta) & \cosh(\lambda A_s \Delta) \end{vmatrix}. \tag{4.6.42}$$

Now

$$\| P_{12}(\lambda;\,\Delta) \|$$

is bounded in the strip (4.6.37) and thus it follows that for every k

$$\| (P_{12}(\lambda;\,\Delta) - (\lambda A_s \Delta)^{-1} \sinh(\lambda A_s \Delta)) \mathbf{e}_k \| \;\to\; 0$$

and hence

$$\| P_{12}(\lambda;\,\Delta) - (\lambda A_s \Delta)^{-1} \sinh(\lambda A_s \Delta) \| \to 0$$

as $|\lambda| \to \infty$ in the strip (4.6.37).

The zeros of

$$\frac{\sinh \gamma_k(\lambda)\Delta}{\gamma_k(\lambda)\Delta}$$

are given by

$$\gamma_k(\lambda_n)\Delta = in\pi.$$

Let

$$\lambda_n' \sqrt{\mu_k}\Delta = in\pi$$

be one sequence of zeros of

$$\det((\lambda A_s \Delta)^{-1} \sinh(\lambda A_s \Delta))$$

corresponding to the eigenvalue

$$\lambda_n' \sqrt{\mu_k} \quad \text{and eigenvector } \mathbf{e}_k.$$

Then

$$\|(\gamma_k(\lambda_n) - \lambda_n \sqrt{\mu_k})\Delta| = \left|(\lambda_n - \lambda_n')\sqrt{\mu_k}\Delta\right| \to 0, \qquad \text{as } n \to \infty$$

and hence

$$\left|\lambda_n - \lambda_n'\right| \to 0, \qquad \text{as } n \to \infty.$$

We say in this case that $\{\lambda_n\}$ is asymptotically equivalent to $\{\lambda_n'\}$ or that the asymptotic zeros of $\det P_{12}(\lambda; \Delta)$ are given by

$$\lambda_{n,k} = \frac{\pm in\pi}{\sqrt{\mu_k}\Delta}, \qquad k = 1, \ldots, 6. \tag{4.6.43}$$

Thus the asymptotic zeros of (4.6.40) are given by

$$\lambda_{n,k,j} = \frac{\pm in\pi}{\sqrt{\mu_k}\Delta_j}, \qquad k = 1, \ldots, 6, \qquad j = 1, \ldots, m-1. \tag{4.6.44}$$

Let us turn now to the asymptotic zeros of $d(\lambda; \alpha)$. Let

$$Q_{21}(\lambda; \Delta) = \lambda P_{21}(\lambda; \Delta), \qquad Q_{12}(\lambda; \Delta) = \lambda P_{12}(\lambda; \Delta),$$

and

$$h(\lambda; \alpha) = \lambda^m \left(q_m(\lambda; \alpha) + \frac{r(\lambda; \alpha)}{\lambda} \right), \tag{4.6.45}$$

where, using (4.6.14),

$$q_m(\lambda; \alpha) = A_2^{-1} \left(M_{b,L} + \frac{\alpha}{\lambda} D_m \right) Q_{12}(\lambda; \Delta_{m-1})$$
$$\cdots Q_{12}(\lambda; \Delta_1) A_2^{-1} \left(M_{b,0} + \frac{\alpha}{\lambda} D_1 \right) \tag{4.6.46}$$

and

$$\|r(\lambda; \alpha)\|$$

is bounded in the strip (4.6.37). Hence we can write

$$d(\lambda; \alpha) = \det q_m(\lambda; \alpha) + \sum_1^6 \frac{1}{\lambda^k} d_k(\lambda; \alpha),$$

where

$$d_k(\lambda;\alpha) \quad \text{are bounded in the strip.} \tag{4.6.47}$$

For large $|\lambda|$ we can use the approximation (4.6.42) and hence obtain

$$\det q_m(\lambda;\alpha) = \prod_{k=1}^{m} \det\left[A_2^{-1}\left(M_{b,L} + \frac{\alpha}{\lambda} D_k\right)\right] \cdot \prod_{k=1}^{m-1} \det[(A_s\Delta_k)^{-1} \sinh \lambda A_s \Delta_k],$$

which for all $|\lambda|$ sufficiently large can be expressed as

$$(\text{constant}) \prod_{k=1}^{m-1}\left(\prod_{i=1}^{6} \frac{\sinh \gamma_i \lambda \Delta_k}{\sqrt{\mu_i}\Delta_k}\right) + \frac{1}{\lambda}\left(d_1(\lambda;\alpha) + \sum_{2}^{6} \frac{1}{\lambda_k} d_k(\lambda;\alpha)\right). \tag{4.6.48}$$

Given $\varepsilon > 0$, we can make $|\lambda|$ large enough so that the second term is less than ε. Thus taking

$$\gamma_k(\lambda)\Delta_k = in\pi + \theta,$$

where

$$|\sinh \theta| < \varepsilon,$$

we get, approximately,

$$\lambda = \frac{in\pi + \theta}{\Delta_k \sqrt{\mu_k}}.$$

We see that there is a value of θ such that (4.6.48) is zero. Since we know that the real part of the eigenvalue must be negative, we have that

$$\operatorname{Re}\theta < 0.$$

We see however that the zeros are again asymptotically the same as that given by (4.6.35). In summary, asymptotically the zeros of $d(\lambda, \alpha)$ for any α are given by the zeros of

$$\det\left(\prod_{k=1}^{m-1} (\Delta_i A_s)^{-1} \sinh \lambda A_s \Delta_i\right).$$

4.6.4 Mode Shape

The (unpurged!) asymptotic mode shapes are determined by the eigenvector y:

$$h_{2m}(\lambda;\alpha)y = 0,$$

where y can be determined as follows. Fix j, and let

$$\lambda_n = \frac{in\pi}{\Delta_{m-1}\sqrt{\mu_j}}.$$

Assuming for simplicity that the Δ_i are distinct, we see that $P_{12}(\lambda_n; \Delta_i)$ are nonsingular for i not equal to $m-1$, and hence we can take

$$y = \left(A_2^{-1} M_{b,m-2} P_{12}(\lambda_n; \Delta_{m-2}) \cdots P_{12}(\lambda_n; \Delta_1) A_2^{-1} M_{b,0}\right)^{-1} \mathbf{e}_j$$

so that

$$P_{12}(\lambda_n; \Delta_{m-1})\mathbf{e}_j = 0.$$

We can repeat this procedure for $\Delta_{m-2}, \ldots, \Delta_1$. The corresponding mode shape is then determined by (4.6.18) where $f(0)$ is now denoted y.

4.7 Modal Expansion

For the undamped structure ($\alpha = 0$), we have developed a "modal expansion" in terms of the eigenfunctions of $\mathcal{A}$ given by (4.4.16). The eigenfunctions are orthogonal and complete. The question arises as to what extent this property holds in the closed-loop case – for the eigenfunctions of

$$\mathcal{A} - \alpha\mathcal{B}\mathcal{B}^*, \qquad \alpha > 0.$$

We have seen that these functions are not orthogonal and hence we need to examine what happens to the modal expansion.

We have seen (cf. (4.5.15)) that the eigenfunctions are of the form

$$Y_k = \mathcal{R}(\lambda_k, \mathcal{A})\mathcal{B}^* u(\lambda_k), \tag{4.7.1}$$

where $\{\lambda_k\}$ are the eigenvalues. For simplicity we shall assume that the dimension of the eigenfunction space is unity. Let Z_k denote the eigenfunction of

$$(\mathcal{A} - \alpha\mathcal{B}\mathcal{B}^*)^* = \mathcal{A}^* - \alpha\mathcal{B}\mathcal{B}^*$$

corresponding to the eigenvalue $\bar{\lambda}_k$. Then since the pertinent properties of $\mathcal{A}^*$ are similar to those of $\mathcal{A}$, we have

$$\begin{aligned} Z_k &= \mathcal{R}(\bar{\lambda}_k, \mathcal{A}^*)\,\mathcal{B}^* v\left(\bar{\lambda}_k\right), \qquad v(\cdot) \in E^{m_c} \\ &= -\mathcal{R}(-\bar{\lambda}_k, \mathcal{A})\,\mathcal{B}^* v(\bar{\lambda}_k). \end{aligned} \tag{4.7.2}$$

The main feature of these eigenfunctions is that they are "biorthogonal":

$$\begin{aligned} &[Y_k, Z_j] = 0, \qquad k \neq j, \\ &[Y_k, Z_k] \neq 0. \end{aligned}$$

4.7.1 Riesz Basis

Recall now that a sequence $\{\Psi_k\}$ of elements in a Hilbert space $\mathcal{H}$ is called a "basis" if every element Y in the space can be expressed as

$$Y = \sum_1^\infty a_k \Psi_k,$$

where

$$\sum_1^\infty |a_k| \, \|\psi_k\| < \infty$$

and

$$0 = \sum_1^\infty a_k \psi_k$$

implies

$$a_k = 0 \quad \text{for every } k.$$

A biorthogonal sequence $\{\Phi_k, \psi_k\}$ is called a Riesz basis if there is a linear bounded operator T on $\mathcal{H}$ into $\mathcal{H}$ with bounded inverse such that $\{T\phi_k\}$ is an orthonormal basis. This implies in particular that we have the expansion

$$Y = \sum_1^\infty [Y, \psi_k]\phi_k = \sum_1^\infty [Y, \Phi_i]\psi_k, \tag{4.7.3}$$

where

$$\sum_1^\infty |[Y, \psi_k]|^2 < \infty, \qquad \sum_1^\infty |[Y, \phi_k]|^2 < \infty,$$

$$T\phi_k = T^{*-1}\psi_k.$$

Also, for Y, Z in $\mathcal{H}$

$$[Y, Z] = \sum_1^\infty [Y, \psi_k][\phi_k, Z]. \tag{4.7.4}$$

The main result in this section is that $\{Y_k, Z_k\}$ upon "normalization" so that

$$[Y_k, Z_k] = 1$$

("normalizing" $u(\lambda_k)$, $v(\bar{\lambda}_k)$ appropriately, which we shall assume from now on) form a Riesz basis. This will follow from Balakrishnan (1996) upon verifying the conditions

(i)

$$\left|\lambda_k + \bar{\lambda}_j\right| \geq \delta > 0, \quad \text{for } k \neq j,$$

(ii)

$$\sum_{1}^{\infty} |\sigma_k| < \infty,$$

and

(iii) (algebraic) multiplicity of each eigenvalue is equal to unity.

Condition (i) follows readily from the asymptotic estimate (4.5.35). Condition (ii) is verified in Section 4.5 (4.5.16). Condition (iii) is automatic since we are assuming the dimension of the eigenfunction space for each $\alpha > 0$ is unity.

We can now proceed to exploit the modal expansion (4.7.3). First we note that if λ_k is an eigenvalue so is $\bar{\lambda}_k$ if

$$\operatorname{Im} \lambda_k \neq 0$$

and there are exactly six real-valued λ_k. Let

$$(\mathcal{A} - \alpha\mathcal{B}\mathcal{B}^*)Y_k = \lambda_k Y_k.$$

Then

$$(\mathcal{A} - \alpha\mathcal{B}\mathcal{B}^*)\bar{Y}_k = \bar{\lambda}_k \bar{Y}_k,$$
$$(\mathcal{A} - \alpha\mathcal{B}\mathcal{B}^*)^* \bar{Z}_k = \lambda_k \bar{Z}_k .$$

Hence numbering so that $\lambda_1, \ldots, \lambda_6$ are real, and the λ_k are in increasing order in $|\lambda_k|$, we can express the modal expansion as

$$Y = \sum_{1}^{6} [Y, Z_k] Y_k + \sum_{7}^{\infty} ([Y, Z_k] Y_k + [Y, \bar{Z}_k] - Y_k). \tag{4.7.5}$$

Correspondingly, the solution of the closed-loop system:

$$\dot{Y}(t) = (\mathcal{A} - \alpha\mathcal{B}\mathcal{B}^*)Y(t), \qquad Y(0) = Y$$

can be expressed as

$$Y(t) = \sum_{1}^{6} [Y, Z_k] e^{-|\sigma_k| t} Y_k + \sum_{7}^{\infty} \left([Y, Z_k] e^{\lambda_k t} Y_k + [Y, \bar{Z}_k] e^{\bar{\lambda}_k t} \bar{Y}_k \right). \tag{4.7.6}$$

We can easily deduce strong stability of the semigroup $S_\alpha(\cdot)$ from (4.7.6) or equivalently from the fact that the eigenfunctions $\{Y_k\}$ are complete in $\mathcal{H}_{\mathrm{E}}$, exploiting the dissipativity

$$\|S_\alpha(t)\| \leq 1.$$

If Y is real-valued, we note that

$$\overline{[Y, Z_k]} = \left[Y, \bar{Z}_k\right]$$

and hence

$$[Y, Z_k]e^{\lambda_k t}Y_k + [Y, \bar{Z}_k]e^{\bar{\lambda}_k t}\bar{Y}_k$$

can be expressed as

$$e^{-|\sigma_k|t}[[Y, Z_k]Y_k + [Y, \bar{Z}_k]\bar{Y}_k]\cos\omega_k t + e^{-|\sigma_k|t}[[Y, Z_k]Y_k - [Y, \bar{Z}_k]\bar{Y}_k]\sin\omega_k t. \tag{4.7.7}$$

Because we must have

$$Y_k = \begin{vmatrix} y_k \\ \lambda_k y_k \end{vmatrix}, \qquad Z_k = \begin{vmatrix} z_k \\ \bar{\lambda}_k z_k \end{vmatrix}$$

we can proceed to develop expansions for $x_1(t)$, $x_2(t)$ where

$$Y(t) = \begin{vmatrix} x_1(t) \\ x_2(t) \end{vmatrix}$$

by going back to (4.5.7) and (4.5.8). We omit the details. From (4.7.6) we can readily deduce (4.5.17).

4.8 Illustrative Example

To illustrate the foregoing theory and concepts, we consider now an example – simplified in the extreme to reduce notational complexity and wholly nonnumeric to avoid computer calculation. Thus we consider beam torsion about a single axis with a control at one end and a lumped mass at the other, and with no interior nodes.

Retaining the nomenclature of Section 4.2 as much as possible but using $\theta(t, s)$ in place of $\phi(t, s)$, the dynamics can be described by

$$m_{44}\ddot{\theta} - c_{66}\theta'' = 0, \qquad 0 < s < L, \qquad 0 < t, \tag{4.8.1}$$

which in the notation of Section 4.2 yields

$$M_0 = m_{44}, \qquad A_2 = c_{66}.$$

The abstract version becomes

$$\mathcal{H} = L_2(0, L) \times E^2,$$

$$x = \begin{vmatrix} f \\ b \end{vmatrix}, \qquad f(\cdot) \in L_2(0, L), \qquad b \in E^2, \qquad b = \begin{vmatrix} b_0 \\ b_L \end{vmatrix}.$$

The stiffness operator A is then given by

$$\text{domain of } A = \left[x = \begin{vmatrix} f \\ b \end{vmatrix}, \qquad f, f', f'' \in L_2(0, L), \qquad b = \begin{vmatrix} f(0) \\ f(L) \end{vmatrix}\right],$$

$$Ax = y, \qquad y = \begin{vmatrix} g \\ c \end{vmatrix},$$

$$g(s) = -c_{66} g''(s), \qquad 0 < s < L,$$

$$c = \begin{vmatrix} -c_{66} f'(0) \\ c_{66} f'(L) \end{vmatrix}.$$

Thus

$$[Ax, x] = -c_{66} \int_0^L f''(s)\overline{f(s)}\, ds + c_{66} f(L)\overline{f''(L)} - c_{66} f(0)\overline{f'(0)}$$

$$= c_{66} \int_0^L |f'(s)|^2\, ds, \tag{4.8.2}$$

yielding the potential energy, as required. There is a rigid-body mode:

$$Ax = 0,$$

where x is of the form

$$x = \begin{vmatrix} f \\ a \\ a \end{vmatrix} \tag{4.8.3}$$

with

$$f(s) = a, \qquad 0 \leq s \leq L.$$

Placing the control at $s = 0$, the control operator B is given by

$$Bu = \begin{vmatrix} 0 \\ B_u u \end{vmatrix}, \qquad B_u u = \begin{vmatrix} u \\ 0 \end{vmatrix}.$$

For x in $\mathcal{D}(A)$,

$$x = \begin{vmatrix} f \\ b \end{vmatrix},$$

we see that

$$B^* x = f(0).$$

Since there is a control at one end, we see that all modes are controllable and that $(A \sim B)$ is controllable.

Finally, the mass operator M is given by

$$Mx = y, \qquad y = \begin{vmatrix} m_{44} \\ M_b b \end{vmatrix}, \qquad M_b = \begin{vmatrix} m_0 & 0 \\ 0 & m_L \end{vmatrix}.$$

Correspondingly, we have the boundary equations

$$\left.\begin{aligned} m_0\ddot{\theta}(t,0) - c_{66}\theta'(t,0) + u(t) = 0 \\ m_L\ddot{\theta}(t,L) + c_{66}\theta'(t,L) = 0 \end{aligned}\right\}. \tag{4.8.1a}$$

The space $\mathcal{H}_1$ (M-orthogonal to the null space of A) consists of elements of the form

$$x = \begin{vmatrix} f(\cdot) \\ b_0 \\ b_L \end{vmatrix},$$

where

$$m_{44}\int_0^L f(s)\,ds + m_0 b_0 + m_L b_L = 0. \tag{4.8.4}$$

The domain of $\sqrt{A}$, is characterized by elements of the form (Balakrishnan, 1990)

$$x = \begin{vmatrix} f(\cdot) \\ f(0) \\ f(L) \end{vmatrix},$$

where $f(\cdot)$ is absolutely continuous and $f'(\cdot) \in L_2(0, L)$. In particular,

$$\|\sqrt{A}x\|^2$$

is the potential energy given by (4.8.2). Thus

$$\mathcal{D}(\sqrt{A}) \cap \mathcal{H}_1 = \left[\begin{vmatrix} f \\ f(0) \\ f(L) \end{vmatrix}, \quad \begin{matrix} f \text{ is absolutely continuous with } f' \\ \text{in } L_2(0,L), \text{ and (4.8.4) holds} \end{matrix} \right].$$

Also,

$$\mathcal{H}_E = (\mathcal{D}(\sqrt{A}) \cap \mathcal{H}_1) \times \mathcal{H}$$

with energy inner product, as in Section 4.2.

The feedback control is

$$u(t) = \alpha B^* \dot{x}(t) = \alpha\dot{\theta}(t,0), \qquad \alpha > 0.$$

4.8.1 Closed-loop Modes

We proceed directly to characterize the closed-loop modes. In the notation of Section 4.6 we have

$$\mathcal{A}(\lambda) = \begin{vmatrix} 0 & 1 \\ \lambda^2\nu^2 & 0 \end{vmatrix},$$

where

$$\nu^2 = \frac{m_{44}}{c_{66}},$$

yielding

$$e^{\mathcal{A}(\lambda)s} = \begin{vmatrix} \cosh \lambda\nu s & \frac{\sinh \lambda\nu s}{\lambda\nu} \\ \lambda\nu \sinh \lambda\nu s & \cosh \lambda\nu s \end{vmatrix}$$

and

$$\begin{aligned} h(\lambda, \alpha) &= \begin{vmatrix} \frac{\lambda^2 m_L}{c_{66}} & 1 \end{vmatrix} e^{\mathcal{A}(\lambda)L} \begin{vmatrix} 1 \\ \frac{\alpha\lambda+\lambda^2 m_0}{c_{66}} \end{vmatrix} \\ &= \lambda(a_1(\lambda) \sinh \lambda\nu L + a_2(\lambda) \cosh \lambda\nu L), \end{aligned} \tag{4.8.5}$$

where

$$a_1(\lambda) = \frac{1}{\nu c_{66}^2}\left(\nu^2 c_{66}^2 + \lambda\alpha m_L + \lambda^2 m_0 m_L\right),$$

$$a_2(\lambda) = \frac{\alpha + \lambda(m_L + m_0)}{c_{66}}.$$

We see that

$$d(\lambda; \alpha) = h(\lambda; \alpha)$$

is an entire function of order one. It has a zero of order one at $\lambda = 0$, for nonzero α, and of order two for $\alpha = 0$. It is of "completely regular growth" in the terminology of Levin (1980):

$$\lim_{r\to\infty} \frac{\log |d(re^{i\theta}, \alpha)|}{r} = \nu L |\cos\theta|$$

and hence (Levin, 1980, p. 169)

$$\lim_{r\to 0} \frac{N(r)}{r} = \frac{1}{2\pi}\int_0^{2\pi} \nu L|\cos\theta|\, d\theta > 0,$$

where $N(r)$ is the number of zeros in the circle of radius r. Hence the number of zeros is not finite. For large $|\lambda|$

$$d(\lambda; \alpha) \sim \lambda^3 m_0 m_L \sinh \lambda\nu L$$

and thus the asymptotic modes are the roots of

$$\sinh \lambda \nu L = 0,$$

or

$$\lambda_n = \frac{\pm i n \pi}{\nu L} \tag{4.8.6}$$

for all $\alpha \geq 0$. Here, however, we can make a more exact calculation. Thus the eigenvalues $\{\lambda_k\}$ are the roots of

$$\tanh \lambda \nu L + b(\lambda; \alpha) = 0, \tag{4.8.7}$$

where

$$b(\lambda; \alpha) = \frac{m_{44}}{\nu} \frac{\alpha + \lambda(m_L + m_0)}{m_{44}c_{66} + \lambda \alpha m_L + \lambda^2 m_0 m_L} \tag{4.8.8}$$

and

$$|b(\lambda; \alpha)| = O\left(\frac{1}{|\lambda|}\right) \quad \text{as } |\lambda| \to \infty.$$

We can rewrite (4.8.4) as

$$\lambda \nu L + \tanh^{-1} b(\lambda; \alpha) = 0,$$

where

$$b(\lambda; \alpha) = \frac{a_2(\lambda)}{a_1(\lambda)},$$

and since

$$\tanh x = \tanh(x \pm 2 i n \pi) \quad (n \text{ integer })$$

we have

$$\lambda \nu L = \pm i n \pi + \frac{1}{2} \log \frac{1 + b(\lambda; \alpha)}{1 - b(\lambda; \alpha)} = 0, \tag{4.8.7a}$$

using the principal value of $\log x$, which is real when x is positive. For $|\lambda|$ large, (4.8.7a) becomes

$$\lambda \nu L = \pm i n \pi + \frac{m_{44}}{\nu} \cdot \left(\frac{m_L + m_0}{m_L m_0}\right) \frac{1}{\lambda} \tag{4.8.9}$$

yielding a slightly better approximation than (4.8.6), for large n. For nonzero α, we can see that (4.8.4) has exactly one real root, approximately

$$\lambda = \frac{-\alpha}{m_{44} L + m_L + m_0}. \tag{4.8.10}$$

4.8.2 Clamped Modes

We can calculate that

$$L(\lambda) = \begin{vmatrix} 1 & 0 \\ \cosh \lambda\nu L & \frac{\sinh \lambda\nu L}{\lambda\nu} \end{vmatrix}$$

and that the clamped modes are the zeros of $\sinh \lambda\nu L$, or

$$\lambda_k = \frac{\pm ik\pi}{\nu L}, \qquad k = 1, 2, \ldots . \tag{4.8.11}$$

4.8.3 Dynamic Stiffness Matrix

We can calculate that

$$K(\lambda) = c_{66} \begin{vmatrix} 0 & -1 \\ \lambda\nu \sinh \lambda\nu L & \cosh \lambda\nu L \end{vmatrix}$$

and hence that

$$T(\lambda) = \frac{\lambda\nu c_{66}}{\sinh \lambda\nu L} \begin{vmatrix} \cosh \lambda\nu L & -1 \\ -1 & \cosh \lambda\nu L \end{vmatrix},$$

which is clearly nonnegative definite for λ real, and nonsingular except for $\lambda = 0$. Also, the inverse of the dynamic stiffness matrix is

$$(\lambda^2 m_b + T(\lambda))^{-1} = \frac{1}{\lambda h(\lambda; 0)}$$
$$\times \begin{vmatrix} \lambda^2 m_L \sinh \lambda\nu L + c_{66}\lambda\nu \cosh \lambda\nu L & -\lambda\nu c_{66} \\ -\lambda\nu c_{66} & \lambda^2 m_0 \sinh \lambda\nu L + c_{66}\lambda\nu \cosh \lambda\nu L \end{vmatrix}.$$

Hence

$$\begin{aligned} \lambda B_u^*(\lambda^2 M_b + T(\lambda))^{-1} Bu &= \frac{1}{h(\lambda; 0)}(\lambda^2 m_L \sinh \lambda\nu L + c_{66}\lambda\nu \cosh \lambda\nu L) \\ &= m(\lambda). \end{aligned} \tag{4.8.12}$$

Hence we can verify that for this example

$$h(\lambda; 0) = h(\lambda, 0) D(\lambda; \alpha).$$

In other words in (4.6.27)

$$p_1(\lambda) = 0,$$

which we may conjecture holds in general.

4.8.4 Root Locus

Beginning first with the limiting eigenvalues as α goes to infinity, given by the roots of

$$m(\lambda) = 0,$$

we have from (4.8.9)

$$\tanh \lambda \nu L + \frac{\nu c_{66}}{\lambda m_L} = 0, \tag{4.8.13}$$

which can also be obtained directly from (4.8.7) by taking the limit as α goes to infinity in (4.8.8). The roots are of course purely imaginary:

$$\lambda_n = i\beta_n, \qquad \beta_n \text{ real,}$$
$$\beta_n \nu L = \pm in\pi + i\delta_n, \qquad |\delta_n| < \pi.$$

These are the modes that satisfy

$$m_L \ddot{\theta}(t, L) + c_{66}\theta'(t, L) = 0, \qquad \theta(t, 0) = 0,$$
$$m_{44}\ddot{\theta}(t, s) \; - \; c_{66}\theta''(t, s) = 0, \qquad 0 < s < L.$$

These are *not* the clamped modes, although they are, asymptotically.

Since we are only interested in the nonzero eigenvalues, let

$$F(\lambda; \alpha) = \left(\nu^2 c_{66}^2 + \lambda \alpha m_L + \lambda^2 m_0 m_L\right) \sinh \lambda \nu L$$
$$+ \nu c_{66}(\alpha + \lambda(m_L + m_0)) \cosh \lambda \nu L$$

whose zeros are the nonzero eigenvalues. Let $\{i\omega_k\}$ denote the zeros for $\alpha = 0$. Fix k. Now

$$F(\lambda(\alpha); \alpha) = 0 \tag{4.8.14}$$

defines an implicit function $\lambda_k(\alpha)$, with

$$\lambda_k(0) = i\omega_k,$$

and we define all derivatives at $\alpha = 0$ using (4.8.14). In particular,

$$\left.\frac{d\lambda_k(\alpha)}{d\alpha}\right|_{\alpha=0} = \left.\frac{-F_\alpha(\lambda_k(\alpha);\ \alpha)}{F_\lambda(\lambda_k(\alpha);\ \alpha)}\right|_{\alpha=0},$$

where the subscripts denote partial derivatives. The main point is that the derivative is real and negative. This shows that the real part is decreasing. Since we know that the real part goes to zero as α goes to infinity, we see that there is a value of α at which its derivative must change sign.

4.8.5 Closed-loop Mode Shapes

Following Section 4.7, the (unpurged) closed-loop mode shape corresponding to the eigenvalue λ_k is given by

$$f_k(s) = |1 \quad 0|\, e^{A(\lambda_k)s} \begin{vmatrix} 1 \\ \frac{\alpha\lambda_k + \lambda_k^2 m_0}{c_{66}} \end{vmatrix} = \left(\cosh \lambda_k \nu s + \frac{(\alpha + \lambda_k m_0)}{\nu c_{66}} \sinh \lambda_k \nu s \right) f(0).$$

Since arbitrary multiplicative constants can be used, we may define the mode shape as

$$f_k(s) = A_k \sinh(\lambda_k \nu s + \theta_k), \qquad 0 \le s \le L,$$

where

$$\tanh \theta_k = \frac{\nu c_{66}}{\alpha + \lambda_k m_0}.$$

For $\lambda_k \neq 0$, the purged version would be (in the notation of Section 4.7)

$$Y_k = \begin{vmatrix} \phi_k \\ \lambda_k \phi_k \end{vmatrix}, \qquad Z_k = \begin{vmatrix} \bar{\phi}_k \\ \bar{\lambda}_k \phi_k \end{vmatrix},$$

where (new notation, not to be confused with that of Section 4.2)

$$\phi_k = \begin{vmatrix} \tilde{f}_k(\cdot) \\ \tilde{f}_k(0) \\ \tilde{f}_k(L) \end{vmatrix},$$

$$\tilde{f}_k(s) = f_k(s) - \left(m_{44} \int_0^L f_k(s)\, ds + m_0 f_k(0) + m_L f_k(L) \right).$$

The constant A_k can be determined to normalize the biorthogonal system as in Section 4.7 and thus to obtain a Riesz basis for $\mathcal{H}_E$.

4.8.6 A Limiting Case

We illustrate finally how to handle the case when one end is clamped. We set $m_L = +\infty$. This results in the boundary condition

$$\theta(t, L) = 0$$

replacing the condition at L in (4.8.1a). We may take

$$\mathcal{H} = L_2(0, L) \times E^1$$

and

$$Ax = y; \quad y = \begin{vmatrix} -c_{66} f''(\cdot) \\ -c_{66} f'(0) \end{vmatrix}, \qquad x = \begin{vmatrix} f(\cdot) \\ f(0) \end{vmatrix}.$$

There are no rigid-body modes and the eigenvalues are roots of

$$(\alpha + \lambda m_0) \sinh \lambda \nu L + \nu c_{66} \cosh \lambda \nu L = 0 \tag{4.8.15}$$

or

$$\lambda_n \nu L = \pm i n\pi - \tanh^{-1} b(\lambda_n),$$
$$b(\lambda) = \frac{\nu c_{66}}{\alpha + \lambda m_0}.$$

The eigenfunctions are

$$Y_k = \begin{vmatrix} \phi_k \\ \lambda_k \phi_k \end{vmatrix}, \qquad \phi_k = \begin{vmatrix} f_k(\cdot) \\ f_k(0) \end{vmatrix}, \qquad f_k(s) = A_k \sinh \lambda_k \nu (L - s).$$

The root-locus problem becomes much simpler than before. Again, we omit the details. For $\alpha = \infty$, the modes are the zeros of $\sinh \lambda \nu L$, or

$$\lambda_n = \frac{\pm i n\pi}{\nu L}.$$

From (4.8.15) we see that

$$\frac{d\lambda_n}{d\alpha} = \frac{-1}{1 - \left(\frac{\nu c_{66}}{\alpha + \lambda_n m_0}\right)^2},$$

which is real negative at $\alpha = 0$ and goes to (-1) as α goes to infinity. For large n we have the approximation

$$\frac{d\lambda_n}{d\alpha} \sim -(\nu c_{66})^2 \frac{1}{(\alpha + \lambda_n m_0)^2}$$

and

$$\operatorname{Re} \frac{d\lambda_n}{d\alpha} = 0$$

implies

$$\alpha_{\text{crit}} = m_0(|\operatorname{Re} \lambda_n| + |\operatorname{Im} \lambda_n|)$$

and shows that the critical value of α increases with mode number.

Acknowledgment

This research was partially supported by NASA grant NCC 2-374.

References

Achieser, N. I., and Glassman, I. M., 1966, *Theory of Linear Operators in a Hilbert Space*, Dover Publications, New York.

Balakrishnan, A. V., 1981, *Applied Functional Analysis*. Springer-Verlag, New York.

Balakrishnan, A. V., 1990, "Damping Operators in Continuum Models of Flexible Structures: Explicit Models for Proportional Damping in Beam Torsion," *Journal of Differential and Integral Equations*, Vol. 3, No. 3, pp. 381–96.

Balakrishnan, A. V., 1991a, "Compensator Design for Stability Enhancement with Collocated Controllers," *IEEE Transactions on Automatic Control*, Vol. 36, pp. 994–1007.

Balakrishnan, A. V., 1991b, "A Continuum Model for Interconnected Lattice Trusses," *Proceedings of the Eighth VPI&SU Symposium on Dynamics and Control of Large Structures, May 6–8, 1991, Blacksburg, Virginia*, pp. 479–89.

Balakrishnan, A. V., 1992, "Combined Structures–Controls Optimization of Lattice Trusses," *Computer Methods in Applied Mechanics and Engineering*, Vol. 94, pp. 131–52.

Balakrishnan, A. V., 1996, "On Superstable Semigroups of Operators," *Dynamic Systems and Applications*, Vol. 5, 1996, pp. 371–84.

Balakrishnan, A. V., and Triggiani, R., 1993, "Lack of Generation of Strongly Continuous Semigroups by the Damped Wave Operator on $H \times H$ (Or: The Little Engine that Couldn't)," *Applied Mathematics Letters*, Vol. 6, pp. 33–37.

Benchimol, C. D., 1978, "A Note on the Stabilizabilty of Contraction Semigroups," *SIAM Journal on Control and Optimization*, Vol. 16, pp. 373–79.

Gohberg, I. C., and Krein, M. G., 1969, *Introduction to the Theory of Linear Non-Self-Adjoint Operators*. Translations of Mathematical Monographs, Vol. 18, American Mathematical Society.

Levin, B. Ja., 1980, *Distribution of Zeros of Entire Functions*, Translations of Math Monographs, American Mathematical Society.

Meirovich, L., 1967, *Analytical Methods in Vibrations*. MacMillan, New York.

Noor, A. K., and Anderson, C. M., 1979, "Analysis of Beamlike Lattice Trusses," *Computer Methods in Applied Mechanics and Engineering*, Vol. 20, pp. 53–70.

Noor, A. K., and Russell, W. C., 1986, "Anisotropic Continuum Models for Beamlike Lattice Trusses," *Computer Methods in Applied Mechanics and Engineering*, Vol. 57, pp. 257–77.

Riesz, F., and Nagy, B. Sz., 1955, *Functional Analysis*. Frederick Ungar Publishing Company, New York.

Timoshenko, S., Young, D. H., and Waver, W., Jr., 1974, *Vibrating Problems in Engineering*, 4th ed., John Wiley & Sons, Inc., New York.

Wang, H. C., 1994, "Distributed Parameter Modeling of Repeated Truss Structures," *NASA Workshop on Distributed Parameter Modeling and Control of Flexible Aerospace Systems*, NASA Conference Publication 3242, pp. 41–63.

Wittrick, W. H., and Williams, F. W., 1971, "A General Algorithm for Computing Natural Frequencies of Elastic Structures," *Quarterly Journal of Mechanics and Applied Math*, Vol. 24, Part 3.

Nomenclature

A	stiffness operator
B	control operator
s	position along beam

(x_1, x_2, x_3)	rectangular coordinates
u	axial displacement (x_1 component)
v	displacement (x_2 component)
w	displacement (x_3 component)
ϕ_1	torsion angle about x_1 axis
ϕ_2	torsion angle about x_2 axis
ϕ_3	torsion angle about x_3 axis
M	mass/inertia operator
M_0	mass/inertia matrix
M_b	composite matrix of mass/inertia at nodes
$\mathcal{H}$	Hilbert space
$\mathcal{H}_1$	space M-orthogonal to null space of A
$L_2(0, L)^6$	L_2-space of 6×1 vector functions over $(0, L)$
R^6, E^6	Euclidean 6-space
m_c	number of control inputs
$[\,,\,]$	inner product
$\mathcal{D}(A)$	domain of operator A
I_0, I_c, I_L	moments of inertia
$u(t)$	control input
λ_k	eigenvalues
σ_k	real part of λ_k
ω_k	angular mode frequencies
Tr	trace
$\mathrm{Re}\, z$	real part of z
$\mathrm{Im}\, z$	imaginary part of z
$\det M$	determinant of M
$\lvert z\rvert$	absolute value of z
$\lVert F\rVert$	norm of vector F; operator norm of matrix F
$\bar{z}$	conjugate of z
A^*	adjoint of A
SCOLE	Spacecraft COntrol Laboratory Experiment

5

Numerical Techniques for Simulation, Parameter Estimation, and Noise Control in Structural Acoustic Systems

H. T. BANKS and R. C. SMITH

Abstract

A model for a 3-D structural acoustic system, currently being used for parameter estimation and control experiments in the Acoustics Division, NASA Langley Research Center, is presented. This system consists of a hard-walled cylinder with a flexible circular plate at one end. An exterior noise source causes vibrations in the plate which in turn lead to unwanted noise inside the cylinder. Control is implemented through the excitation of piezoceramic patches bonded to the plate which generate in-plane forces and/or bending moments in response to an input voltage.

The plate and interior acoustic wave dynamics are approximated with expansions involving Fourier components in the circular direction and spline and spectral elements in the radial and axial directions. To guarantee uniqueness and differentiability at the coordinate singularity as well as to ensure stability and the expected convergence rate, the radial basis functions for both the wave and plate components are constructed in a manner that incorporates the Bessel or analytic behavior near the singularity while retaining sufficient generality so as to provide an approximation technique for discretizing complex coupled systems involving circular geometries.

5.1 Introduction

A growing area of research in the structural acoustics community concerns the problem of reducing structure-borne noise levels within an acoustic cavity. A specific example of this is motivated by the development of a new class of turboprob and turbofan engines that are very fuel efficient but also very loud. The low-frequency high-magnitude acoustic fields produced by these engines cause vibrations in the fuselage which in turn generate unwanted interior noise. The passive control techniques that were initially employed to reduce the acoustic

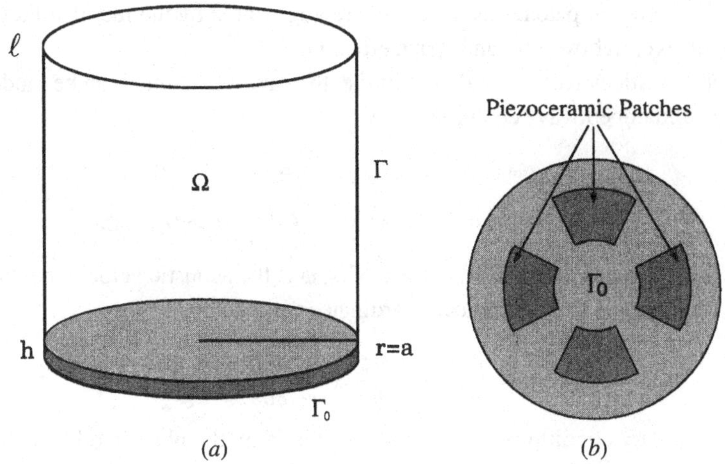

Fig. 5.1. (*a*) The cylindrical acoustic cavity. (*b*) The circular plate with patches.

pressure levels are in general undesirable since the increased weight offsets the advantages gained through the use of the new engines and lighter building materials. This then led to the study of active control techniques for this problem both in a frequency domain setting and from a time domain or partial differential equation (PDE) approach.

In this work we develop the numerical methods needed to extend previous 2-D time domain results (Banks, Fang, Silcox, and Smith, 1993; Banks, Silcox, and Smith, 1994) to a 3-D geometry in which the parameter estimation and control techniques can be experimentally tested. The domain Ω of interest consists of a cylinder of length ℓ and radius a as pictured in Fig. 5.1. At one end of the cylinder is a flexible plate of thickness h, which is assumed to have Kelvin–Voigt damping. It is also assumed that the edges of the plate are fixed. The other end of the cylinder is closed, and hard-wall boundary conditions are assumed on all walls of the cylinder other than the flexible plate. The choice of this geometry and configuration results from the setup of an experimental apparatus currently being used in the Acoustics Division at NASA Langley Research Center, which consists of a concrete pipe with a clamped aluminum plate at one end.

Control is implemented in the model via piezoceramic patches on the plate (see Figure 5.1) which produce pure bending moments and/or extensional forces when a voltage is applied (Banks, Smith, and Wang, 1995). Because of the coupling between the interior acoustic field and the structural vibrations, this provides a means of controlling the interior acoustic pressure levels through the control of the bounding structure dynamics. The benefits of using the piezoceramic patches as actuators are augmented by the fact that they are inexpensive, lightweight, and space efficient.

For acoustic waves of small amplitude, the cavity dynamics can be modeled by the undamped wave equation

$$\begin{aligned} \phi_{tt} &= c^2 \Delta\phi, \qquad (r,\theta,z) \in \Omega, \quad t > 0, \\ \nabla\phi \cdot \hat{n} &= 0, \qquad (r,\theta,z) \in \Gamma, \quad t > 0, \end{aligned}$$

where c is the speed of sound in the cavity, ϕ is the acoustic velocity potential, and the Laplacian in cylindrical coordinates is given by

$$\Delta\phi = \frac{\partial^2\phi}{\partial r^2} + \frac{1}{r}\frac{\partial\phi}{\partial r} + \frac{1}{r^2}\frac{\partial^2\phi}{\partial\theta^2} + \frac{\partial^2\phi}{\partial z^2}.$$

The boundary conditions on Γ (all walls other than the plate) model the hard-wall conditions of the experimental concrete cylinder. Acoustic damping is omitted due to the relatively small dimensions of the experimental cavity ($a = 9\,\text{in}$, $\ell = 42\,\text{in}$). Finally, we point out that the acoustic pressure p is related to the potential through the relationship $p = \rho_f \phi_t$, where ρ_f is the equilibrium density of the interior cavity fluid (air).

The motion of the plate is influenced by a forcing function f which models the exterior noise source, the moments generated by the patches, and the back-pressure due to pressure oscillations within the cavity. In formulating the patch contributions, we have assumed that s piezoceramic patches have been bonded to the plate as depicted in Fig. 5.1. As detailed in Banks and Smith (1994), the strong form of the equations of motion for a fixed-edge, damped circular plate that is subjected to these moments and forces is

$$\begin{aligned} &\rho_p h w_{tt} + \frac{\partial^2 \mathcal{M}_r}{\partial r^2} + \frac{2}{r}\frac{\partial \mathcal{M}_r}{\partial r} - \frac{1}{r}\frac{\partial \mathcal{M}_\theta}{\partial r} + \frac{2}{r}\frac{\partial^2 \mathcal{M}_{r\theta}}{\partial r \partial\theta} + \frac{2}{r^2}\frac{\partial \mathcal{M}_{r\theta}}{\partial\theta} + \frac{1}{r^2}\frac{\partial^2 \mathcal{M}_\theta}{\partial\theta^2} \\ &= -\rho_f \phi_t(t,r,\theta,w(t,r,\theta)) + f(t,r,\theta), \end{aligned}$$

$$w(t,a,\theta) = \frac{\partial w}{\partial r}(t,a,\theta) = 0,$$

where w is the transverse displacement and ρ_p is the density of the plate. The general moments are given by

$$\begin{aligned} \mathcal{M}_r &= M_r - (M_r)_{pe}, \\ \mathcal{M}_\theta &= M_\theta - (M_\theta)_{pe}, \\ \mathcal{M}_{r\theta} &= M_{r\theta}, \end{aligned}$$

where

$$M_r = D\left(\frac{\partial^2 w}{\partial r^2} + \frac{\nu}{r}\frac{\partial w}{\partial r} + \frac{\nu}{r^2}\frac{\partial^2 w}{\partial \theta^2}\right) + c_D\left(\frac{\partial^3 w}{\partial r^2 \partial t} + \frac{\nu}{r}\frac{\partial^2 w}{\partial r \partial t} + \frac{\nu}{r^2}\frac{\partial^3 w}{\partial \theta^2 \partial t}\right),$$

$$M_\theta = D\left(\frac{1}{r}\frac{\partial w}{\partial r} + \frac{1}{r^2}\frac{\partial^2 w}{\partial \theta^2} + \nu\frac{\partial^2 w}{\partial r^2}\right) + c_D\left(\frac{1}{r}\frac{\partial^2 w}{\partial r \partial t} + \frac{1}{r^2}\frac{\partial^3 w}{\partial \theta^2 \partial t} + \nu\frac{\partial^3 w}{\partial r^2 \partial t}\right),$$

$$M_{r\theta} = D(1-\nu)\left(\frac{1}{r}\frac{\partial^2 w}{\partial r \partial \theta} - \frac{1}{r^2}\frac{\partial w}{\partial \theta}\right) + c_D(1-\nu)\left(\frac{1}{r}\frac{\partial^3 w}{\partial r \partial \theta \partial t} - \frac{1}{r^2}\frac{\partial^2 w}{\partial \theta \partial t}\right) \tag{5.1}$$

are the internal plate moments, and

$$(M_r)_{pe} = (M_\theta)_{pe} = \sum_{i=1}^{s} \mathcal{K}_i^B u_i(t)[H(r - r_{i1}) - H(r - r_{i2})] \times [H(\theta - \theta_{i1}) - H(\theta - \theta_{i2})]$$

are the applied patch moments for patches located at $r_{i1} \le r \le r_{i2}, \theta_{i1} \le \theta \le \theta_{i2}$. With E denoting the Young's modulus, the parameters $D = \frac{Eh^3}{12(1-\nu^2)}$, c_D, and ν represent the flexural rigidity, damping coefficient, and Poisson's ratio for the plate. Here H denotes the Heaviside function, $u_i(t)$ is the voltage into the ith patch, and $\mathcal{K}_i^B$ is a parameter that depends on the geometry, piezoceramic and plate material properties, and piezoelectric strain constant (see Banks, Smith, and Wang (1995) for details). The piezoceramic material parameters $\mathcal{K}_i^B, i = 1, \ldots, s$ as well as the plate parameters ρ_p, D, c_D, and ν must ultimately be considered to be unknown and will in actual applications be obtained using parameter estimation techniques.

The final coupling equation is the continuity of velocity condition

$$\frac{\partial \phi}{\partial z}(t, r, \theta, w(t, r, \theta)) = -w_t(t, r, \theta), \qquad (r, \theta) \in \Gamma_0, \qquad t > 0,$$

which results from the assumption that the plate is impenetrable to air. We note that both the backpressure and velocity (momentum) coupling conditions are in general nonlinear due to the fact that they occur on the surface of the plate.

Under the assumption of small displacements inherent in the Love–Kirchhoff plate formulation, these nonlinear coupling terms are replaced by their linear approximations, which yields the approximate system model:

$$\phi_{tt} = c^2 \Delta \phi, \qquad (r, \theta, z) \in \Omega, t > 0,$$

$$\nabla \phi \cdot \hat{n} = 0, \qquad (r, \theta, z) \in \Gamma, t > 0,$$

$$\frac{\partial \phi}{\partial z}(t, r, \theta, 0) = -w_t(t, r, \theta), \qquad (r, \theta) \in \Gamma_0, t > 0,$$

$$\rho_p h w_{tt} + \frac{\partial^2 M_r}{\partial r^2} + \frac{2}{r}\frac{\partial M_r}{\partial r} - \frac{1}{r}\frac{\partial M_\theta}{\partial r} + \frac{2}{r}\frac{\partial^2 M_{r\theta}}{\partial r \partial \theta} + \frac{2}{r^2}\frac{\partial M_{r\theta}}{\partial \theta} + \frac{1}{r^2}\frac{\partial^2 M_\theta}{\partial \theta^2}$$

$$= \frac{\partial^2 (M_r)_{pe}}{\partial r^2} + \frac{2}{r}\frac{\partial (M_r)_{pe}}{\partial r} - \frac{1}{r}\frac{\partial (M_\theta)_{pe}}{\partial r} + \frac{1}{r^2}\frac{\partial^2 (M_\theta)_{pe}}{\partial \theta^2}$$

$$- \rho_f \phi_t(t, r, \theta, 0) + f(t, r, \theta), \tag{5.2}$$

$$w(t, a, \theta) = \frac{\partial w}{\partial r}(t, a, \theta) = 0,$$

$$\phi(0, r, \theta, z) = \phi_0(r, \theta, z), \qquad w(0, r, \theta) = w_0(r, \theta),$$

$$\phi_t(0, r, \theta, z) = \phi_1(r, \theta, z), \qquad w_t(0, r, \theta) = w_1(r, \theta).$$

In the system model (5.2), the plate and acoustic equations are in strong form, which leads to difficulties in the control problem since it involves the differentiation of the Heaviside function and the Dirac delta, which then yields an unbounded (discontinuous) control input term. Moreover, because of the presence and differing material properties of the piezoceramic patches, it is assumed that ρ_p, D, c_D, and ν for the combined structure are piecewise constant in nature (Banks, Wang, Inman, and Slater, 1992). Hence these parameters will be expanded in terms of a Heaviside basis with the edges of the patches defining the support of the basis functions.This also leads to problems in the strong form, however, since it necessitates the differentiation of discontinuous material parameters. To avoid these difficulties, it is advantageous to formulate the problem in a weak or variational form and so throughout the discussion that follows, a weak form will be used when approximating the various components of the problem.

The remainder of this chapter is devoted to the development of an infinite-dimensional formulation and approximation framework that is amenable to the application of LQR and H^∞/MinMax optimal control techniques as well as the estimation of physical parameters. From the standpoint of performing forward simulations and estimating parameters, it is desirable to have a scheme that is accurate, robust, efficient, and easily implemented. From a control perspective, it is desirable to have a scheme that uniformly preserves stability margins as the dimension of the approximating system increases (Banks, Ito, and Wang, 1991). This last criterion is especially important in weakly damped systems having hyperbolic components. Because the only damping in our system is in the plate, the system is weakly damped and care must be taken to choose an approximation scheme that uniformly preserves stability and stabilizability margins. Finally, all of these criteria must be satisfied in the presence of the coordinate singularity at the origin (the careful handling of the singularity is especially important when using spectral basis functions (Bouaoudia and Marcus, 1991; Gottlieb and Orszag, 1977; Orszag, 1974; Orszag and Patera, 1983; Patera and Orszag, 1981; Zang, Streett and Hussaini, 1989)).

Section 5.2 contains a discussion of the wave equation with Neumann boundary conditions on a 2-D circular domain. Approximation techniques for this problem are developed and several test examples are presented that highlight the strengths and weaknesses of the various proposed methods. In the third section, the methods are then extended to the wave equation in a 3-D cylindrical domain. The discretization of the circular plate equations is examined in Section 5.4. Again, several test examples are examined as well as the problem of determining the natural frequencies and mode shapes of a plate having dimensions consistent with those in the experimental setup. The approximation techniques developed in Sections 5.3 and 5.4 are extended in Section 5.5 to the coupled system, and an example demonstrating the accuracy of the method is presented. The problem of employing these numerical techniques in a PDE-based controller to reduce sound pressure levels in the structural acoustic system is considered in Section 5.6. Appropriate LQR optimal control theory is outlined and a detailed numerical example demonstrating the reduction of acoustic sound pressure levels through the activation of piezoceramic patches bonded to the plate is given. These results demonstrate that with careful choices of approximation methods, PDE-based control methods provide a viable means of controlling noise in structural acoustic systems using piezoceramic actuators.

5.2 2-D Wave Equation in Polar Coordinates

In order to develop an efficient and accurate numerical scheme suitable for discretizing the wave portion of the coupled system, we consider first the 2-D acoustic wave equation on a disk of radius a as shown in Fig. 5.2.

For Neumann boundary conditions, the equations of motion are

$$
\begin{gathered}
\phi_{tt} = c^2\Delta\phi + g(t,r,\theta), \qquad (r,\theta) \in \Omega, t > 0, \\
\nabla\phi \cdot \hat{r} = 0, \qquad (r,\theta) \in \Gamma, \qquad t > 0, \\
\phi(0,r,\theta) = \phi_0(r,\theta), \\
\phi_t(0,r,\theta) = \phi_1(r,\theta),
\end{gathered}
\tag{5.3}
$$

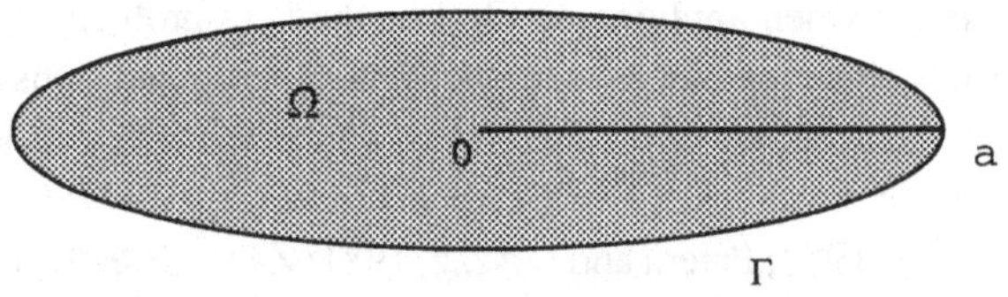

Fig. 5.2. Domain for the 2-D acoustic wave equation.

where ϕ denotes the velocity potential and c is the speed of sound. These boundary conditions were chosen so as to be consistent with the hard-wall conditions in the coupled problem of interest.

To pose the problem in a form conducive to approximation, the state is taken to be ϕ in the space $H = \bar{L}^2(\Omega)$ with the weighted inner product

$$\langle \phi, \xi \rangle_H = \int_\Omega \frac{1}{c^2} \phi \bar{\xi}\, d\omega.$$

Here $\bar{L}^2(\Omega)$ is the quotient space of L^2 over the constant functions (the use of the quotient space results from the fact that the potentials are determined only up to a constant). To provide a class of functions to be considered when defining a variational form of the problem, we also define the Hilbert space $V = \bar{H}^1(\Omega)$ with the inner product

$$\langle \phi, \xi \rangle_V = \int_\Omega \nabla \phi \cdot \overline{\nabla \xi}\, d\omega,$$

where again we interpret this as a quotient space of H^1 over constant functions.

Now, consider the Gelfand triple $V \hookrightarrow H \hookrightarrow V^*$ with pivot space H and define the sesquilinear form $\sigma : V \times V \to \mathbb{C}$ by

$$\sigma(\phi, \xi) = \int_\Omega \nabla \phi \cdot \overline{\nabla \xi}\, d\omega.$$

The system can then be written in weak or variational form as

$$\langle \phi_{tt}, \xi \rangle_{V^*,V} + \sigma(\phi, \xi) = \langle g, \xi \rangle_{V^*,V} \qquad (5.4)$$

for ξ in V. The duality product $\langle \cdot, \cdot \rangle_{V^*,V}$ is the unique extension by continuity of the inner product $\langle \cdot, \cdot \rangle_H$ from $H \times V$ to $V^* \times V$.

To approximate the solution to Eq. (5.4), let $\{B_k^{\mathcal{M}}\}_{k=1}^{\mathcal{M}}$ denote the 2-D basis functions used on the disk and let $H^{\mathcal{M}} = \text{span}\{B_k^{\mathcal{M}}\}$ be the approximating subspace. The product space for the first-order system is $\mathcal{H}^{\mathcal{M}} = H^{\mathcal{M}} \times H^{\mathcal{M}}$. As discussed in Canuto, Hussaini, Quarteroni, and Zang (1988, pp. 90–91), care must be taken when applying spectral methods to problems with coordinate singularities since the incorrect application of pole conditions can significantly degrade the accuracy of the method as well as introduce strong instabilities (Bouaoudia and Marcus, 1991; Gottlieb and Orszag, 1977; Orszag, 1974; Orszag and Patera, 1983; Patera and Orszag, 1981; Zang, Streett, and Hussaini, 1989).

5.2.1 Approximation Techniques

A Fourier–Galerkin expansion in θ yields the approximate solution

$$\phi^{\mathcal{M}}(t, r, \theta) = \sum_{m=-M}^{M} \tilde{\phi}_m(t, r) e^{im\theta}. \tag{5.5}$$

We point out that the use of the complex Fourier expansion simplifies the following discussion both in describing the form of the approximate solution and the construction of the system matrices. However, when combining these wave results with those of the circular plate to yield an approximation scheme for the coupled system, it is easier to use a real Fourier expansion when performing the actual computations (this is due to the presence of the piezoceramic patches on the plate). The interchange between the two expansions is straightforward, and hence details concerning the implementation of the real Fourier scheme are left to the reader.

Several possibilities exist for Legendre or Chebyshev expansions of $\tilde{\phi}_m(t, r)$ both in a collocation and Galerkin setting. These include direct expansions that maintain the parity of the solution, refinements to incorporate the decay of the solution at the origin, and mapped expansions that use all the polynomials and yield better center resolution. These expansions must satisfy the condition

$$\frac{\partial \phi^{\mathcal{M}}}{\partial \theta} = 0$$

at the origin thereby guaranteeing the uniqueness of the solution. This yields the requirement

$$\tilde{\phi}_m(t, r) = 0 \qquad \text{at } r = 0, m \neq 0. \tag{5.6}$$

To guarantee differentiability at the origin, it is appropriate to require that the remaining component satisfies

$$\frac{\partial \tilde{\phi}_0}{\partial r} = 0 \qquad \text{at } r = 0. \tag{5.7}$$

One expansion that satisfies Eqs. (5.6) and (5.7) as well as guarantees that the numerical solution will have the same parity, $\tilde{\phi}_m(t, -r) = (-1)^m \tilde{\phi}_m(t, r)$, as the analytic one is

$$\tilde{\phi}_m(t, r) = \sum_{\substack{n=0 \\ m+n \text{ even}}}^{N} \phi_{mn}(t) r^{|m|} \tilde{P}_n(r), \tag{5.8}$$

where $\tilde{P}_n(r)$ is the nth Legendre or Chebyshev polynomial on the interval $(-a, a)$ (see Orszag and Patera (1983) and Patera and Orszag (1981) for examples where this type of expansion is employed in a CFD setting). The inclusion of the weighting term $r^{|m|}$ is motivated by the observation that the analytic solution decays like $r^{\pm m}$ as $r \to 0$ due to the asymptotic properties of the Bessel functions (see Example 3 or Abramowitz and Stegan (1972)). The choice $|m|$ results from the fact that we are interested in the bounded solution.

Combining the Fourier and polynomial components then yields the basis

$$B_k(r, \theta) = e^{im\theta} r^{|m|} \tilde{P}_n(r). \tag{5.9}$$

We point out that the approximate solution constructed with this basis satisfies Eq. (5.7) because of the property that $\tilde{P}_n'(0) = 0$ for even n.

If the analytic solution is C^∞ for all time, then for all m, the terms $\tilde{\phi}_m$ behave as

$$\tilde{\phi}_m(t, r) = \mathcal{O}\left(r^{|m|}\right) \qquad (r \to 0)$$

(Bouaoudia and Marcus, 1991) and exponential convergence can be expected. As demonstrated in Example 1, this convergence rate is also attained in many problems with less smoothness; however, in problems in which the analytic solution is only $C^{\mathcal{P}}$ with $\mathcal{P} < \mathcal{M}$, the addition of the term $r^{|m|}$ with the parity-preserving polynomials can in some cases force the approximate solution to have more decay at the origin than does the analytic one. Depending on the choice of basis, this can have the effect of drastically slowing the convergence rate (see Example 2).

A second expansion for $\tilde{\phi}_m(t, r)$ that uses all the polynomials is

$$\tilde{\phi}_m(t, r) = \sum_{n=0}^{N} \phi_{mn}(t) r^{|m|} P_n(r),$$

where $P_n(r)$ is the nth Legendre or Chebyshev polynomial that has been mapped to the interval $(0, a)$. Although this expansion does not explicitly enforce the condition (5.7), as shown in Examples 1 and 2 it does yield an accurate approximation to the solution.

A third possibility for the basis includes a Bessel function expansion in radius coupled with the above Fourier basis in θ. As discussed on page 156 of Gottlieb and Orszag (1977), the Legendre or Chebyshev bases are probably preferable to a Bessel expansion since the Legendre and Chebyshev series converge more rapidly to general functions, regardless of the boundary conditions (the Bessel expansions can be unstable with respect to boundary conditions).

The approximating polynomial sets $\{\tilde{P}_n(r)\}$ and $\{P_n(r)\}$ must also be chosen so as to be suitable as a basis for the quotient space. It is tempting to simply impose the condition $|m| + n \neq 0$, which eliminates the constant function as was done in the 2-D case. As demonstrated by results summarized in Banks and Smith (1994), however, this leads to incorrect natural frequencies for the system. A second means of generating a suitable quotient space basis is to impose the condition

$$\int_{\Omega} \phi^{\mathcal{M}}(t, r, \theta)\, d\omega = 0$$

(note that this technique was used when testing the linear spline and finite element schemes in the rectangular 2-D problem (Banks, Fang, Silcox, and Smith, 1993; Banks, Silcox, and Smith, 1994)).

Demonstrating this condition for the translated set of Legendre polynomials, we find this yields the requirement that

$$\phi_{00}(t) = -\frac{2}{a^2} \sum_{n=1}^{N} \phi_{0n}(t) a_n,$$

where $a_n = \int_0^a r P_n(r) dr$. By noting that this latter term can be written as

$$a_n = \int_0^a \left[\frac{a}{2} P_1(r) + \frac{a}{2} P_0(r) \right] P_n(r)\, dr$$

and using the orthogonality properties of the Legendre polynomials, this leads to the polynomial set $\{P_n^m(r)\}$ where

$$P_n^m(r) = \begin{cases} P_1(r) - 1/3, & m = 0, n = 1 \\ P_n(r), & \text{otherwise} \end{cases} \tag{5.10}$$

if one is using the polynomials that have been mapped to the interval $(0, a)$. This yields the quotient space basis $\{B_k^{\mathcal{M}}\}$, $\mathcal{M} = (2M + 1) \cdot (N + 1) - 1$, where

$$B_k^{\mathcal{M}}(r, \theta) = r^{|m|} P_n^m(r) e^{im\theta}. \tag{5.11}$$

Similarly, if the parity-preserving polynomials $\{\tilde{P}_n(r)\}$ are employed, the set satisfying the quotient space condition has the components

$$\tilde{P}_n^m(r) = \begin{cases} \tilde{P}_1(r) - 1/3, & m = 0, n = 1 \\ \tilde{P}_n(r), & \text{otherwise} \end{cases} \tag{5.12}$$

with a resulting quotient space basis similar to that defined in Eq. (5.11). In this case, $\mathcal{M} = (2M + 1) \cdot (N + 1)/2 - 1$.

One further refinement that can be made on the basis is to reduce the decay of $r^{|m|}$ by instead using the terms

$$B_k^{\mathcal{M}}(r, \theta) = r^{|\hat{m}|} P_n^m(r)\, e^{im\theta}, \tag{5.13}$$

where

$$\hat{m} = \begin{cases} m, & |m| = 0, \ldots, 5 \\ 5, & |m| = 6, \ldots, M. \end{cases}$$

The use of this basis results in a slight loss of accuracy when approximating high-frequency responses. The advantage, however, is that it yields a better conditioned mass matrix than does the choice (5.11). For high-frequency approximations, this advantage appears to outweigh any disadvantages associated with the slight accuracy loss due to truncating the power on the weighting term. Finally, the above discussion has been for the polynomials that have been mapped to the interval $(0, a)$, and similar expressions for the basis can be written in terms of the parity-preserving polynomials $\{\tilde{P}_n(r)\}$ that are defined on the interval $(-a, a)$.

In terms of the basis elements defined in Eqs. (5.11) or (5.13), the approximate solution can then be written as

$$\phi^{\mathcal{M}}(t, r, \theta) = \sum_{k=1}^{\mathcal{M}} \phi_k(t) B_k^{\mathcal{M}}(r, \theta),$$

where, again, $\mathcal{M} = (2M + 1) \cdot (N + 1) - 1$ and N is the number of Legendre polynomials. The ordering of indices is assumed to be that defined by the Fourier expansion (5.5).

5.2.2 Matrix System

The restriction of the infinite-dimensional system (5.4) to the space $H^{\mathcal{M}} \times H^{\mathcal{M}}$ then yields

$$\langle \phi_{tt}^{\mathcal{M}}, \xi \rangle_H + \sigma(\phi^{\mathcal{M}}, \xi) = \langle g, \xi \rangle_H$$

or

$$\int_\Omega \frac{1}{c^2} \phi_{tt}^{\mathcal{M}} \bar{\xi}\, d\omega + \int_\Omega \nabla\phi^{\mathcal{M}} \cdot \overline{\nabla\xi}\, d\omega = \int_\Omega \frac{1}{c^2} g \bar{\xi}\, d\omega$$

for ξ in $H^{\mathcal{M}}$. The corresponding matrix system is

$$\begin{aligned} M^{\mathcal{M}} \dot{y}^{\mathcal{M}}(t) &= \tilde{A}^{\mathcal{M}} y^{\mathcal{M}}(t) + \tilde{G}^{\mathcal{M}}(t), \\ M^{\mathcal{M}} y^{\mathcal{M}}(0) &= \tilde{y}_0^{\mathcal{M}}, \end{aligned}$$

where

$$y^{\mathcal{M}}(t) = \begin{pmatrix} \vartheta^{\mathcal{M}}(t) \\ \dot{\vartheta}^{\mathcal{M}}(t) \end{pmatrix}.$$

Here $\vartheta^{\mathcal{M}}(t) = [\phi_1(t), \phi_2(t), \ldots, \phi_{\mathcal{M}}(t)]^T$ denotes the $\mathcal{M} \times 1$ vector containing the approximate state coefficients. The full system has the form

$$\begin{bmatrix} K_{r\theta}^{\mathcal{M}} & 0 \\ 0 & M_{r\theta}^{\mathcal{M}} \end{bmatrix} \begin{bmatrix} \dot{\vartheta}^{\mathcal{M}}(t) \\ \ddot{\vartheta}^{\mathcal{M}}(t) \end{bmatrix} = \begin{bmatrix} 0 & K_{r\theta}^{\mathcal{M}} \\ -K_{r\theta}^{\mathcal{M}} & 0 \end{bmatrix} \begin{bmatrix} \vartheta^{\mathcal{M}}(t) \\ \dot{\vartheta}^{\mathcal{M}}(t) \end{bmatrix} + \begin{bmatrix} 0 \\ \tilde{G}^{\mathcal{M}}(t) \end{bmatrix},$$

$$\begin{bmatrix} K_{r\theta}^{\mathcal{M}} & 0 \\ 0 & M_{r\theta}^{\mathcal{M}} \end{bmatrix} \begin{bmatrix} \vartheta^{\mathcal{M}}(0) \\ \dot{\vartheta}^{\mathcal{M}}(0) \end{bmatrix} = \begin{bmatrix} g_1^{\mathcal{M}} \\ g_2^{\mathcal{M}} \end{bmatrix}.$$

The component matrices and vectors are given by

$$
\begin{aligned}
\left[K_{r\theta}^{\mathcal{M}}\right]_{\ell,k} &= \int_{\Omega} \nabla B_k^{\mathcal{M}} \cdot \overline{\nabla B_\ell^{\mathcal{M}}}\, d\omega, \\
\left[g_1^{\mathcal{M}}\right]_{\ell} &= \int_{\Omega} \nabla \phi_0 \cdot \overline{\nabla B_\ell^{\mathcal{M}}} d\omega = \langle \phi_0, B_\ell^{\mathcal{M}} \rangle_V, \\
\left[M_{r\theta}^{\mathcal{M}}\right]_{\ell,k} &= \int_{\Omega} \frac{1}{c^2} B_k^{\mathcal{M}} \overline{B_\ell^{\mathcal{M}}}\, d\omega, \qquad (5.14) \\
\left[g_2^{\mathcal{M}}\right]_{\ell} &= \int_{\Omega} \frac{1}{c^2} \phi_1 \overline{B_\ell^{\mathcal{M}}} d\omega = \langle \phi_1, B_\ell^{\mathcal{M}} \rangle_H, \\
\left[\tilde{G}_2^{\mathcal{M}}(t)\right]_{\ell} &= \int_{\Omega} \frac{1}{c^2} g \overline{B_\ell^{\mathcal{M}}}\, d\omega,
\end{aligned}
$$

where the index ranges are $k, \ell = 1, \ldots, \mathcal{M}$. The reader is referred to Banks and Smith (1994) for explicit details concerning the construction of these matrices.

5.2.3 Example 1

Consider the forced wave equation

$$
\begin{aligned}
\phi_{tt} &= \Delta\phi + g(t, r, \theta), \qquad (r, \theta) \in \Omega, t > 0, \\
\nabla\phi \cdot \hat{r} &= 0, \qquad (r, \theta) \in \Gamma, t > 0, \qquad (5.15) \\
\phi(0, r, \theta) &= 0, \qquad \phi_t(0, r, \theta) = 0,
\end{aligned}
$$

where

$$
\begin{aligned}
g(t, r, \theta) = \Big[& 2(\cos(2\pi r/a) - 1) + t^2 \cdot 4\pi^2/a^2 \cos(2\pi r/a) \\
& + t^2 \cdot 2\pi/a \cdot \frac{1}{r} \sin(2\pi r/a) + t^2 \cdot \frac{1}{r^2}(\cos(2\pi r/a) - 1) \Big] \sin\theta.
\end{aligned}
$$

The hard-wall boundary conditions were chosen so as to be consistent with the experimental setup described in the next section. The true solution is $\phi(t, r, \theta) = t^2(\cos(2\pi r/a) - 1)\sin\theta$. The absolute and relative errors on a uniform grid at the time $T = 1$ are shown in Tables 5.1 and 5.2. The results in Table 5.1 were obtained with the basis $\{e^{im\theta} r^{|\hat{m}|} \tilde{P}_n^m(r)\}$, where $\tilde{P}_n^m(r)$ is given by (5.12) and hence consists of Legendre polynomials on the interval $(-a, a)$ (in this case, $a = 1$). Figures 5.3 and 5.4 show the true and approximate solutions with $N = 5, 7$, and 9. Corresponding results for the basis $\{e^{im\theta} r^{|\hat{m}|} P_n^m(r)\}$, where $P_n^m(r)$ is given by (5.10), are found in Table 5.2. Here the radial basis consists of Legendre polynomials that have been mapped to the interval $(0, a)$ where, again, $a = 1$. As demonstrated by these results, both the parity-preserving and mapped basis functions work quite well for this example.

Table 5.1. *Absolute and relative errors with the Fourier–Legendre basis* $\{e^{im\theta}r^{|\hat{m}|}\tilde{P}_n^m(r)\}$ *on the radial interval (0,1).*

M	N	size($\tilde{A}^{\mathcal{M}}$)	$\lVert\phi_{true}-\phi_{app}\rVert$	$\frac{\lVert\phi_{true}-\phi_{app}\rVert}{\lVert\phi_{true}\rVert}$
1	5	16 × 16	.4083 − 0	.2041 − 0
1	7	22 × 22	.6258 − 1	.3129 − 1
1	9	28 × 28	.5632 − 2	.2816 − 2
1	11	34 × 34	.3685 − 3	.1843 − 3
1	13	40 × 40	.1460 − 4	.7298 − 5

Table 5.2. *Absolute and relative errors with the Fourier–Legendre basis* $\{e^{im\theta}r^{|\hat{m}|}P_n^m(r)\}$ *on the radial interval (0,1).*

M	N	size($\tilde{A}^{\mathcal{M}}$)	$\lVert\phi_{true}-\phi_{app}\rVert$	$\frac{\lVert\phi_{true}-\phi_{app}\rVert}{\lVert\phi_{true}\rVert}$
1	3	22 × 22	.4269 − 1	.2158 − 1
1	6	40 × 40	.2568 − 2	.1298 − 2
1	9	58 × 58	.9985 − 6	.5048 − 6

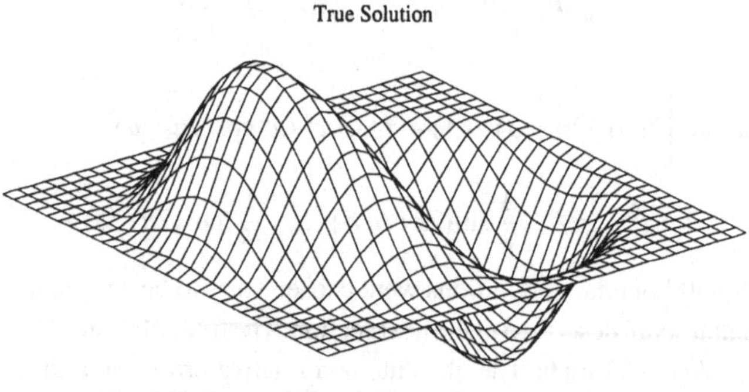

Fig. 5.3. True solution to Example 1 at time $T = 1$.

5.2.4 Example 2

Consider now the problem (5.15) with the forcing function

$$g(t,r,\theta) = \Big[2(\cos(2\pi r)-1) + t^2\cdot 4\pi^2\cos(2\pi r) + t^2\cdot 2\pi\frac{1}{r}\sin(2\pi r) + t^2\cdot\frac{1}{r^2}\cdot 9(\cos(2\pi r)-1)\Big]\sin 3\theta.$$

Table 5.3. *Absolute and relative errors with the parity preserving basis* $\{e^{im\theta} r^{|\hat{m}|} \tilde{P}_n^m(r)\}$.

M	N	size($\tilde{A}^{\mathcal{M}}$)	$\|\phi_{true} - \phi_{app}\|$	$\frac{\|\phi_{true}-\phi_{app}\|}{\|\phi_{true}\|}$
3	9	68 × 68	.6624 − 0	.3349 − 0
3	13	96 × 96	.3795 − 0	.1919 − 0
3	17	124 × 124	.2342 − 0	.1184 − 0

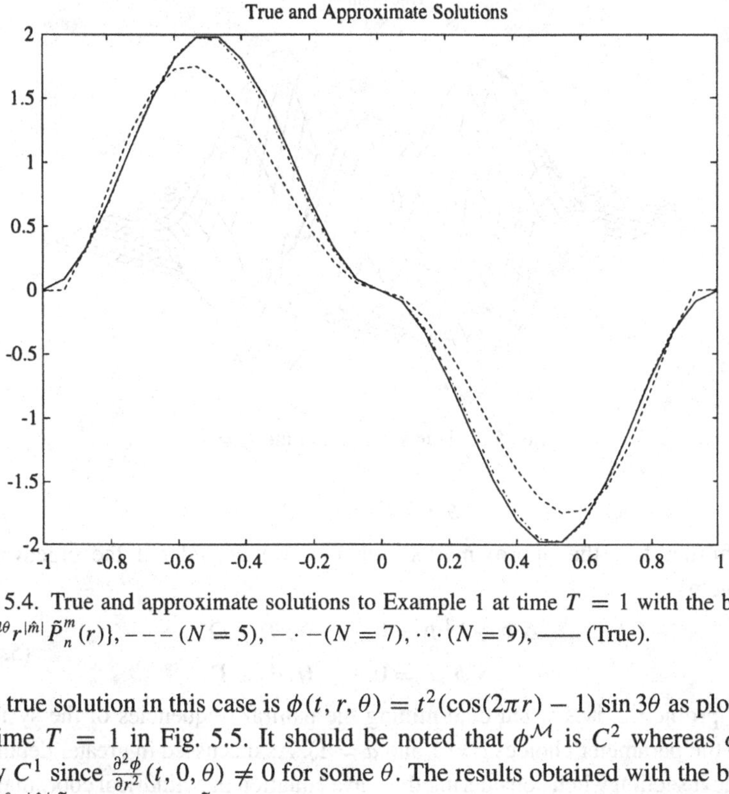

Fig. 5.4. True and approximate solutions to Example 1 at time $T = 1$ with the basis $\{e^{im\theta} r^{|\hat{m}|} \tilde{P}_n^m(r)\}$, − − − ($N = 5$), − · − ($N = 7$), · · · ($N = 9$), —— (True).

The true solution in this case is $\phi(t, r, \theta) = t^2(\cos(2\pi r) - 1) \sin 3\theta$ as plotted at time $T = 1$ in Fig. 5.5. It should be noted that $\phi^{\mathcal{M}}$ is C^2 whereas ϕ is only C^1 since $\frac{\partial^2 \phi}{\partial r^2}(t, 0, \theta) \neq 0$ for some θ. The results obtained with the basis $\{e^{im\theta} r^{|\hat{m}|} \tilde{P}_n^m(r)\}$ (with $\tilde{P}_n^m(r)$ given by (5.12) and hence containing polynomials on the interval $(-a, a)$) are shown in Table 5.3 and Fig. 5.6. As shown in this figure, the approximate solution has too much decay at the origin. This also correlates with the results in the table that indicate a drastically reduced convergence rate. The corresponding results obtained with the translated basis $\{e^{im\theta} r^{|\hat{m}|} P_n^m(r)\}$, with $P_n^m(r)$ given by (5.10), are given in Table 5.4 and Fig. 5.7. Both the table and the figure indicate a substantial improvement in the convergence with the translated basis.

Table 5.4. *Absolute and relative errors with the mapped basis* $\{e^{im\theta} r^{|\hat{m}|} P_n^m(r)\}$.

M	N	size$(\tilde{A}^{\mathcal{M}})$	$\|\phi_{true} - \phi_{app}\|$	$\frac{\|\phi_{true}-\phi_{app}\|}{\|\phi_{true}\|}$
3	3	54×54	.1637 − 0	.8278 − 1
3	6	96×96	.3104 − 1	.1569 − 1
3	9	138×138	.1152 − 1	.5825 − 2

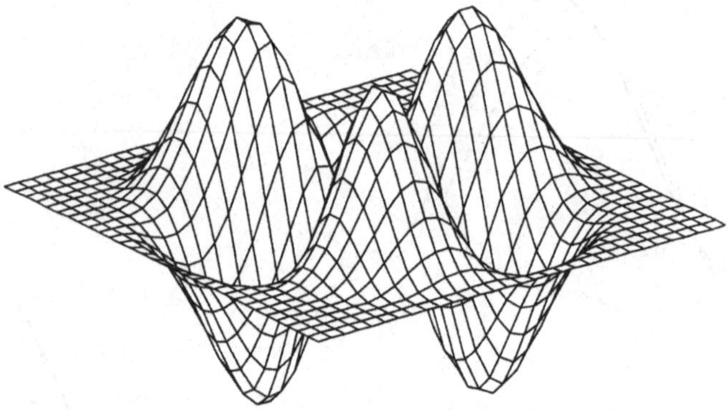

Fig. 5.5. True solution at time $T = 1$.

5.2.5 Example 3

To further test the approximation scheme, we considered the eigenvalue problem

$$\begin{aligned} \Delta\Phi + \lambda^2\Phi &= 0, \qquad (r,\theta) \in \Omega, \\ \nabla\phi \cdot \hat{r} &= 0, \qquad (r,\theta) \in \Gamma \end{aligned} \tag{5.16}$$

(this problem arises when determining the natural frequencies of the system with the parameter choices $c = 1$ and $a = 1$). As discussed in greater detail in the next section when considering the wave equation in cylindrical coordinates, the solution to (5.16) is determined by the nonlinear equation

$$\frac{dJ_m(\lambda)}{dr} = 0 \tag{5.17}$$

when the boundary constraint $\frac{\partial\phi}{\partial r}\Big|_{a=1} = 0$ is enforced. Here J_m denotes the mth Bessel function of the first kind. Several solutions, λ_{mn}, to (5.17) for various values of n are given in Table 5.5.

To compare these results with those obtained via the Fourier–Galerkin expansion, it is noted that under approximation, the eigenvalue problem (5.16)

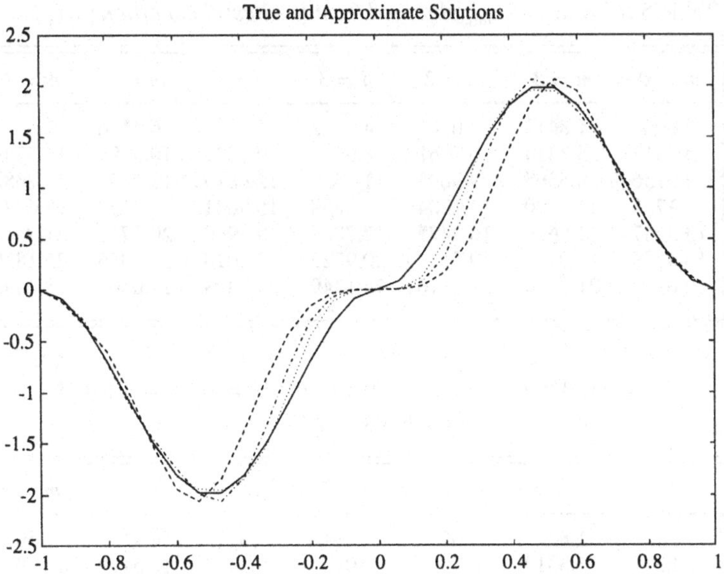

Fig. 5.6. True and approximate solutions at time $T = 1$ with the parity preserving basis $\{e^{im\theta} r^{|\hat{m}|} \tilde{P}_n^m(r)\}$, – – – ($N = 9$), – · –($N = 13$), · · · ($N = 17$), —— (True).

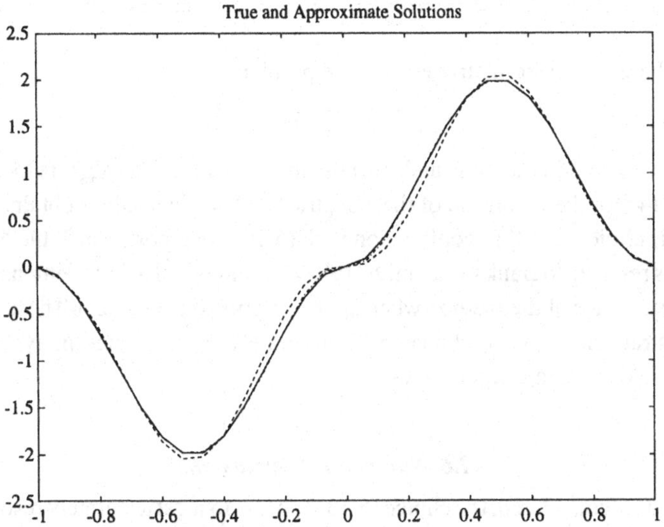

Fig. 5.7. True and approximate solutions at time $T = 1$ with the mapped basis $\{e^{im\theta} r^{|\hat{m}|} P_n^m(r)\}$, – – – ($N = 3$), – · –($N = 6$), · · · ($N = 9$), —— (True).

Table 5.5. *Values of λ_{mn} obtained from the Bessel condition (5.17).*

n	$m=0$	$m=1$	$m=2$	$m=3$	$m=4$	$m=5$	$m=6$
0	0.0000	1.8412	3.0542	4.2012	5.3176	6.4156	7.5013
1	3.8317	5.3314	6.7061	8.0152	9.2824	10.5199	11.7349
2	7.0156	8.5363	9.9695	11.3459	12.6819	13.9872	15.2682
3	10.1735	11.7060	13.1704	15.5858	15.9641	17.3128	18.6374
4	13.3237	14.8636	16.3475	17.7887	19.1960	20.5755	21.9317
5	16.4706	18.0155	19.5129	20.9725	22.4010	23.8036	25.1839
6	19.6159	21.1644	22.6716	24.1449	25.5898	27.0103	28.4098

Table 5.6. *Values of λ_{mn} obtained with $M = 6$, $N = 12$, and the basis (5.13).*

n	$m=0$	$m=1$	$m=2$	$m=3$	$m=4$	$m=5$	$m=6$
0		1.8412	3.0542	4.2012	5.3176	6.4156	7.5013
1	3.8317	5.3314	6.7061	8.0152	9.2824	10.5199	11.7349
2	7.0156	8.5363	9.9695	11.3459	12.6819	13.9872	15.2682
3	10.1735	11.7060	13.1704	14.5858	15.9641	17.3128	18.6374
4	13.3237	14.8636	16.3475	17.7888	19.1960	20.5755	21.9317
5	16.4717	18.0159	19.5131	20.9730	22.4018	23.8043	25.1860
6	19.6250	21.1860	22.6929	24.1559	25.5937	27.0152	28.4223

yields the generalized matrix eigenvalue problem

$$K_{r\theta}^{\mathcal{M}} \vartheta^{\mathcal{M}} = \lambda^2 M_{r\theta}^{\mathcal{M}} \vartheta^{\mathcal{M}}$$

with $c^2 = a = 1$ in the mass and stiffness matrices $M_{r\theta}^{\mathcal{M}}$ and $K_{r\theta}^{\mathcal{M}}$, respectively (see (5.14) for the definition of these matrices). The eigenvalues obtained with the basis choice (5.13) in conjunction with (5.10) are reported in Table 5.6 (the reader is referred to Banks and Smith (1994) for numerical results demonstrating the convergence of the method when approximating eigenvalues). These results demonstrate that the basis choice (5.13) accurately approximates the system and hence yields accurate eigenvalues.

5.2.6 Numerical Conclusions

Of the bases considered, the choice (5.13) appears to most accurately resolve the behavior of the solution at the origin in problems in which the forcing function has limited continuity. This consideration is important since this will be the case when piezoceramic patches are used as actuators in coupled structural acoustics problems. As seen in Example 2, the basis $\{e^{im\theta} r^{|\hat{m}|} \tilde{P}_n(r)\}$, which was chosen to

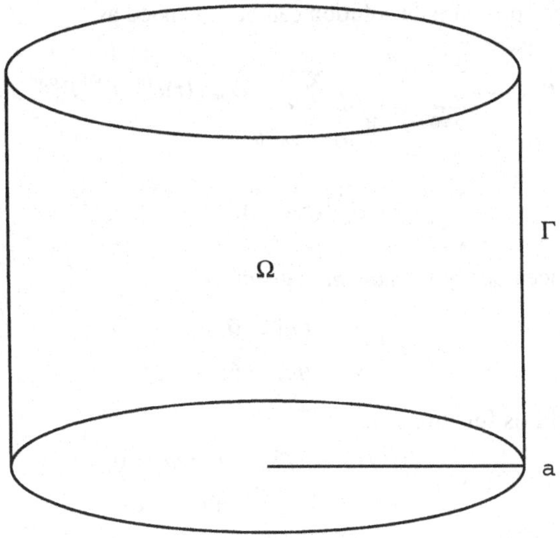

Fig. 5.8. Domain for the 3-D acoustic wave equation.

maintain parity, can in some cases have too much decay at the origin. This latter choice is also more difficult to implement than the basis containing the mapped Legendre polynomials. Hence the choice (5.13) will be used when discretizing the cylindrical wave problem and the coupled acoustic/structure model.

5.3 3-D Wave Equation in Cylindrical Coordinates

We next consider the full 3-D acoustic wave equation cylindrical domain having radius a and length ℓ as shown in Fig. 5.8. For Neumann boundary conditions, the equations of motion are

$$\begin{aligned} \phi_{tt} &= c^2\Delta\phi + g(t,r,\theta,z), \qquad (r,\theta,z) \in \Omega, t > 0, \\ \nabla\phi\cdot\hat{n} &= 0, \qquad (r,\theta,z) \in \Gamma, t > 0, \\ \phi(0,r,\theta,z) &= \phi_0(r,\theta,z), \\ \phi_t(0,r,\theta,z) &= \phi_1(r,\theta,z), \end{aligned} \tag{5.18}$$

where, again, ϕ is the velocity potential. The boundary conditions were chosen so as to be consistent with those arising in the wave portion of the coupled system.

5.3.1 Approximate Solution

The theoretical framework for the problem is analogous to that of the 2-D problem in the last section with the state space again taken to be $H = \bar{L}^2(\Omega)$.

In this case, the approximate solution can be expanded as

$$\phi^{\mathcal{M}}(t,r,\theta,z) = \sum_{p=0}^{P}\sum_{m=-M}^{M}\sum_{\substack{n=0\\ p+|m|+n\neq 0}}^{N^{p,m}} \phi_{pmn}(t)e^{im\theta}r^{|\hat{m}|}P_n^{p,m}(r)P_p(z)$$

$$= \sum_{k=1}^{\mathcal{M}} \phi_k(t)B_k^{\mathcal{M}}(r,\theta,z),$$

where the reduced decay variable $\hat{m}$ is given by

$$\hat{m} = \begin{cases} m, & |m| = 0,\ldots,5 \\ 5, & |m| = 6,\ldots,M \end{cases}$$

and the radial basis functions are

$$P_n^{p,m}(r) = \begin{cases} P_1(r) - 1/3, & p = m = 0, n = 1 \\ P_n(r), & \text{otherwise} \end{cases}$$

(see (5.13)). Here $P_n(r)$ and $P_p(z)$ are the nth and pth Legendre polynomials that have been mapped to the intervals $(0,a)$ and $(0,\ell)$, respectively. The term $P_1(r) - 1/3$ when $p = m = 0, n = 1$ results from the orthogonality properties of the Legendre polynomials and arises when enforcing the condition

$$\int_\Omega \phi^{\mathcal{M}}(t,r,\theta,z)\,d\omega = 0$$

so as to guarantee that the functions are suitable as a basis for the quotient space. The inclusion of the weight $r^{|\hat{m}|}$ again incorporates the decay of the analytic solution near the origin while ensuring its uniqueness at that point. Finally, we note that the limit $N^{p,m}$ is given by $N^{p,m} = N+1$ when $p+|m| \neq 0$ and $N^{p,m} = N$ when $p = m = 0$, which implies that $\mathcal{M} = (2M+1)(N+1)(P+1) - 1$ basis functions are used in the wave expansion.

The $\mathcal{M}$-dimensional approximating subspace is then taken to be $H^{\mathcal{M}} = \text{span}\{B_k^{\mathcal{M}}\}$ with the 3-D basis

$$B_k(r,\theta,z) = e^{im\theta}r^{|\hat{m}|}P_n^{p,m}(r)P_p(z).$$

5.3.2 Matrix System

The restriction of the infinite-dimensional system (5.18) to the space $H^{\mathcal{M}} \times H^{\mathcal{M}}$ then yields

$$\langle \phi_{tt}^{\mathcal{M}}, \xi\rangle_H + \sigma(\phi^{\mathcal{M}}, \xi) = \langle G, \xi\rangle_H$$

or

$$\int_\Omega \frac{1}{c^2}\phi_{tt}^{\mathcal{M}}\bar{\xi}d\omega + \int_\Omega \nabla\phi^{\mathcal{M}}\cdot\overline{\nabla\xi}\,d\omega = \int_\Omega \frac{1}{c^2}g\bar{\xi}\,d\omega$$

for ξ in $H^{\mathcal{M}}$. The corresponding matrix system is

$$M^{\mathcal{M}}\dot{y}^{\mathcal{M}}(t) = \tilde{A}^{\mathcal{M}}y^{\mathcal{M}}(t) + \tilde{G}^{\mathcal{M}}(t),$$
$$M^{\mathcal{M}}y^{\mathcal{M}}(0) = \tilde{y}_0^{\mathcal{M}},$$

where

$$y^{\mathcal{M}}(t) = \begin{pmatrix} \vartheta^{\mathcal{M}}(t) \\ \dot{\vartheta}^{\mathcal{M}}(t) \end{pmatrix}$$

with the approximate state vector $\vartheta^{\mathcal{M}}(t) = [\phi_1(t), \phi_2(t), \ldots, \phi_{\mathcal{M}}(t)]^T$. The full system has the form

$$\begin{bmatrix} K_{r\theta z}^{\mathcal{M}} & 0 \\ 0 & M_{r\theta z}^{\mathcal{M}} \end{bmatrix} \begin{bmatrix} \dot{\vartheta}^{\mathcal{M}}(t) \\ \ddot{\vartheta}^{\mathcal{M}}(t) \end{bmatrix} = \begin{bmatrix} 0 & K_{r\theta z}^{\mathcal{M}} \\ -K_{r\theta z}^{\mathcal{M}} & 0 \end{bmatrix} \begin{bmatrix} \vartheta^{\mathcal{M}}(t) \\ \dot{\vartheta}^{\mathcal{M}}(t) \end{bmatrix} + \begin{bmatrix} 0 \\ \tilde{G}^{\mathcal{M}}(t) \end{bmatrix},$$

$$\begin{bmatrix} K_{r\theta z}^{\mathcal{M}} & 0 \\ 0 & M_{r\theta z}^{\mathcal{M}} \end{bmatrix} \begin{bmatrix} \vartheta^{\mathcal{M}}(0) \\ \dot{\vartheta}^{\mathcal{M}}(0) \end{bmatrix} = \begin{bmatrix} g_1^{\mathcal{M}} \\ g_2^{\mathcal{M}} \end{bmatrix}.$$

The component matrices $M_{r\theta z}^{\mathcal{M}}$ and $K_{r\theta z}^{\mathcal{M}}$ can be succinctly described as follows. Let the fundamental $(P+1) \times (P+1)$ matrices M_z and K_z be defined as

$$[M_z]_{ij} = \int_0^{\ell} P_i(z)P_j(z)\,dz,$$

$$[K_z]_{ij} = \int_0^{\ell} P_i'(z)P_j'(z)\,dz.$$

By using the tensor properties of the 3-D basis, the matrices $M_{r\theta z}^{\mathcal{M}}$ and $K_{r\theta z}^{\mathcal{M}}$ can then be formed as the tensor products

$$\begin{aligned} M_{r\theta z}^{\mathcal{M}} &= M_z^{\mathcal{M}} \otimes M_{r\theta}^{\mathcal{M}}, \\ K_{r\theta z}^{\mathcal{M}} &= M_z^{\mathcal{M}} \otimes K_{r\theta}^{\mathcal{M}} + K_z^{\mathcal{M}} \otimes M_{r\theta}^{\mathcal{M}}, \end{aligned} \tag{5.19}$$

where $M_{r\theta}^{\mathcal{M}}$ and $K_{r\theta}^{\mathcal{M}}$ are given in (5.14). The construction is completed by updating the row and column of $M_{r\theta z}^{\mathcal{M}}$ and $K_{r\theta z}^{\mathcal{M}}$ affected by the alterations used to guarantee that the functions are a basis for the quotient space. The vectors $\tilde{G}^{\mathcal{M}}(t)$, $g_1^{\mathcal{M}}$, and $g_2^{\mathcal{M}}$ are defined in a manner similar to that in (5.14).

5.3.3 Example 4

Consider the problem

$$\phi_{tt} = \Delta\phi + g(t, r, \theta, z), \qquad (r, \theta, z) \in \Omega, t > 0,$$

$$\nabla\phi \cdot \hat{n} = 0, \qquad (r, \theta, z) \in \Gamma, t > 0,$$

$$\phi(0, r, \theta, z) = 0, \qquad \phi_t(0, r, \theta, z) = 0,$$

Table 5.7. *Absolute and relative errors with* $a = .6$, $\ell = 1$.

M	N	P	size$(\tilde{A}^{\mathcal{M}})$	$\lVert\phi_{true} - \phi_{app}\rVert$	$\frac{\lVert\phi_{true}-\phi_{app}\rVert}{\lVert\phi_{true}\rVert}$
1	3	3	94 × 94	.2216 − 0	.2792 − 0
1	6	6	292 × 292	.3168 − 3	.3992 − 3
1	9	9	598 × 598	.4084 − 4	.5146 − 4

where

$$\begin{aligned} g(t,r,\theta) = &\Bigg[\bigg(2(\cos(2\pi r/a) - 1) + t^2 \cdot 4\pi^2/a^2 \cos(2\pi r/a) + t^2 \cdot 2\pi/a \\ &\cdot \frac{1}{r}\sin(2\pi r/a) + t^2 \cdot \frac{1}{r^2}(\cos(2\pi r/a) - 1)\bigg)\bigg(\sin^2(\pi z)\frac{1}{2}\bigg) \\ &- t^2 \cdot (\cos(2\pi r/a) - 1) \cdot 2\pi^2(\cos^2(\pi z) - \sin^2(\pi z))\Bigg]\sin\theta. \end{aligned}$$

The true solution here is $\phi(t,r,\theta,z) = t^2(\cos(2\pi r/a)-1)(\sin^2(\pi z)-\frac{1}{2})\sin\theta$. The absolute and relative errors on a uniform grid at the time $T = 1$ are shown in Table 5.7. The cylinder in this case had axial length $\ell = 1$ and radius $a = .6$. These results demonstrate the convergence of the method for this example.

5.3.4 Example 5

To further test the wave discretization and to find the natural frequencies for the wave equation in a cylinder having the same dimensions as those used in the experimental setup, we consider the problem

$$\begin{aligned} \phi_{tt} &= c^2\Delta\phi, \qquad (r,\theta,z) \in \Omega, t > 0, \\ \nabla\phi \cdot \hat{n} &= 0, \qquad (r,\theta,z) \in \Gamma, t > 0, \\ \phi(0,r,\theta,z) &= \phi_0(r,\theta,z), \\ \phi_t(0,r,\theta,z) &= \phi_1(t,\theta,z), \end{aligned}$$

where c as usual denotes the speed of sound in the medium. Through the separation of variables $\phi(t,r,\theta,z) = T(t)\Phi(r,\theta,z)$, one arrives at Helmholtz's equation

$$\begin{aligned} \Delta\Phi + \gamma^2\Phi &= 0, \qquad (r,\theta,z) \in \Omega, \\ \nabla\Phi \cdot \hat{n} &= 0, \qquad (r,\theta,z) \in \Gamma \end{aligned} \tag{5.20}$$

and the relation

$$T'' + \omega^2 T = 0, \qquad t > 0.$$

The separation constant here is $\gamma = \omega/c$ where ω is the circular frequency with units of radians/sec. To find Φ, we separate variables once more. Letting $\Phi(r,\theta,z) = R(r)\Theta(\theta)Z(z)$, the expansion of Helmholtz's equation yields

$$\frac{1}{rR}\frac{d}{dr}\left(r\frac{dR}{dr}\right) + \frac{1}{r^2\Theta}\frac{d^2\Theta}{d\theta^2} + \frac{1}{Z}\frac{\partial^2 Z}{dz^2} + \gamma^2 = 0,$$

which implies that R, Θ, and Z must satisfy the differential equations

$$\frac{1}{rR}\frac{d}{dr}\left(r\frac{dR}{dr}\right) - \frac{m^2}{r^2} - k^2 + \gamma^2 = 0, \qquad \frac{dR(a)}{dr} = 0, \tag{5.21}$$

$$\frac{d^2\Theta}{d\theta^2} + m^2\Theta = 0, \qquad \Theta(0) = \Theta(2\pi) \tag{5.22}$$

and

$$\frac{d^2Z}{dz^2} + k^2 Z = 0, \qquad \frac{dZ(0)}{dz} = \frac{dZ(\ell)}{dz} = 0, \tag{5.23}$$

respectively.

The general solution to (5.23) is

$$Z(z) = A_1\cos(kz) + B_1\sin(kz),$$

and by enforcing the boundary conditions, it is found that $B_1 = 0$ and the frequencies are $k_p = \frac{p\pi}{\ell}$, $p = 0, 1, 2, \ldots$. Similarly, the general solution to (5.22) is periodic and is given by $\Theta(\theta) = A_2\cos(m\theta) + B_2\sin(m\theta)$. Finally, by enforcing differentiability constraints, the general solution to (5.21) is found to be

$$R(r) = A_3 J_m(\Lambda r),$$

where $\Lambda^2 = -k^2 + \gamma^2$ and J_m is the mth Bessel function of the first kind. The eigenvalues of (5.21) are then determined by applying the boundary condition and solving for the zeros of the nonlinear equation

$$\frac{dJ_m(\lambda)}{dr} = 0, \tag{5.24}$$

where $\lambda = \Lambda a$. This then yields the set of eigenvalues

$$\gamma_{mnp}^2 = \left(\frac{p\pi}{\ell}\right)^2 + \left(\frac{\lambda_{mn}}{a}\right)^2, \tag{5.25}$$

where, again, λ_{mn} solves (5.24).

To determine the natural frequencies for the problem, we recall that $f = \frac{1}{2\pi}\omega$, where f has units of hertz. By combining this with (5.25) and the fact that $\gamma = \omega/c$, the natural frequencies can be expressed as

$$f_{mnp} = \frac{1}{2\pi}c\sqrt{\left(\frac{p\pi}{\ell}\right)^2 + \left(\frac{\lambda_{mn}}{a}\right)^2},$$

Table 5.8. *Radial and circular frequencies* $f_{mn0} = \frac{\lambda_{mn}c}{2\pi a}$.

n	$m=0$	$m=1$	$m=2$	$m=3$	$m=4$
0		439.6823	729.3492	1003.2551	1269.8537
1	915.0178	1273.1492	1601.4303	1914.0461	2216.6560
2	1675.3396	2038.4858	2380.7369	2709.4240	3028.4635
3	2429.4525	2795.4166	3145.1183	3483.1187	3812.2596

Table 5.9. *Axial frequencies* $f_{00p} = \frac{pc}{2\ell}$.

	$p=1$	$p=2$	$p=3$	$p=4$	$p=5$	$p=6$	$p=7$
k_p	160.7612	321.5223	482.2835	643.0446	803.8058	964.5669	1125.3281

Table 5.10. *Cross frequencies* $f_{0np} = \frac{1}{2\pi}c\sqrt{\left(\frac{p\pi}{\ell}\right)^2 + \left(\frac{\lambda_{0n}}{a}\right)^2}$.

n	$p=1$	$p=2$	$p=3$	$p=4$	$p=5$	$p=6$	$p=7$
0	160.7612	321.5223	482.2835	643.0446	803.8058	964.5669	1125.3281
1	929.0326	969.8629	1034.3379	1118.3756	1217.9332	1329.5288	1450.3864
2	1683.0350	1705.9130	1743.3760	1794.5108	1858.1890	1933.1715	2018.1987

where $p = 0, 1, 2, \ldots, m = 0, 1, 2, \ldots$, and $n = 1, 2, \ldots$. For the parameter choices $a = .2286$ m (9 in), $\ell = 1.0668$ m (42 in), and $c = 343$ m/sec being used in the experiments, several frequencies deriving from this expression are reported in Tables 5.8–10. The results in Table 5.8 are the radial and circular frequencies ($p = 0$) while those in Table 5.9 are the axial values ($\lambda_{mn} = 0$). Finally, Table 5.10 contains the cross frequencies obtained when $m = 0$.

To compare these results with those obtained via the Fourier–Galerkin expansion, it is noted that under approximation, Helmholtz's equation (5.20) yields the matrix eigenvalue problem

$$K_{r\theta z}^{\mathcal{M}} \vartheta^M = \gamma^2 M_{r\theta z}^{\mathcal{M}} \vartheta^M, \tag{5.26}$$

where $M_{r\theta z}^{\mathcal{M}}$ and $K_{r\theta z}^{\mathcal{M}}$ are the mass and stiffness matrices given by (5.19). The frequencies obtained by solving (5.26) with the basis limits $M = 1$, $N = 15$, and $P = 15$ are reported in Tables 5.11–13 (the choice of the Fourier number $M = 1$ was dictated by limitations encountered when solving the generalized eigenvalue problem).

Table 5.11. *Radial and circular frequencies obtained with $M = 1$, $N = 15$, and $P = 15$ basis functions.*

n	$m = 0$	$m = 1$
0		439.6784
1	915.0192	1273.1594
2	1675.3364	2038.4897
3	2429.4449	2795.4178

Table 5.12. *Axial frequencies obtained with $M = 1$, $N = 15$, and $P = 15$.*

	$p = 1$	$p = 2$	$p = 3$	$p = 4$	$p = 5$	$p = 6$	$p = 7$
k_p	160.7612	321.5223	482.2835	643.0446	803.8058	964.5735	1125.3657

Table 5.13. *Cross frequencies obtained with $M = 1$, $N = 15$, and $P = 15$.*

n	$p = 1$	$p = 2$	$p = 3$	$p = 4$	$p = 5$	$p = 6$	$p = 7$
0	160.7612	321.5223	482.2835	643.0446	803.8058	964.5735	1125.3657
1	929.0341	969.8643	1034.3391	1118.3767	1217.9343	1329.5346	1450.4165
2	1683.0318	1705.9099	1743.3730	1794.5078	1858.1861	1933.1720	2018.2170

By comparing the analytical results in Tables 5.8–10 with those in Tables 5.11–13, it can be seen that the Fourier–Galerkin method is performing well for this problem. Although omitted here, numerical results obtained with $M = 1$, $N = 12$, and $P = 12$ demonstrate that the method is also converging as expected. Hence this expansion will be used when discretizing the wave component of the coupled system.

5.4 Discretization of the Circular Plate Equations

The second component of the system that must be discretized is the circular plate, and the development of an efficient and accurate numerical scheme for approximating the motion of the plate is considered in this section. The circular plate is assumed to have thickness h and radius a as shown in Fig. 5.9.

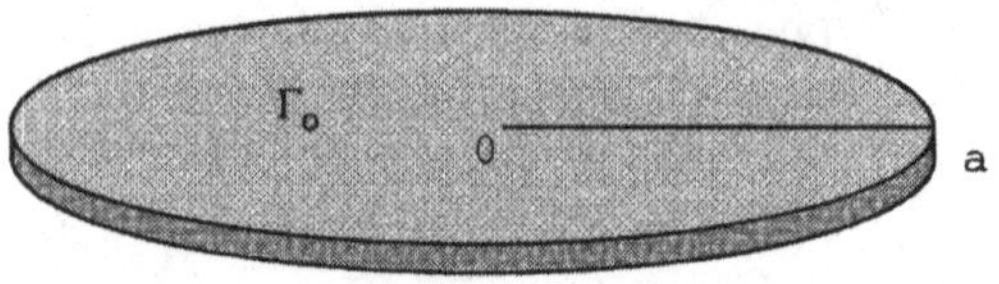

Fig. 5.9. Circular plate of radius a.

For a fixed plate with no control input, moment and force balancing yields

$$\rho_p h w_{tt} + \frac{\partial^2 M_r}{\partial r^2} + \frac{2}{r}\frac{\partial M_r}{\partial r} - \frac{1}{r}\frac{\partial M_\theta}{\partial r} + \frac{2}{r}\frac{\partial^2 M_{r\theta}}{\partial r \partial \theta} + \frac{2}{r^2}\frac{\partial M_{r\theta}}{\partial \theta} + \frac{1}{r^2}\frac{\partial^2 M_\theta}{\partial \theta^2}$$
$$= f(t, r, \theta),$$

where the moments are

$$M_r = D\left(\frac{\partial^2 w}{\partial r^2} + \frac{\nu}{r}\frac{\partial w}{\partial r} + \frac{\nu}{r^2}\frac{\partial^2 w}{\partial \theta^2}\right) + c_D\left(\frac{\partial^3 w}{\partial r^2 \partial t} + \frac{\nu}{r}\frac{\partial^2 w}{\partial r \partial t} + \frac{\nu}{r^2}\frac{\partial^3 w}{\partial \theta^2 \partial t}\right),$$
$$M_\theta = D\left(\frac{1}{r}\frac{\partial w}{\partial r} + \frac{1}{r^2}\frac{\partial^2 w}{\partial \theta^2} + \nu\frac{\partial^2 w}{\partial r^2}\right) + c_D\left(\frac{1}{r}\frac{\partial^2 w}{\partial r \partial t} + \frac{1}{r^2}\frac{\partial^3 w}{\partial \theta^2 \partial t} + \nu\frac{\partial^3 w}{\partial r^2 \partial t}\right),$$

and

$$M_{r\theta} = D(1-\nu)\left(\frac{1}{r}\frac{\partial^2 w}{\partial r \partial \theta} - \frac{1}{r^2}\frac{\partial w}{\partial \theta}\right) + c_D(1-\nu)\left(\frac{1}{r}\frac{\partial^3 w}{\partial r \partial \theta \partial t} - \frac{1}{r^2}\frac{\partial^2 w}{\partial \theta \partial t}\right)$$

(see Banks and Smith (1994) for a derivation of the strong form of the circular plate equations). Here w denotes the transverse displacement of the plate, ρ_p denotes the density, ν is Poisson's ratio, c_D is the damping coefficient, and $D = \frac{Eh^3}{12(1-\nu^2)}$ is the flexural rigidity.

The boundary and initial conditions are

$$w(t, a, \theta) = \frac{\partial w}{\partial r}(t, a, \theta) = 0$$

and

$$w(0, r, \theta) = w_0(r, \theta),$$
$$w_t(0, r, \theta) = w_1(r, \theta),$$

respectively. We point out that the boundary conditions are chosen so as to be consistent with those of the plate in the coupled system.

To pose the problem abstractly and provide a setting in which to consider the variational form, let the state space be $H = L^2(\Gamma_0)$ with the inner product

$$\langle w, \eta \rangle_H = \int_{\Gamma_0} \rho_p h w \bar{\eta}\, d\gamma.$$

Also, define the Hilbert space $V = H_0^2(\Gamma_0) \equiv \{\psi \in H^2(\Gamma_0) : \psi(x) = \psi'(x) = 0$ at $x = 0, a\}$. The V inner product is taken to be

$$\langle w, \eta \rangle_V = \left\langle M_r, \frac{\partial^2 \eta}{\partial r^2} \right\rangle + \left\langle \frac{1}{r} M_\theta, \frac{\partial \eta}{\partial r} \right\rangle + \left\langle \frac{1}{r^2} M_\theta, \frac{\partial^2 \eta}{\partial \theta^2} \right\rangle + 2\left\langle \frac{1}{r} M_{r\theta}, \frac{\partial^2 \eta}{\partial r \partial \theta} \right\rangle - 2\left\langle \frac{1}{r^2} M_{r\theta}, \frac{\partial \eta}{\partial \theta} \right\rangle,$$

where $\langle F, G \rangle = \int_{\Gamma_0} F \bar{G} d\gamma$ with $d\gamma = r dr d\theta$. As shown in Banks and Smith (1994), this definition follows from the use of the strain energy when deriving the weak form of the circular plate equations. We note that throughout this and following sections, subscripts r and θ will be used to denote both the sense of moments and differentiation with respect to these variables. The use of this notation to denote differentiation will be limited to the state and test functions and Leibniz differential notation will be used in all expressions involving the moments M_r, M_θ, and $M_{r\theta}$.

From energy considerations, Banks and Smith (1994) showed that, for an undamped plate, the weak or variational form of the equations of motion is

$$\begin{aligned}\langle \rho_p h w_{tt} - f, \eta \rangle &+ \left\langle D\left(w_{rr} + \frac{\nu}{r} w_r + \frac{\nu}{r^2} w_{\theta\theta} \right), \eta_{rr} \right\rangle \\ &+ \left\langle D\left(\frac{1}{r^2} w_r + \frac{1}{r^3} w_{\theta\theta} + \frac{\nu}{r} w_{rr} \right), \eta_r \right\rangle \\ &+ \left\langle D\left(\frac{1}{r^3} w_r + \frac{1}{r^4} w_{\theta\theta} + \frac{\nu}{r^2} w_{rr} \right), \eta_{\theta\theta} \right\rangle \\ &+ \left\langle 2(1-\nu) D\left(\frac{1}{r^2} w_{\theta r} - \frac{1}{r^3} w_\theta \right), \eta_{\theta r} \right\rangle \\ &+ \left\langle 2(1-\nu) D\left(-\frac{1}{r^3} w_{\theta r} + \frac{1}{r^4} w_\theta \right), \eta_\theta \right\rangle = 0 \qquad (5.27)\end{aligned}$$

for all $\eta \in H_0^2(\Gamma_0)$. Similar expressions arise when damping is incorporated in the model, and in terms of the plate moments, the weak form of the damped plate equation can be rewritten as

$$\langle \rho_p h w_{tt} - f, \eta \rangle + \left\langle M_r, \frac{\partial^2 \eta}{\partial r^2} \right\rangle + \left\langle \frac{1}{r} M_\theta, \frac{\partial \eta}{\partial r} \right\rangle + \left\langle \frac{1}{r^2} M_\theta, \frac{\partial^2 \eta}{\partial \theta^2} \right\rangle + 2\left\langle \frac{1}{r} M_{r\theta}, \frac{\partial^2 \eta}{\partial r \partial \theta} \right\rangle - 2\left\langle \frac{1}{r^2} M_{r\theta}, \frac{\partial \eta}{\partial \theta} \right\rangle = 0$$

for all $\eta \in H_0^2(\Gamma_0)$.

In order to write the system in abstract form, we begin by defining the sesquilinear forms $\sigma_i : V \times V \to \mathbb{C}, i = 1, 2$ by

$$\begin{aligned}
\sigma_1(w,\eta) &= \int_{\Gamma_0} D\left(w_{rr} + \frac{\nu}{r}w_r + \frac{\nu}{r^2}w_{\theta\theta}\right)\overline{\eta_{rr}}\,d\gamma \\
&+ \int_{\Gamma_0} D\left(\frac{1}{r^2}w_r + \frac{1}{r^3}w_{\theta\theta} + \frac{\nu}{r}w_{rr}\right)\overline{\eta_r}\,d\gamma \\
&+ \int_{\Gamma_0} D\left(\frac{1}{r^3}w_r + \frac{1}{r^4}w_{\theta\theta} + \frac{\nu}{r^2}w_{rr}\right)\overline{\eta_{\theta\theta}}\,d\gamma \\
&+ \int_{\Gamma_0} 2(1-\nu)D\left(\frac{1}{r^2}w_{\theta r} - \frac{1}{r^3}w_\theta\right)\overline{\eta_{r\theta}}\,d\gamma \\
&+ \int_{\Gamma_0} 2(1-\nu)D\left(-\frac{1}{r^3}w_{\theta r} + \frac{1}{r^4}w_\theta\right)\overline{\eta_\theta}\,d\gamma
\end{aligned}$$

and

$$\begin{aligned}
\sigma_2(w,\eta) &= \int_{\Gamma_0} c_D\left(w_{rr} + \frac{\nu}{r}w_r + \frac{\nu}{r^2}w_{\theta\theta}\right)\overline{\eta_{rr}}\,d\gamma \\
&+ \int_{\Gamma_0} c_D\left(\frac{1}{r^2}w_r + \frac{1}{r^3}w_{\theta\theta} + \frac{\nu}{r}w_{rr}\right)\overline{\eta_r}\,d\gamma \\
&+ \int_{\Gamma_0} c_D\left(\frac{1}{r^3}w_r + \frac{1}{r^4}w_{\theta\theta} + \frac{\nu}{r^2}w_{rr}\right)\overline{\eta_{\theta\theta}}\,d\gamma \\
&+ \int_{\Gamma_0} 2(1-\nu)c_D\left(\frac{1}{r^2}w_{\theta r} - \frac{1}{r^3}w_\theta\right)\overline{\eta_{r\theta}}\,d\gamma \\
&+ \int_{\Gamma_0} 2(1-\nu)c_D\left(-\frac{1}{r^3}w_{\theta r} + \frac{1}{r^4}w_\theta\right)\overline{\eta_\theta}\,d\gamma,
\end{aligned}$$

where w and η are in V. With $\langle\cdot,\cdot\rangle_{V^*,V}$ denoting the usual duality product and F given by $F = \frac{f}{\rho_p h}$, the system can then be written as

$$\langle w_{tt}(t),\eta\rangle_{V^*,V} + \sigma_2(w_t(t),\eta) + \sigma_1(w(t),\eta) = \langle F,\eta\rangle_{V^*,V} \tag{5.28}$$

for η in V.

5.4.1 Approximate Solution

To approximate the solution to (5.28), let $\{B_k^{\mathcal{N}}\}_{k=1}^{\mathcal{N}}$ denote the 2-D basis functions used on the plate and let $H^{\mathcal{N}} = \text{span}\{B_k^{\mathcal{N}}\}$ be the approximating subspace. The product space for the first-order system is $\mathcal{H}^{\mathcal{N}} = H^{\mathcal{N}} \times H^{\mathcal{N}}$.

A Fourier–Galerkin expansion in θ then yields the approximate solution

$$w^{\mathcal{N}}(t,r,\theta) = \sum_{m=-M}^{M} \tilde{w}_m(t,r)e^{im\theta}. \tag{5.29}$$

We remark that the complex Fourier expansion will be used while describing the method since it simplifies the notation. Due to the presence of the piezoceramic patches on the plate, however, the method is more easily implemented using the real (trigonometric) expansion (with patches present, the complex expansion leads to a complex system matrix that proves troublesome when solving the Riccati equation). We have omitted details concerning the real expansions used in the implementation since it is straightforward to interchange between the two expansions.

We now have various choices for $\tilde{w}_m(t,r)$. To aid in determining an appropriate expansion for $\tilde{w}_m(t,r)$, it is first noted that the approximate solution must satisfy the condition

$$\frac{\partial w^{\mathcal{N}}}{\partial \theta} = 0$$

at the origin to guarantee the uniqueness of the solution. This yields the requirement

$$\tilde{w}_m(t,r) = 0 \qquad \text{at } r = 0,\, m \neq 0. \tag{5.30}$$

To guarantee differentiability at the origin, it is appropriate to require that the remaining component satisfies

$$\frac{\partial \tilde{w}_0}{\partial r} = 0 \qquad \text{at } r = 0. \tag{5.31}$$

As will be seen when the sesquilinear forms are expanded in terms of the basis, these properties are necessary for guaranteeing the convergence of the various integrals.

One expansion that satisfies Eqs. (5.30) and (5.31) is

$$\tilde{w}_m(t,r) = \sum_{n=1}^{N} w_{mn}(t) r^{|m|} B_n(r), \tag{5.32}$$

where

$$B_n(a) = \frac{dB_n(a)}{dr} = \frac{dB_n(0)}{dr} = 0.$$

As in the case of the circular wave equation, the inclusion of the weighting term $r^{|m|}$ is motivated by the asymptotic behavior of the Bessel functions as $r \to 0$ (see Examples 7 and 8 for a more detailed discussion concerning the analytic behavior of the plate dynamics and Abramowitz and Stegan (1972) for a summary of the asymptotic properties of the Bessel functions). It is noted that because of the inclusion of the weighting term in the basis, this expansion does allow for nonzero derivative values of the approximate solution at $r = 0$. This choice of basis is slightly more restrictive than necessary, however, since the condition (5.31) requires $\frac{dB_n(0)}{dr} = 0$ only when $m = 0$ whereas the expansion (5.32) enforces this for all m.

A second expansion that satisfies (5.30) and (5.31) and contains a more complete set of cubic spline basis functions is

$$\tilde{w}_m(t,r) = \sum_{n=1}^{N^m} w_{mn}(t) r^{|m|} B_n^m(r), \tag{5.33}$$

where

$$N^m = \begin{cases} N+1, & m \neq 0 \\ N, & m = 0. \end{cases}$$

Here $B_n^m(r)$ is the nth modified cubic spline satisfying

$$B_n^m(a) = \frac{dB_n^m(a)}{dr} = 0 \tag{5.34}$$

with the condition

$$\frac{dB_n^m(0)}{dr} = 0 \tag{5.35}$$

being enforced only when $m = 0$. We note that, in this case, the total number of basis functions is $\mathcal{N} = (2M+1)(N+1) - 1$.

A difficulty that occurs with both of the expansions (5.32) and (5.33) is that as m increases in magnitude, the term $r^{|m|}$ rapidly decays to 0 at the origin thus causing the mass and stiffness matrices to become singular (this is due to the local support of the basis functions and does not occur in the analytic solution owing to the global nature of the Bessel functions). To remedy this, a further refinement that can be made on the basis is to use the expansion

$$\tilde{w}_m(t,r) = \sum_{n=1}^{N^m} w_{mn}(t) r^{|\hat{m}|} B_n^m(r), \tag{5.36}$$

where $\hat{m}$ is given by

$$\hat{m} = \begin{cases} 0, & m = 0 \\ 1, & m \neq 0 \end{cases}$$

and N^m and $B_n^m(r)$ are defined as above.

All three expansions were tested when running the numerical examples and it was found that, depending on the test example, all yielded very good results. The discrete system obtained with the expansion (5.32) is slightly smaller

($\mathcal{N} = (2M+1)N$ versus $\mathcal{N} = (2M+1)(N+1) - 1$) and easier to construct than that of (5.33). This is because, in the former case, the submatrices obtained for various values of m are uniform in size whereas the $m = 0$ submatrix in the latter expansion must be constructed separately and has dimension that is one less than its neighbors. On the other hand, the expansion (5.33) yields an approximate solution that is slightly more accurate in many problems than that obtained with the more restrictive expansion (5.32). As mentioned previously and demonstrated in examples reported in Banks and Smith (1994), the expansions (5.32) and (5.33) can lead to singular mass and stiffness matrices for large values of m, and this problem is alleviated by the expansion (5.36).

5.4.2 Matrix System

To simplify notation, the approximate solution is written as

$$w^{\mathcal{N}}(t,r,\theta) = \sum_{k=1}^{\mathcal{N}} w_k(t) B_k^{\mathcal{N}}(r,\theta).$$

The restriction of the infinite-dimensional system (5.28) to the space $H^{\mathcal{N}} \times H^{\mathcal{N}}$ then yields

$$\left\langle w_{tt}^{\mathcal{N}}(t), \xi \right\rangle_H + \sigma_2\left(w_t^{\mathcal{N}}(t), \xi\right) + \sigma_1(w^{\mathcal{N}}(t), \xi) = \langle F, \xi \rangle_H$$

for ξ in $H^{\mathcal{N}}$, and the corresponding matrix system is

$$\begin{aligned} M^{\mathcal{N}} \dot{y}^{\mathcal{N}}(t) &= \tilde{A}^{\mathcal{N}} y^{\mathcal{N}}(t) + \tilde{F}^{\mathcal{N}}(t), \\ M^{\mathcal{N}} y^{\mathcal{N}}(0) &= \tilde{y}_0^{\mathcal{N}}, \end{aligned}$$

where

$$y^{\mathcal{N}}(t) = \begin{pmatrix} \vartheta^{\mathcal{N}}(t) \\ \dot{\vartheta}^{\mathcal{N}}(t) \end{pmatrix}.$$

Here $\vartheta^{\mathcal{N}}(t) = [w_1(t), w_2(t), \ldots, w_{\mathcal{N}}(t)]^T$ denotes the $\mathcal{N} \times 1$ vector containing the approximate state coefficients. The full system has the form

$$\begin{bmatrix} K^{\mathcal{N}} & 0 \\ 0 & M^{\mathcal{N}} \end{bmatrix} \begin{bmatrix} \dot{\vartheta}^{\mathcal{N}}(t) \\ \ddot{\vartheta}^{\mathcal{N}}(t) \end{bmatrix} = \begin{bmatrix} 0 & K^{\mathcal{N}} \\ -K_D^{\mathcal{N}} & -K_{CD}^{\mathcal{N}} \end{bmatrix} \begin{bmatrix} \vartheta^{\mathcal{N}}(t) \\ \dot{\vartheta}^{\mathcal{N}}(t) \end{bmatrix} + \begin{bmatrix} 0 \\ \tilde{F}^{\mathcal{N}}(t) \end{bmatrix},$$

$$\begin{bmatrix} K^{\mathcal{N}} & 0 \\ 0 & M^{\mathcal{N}} \end{bmatrix} \begin{bmatrix} \vartheta^{\mathcal{N}}(0) \\ \dot{\vartheta}^{\mathcal{N}}(0) \end{bmatrix} = \begin{bmatrix} g_1^{\mathcal{N}} \\ g_2^{\mathcal{N}} \end{bmatrix}.$$

The component matrices and vectors are given by

$$
\begin{aligned}
K^{\mathcal{N}} &= K_1 + K_2 + K_3 + K_4 + K_5, \\
K_D^{\mathcal{N}} &= K_{D1} + K_{D2} + K_{D3} + K_{D4} + K_{D5}, \\
K_{c_D}^{\mathcal{N}} &= K_{c_D 1} + K_{c_D 2} + K_{c_D 3} + K_{c_D 4} + K_{c_D 5}, \\
[M^{\mathcal{N}}]_{\ell,k} &= \int_{\Gamma_0} \rho_p h B_k^{\mathcal{N}} \overline{B_\ell^{\mathcal{N}}} \, d\gamma, \\
[\tilde{F}^{\mathcal{N}}(t)]_\ell &= \int_{\Gamma_0} f \overline{B_\ell^{\mathcal{N}}} \, d\gamma, \\
[g_1^{\mathcal{N}}]_\ell &= \langle w_0, B_\ell^{\mathcal{N}} \rangle_V, \\
[g_2^{\mathcal{N}}]_\ell &= \int_{\Gamma_0} w_1 \overline{B_\ell^{\mathcal{N}}} \, d\gamma,
\end{aligned}
\tag{5.37}
$$

where

$$[K_1]_{\ell,k} = \int_{\Gamma_0} \left[\frac{\partial^2 B_k^{\mathcal{N}}}{\partial r^2} + \frac{\nu}{r} \frac{\partial B_k^{\mathcal{N}}}{\partial r} + \frac{\nu}{r^2} \frac{\partial^2 B_k^{\mathcal{N}}}{\partial \theta^2} \right] \frac{\partial^2 \overline{B_\ell^{\mathcal{N}}}}{\partial r^2} \, d\gamma,$$

$$[K_2]_{\ell,k} = \int_{\Gamma_0} \left[\frac{1}{r^2} \frac{\partial B_k^{\mathcal{N}}}{\partial r} + \frac{1}{r^3} \frac{\partial^2 B_k^{\mathcal{N}}}{\partial \theta^2} + \frac{\nu}{r} \frac{\partial^2 B_k^{\mathcal{N}}}{\partial r^2} \right] \frac{\partial \overline{B_\ell^{\mathcal{N}}}}{\partial r} \, d\gamma,$$

$$[K_3]_{\ell,k} = \int_{\Gamma_0} \left[\frac{1}{r^3} \frac{\partial B_k^{\mathcal{N}}}{\partial r} + \frac{1}{r^4} \frac{\partial^2 B_k^{\mathcal{N}}}{\partial \theta^2} + \frac{\nu}{r^2} \frac{\partial^2 B_k^{\mathcal{N}}}{\partial r^2} \right] \frac{\partial^2 \overline{B_\ell^{\mathcal{N}}}}{\partial \theta^2} \, d\gamma,$$

$$[K_4]_{\ell,k} = 2 \int_{\Gamma_0} (1 - \nu) \left[\frac{1}{r^2} \frac{\partial^2 B_k^{\mathcal{N}}}{\partial r \partial \theta} - \frac{1}{r^3} \frac{\partial B_k^{\mathcal{N}}}{\partial \theta} \right] \frac{\partial^2 \overline{B_\ell^{\mathcal{N}}}}{\partial r \partial \theta} \, d\gamma,$$

$$[K_5]_{\ell,k} = 2 \int_{\Gamma_0} (1 - \nu) \left[-\frac{1}{r^3} \frac{\partial^2 B_k^{\mathcal{N}}}{\partial r \partial \theta} + \frac{1}{r^4} \frac{\partial B_k^{\mathcal{N}}}{\partial \theta} \right] \frac{\partial \overline{B_\ell^{\mathcal{N}}}}{\partial \theta} \, d\gamma.$$

The matrices $K_{D1}, \ldots, K_{D5}$ and $K_{c_D 1}, \ldots, K_{c_D 5}$ are defined similarly with the inclusion of the parameters D and c_D in the various integrals.The index ranges are $k, \ell = 1, \ldots, \mathcal{N}$.

5.4.3 Example 6

Consider the steady state problem

$$\frac{\partial^2 M_r}{\partial r^2} + \frac{2}{r} \frac{\partial M_r}{\partial r} - \frac{1}{r} \frac{\partial M_\theta}{\partial r} + \frac{2}{r} \frac{\partial^2 M_{r\theta}}{\partial r \partial \theta} + \frac{2}{r^2} \frac{\partial M_{r\theta}}{\partial \theta} + \frac{1}{r^2} \frac{\partial^2 M_\theta}{\partial \theta^2} = f(r, \theta)$$

Table 5.14. *Absolute and relative errors with $m = 1$ and $a = .6$.*

M	N	size($K^{\mathcal{N}}$)	$\|w_{true} - w_{app}\|$	$\frac{\|w_{true} - w_{app}\|}{\|w_{true}\|}$	Asym. Error
1	5	17×17	$.4290 - 2$	$.2169 - 2$	
1	10	32×32	$.2674 - 3$	$.1352 - 3$	$.1356 - 3$
1	20	62×62	$.1588 - 4$	$.8026 - 5$	$.8450 - 5$
1	40	122×122	$.1005 - 5$	$.5090 - 6$	$.5016 - 6$

with the moments

$$M_r = D\left(\frac{\partial^2 w}{\partial r^2} + \frac{\nu}{r}\frac{\partial w}{\partial r} + \frac{\nu}{r^2}\frac{\partial^2 w}{\partial \theta^2}\right),$$

$$M_\theta = D\left(\frac{1}{r}\frac{\partial w}{\partial r} + \frac{1}{r^2}\frac{\partial^2 w}{\partial \theta^2} + \nu\frac{\partial^2 w}{\partial r^2}\right),$$

$$M_{r\theta} = D(1-\nu)\left(\frac{1}{r}\frac{\partial^2 w}{\partial r \partial \theta} - \frac{1}{r^2}\frac{\partial w}{\partial \theta}\right).$$

The parameter ν is taken to be $1/2$, which yields $D = 4/3$ when $EI \equiv 1$. For a plate of radius a, the true solution $w(r, \theta) = (\cos(2\pi r/a) - 1)\sin(m\theta)$ yields the forcing function

$$\begin{aligned} f(r,\theta) = {} & \frac{1}{r^4}[(4m^2 - m^4) + (-2\pi r/a - 4\pi r m^2/a + 16\pi^3 r^3/a^3) \\ & \times \sin(2\pi r/a)(-4m^2 + m^4 + 4\pi^2 r^2/a^2 + 8\pi^2 r^2 m^2/a^2 \\ & + 16\pi^4 r^4/a^4)\cos(2\pi r/a)]\sin(m\theta). \end{aligned}$$

The absolute and relative errors with $m = 1$, $a = .6$ and the expansion (5.33) (or (5.36)) are reported in Table 5.14. The asymptotic error is obtained by dividing the previous relative error by 16. Since the number of radial basis functions is doubled each time, this provides a means of checking whether or not the expected $\mathcal{O}(h^4)$ convergence rate is being maintained (the results demonstrate that it is). Note that the system for the steady-state problem is simply

$$K^{\mathcal{N}}\vartheta^{\mathcal{N}} = \tilde{F}^{\mathcal{N}},$$

where $K^{\mathcal{N}}$ and $\tilde{F}^{\mathcal{N}}$ are given in (5.37).

5.4.4 Example 7

Consider the eigenvalue problem

$$\nabla^4 W = \gamma^4 W,$$

Table 5.15. *Values of λ_{mn}^2 as reported in Leissa (1969).*

n	$m = 0$	$m = 1$	$m = 2$	$m = 3$	$m = 4$	$m = 5$
0	10.2158	21.26	34.88	51.04	69.6659	90.7390
1	39.771	60.82	84.58	111.01	140.1079	171.8029
2	89.104	120.08	153.81	190.30	229.5186	271.4283
3	158.183	199.06	242.71	289.17	338.4113	390.3896
4	247.005	297.77	351.38	407.72		
5	355.568	416.20	479.65	545.97		

Table 5.16. *Values of λ_{mn}^2 obtained with $M = 5$, $N = 24$, and the expansion (5.36).*

n	$m = 0$	$m = 1$	$m = 2$	$m = 3$	$m = 4$	$m = 5$
0	10.2158	21.2604	34.8770	51.0301	69.6659	90.7391
1	39.7712	60.8288	84.5830	111.0220	140.1089	171.8045
2	89.1051	120.0808	153.8181	190.3088	229.5261	271.4388
3	158.1095	199.0627	242.7364	289.2037	338.4446	390.4345
4	247.0331	297.7975	351.3938	407.8114		
5	355.6562	416.3192	479.8449	546.2140		

where

$$\nabla^4 W = \left(\frac{\partial^2}{\partial r^2} + \frac{1}{r}\frac{\partial}{\partial r} + \frac{1}{r^2}\frac{\partial^2}{\partial \theta^2}\right)\left(\frac{\partial^2 W}{\partial r^2} + \frac{1}{r}\frac{\partial W}{\partial r} + \frac{1}{r^2}\frac{\partial^2 W}{\partial \theta^2}\right)$$

(this type of problem arises when determining the natural frequencies of the plate as detailed further in the next example). Note that, under approximation, this yields the matrix eigenvalue problem

$$K^{\mathcal{N}}\vartheta^{\mathcal{N}} = \gamma^4 M^{\mathcal{N}}\vartheta^{\mathcal{N}}$$

with $\rho_p h = 1$ in the mass matrix $M^{\mathcal{N}}$ (see (5.37) for matrix definitions).

To normalize dimensions, we consider $\lambda \equiv \gamma a$. For $a = 1$, the eigenvalues $\lambda_{mn}^2 = \gamma_{mn}^2 a^2 = \gamma_{mn}^2$, as reported in Leissa (1969), are given in Table 5.15. Those obtained with $M = 5$ and $N = 24$ and the expansion (5.36) are given in Table 5.16. The corresponding modes are plotted in Figs. 5.10 and 5.11. In the tables and figures, m refers to the number of nodal diameters and n denotes the number of nodal circles not including the boundary.

To test the modes obtained with this Fourier–Galerkin expansion, a comparison was made with modes obtained with Bessel expansions, as discussed in Leissa (1969) and Morse and Ingard (1968). For $m = 1, \ldots, \infty$, a typical mode

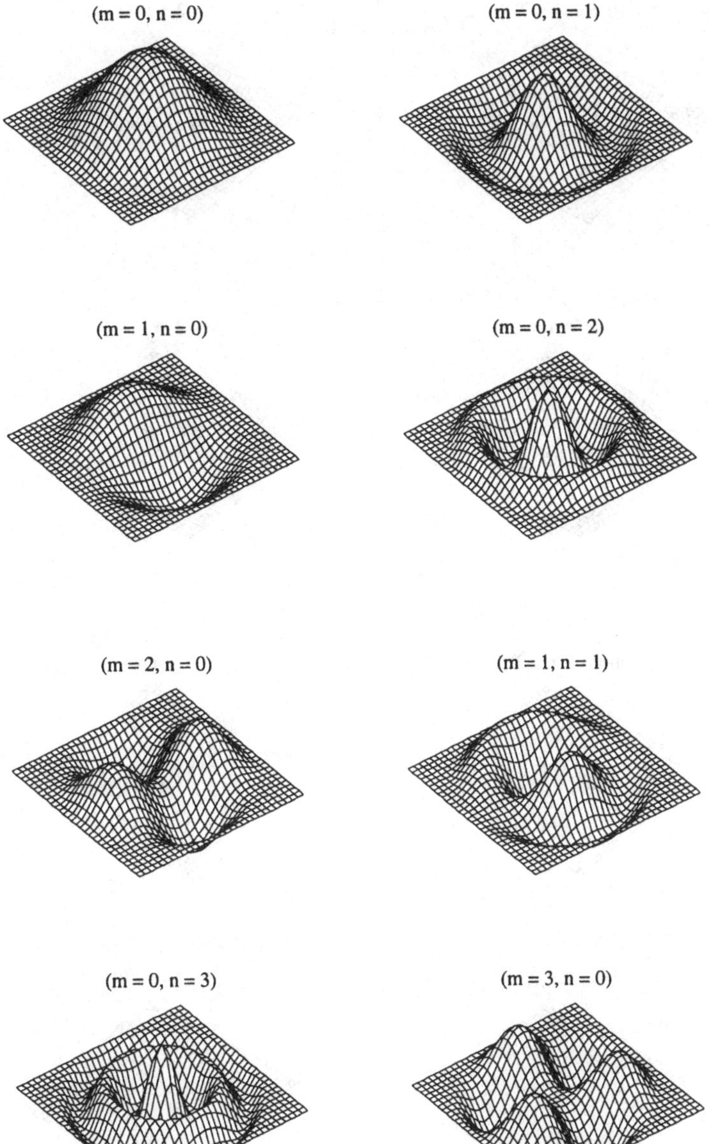

Fig. 5.10. The (0, 0) through (3, 0) modes obtained via the Fourier–Galerkin scheme with $M = 5$ and $N = 16$.

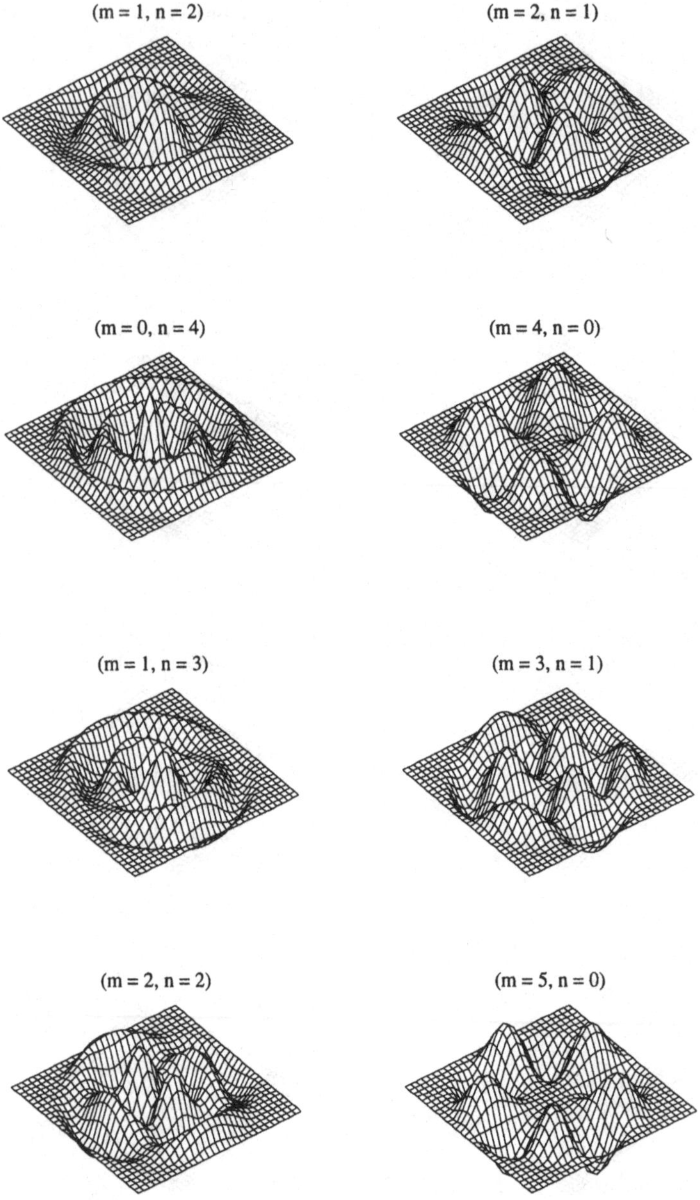

Fig. 5.11. The (1, 2) through (5, 0) modes obtained via the Fourier–Galerkin scheme with $M = 5$ and $N = 16$.

Table 5.17. *Absolute and relative errors for the (1, 2) through (5, 0) modes with $M = 5$, $N = 16$ basis functions.*

Mode	$\|\Psi_{app} - \Psi_{bes}\|$	$\frac{\|\Psi_{app}-\Psi_{bes}\|}{\|\Psi_{bes}\|}$	Mode	$\|\Psi_{app} - \Psi_{bes}\|$	$\frac{\|\Psi_{app}-\Psi_{bes}\|}{\|\Psi_{bes}\|}$
(1, 2)	.1123 − 3	.2016 − 3	(1, 3)	.3427 − 3	.5893 − 3
(2, 1)	.2223 − 2	.4574 − 2	(3, 1)	.1483 − 3	.3416 − 3
(0, 4)	.1197 − 2	.1197 − 2	(2, 2)	.3992 − 2	.8360 − 2
(4, 0)	.1653 − 2	.4088 − 2	(5, 0)	.1976 − 4	.5226 − 4

is given by

$$\Psi_m(r, \theta) = \cos(m\theta)[A_m J_m(\gamma_{mn} r) + B_m I_m(\gamma_{mn} r)]$$

(the sine terms can be dropped due to the symmetry of the boundary conditions). Here J_m is the mth Bessel function of the first kind and $I_m(z) = i^{-m} J_m(iz)$ is the modified Bessel function of the first kind. The coefficients A_m and B_m are solved for from the boundary conditions and determine the mode shapes. The condition $w(a) = w(1) = 0$ yields

$$A_m = -B_m \frac{I_m(\lambda_{mn})}{J_m(\lambda_{mn})},$$

where, again, $\lambda \equiv \gamma a$. We add that γ is related to the circular frequence ω through the expression $\gamma^4 = \frac{\rho_p h \omega^2}{D}$ (hence the eigenvalues λ_{mn} determine the frequencies ω of the plate). The derivative boundary condition $\frac{\partial w(a)}{\partial r} = \frac{\partial w(1)}{\partial r} = 0$ is enforced by requiring that λ satisfies

$$I_m(\lambda) \frac{dJ_m(\lambda)}{dr} - J_m(\lambda) \frac{dI_m(\lambda)}{dr} = 0,$$

thus yielding the eigenvalues λ_{mn}. Several solutions to this equation are listed in Table 5.15 as reported in Leissa (1969).

The eigenfunctions or modes are then given by

$$\Psi_m(r, \theta) = \cos(m\theta) \left[J_m\left(\frac{\lambda_{mn} r}{a}\right) - \frac{J_m(\lambda_{mn})}{I_m(\lambda_{mn})} I_m\left(\frac{\lambda_{mn} r}{a}\right) \right].$$

The (0, 1), (1, 1), (2, 2), and (5, 0) modes obtained with this expression are plotted in Fig. 5.12 (the ordering here is (m, n)). The absolute and relative errors between the modes obtained via the Fourier–Galerkin approximations (5.33) with $M = 5$, $N = 16$ and the Bessel expansions are reported in Table 5.17 (the modal results obtained with the expansion (5.36) were identical to those reported). Note that in order to compare the two sets of eigenfunctions, they had to be normalized to the same scale. The results in Table 5.17 and Fig. 5.12 both indicate a very good agreement between the modes obtained with the

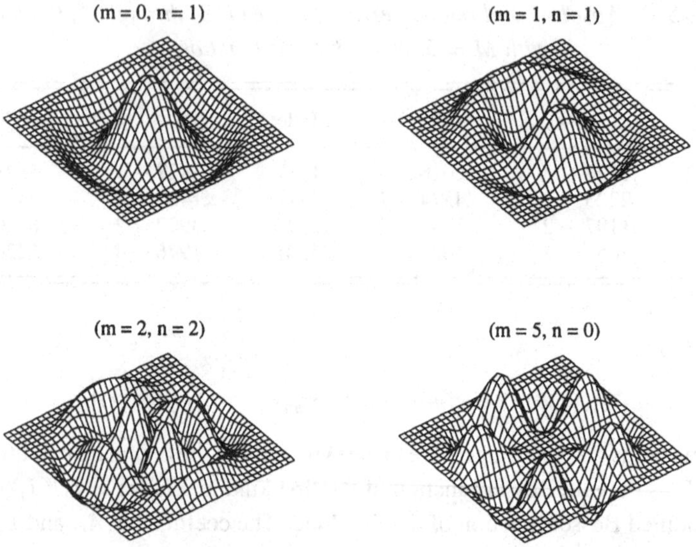

Fig. 5.12. The (0, 1), (1, 1), (2, 2), and (5, 0) modes obtained with Bessel expansions.

Fourier–Galerkin approximation and the Bessel expansions, thus demonstrating the feasibility of the Fourier–Galerkin scheme.

5.4.5 Example 8

To further test the accuracy and efficiency of the plate expansion as well as to determine the natural frequencies of an uncoupled plate having dimensions consistent with those in the experimental setup, we consider the equation of motion for an undamped circular plate with constant stiffness and density. As discussed in Banks and Smith (1994), the motion is modeled by the equation

$$\rho_p h \frac{\partial^2 w}{\partial t^2} + D\nabla^4 w = 0,$$

where $D = \frac{Eh^3}{12(1-\nu^2)}$ is the flexural rigidity, E is Young's modulus, ν is Poisson's ratio, h is the thickness, and ρ_p is the mass density of the plate. To determine the analytic frequencies, free vibrations are assumed, and the displacements are taken to be of the form

$$w = W\cos(\omega t).$$

Here ω is the circular frequency (with units of radians/sec) and W contains the remaining spatial contributions. The substitution of this expression into the equation of motion yields the eigenvalue problem

$$\nabla^4 W = \gamma^4 W, \tag{5.38}$$

Table 5.18. *Natural frequencies deriving from the Bessel expansions (in hertz).*

m	$n=0$	$n=1$	$n=2$	$n=3$	$n=4$	$n=5$	$n=6$
0	61.96	128.95	211.56	309.58	422.56	550.38	692.75
1	241.23	368.90	513.02	673.33	849.82	1042.07	1249.92
2	540.46	728.34	932.93	1154.26	1392.14	1646.34	1916.70
3	959.46	1207.39	1472.15	1753.95	2052.63	2367.90	
4	1498.20	1806.12	2131.29	2473.02			
5	2156.69	2524.45	2909.31	3311.57			
6	2934.91	3362.52	3807.60	4269.79			

where $\gamma^4 = \rho_p h\omega^2/D$. As noted in the previous example, the frequencies ω for a fixed plate are determined by solving the nonlinear problem

$$I_m(\lambda)\frac{dJ_m(\lambda)}{dr} - J_m(\lambda)\frac{dI_m(\lambda)}{dr} = 0,$$

where J_m is the mth Bessel function of the first kind, $I_m(z) = i^{-m}J_m(iz)$ is the modified Bessel function of the first kind, and, again, λ is related to γ by $\lambda = \gamma a$.

By noting the relationship $f = \frac{1}{2\pi}\omega$, where f is the frequency expressed in hertz, the natural frequencies of the fixed circular plate can be written as

$$f = \frac{1}{2\pi}\left(\frac{\lambda}{a}\right)^2\sqrt{\frac{D}{\rho_p h}}.$$

The dimensions and parameters of the plate were taken to be $a = .2286$ m (9 in), $\rho_p = 2{,}700$ kg/m^3, $h = .00127$ m (.05 in), $E = 7.1 \times 10^{10}$ N/m^2, and $\nu = .33$ so as to be consistent with those in the experimental setup. This then yields the flexural rigidity $D = 13.6007$ N·m. For these values, several frequencies deriving from the Bessel solutions λ^2 reported in Leissa (1969) are given in Table 5.18.

To compare the results reported in Table 5.18 with those obtained via the Fourier–Galerkin scheme, it is noted that, under approximation, the infinite-dimensional eigenvalue problem (5.38) yields the matrix eigenvalue problem

$$K^{\mathcal{N}}\vartheta^{\mathcal{N}} = \gamma^4 M^{\mathcal{N}}\vartheta^{\mathcal{N}},$$

where $M^{\mathcal{N}}$ and $K^{\mathcal{N}}$ are the mass and stiffness matrices for the undamped plate (see (5.37)). The frequencies obtained by solving this generalized matrix eigenvalue problem are reported in Table 5.19. In this case, the basis limits $M = 6, N = 24$ were used along with the expansion in (5.36). By comparing the Bessel and Galerkin results in Tables 5.18 and 5.19, respectively, it can be

Table 5.19. *Natural frequencies obtained via the Fourier–Galerkin scheme with $M = 6$, $N = 24$ basis functions and the expansion (5.36) (in hertz).*

m	$n = 0$	$n = 1$	$n = 2$	$n = 3$	$n = 4$	$n = 5$	$n = 6$
0	61.96	128.95	211.55	309.52	422.56	550.38	692.75
1	241.23	368.96	513.04	673.40	849.83	1042.08	1249.93
2	540.46	728.35	932.98	1154.31	1392.18	1646.40	1916.79
3	959.50	1207.41	1472.31	1754.16	2052.83	2368.17	
4	1498.37	1806.28	2131.37	2473.57			
5	2157.22	2525.17	2910.49	3313.05			
6	2936.35	3364.46	3810.17	4273.26			

seen that the Fourier–Galerkin scheme does a very good job of approximating the system and hence determining the natural frequencies.

5.4.6 Numerical Conclusions

As demonstrated by the results in the examples, the expansion (5.36) with modified cubic splines satisfying (5.34) and (5.35) yields good results both in contrived test problems and in the physically realistic problem of determining the natural frequencies of an undamped plate with constant density and stiffness parameters. Due to its success, this expansion will be used when discretizing the coupled acoustic/structure problem.

5.5 The Fluid/Structure Interaction Problem

In this section we combine the results from the last three sections to yield a numerical method for approximating the dynamics of the fully coupled system described in the introduction. As described there, the 3-D cavity consists of a cylinder of length ℓ and radius a (see Fig. 5.13). At one end of the cylinder is a flexible plate of thickness h, which is assumed to have Kelvin–Voigt damping. It is also assumed that the edges of the plate are fixed. The other end of the cylinder is closed and hard-wall boundary conditions are assumed on all walls of the cylinder other than the flexible plate.

The strong form of the equations of motion for this system are given in (5.2). As noted in the introduction, however, the use of the strong form leads to difficulties as a result of discontinuities in the control input term and physical patch parameters. To avoid these difficulties, it is advantageous to formulate the problem in weak or variational form.

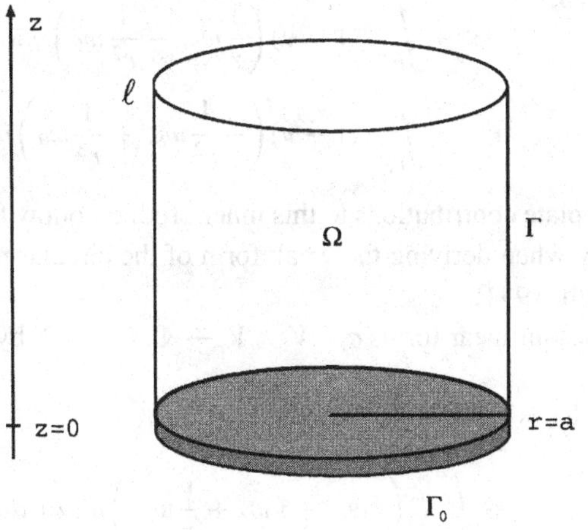

Fig. 5.13. The cylindrical acoustic cavity.

To pose the problem in a manner conducive to approximation, parameter estimation, and control, the state is taken to be $z = (\phi, w)$ in the Hilbert space $H = \bar{L}^2(\Omega) \times L^2(\Gamma_0)$ with the energy inner product

$$\left\langle \begin{pmatrix} \phi \\ w \end{pmatrix}, \begin{pmatrix} \xi \\ \eta \end{pmatrix} \right\rangle_H = \int_\Omega \frac{\rho_f}{c^2} \phi \bar{\xi}\, d\omega + \int_{\Gamma_0} \rho_p h w \bar{\eta}\, d\gamma$$

(here $d\omega = r\,dr\,d\theta\,dz$ and $d\gamma = r\,dr\,d\theta$). The choice of the quotient space $\bar{L}^2(\Omega)$ results from the fact that the potentials are determined only up to a constant.

To provide a class of functions to be considered when defining a variational form of the problem, we also define the Hilbert space $V = \bar{H}^1(\Omega) \times H_0^2(\Gamma_0)$, where $\bar{H}^1(\Omega)$ is the quotient space of H^1 over the constant functions and $H_0^2(\Gamma_0) = \{\psi \in H^2(\Gamma_0) : \psi = \psi_r = 0 \text{ at } r = a\}$. The V inner product is

$$\begin{aligned} \left\langle \begin{pmatrix} \phi \\ w \end{pmatrix}, \begin{pmatrix} \xi \\ \eta \end{pmatrix} \right\rangle_V &= \int_\Omega \nabla\phi \cdot \overline{\nabla\xi}\, d\omega \\ &+ \int_{\Gamma_0} \left(r w_{rr} + \nu w_r + \frac{\nu}{r} w_{\theta\theta} \right) \overline{\eta_{rr}}\, dr\, d\theta \\ &+ \int_{\Gamma_0} \left(\frac{1}{r} w_r + \frac{1}{r^2} w_{\theta\theta} + \nu w_{rr} \right) \overline{\eta_r}\, dr\, d\theta \\ &+ \int_{\Gamma_0} \left(\frac{1}{r^2} w_r + \frac{1}{r^3} w_{\theta\theta} + \frac{\nu}{r} w_{rr} \right) \overline{\eta_{\theta\theta}}\, dr\, d\theta \end{aligned}$$

$$+\int_{\Gamma_0} 2(1-\nu)\left(\frac{1}{r}w_{\theta r}-\frac{1}{r^2}w_\theta\right)\overline{\eta_{r\theta}}\,dr\,d\theta$$
$$+\int_{\Gamma_0} 2(1-\nu)\left(-\frac{1}{r^2}w_{\theta r}+\frac{1}{r^3}w_\theta\right)\overline{\eta_\theta}\,dr\,d\theta.$$

The form of the plate contributions to this inner product follow from the use of the strain energy when deriving the weak form of the circular plate equations (Banks and Smith, 1994).

Now, define sesquilinear forms $\sigma_i : V \times V \to \mathbb{C}$, $i = 1, 2$, by

$$\begin{aligned}\sigma_1(\Phi,\Psi) = &\int_\Omega \rho_f \nabla\phi\cdot\overline{\nabla\xi}\,d\omega\\
&+\int_{\Gamma_0} D\left(rw_{rr}+\nu w_r+\frac{\nu}{r}w_{\theta\theta}\right)\overline{\eta_{rr}}\,dr\,d\theta\\
&+\int_{\Gamma_0} D\left(\frac{1}{r}w_r+\frac{1}{r^2}w_{\theta\theta}+\nu w_{rr}\right)\overline{\eta_r}\,dr\,d\theta\\
&+\int_{\Gamma_0} D\left(\frac{1}{r^2}w_r+\frac{1}{r^3}w_{\theta\theta}+\frac{\nu}{r}w_{rr}\right)\overline{\eta_{\theta\theta}}\,dr\,d\theta\\
&+\int_{\Gamma_0} 2(1-\nu)D\left(\frac{1}{r}w_{\theta r}-\frac{1}{r^2}w_\theta\right)\overline{\eta_{r\theta}}\,dr\,d\theta\\
&+\int_{\Gamma_0} 2(1-\nu)D\left(-\frac{1}{r^2}w_{\theta r}+\frac{1}{r^3}w_\theta\right)\overline{\eta_\theta}\,dr\,d\theta\end{aligned}$$

and

$$\begin{aligned}\sigma_2(\Phi,\Psi) = &\int_{\Gamma_0} c_D\left(rw_{rr}+\nu w_r+\frac{\nu}{r}w_{\theta\theta}\right)\overline{\eta_{rr}}\,dr\,d\theta\\
&+\int_{\Gamma_0} c_D\left(\frac{1}{r}w_r+\frac{1}{r^2}w_{\theta\theta}+\nu w_{rr}\right)\overline{\eta_r}\,dr\,d\theta\\
&+\int_{\Gamma_0} c_D\left(\frac{1}{r^2}w_r+\frac{1}{r^3}w_{\theta\theta}+\frac{\nu}{r}w_{rr}\right)\overline{\eta_{\theta\theta}}\,dr\,d\theta\\
&+\int_{\Gamma_0} 2(1-\nu)c_D\left(\frac{1}{r}w_{\theta r}-\frac{1}{r^2}w_\theta\right)\overline{\eta_{r\theta}}\,dr\,d\theta\\
&+\int_{\Gamma_0} 2(1-\nu)c_D\left(-\frac{1}{r^2}w_{\theta r}+\frac{1}{r^3}w_\theta\right)\overline{\eta_\theta}\,dr\,d\theta\\
&+\int_{\Gamma_0} \rho_f(\phi\bar{\eta}-w\bar{\xi})r\,dr\,d\theta,\end{aligned}$$

where, again, $\Phi = (\phi, w)$ and $\Psi = (\xi, \eta)$ are in V. Note that the sesquilinear forms satisfy the continuity and coercivity conditions

$$\begin{aligned}
&\text{Re}\,\sigma_1(\Phi,\Phi) \geq c_1|\Phi|_V^2,\\
&|\sigma_1(\Phi,\Psi)| \leq c_2|\Phi|_V|\Psi|_V,\\
&\text{Re}\,\sigma_2(\Phi,\Phi) \geq c_3\langle\nabla^2 w, \nabla^2 w\rangle_{L^2(\Gamma_0)} = c_3|w|^2_{H_0^2(\Gamma_0)},\\
&|\sigma_2(\Phi,\Psi)| \leq c_4|\Phi|_V|\Psi|_V.
\end{aligned}$$

Furthermore, we define a control operator $B \in \mathcal{L}(U, V^*)$ by

$$\langle Bu, \Psi\rangle_{V^*,V} = \int_{\Gamma_0}\left([H_1(r) - H_2(r)]\sum_{i=1}^{s}\mathcal{K}_i^B u_i(t)[H_{i1}(\theta) - H_{i2}(\theta)]\right) \times \overline{\nabla^2\eta} r\, dr\, d\theta$$

for Ψ in V. Finally, for $F = (0, \frac{f}{\rho_p h})$ we can write the control system in weak or variational form

$$\langle z_{tt}(t), \Psi\rangle_H + \sigma_2(z_t(t), \Psi) + \sigma_1(z(t), \Psi) = \langle Bu(t) + F, \Psi\rangle_{V^*,V}$$

for Ψ in V. We reiterate that the state is given by $z(t) = (\phi(t,\cdot,\cdot), w(t,\cdot))$ in $V \hookrightarrow H$.

Since σ_1 and σ_2 are bounded, we can define operators $A_1, A_2 \in \mathcal{L}(V, V^*)$ by

$$\langle A_i\Phi, \Psi\rangle_{V^*,V} = \sigma_i(\Phi, \Psi)$$

for $i = 1, 2$. This then yields the second-order system

$$z_{tt}(t) + A_2 z_t(t) + A_1 z(t) = Bu(t) + F$$

in V^*.

We now want to write the system in first-order form. To do that, we define the product spaces $\mathcal{V} = V \times V$ and $\mathcal{H} = V \times H$ with the norms

$$|(\Phi, \Psi)|^2_{\mathcal{H}} = |\Phi|^2_V + |\Psi|^2_H$$

and

$$|(\Phi, \Psi)|^2_{\mathcal{V}} = |\Phi|^2_V + |\Psi|^2_V.$$

For $\chi = (\Phi, \Psi)$ and $\Theta = (\Upsilon, \Lambda)$, the sesquilinear form $\sigma : \mathcal{V} \times \mathcal{V} \to \mathbb{C}$ is then defined by

$$\sigma((\Upsilon, \Lambda), (\Phi, \Psi)) = -\langle\Lambda, \Phi\rangle_V + \sigma_1(\Upsilon, \Psi) + \sigma_2(\Lambda, \Psi).$$

Since the duality product $\langle\cdot,\cdot\rangle_{V^*,V}$ is the unique extension by continuity of the scalar product $\langle\cdot,\cdot\rangle_H$ from $H \times V$ to $V^* \times V$, it follows that for appropriate restrictions on Θ we can write

$$\begin{aligned}\sigma(\Theta,\chi)=\sigma((\Upsilon,\Lambda),(\Phi,\Psi)) &= -\langle\Lambda,\Phi\rangle_V+\langle A_1\Upsilon,\Psi\rangle_{V^*,V}+\langle A_2\Lambda,\Psi\rangle_{V^*,V}\\ &= -\langle\Lambda,\Phi\rangle_V+\langle A_1\Upsilon+A_2\Lambda,\Psi\rangle_H\\ &= \langle(-\Lambda,A_1\Upsilon+A_2\Lambda),(\Phi,\Psi)\rangle_{\mathcal{H}}\\ &= \langle -\mathcal{A}\Theta,\chi\rangle_{\mathcal{H}}.\end{aligned}$$

The operator $\mathcal{A}:\mathcal{H}\to\mathcal{H}$ is given by

$$\mathcal{A}=\begin{bmatrix}0 & I\\ -A_1 & -A_2\end{bmatrix}, \tag{5.39}$$

where $\mathrm{dom}\,\mathcal{A}=\{\Theta=(\Upsilon,\Lambda)\in\mathcal{H}:\Lambda\in V, A_1\Upsilon+A_2\Lambda\in H\}$, A_1 and A_2 are the operators defined by σ_1 and σ_2, respectively, and the above calculations hold for $\Theta\in\mathrm{dom}\,\mathcal{A}$.

To write the first-order system in weak or variational form, let $\mathcal{Z}(t)=(z(t),z_t(t))$, $\mathcal{F}(t)=(0,F(t))$, and $\mathcal{B}u(t)=(0,Bu(t))$. The weak form of the system is then

$$\langle\mathcal{Z}_t(t),\chi\rangle_{\mathcal{V}^*,\mathcal{V}}+\sigma(\mathcal{Z}(t),\chi)=\langle\mathcal{B}u(t)+\mathcal{F}(t),\chi\rangle_{\mathcal{V}^*,\mathcal{V}} \tag{5.40}$$

for $\chi\in\mathcal{V}$. Formally, this is equivalent to the system

$$\mathcal{Z}_t(t)=\mathcal{A}\mathcal{Z}(t)+\mathcal{B}u(t)+\mathcal{F}(t) \tag{5.41}$$

in $\mathcal{H}$, where $\mathcal{A}$ is given in (5.39) (see Banks, Ito, and Wang (1995) and Banks and Smith (1995b) for details concerning the formulation and well-posedness of systems of this type with unbounded input operators).

5.5.1 Finite-dimensional Approximation

To approximate the solution to the coupled system, suitable expansions must be chosen for the state variables w and ϕ. The plate displacement is approximated by

$$w^{\mathcal{N}}(t,r,\theta)=\sum_{m=-M_p}^{M_p}\sum_{n=1}^{N_p^m} w_{mn}(t)e^{im\theta}r^{|\hat{m}|}B_n^m(r),\quad N_p^m=\begin{cases}N_p+1, & m\neq 0\\ N_p, & m=0,\end{cases}$$

where $B_n(r)$ is the nth modified cubic spline satisfying $B_n(a)=\frac{dB_n(a)}{dr}=0$ with the condition $\frac{dB_n(0)}{dr}=0$ being enforced when $m=0$ (this latter condition guarantees differentiability at the origin). As discussed in Section 4.4, the inclusion of the weighting term $r^{|\hat{m}|}$ with

$$\hat{m}=\begin{cases}0, & m=0\\ 1, & m\neq 0\end{cases}$$

is motivated by the asymptotic behavior of the Bessel functions (which make up the analytic plate solution) as $r \to 0$. It also serves to ensure the uniqueness of the solution at the origin. The Fourier coefficient in the weight is truncated to control the conditioning of the mass and stiffness matrices (see the examples in Section 4.4).

A suitable Fourier–Galerkin expansion of the potential is

$$\phi^{\mathcal{M}}(t,r,\theta,z) = \sum_{p=0}^{P_c} \sum_{m=-M_c}^{M_c} \sum_{\substack{n=0 \\ p+|m|+n\neq 0}}^{N_c^{p,m}} \phi_{pmn}(t) e^{im\theta} r^{|\hat{m}|} P_n^{p,m}(r) P_p(z),$$

where

$$\hat{m} = \begin{cases} m, & |m| = 0, \ldots, 5 \\ 5, & |m| = 6, \ldots, M \end{cases}$$

and

$$P_n^{p,m}(r) = \begin{cases} P_1(r) - 1/3, & p = m = 0, n = 1 \\ P_n(r), & \text{otherwise.} \end{cases}$$

Here $P_n(r)$ and $P_p(z)$ are the nth and pth Legendre polynomials that have been mapped to the intervals $(0, a)$ and $(0, \ell)$, respectively. The term $P_1(r) - 1/3$ when $p = m = 0, n = 1$ results from the orthogonality properties of the Legendre polynomials and arises when enforcing the condition $\int_\Omega \phi^{\mathcal{M}}(t,r,\theta,z)\,d\omega = 0$ so as to guarantee that the functions are suitable as a basis for the quotient space. The inclusion of the weight $r^{|\hat{m}|}$ again incorporates the decay of the analytic solution near the origin while ensuring its uniqueness at that point. Finally, we note that the limit $N_c^{p,m}$ is given by $N_c^{p,m} = N_c + 1$ when $p + |m| \neq 0$ and $N_c^{p,m} = N_c$ when $p = m = 0$.

To simplify notation, the above expansions are then expressed as

$$w^{\mathcal{N}}(t,r,\theta) = \sum_{j=1}^{\mathcal{N}} w_j(t) B_j^{\mathcal{N}}(r,\theta) \tag{5.42}$$

and

$$\phi^{\mathcal{M}}(t,r,\theta,z) = \sum_{j=1}^{\mathcal{M}} \phi_j(t) B_j^{\mathcal{M}}(r,\theta,z), \tag{5.43}$$

where $\mathcal{N} = (2M_p+1)(N_p+1)-1$ and $\mathcal{M} = (2M_c+1)(N_c+1)(P_c+1)-1$. The plate and potential bases thus have the components $B_j^{\mathcal{N}}(r,\theta) = e^{im\theta} r^{|\hat{m}|} B_n(r)$ and $B_j^{\mathcal{M}}(r,\theta,z) = e^{im\theta} r^{|\hat{m}|} P_n(r) P_p(z)$, respectively.

5.5.2 Matrix System

The $\mathcal{N}$- and $\mathcal{M}$-dimensional approximating plate and cavity subspaces are taken to be $H_p^{\mathcal{N}} = \text{span}\{B_i^{\mathcal{N}}\}_{i=1}^{\mathcal{N}}$ and $H_c^{\mathcal{M}} = \text{span}\{B_i^{\mathcal{M}}\}_{i=1}^{\mathcal{M}}$, respectively. Defining $\mathcal{P} = \mathcal{N} + \mathcal{M}$, the approximating state space is $H^{\mathcal{P}} = H_c^{\mathcal{M}} \times H_p^{\mathcal{N}}$ and the product space for the first-order system is $\mathcal{H}^{\mathcal{P}} = H^{\mathcal{P}} \times H^{\mathcal{P}}$. The finite-dimensional approximation is then determined by restricting σ to $\mathcal{H}^{\mathcal{P}} \times \mathcal{H}^{\mathcal{P}}$. This yields the operator $\mathcal{A}^{\mathcal{P}} : \mathcal{H}^{\mathcal{P}} \to \mathcal{H}^{\mathcal{P}}$, where

$$\mathcal{A}^{\mathcal{P}} = \begin{bmatrix} 0 & I \\ -A_1^{\mathcal{P}} & -A_2^{\mathcal{P}} \end{bmatrix}$$

and $A_1^{\mathcal{P}}$ and $A_2^{\mathcal{P}}$ are obtained by restricting σ_1 and σ_2 to $H^{\mathcal{P}} \times H^{\mathcal{P}}$. Note that the restriction of the infinite-dimensional system to the space $H^{\mathcal{P}} \times H^{\mathcal{P}}$ yields, for $\Psi = (\xi, \eta) \in H^{\mathcal{P}}$,

$$\begin{aligned}
&\langle z_{tt}^{\mathcal{P}}(t), \Psi\rangle_H + \sigma_2(z_t^{\mathcal{P}}(t), \Psi) + \sigma_1(z^{\mathcal{P}}(t), \Psi) \\
&\quad = \int_{\Gamma_0} \left([H_1(r) - H_2(r)] \sum_{i=1}^{s} \mathcal{K}_i^B u_i(t)[H_{i1}(\theta) - H_{i2}(\theta)] \right) \overline{\nabla^2 \eta}\, d\gamma \\
&\qquad + \int_{\Gamma_0} f \bar{\eta}\, d\gamma.
\end{aligned}$$

When Ψ is chosen in $H^{\mathcal{P}}$ and the approximate plate and cavity solutions are taken to be (5.42) and (5.43), respectively, this yields the system

$$\begin{aligned}
M^{\mathcal{P}} \dot{y}^{\mathcal{P}}(t) &= \tilde{A}^{\mathcal{P}} y^{\mathcal{P}}(t) + \tilde{B}^{\mathcal{P}} u(t) + \tilde{F}^{\mathcal{P}}(t), \\
M^{\mathcal{P}} y^{\mathcal{P}}(0) &= \tilde{y}_0^{\mathcal{P}},
\end{aligned} \tag{5.44}$$

where

$$y^{\mathcal{P}}(t) = \begin{pmatrix} \vartheta^{\mathcal{P}}(t) \\ \dot{\vartheta}^{\mathcal{P}}(t) \end{pmatrix}.$$

Here $\vartheta^{\mathcal{P}}(t) = [\phi_1^{\mathcal{P}}(t), \phi_2^{\mathcal{P}}(t), \ldots, \phi_{\mathcal{M}}^{\mathcal{P}}(t), w_1^{\mathcal{P}}(t), w_2^{\mathcal{P}}(t), \ldots, w_{\mathcal{N}}^{\mathcal{P}}(t)]^T$ denotes the approximate state vector coefficients while $u(t) = [u_1(t), \ldots, u_s(t)]^T$ contains the s control variables. The full system has the form

$$\begin{bmatrix} M_1^{\mathcal{P}} & 0 \\ 0 & M_2^{\mathcal{P}} \end{bmatrix} \begin{bmatrix} \dot{\vartheta}^{\mathcal{P}}(t) \\ \ddot{\vartheta}^{\mathcal{P}}(t) \end{bmatrix} = \begin{bmatrix} 0 & M_1^{\mathcal{P}} \\ -A_1^{\mathcal{P}} & -A_2^{\mathcal{P}} \end{bmatrix} \begin{bmatrix} \vartheta^{\mathcal{P}}(t) \\ \dot{\vartheta}^{\mathcal{P}}(t) \end{bmatrix} + \begin{bmatrix} 0 \\ \hat{B}^{\mathcal{P}} \end{bmatrix} u(t) + \begin{bmatrix} 0 \\ \hat{F}^{\mathcal{P}}(t) \end{bmatrix},$$

$$\begin{bmatrix} M_1^{\mathcal{P}} & 0 \\ 0 & M_2^{\mathcal{P}} \end{bmatrix} \begin{bmatrix} \vartheta^{\mathcal{P}}(0) \\ \dot{\vartheta}^{\mathcal{P}}(0) \end{bmatrix} = \begin{bmatrix} g_1^{\mathcal{P}} \\ g_2^{\mathcal{P}} \end{bmatrix}$$

with

$$M_1^{\mathcal{P}} = \text{diag}\left[M_{11}^{\mathcal{P}}, M_{12}^{\mathcal{P}}\right],$$

$$M_2^{\mathcal{P}} = \text{diag}\left[M_{21}^{\mathcal{P}}, M_{22}^{\mathcal{P}}\right], \qquad A_2^{\mathcal{P}} = \begin{bmatrix} 0 & A_{31}^{\mathcal{P}} \\ A_{32}^{\mathcal{P}} & A_{22}^{\mathcal{P}} \end{bmatrix}$$

$$A_1^{\mathcal{P}} = \text{diag}\left[A_{11}^{\mathcal{P}}, A_{12}^{\mathcal{P}}\right],$$

and

$$\hat{B}^{\mathcal{P}} = \left[0, \tilde{B}_2^{\mathcal{P}}\right]^T, \qquad \hat{F}^{\mathcal{P}}(t) = \left[0, \tilde{F}_2^{\mathcal{P}}(t)\right]^T.$$

The matrices deriving from the wave contributions have the components

$$\left[M_{11}^{\mathcal{P}}\right]_{\ell,k} = \int_\Omega \nabla B_k^{\mathcal{M}} \cdot \overline{\nabla B_\ell^{\mathcal{M}}} d\omega,$$

$$\left[M_{21}^{\mathcal{P}}\right]_{\ell,k} = \int_\Omega \frac{\rho_f}{c^2} B_k^{\mathcal{M}} \overline{B_\ell^{\mathcal{M}}} d\omega,$$

$$\left[A_{11}^{\mathcal{P}}\right]_{\ell,k} = \int_\Omega \rho_f \nabla B_k^{\mathcal{M}} \cdot \overline{\nabla B_\ell^{\mathcal{M}}} d\omega.$$

We point out that $M_{21}^{\mathcal{P}}$ and $A_{11}^{\mathcal{P}}$ are the mass and stiffness matrices that arise when solving the uncoupled wave equation with Neumann boundary conditions. The form of the matrix $M_{11}^{\mathcal{P}}$ results from the choice of V inner product.

The plate matrices are given by

$$\left[M_{22}^{\mathcal{P}}\right]_{p,i} = \int_{\Gamma_0} \rho_p h B_i^{\mathcal{N}} \overline{B_p^{\mathcal{N}}} d\gamma$$

and

$$M_{12}^{\mathcal{P}} = K_1 + K_2 + K_3 + K_4 + K_5,$$
$$A_{12}^{\mathcal{P}} = K_{D1} + K_{D2} + K_{D3} + K_{D4} + K_{D5},$$
$$A_{22}^{\mathcal{P}} = K_{c_D1} + K_{c_D2} + K_{c_D3} + K_{c_D4} + K_{c_D5},$$

where

$$[K_1]_{p,i} = \int_{\Gamma_0} \left[\frac{\partial^2 B_i^{\mathcal{N}}}{\partial r^2} + \frac{\nu}{r}\frac{\partial B_i^{\mathcal{N}}}{\partial r} + \frac{\nu}{r^2}\frac{\partial^2 B_i^{\mathcal{N}}}{\partial \theta^2}\right] \frac{\partial^2 \overline{B_p^{\mathcal{N}}}}{\partial r^2} d\gamma,$$

$$[K_2]_{p,i} = \int_{\Gamma_0} \left[\frac{1}{r^2}\frac{\partial B_i^{\mathcal{N}}}{\partial r} + \frac{1}{r^3}\frac{\partial^2 B_i^{\mathcal{N}}}{\partial \theta^2} + \frac{\nu}{r}\frac{\partial^2 B_i^{\mathcal{N}}}{\partial r^2}\right] \frac{\partial \overline{B_p^{\mathcal{N}}}}{\partial r} d\gamma,$$

$$[K_3]_{p,i} = \int_{\Gamma_0} \left[\frac{1}{r^3}\frac{\partial B_i^{\mathcal{N}}}{\partial r} + \frac{1}{r^4}\frac{\partial^2 B_i^{\mathcal{N}}}{\partial \theta^2} + \frac{\nu}{r^2}\frac{\partial^2 B_i^{\mathcal{N}}}{\partial r^2}\right] \frac{\partial^2 \overline{B_p^{\mathcal{N}}}}{\partial \theta^2} d\gamma,$$

$$[K_4]_{p,i} = 2\int_{\Gamma_0}(1-\nu)\left[\frac{1}{r^2}\frac{\partial^2 B_i^{\mathcal{N}}}{\partial r\partial\theta} - \frac{1}{r^3}\frac{\partial B_i^{\mathcal{N}}}{\partial\theta}\right]\frac{\partial^2 \overline{B_p^{\mathcal{N}}}}{\partial r\partial\theta}d\gamma,$$

$$[K_5]_{p,i} = 2\int_{\Gamma_0}(1-\nu)\left[-\frac{1}{r^3}\frac{\partial^2 B_i^{\mathcal{N}}}{\partial r\partial\theta} + \frac{1}{r^4}\frac{\partial B_i^{\mathcal{N}}}{\partial\theta}\right]\frac{\partial \overline{B_p^{\mathcal{N}}}}{\partial\theta}d\gamma.$$

The matrices $K_{D1},\ldots,K_{D5}$ and $K_{c_D1},\ldots,K_{c_D5}$ are defined similarly with the inclusion of the parameters D and c_D in the various integrals. Note that the matrices $M_{22}^{\mathcal{P}}$, $A_{12}^{\mathcal{P}}$, and $A_{22}^{\mathcal{P}}$ are the mass, stiffness, and damping matrices that arise when solving the damped plate equation with fixed boundary conditions. The form of the matrix $M_{12}^{\mathcal{P}}$ is again due to the choice of the V inner product.

The contributions from the coupling terms are contained in the matrices

$$\left[A_{31}^{\mathcal{P}}\right]_{\ell,i} = -\rho_f\int_{\Gamma_0}B_i^{\mathcal{N}}\overline{B_\ell^{\mathcal{M}}}d\gamma, \qquad \left[A_{32}^{\mathcal{P}}\right]_{p,k} = \rho_f\int_{\Gamma_0}B_k^{\mathcal{M}}\overline{B_p^{\mathcal{N}}}d\gamma$$

while the control and forcing terms are contained in $\tilde{B}_2^{\mathcal{P}}$ and $\tilde{F}_2^{\mathcal{P}}(t)$, which are given by

$$\left[\tilde{B}_2^{\mathcal{P}}\right]_{p,j} = \int_{j^{th}\text{patch}}\mathcal{K}_j^B\overline{\nabla^2 B_p^{\mathcal{N}}}d\gamma, \qquad \left[\tilde{F}_2^{\mathcal{P}}(t)\right]_p = \int_{\Gamma_0}f\overline{B_p^{\mathcal{N}}}d\gamma.$$

In all cases, the index ranges are $k,\ell = 1,\ldots,\mathcal{M}$ and $i,p = 1,\ldots,\mathcal{N}$. The patch index j ranges from 1 to s. Finally, the vectors $g_1^{\mathcal{P}}$ and $g_2^{\mathcal{P}}$ contain the projections of the initial values into the approximating finite-dimensional subspaces. It should be noted that the construction of the matrices $A_{11}^{\mathcal{P}}$, $M_{11}^{\mathcal{P}}$, and $M_{21}^{\mathcal{P}}$ can be facilitated by taking advantage of the tensor nature of the cavity basis (see Section 5.3 containing the cylindrical wave discussion).

5.5.3 Example 9

In order to test the approximation scheme, a forcing function was incorporated in the wave equation in a manner analogous to that used in the plate model. We then consider the uncontrolled problem

$$\frac{\rho_f}{c^2}\phi_{tt} = \rho_f\Delta\phi + g(t,r,\theta,z), \qquad (r,\theta,z)\in\Omega, t>0,$$

$$\nabla\phi\cdot\hat{n} = 0, \qquad (r,\theta,z)\in\Gamma,$$

$$\frac{\partial\phi}{\partial z}(t,r,\theta,0) = -w_t(t,r,\theta),$$

$$\rho_b h w_{tt} + \frac{\partial^2 M_r}{\partial r^2} + \frac{2}{r}\frac{\partial M_r}{\partial r} - \frac{1}{r}\frac{\partial M_\theta}{\partial r} + \frac{2}{r}\frac{\partial^2 M_{r\theta}}{\partial r\partial\theta} + \frac{2}{r^2}\frac{\partial M_{r\theta}}{\partial\theta} + \frac{1}{r^2}\frac{\partial^2 M_\theta}{\partial\theta^2}$$
$$= -\rho_f\phi_t(t,r,\theta,0) + f(t,r,\theta),$$

$$w(t,a,\theta) = \frac{\partial w}{\partial r}(t,a,\theta) = 0,$$
$$\phi(0,r,\theta,z) = \phi_t(0,r,\theta,z) = w(0,r,\theta) = w_t(0,r,\theta) = 0$$

on the cylindrical domain Ω with dimensions $a = .6$, $h = .00127$, $\ell = 1.1$, and parameters $\rho_p = 2700$, $\rho_f = 1.21$, and $c = 343$. The moments are

$$M_r = D\left(\frac{\partial^2 w}{\partial r^2} + \frac{\nu}{r}\frac{\partial w}{\partial r} + \frac{\nu}{r^2}\frac{\partial^2 w}{\partial \theta^2}\right) + c_D\left(\frac{\partial^3 w}{\partial r^2 \partial t} + \frac{\nu}{r}\frac{\partial^2 w}{\partial r \partial t} + \frac{\nu}{r^2}\frac{\partial^3 w}{\partial \theta^2 \partial t}\right),$$

$$M_\theta = D\left(\frac{1}{r}\frac{\partial w}{\partial r} + \frac{1}{r^2}\frac{\partial^2 w}{\partial \theta^2} + \nu\frac{\partial^2 w}{\partial r^2}\right) + c_D\left(\frac{1}{r}\frac{\partial^2 w}{\partial r \partial t} + \frac{1}{r^2}\frac{\partial^3 w}{\partial \theta^2 \partial t} + \nu\frac{\partial^3 w}{\partial r^2 \partial t}\right),$$

$$M_{r\theta} = D(1-\nu)\left(\frac{1}{r}\frac{\partial^2 w}{\partial r \partial \theta} - \frac{1}{r^2}\frac{\partial w}{\partial \theta}\right) + c_D(1-\nu)\left(\frac{1}{r}\frac{\partial^3 w}{\partial r \partial \theta \partial t} - \frac{1}{r^2}\frac{\partial^2 w}{\partial \theta \partial t}\right)$$

with $\nu = .33$, $D = \frac{Eh^3}{12(1-\nu^2)} = 13.6007$, and $c_D = .00011222$. The forcing functions

$$\begin{aligned} g(t,r,\theta,z) = -\frac{3}{\ell^2}\Bigg[\Bigg(&2\frac{\rho_f}{c}(\cos(2\pi r/a) - 1) + t^2 \cdot 4\pi^2/a^2 \cos(2\pi r/a) \\ &+ t^2 \cdot 2\pi/a \cdot \frac{1}{r}\sin(2\pi r/a) + t^2 \cdot \frac{1}{r^2}(\cos(2\pi r/a) - 1)\Bigg) \\ &\times z(\ell - z)^2 - t^2 \cdot (\cos(2\pi r/a) - 1)(6z - 4\ell)\Bigg] \sin\theta, \end{aligned}$$

$$\begin{aligned} f(t,r,\theta) = {}& 6\rho_p h t(\cos(2\pi r/a) - 1)\sin\theta \\ &+ (t^3 D + 3t^2 c_D) \cdot \frac{1}{r^4}[3 + (-6\pi r/a + 16\pi^3 r^3/a^3)\sin(2\pi r/a) \\ &+ (-3 + 12\pi^2 r^2/a^2 + 16\pi^4 r^4/a^4)\cos(2\pi r/a)]\sin\theta \end{aligned}$$

result from the true solutions

$$w(r,\theta) = t^3(\cos(2\pi r/a) - 1)\sin\theta,$$
$$\phi(t,r,\theta,z) = -\frac{3}{\ell^2}t^2(\cos(2\pi r/a) - 1)z(\ell - z)^2\sin\theta$$

(the results from this example can be compared with those in the plate and wave sections). We note that the addition of the acoustic source g leads to the modification $\tilde{F}^{\mathcal{P}}(t) = [\tilde{G}_2^{\mathcal{P}}(t), \tilde{F}_2^{\mathcal{P}}(t)]$, $[\tilde{G}_2^{\mathcal{P}}(t)]_p = \int_\Omega g\overline{B_p^{\mathcal{M}}}d\omega$ in the right-hand side vector $\mathcal{F}^{\mathcal{P}}(t)$ of (5.44). The absolute and relative errors in the plate displacement, potential, and pressure at time $T = .1$ are reported in Tables 5.20–22. The subscripts p and c in the tables again refer to the plate and wave indices, respectively, and the measurements were made on a 30×30 grid on the plate and a $10 \times 10 \times 10$ grid in the cavity. The true and approximate plate

Table 5.20. *Absolute and relative errors in plate displacement at $T = .1$.*

M_p	N_p	M_c	N_c	P_c	size($\tilde{A}^{\mathcal{P}}$)	$\|w_{true} - w_{app}\|$	$\frac{\|w_{true}-w_{app}\|}{\|w_{true}\|}$
1	5	1	2	2	86 × 86	.4366 − 5	.2207 − 2
1	5	1	4	4	182 × 182	.4294 − 5	.2171 − 2
1	10	1	6	6	356 × 356	.2674 − 6	.1352 − 3
1	20	1	8	8	608 × 608	.1588 − 7	.8030 − 5

Table 5.21. *Absolute and relative errors in potential at $T = .1$.*

M_p	N_p	M_c	N_c	P_c	size($\tilde{A}^{\mathcal{P}}$)	$\|\phi_{true} - \phi_{app}\|$	$\frac{\|\phi_{true}-\phi_{app}\|}{\|\phi_{true}\|}$
1	5	1	2	2	86 × 86	.3132 − 2	.3569 − 0
1	5	1	4	4	182 × 182	.1545 − 3	.1760 − 1
1	10	1	6	6	356 × 356	.9554 − 5	.1089 − 2
1	20	1	8	8	608 × 608	.2251 − 6	.2565 − 4

Table 5.22. *Absolute and relative errors in pressure at $T = .1$.*

M_p	N_p	M_c	N_c	P_c	size($\tilde{A}^{\mathcal{P}}$)	$\|p_{true} - p_{app}\|$	$\frac{\|p_{true}-p_{app}\|}{\|p_{true}\|}$
1	5	1	2	2	86 × 86	.7573 − 1	.3566 − 0
1	5	1	4	4	182 × 182	.3739 − 2	.1761 − 1
1	10	1	6	6	356 × 356	.2312 − 3	.1089 − 2
1	20	1	8	8	608 × 608	.5453 − 5	.2568 − 4

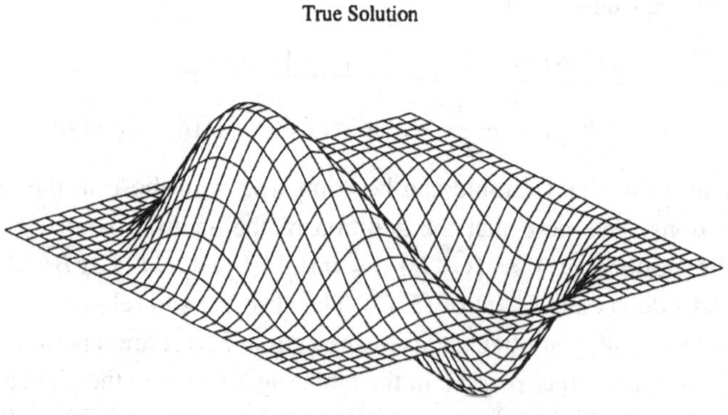

Fig. 5.14. True plate displacement at time $T = .1$.

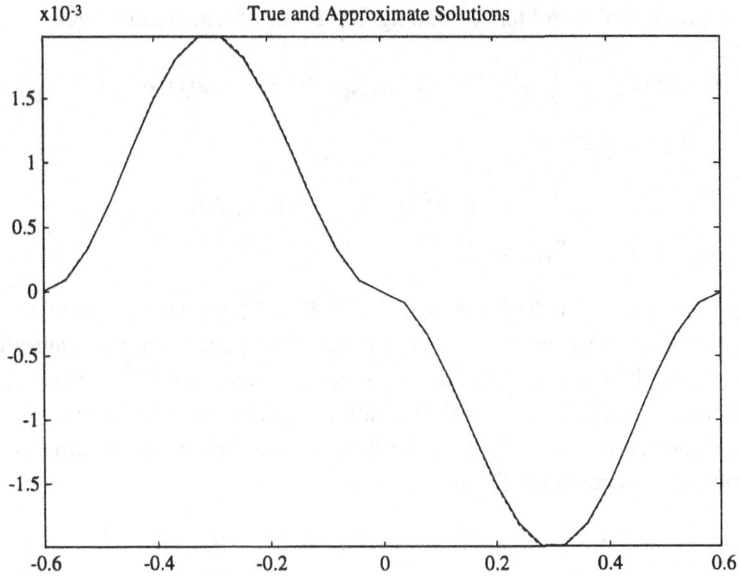

Fig. 5.15. True and approximate plate displacements at time $T = .1$; $- - -(N_c = 2)$, —— (True).

solutions at time $T = .1$ are plotted in Figs. 5.14 and 5.15. The latter figure in conjunction with the error results in the tables demonstrates the convergence of the method for this problem.

5.6 Noise Control in the Coupled System

In this section, the problem of controlling sound pressure levels in the previously described structural acoustic system will be considered. Again, control in this system is attained through the excitation of piezoceramic patches bonded to the vibrating end plate. A brief discussion of applicable optimal control theory is given followed by simulation results demonstrating the manner in which control is obtained using PDE-based techniques.

5.6.1 The Control Problem

The goal in the control problem is to determine voltages to the patches that affect the plate vibrations in a manner that ultimately leads to reduced interior sound pressure levels. For the finite-dimensional problem in which the exterior force is periodic with period τ, the problem of determining a controlling voltage can be posed as the problem of finding $u \in L^2(0,\tau)$ which minimizes

$$J^{\mathcal{P}}(u) = \frac{1}{2}\int_0^\tau \{\langle Q^{\mathcal{P}} y^{\mathcal{P}}(t), y^{\mathcal{P}}(t)\rangle_{\mathbf{R}^{\mathcal{P}}} + \langle Ru(t), u(t)\rangle_{\mathbf{R}^s}\}\,dt, \tag{5.45}$$

where $y^{\mathcal{P}}$ solves

$$\begin{aligned}\dot{y}^{\mathcal{P}}(t) &= A^{\mathcal{P}} y^{\mathcal{P}}(t) + B^{\mathcal{P}} u(t) + F^{\mathcal{P}}(t),\\ y^{\mathcal{P}}(0) &= y_0^{\mathcal{P}}\end{aligned}$$

(see (5.44). In the penalty functional (5.45), $R = \text{diag}[r_{ii}]$ is an $s \times s$ diagonal matrix and $r_{ii} > 0, i = 1, \ldots, s$, is the weight or penalty on the controlling voltage into the ith patch. The nonnegative definite matrix $Q^{\mathcal{P}}$ is chosen in a manner so as to emphasize the minimization of particular state variables. From energy considerations as discussed in Banks, Fang, Silcox, and Smith (1993), an appropriate choice for $Q^{\mathcal{P}}$ in this case is

$$Q^{\mathcal{P}} = M^{\mathcal{P}}\mathcal{D} = M^{\mathcal{P}}\text{diag}[d_1 I^{\mathcal{M}}, d_2 I^{\mathcal{N}}, d_3 I^{\mathcal{M}}, d_4 I^{\mathcal{N}}],$$

where $M^{\mathcal{P}}$ is the mass matrix (see Eq. (5.44)), $I^k, k = \mathcal{N}, \mathcal{M}$, denote $k \times k$ identity matrices, and d_i are parameters chosen to enhance stability and performance of the feedback.

The optimal control is then given by

$$u^{\mathcal{P}}(t) = R^{-1}(B^{\mathcal{P}})^T[r^{\mathcal{P}}(t) - \Pi^{\mathcal{P}} y^{\mathcal{P}}(t)], \tag{5.46}$$

where $\Pi^{\mathcal{P}}$ is the solution to the algebraic Riccati equation

$$(A^{\mathcal{P}})^T\Pi^{\mathcal{P}} + \Pi^{\mathcal{P}}A^{\mathcal{P}} - \Pi^{\mathcal{P}}B^{\mathcal{P}}R^{-1}(B^{\mathcal{P}})^T\Pi^{\mathcal{P}} + Q^{\mathcal{P}} = 0.$$

For the regulator problem with periodic forcing function $F^{\mathcal{P}}(t)$, the tracking variable $r^{\mathcal{P}}(t)$ must satisfy the linear differential equation

$$\begin{aligned}\dot{r}^{\mathcal{P}}(t) &= -[A^{\mathcal{P}} - B^{\mathcal{P}}R^{-1}(B^{\mathcal{P}})^T\Pi^{\mathcal{P}}]^T r^{\mathcal{P}}(t) + \Pi^{\mathcal{P}}F^{\mathcal{P}}(t),\\ r^{\mathcal{P}}(0) &= r^{\mathcal{P}}(\tau),\end{aligned}$$

while the optimal trajectory is the solution to the linear differential equation

$$\begin{aligned}\dot{y}^{\mathcal{P}}(t) &= [A^{\mathcal{P}} - B^{\mathcal{P}}R^{-1}(B^{\mathcal{P}})^T\Pi^{\mathcal{P}}]y^{\mathcal{P}}(t) + B^{\mathcal{P}}R^{-1}(B^{\mathcal{P}})^T r^{\mathcal{P}}(t) + F^{\mathcal{P}}(t),\\ y^{\mathcal{P}}(0) &= y^{\mathcal{P}}(\tau).\end{aligned}$$

Details regarding the formulation of the finite-dimensional periodic control problem for a reduced geometry as well as a discussion about the underlying infinite-dimensional control problem can be found in Banks, Fang, Silcox, and Smith (1993).

5.6.2 Simulation Results

To numerically demonstrate the previously described control methodology, simulations of the uncontrolled and controlled system were performed with the dimensions and physical parameters in the model chosen so as to be consistent with those of the experimental setup described in the introduction. The length and radius of the cavity were taken to be 1.0668 m (42″) and $a = .2286$ m (9″), respectively, with a plate having thickness $h = .00127$ m (.05″) mounted at one end. A pair of circular piezoceramic patches having thickness $T = .0001778$ m (.007″) and radius $rad = .01905$ m (.75″) were located at the center of the plate (see Fig. 5.16).

The physical parameters that were chosen for the structure and acoustic cavity are summarized in Table 5.23. The flexural rigidity D for the plate was obtained using the "handbook" value $E = 7.1 \times 10^{10}$ N/m^2 for the Young's modulus of aluminum. The remaining choices are comparable to values found when estimating parameters for the isolated plate with a similar patch configuration (Banks, Brown, Metcalf, Silcox, Smith, and Wang, 1994). We reemphasize that a first step when determining gains to be used during experimental implementation of the scheme is the estimation of these parameters for the system in the form in which it is going to be controlled. Although robustness of a control scheme might allow for some leeway in these values, the results will be degraded and potentially destabilized by the use of overly inaccurate system parameters.

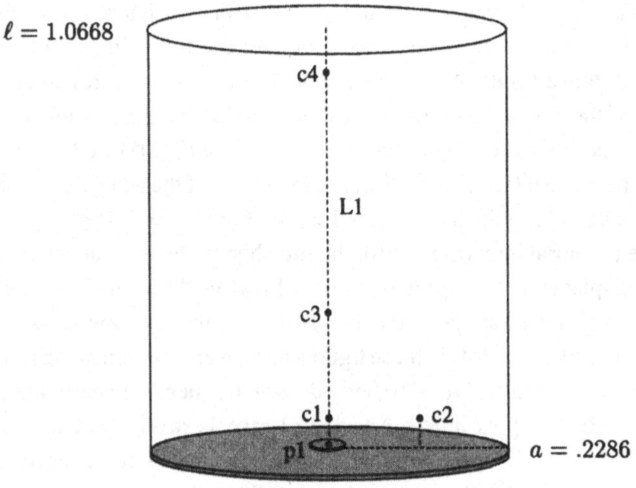

Fig. 5.16. The acoustic cavity with a pair of centered circular patches and the observation points $p1 = (0, 0)$, $c1 = (0, 0, .05)$, $c2 = (a/2, \pi/2, .05)$, $c3 = (0, 0, .35)$, and $c4 = (0, 0, 1.0)$.

Table 5.23. *Physical parameters for the structure and acoustic cavity.*

Structure			Acoustic cavity	
Parameter	Plate	Plate + Pzt	Parameter	Cavity
$\rho_p \cdot$ *Thickness* (kg/m^2)	3.429	3.489	ρ_f (kg/m^3)	1.21
D (N·m)	13.601	13.901		
c_D (N·m·sec)	1.150–4	2.250–4		
ν	.33	.32		
$\mathcal{K}^B$ (N/V)		.0267	c (m/sec)	343

5.6.2.1 *Uncontrolled and Controlled System Dynamics*

To demonstrate the effects of the controlling voltage on the dynamics of the system, the forcing function

$$g(t, r, \theta) = 28.8 \sin(500\pi t)$$

modeling a periodic plane wave having a root mean square (rms) sound pressure level of 120 dB was applied to the plate. The excitation frequency of 250 hertz is close to the natural frequency, 238 hertz, of the third system mode (see Banks, Silcox, and Smith (1993) for examples determining the natural frequencies of the system).

In approximating the dynamics of the system, it was found that the choices $\mathcal{N} = 12$ and $\mathcal{M} = 99$ (see (5.42) and (5.43)), were sufficient for resolving the range of frequencies under consideration, and the following results were obtained with those limits. For both the uncontrolled and controlled cases, time histories of the system response were calculated at the plate point $p1 = (0, 0)$ and cavity points $c1 = (0, 0, .05)$, $c2 = (a/2, \pi/2, .05)$, $c3 = (0, 0, .35)$, and $c4 = (0, 0, 1.0)$ (see Fig. 5.16). The rms values of the acoustic pressure were also calculated along the axial line $L1 = \{(r, \theta, z) : r = 0, 0 \leq z \leq 1.0668\}$.

For the temporal interval [0, .16], the time history and frequency response of the plate displacement are plotted in Fig. 5.17 while the acoustic pressure levels at the points $c1$ and $c3$ are plotted in Fig. 5.18 (note the differing scales). It can be seen from the spectral plots in these figures that several system modes are excited by the transient startup of the 250 hertz driving frequency. In particular, the plot of the uncontrolled pressure observed at $c1$ reveals strong system responses at 61, 165, 238, and 323 hertz along with weaker responses at higher frequencies. As discussed in Banks, Silcox, and Smith (1993), the responses at 61 and 238 hertz correspond to the first two frequencies of the isolated, undamped plate whereas the responses at 165 and 323 hertz correspond to the first two axial frequencies for an uncoupled cavity of this size. We emphasize that due to the

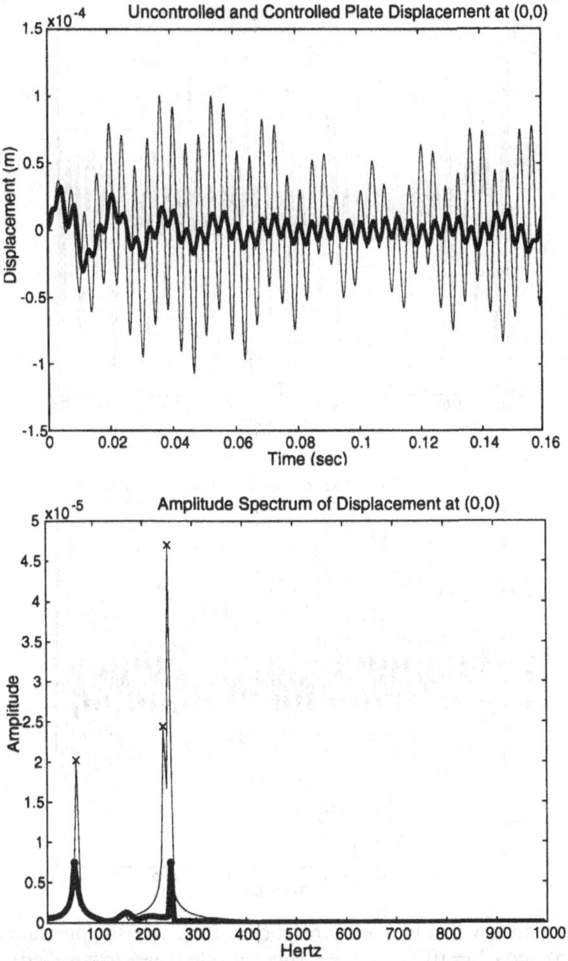

Fig. 5.17. Time history and frequency response of the plate displacement at the point (0, 0): — (uncontrolled), — (controlled).

coupling between the acoustic and structural components, the modes here are truly *system* modes and although comparisons can be made with the modes of the isolated components, the *system* modes differ slightly in frequency and shape due to the interaction between the structural and acoustic dynamics as well as the internal damping in the plate. Finally, we point out that the strong responses at system frequencies are transient, and when simulated sufficiently far in time, the system dynamics eventually reflect only the driving frequencies.

The rms sound pressure levels along the line $L1$ are plotted in Fig. 5.19 (both linear and decibel scales are given) while the rms sound pressure levels at the cavity points $c1$–$c4$ and the rms beam displacement at the central point $p1$ are

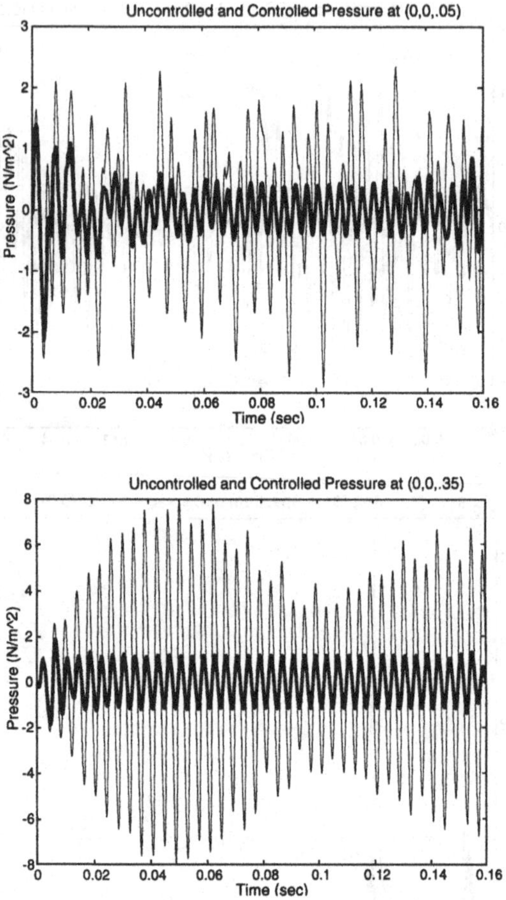

Fig. 5.18. Time history and frequency response of the acoustic pressure at the points $c1 = (0, 0, .05)$ and $c3 = (0, 0, .35)$: — (uncontrolled), — (controlled).

summarized in Table 5.24. The strong responses at the interior points $c3$ and $c4$ are due to the excitation of the system mode corresponding to the first cavity mode. At these points, the sound pressure levels are approximately 12–14 dB less than those of the driving exterior field.

Control in the system was implemented by calculating the voltage (5.46) and applying this to the centered patches. The quadratic cost functional parameters were taken to be $d_1 = d_2 = d_3 = d_4 = 10^{-8}$ and $R = 10^{-4}$ (although several values were tried, this choice provided good control authority with reasonable voltages and system conditioning). The Riccati matrix and tracking solution were computed off-line and then treated as filters in the manner discussed in

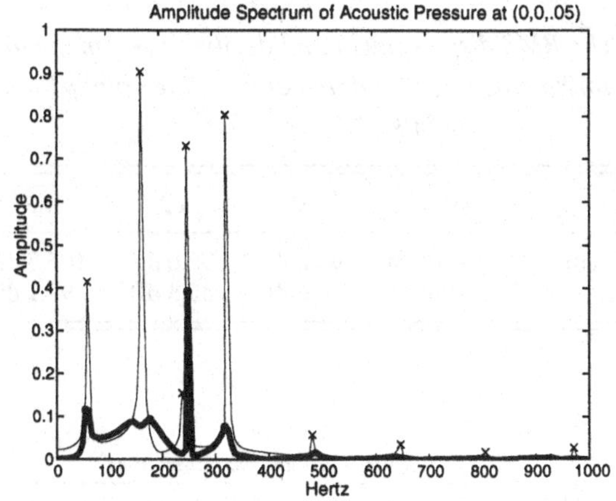

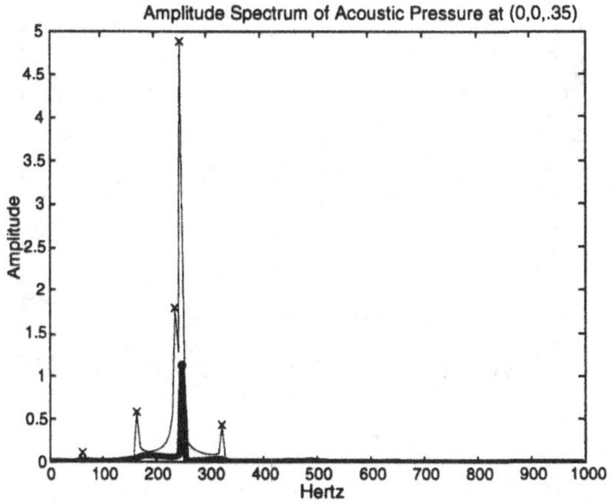

Fig. 5.18. *(cont.)*

Banks and Smith (1995a). In this manner, the feedback of information could be performed without significant delay in the process.

By comparing the time histories and frequency responses of the controlled trajectories with those of the uncontrolled system (see Figs. 5.17, 5.18, and 5.19), it can be seen that, following an initial transient interval, the controlled displacement and pressure are substantially reduced and maintained at a low level throughout the remainder of the period. To quantify the reduction, the rms levels of the controlled displacement and acoustic pressure were calculated and tabulated in Table 5.24. When compared with the uncontrolled levels, it can be seen that reductions of 3.4×10^{-5} m, 8.8 dB, 10.7 dB, 13.6 dB, and 12.9 dB

Table 5.24. *The RMS displacement and decibel levels in the uncontrolled and controlled cases as calculated at the plate point p1 and cavity points c1, c2, c3, and c4.*

	$p1$	$c1$	$c2$	$c3$	$c4$
Uncontrolled system	4.4e − 05 m	94.8 dB	96.0 dB	105.7 dB	107.8 dB
Controlled system	9.6e − 06 m	86.0 dB	85.3 dB	92.1 dB	94.9 dB

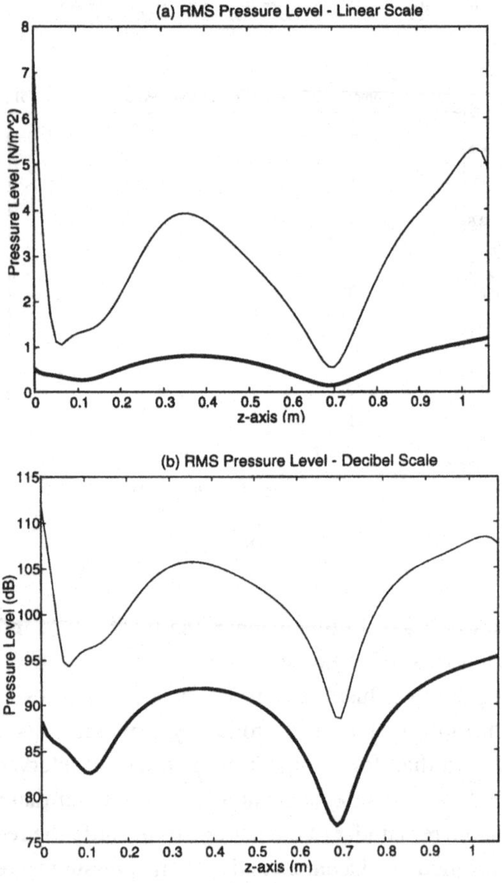

Fig. 5.19. RMS sound pressure levels along the line $L1$: (*a*) linear and (*b*) decibel pressure scales; — (uncontrolled), — (controlled).

are realized at the points $p1, c1, c2, c3$, and $c4$, respectively. Qualitatively, it can be seen that the primary response in the controlled trajectories is at the 250 hertz exciting frequency with the highest levels found near the back of the cavity (see Fig. 5.19). By comparing the rms levels of the uncontrolled and controlled pressure along the axial line $L1$, it can be seen that the very large pressure levels adjacent to the plate in the uncontrolled case are greatly reduced when control is implemented (there is an approximately 25 dB reduction in acoustic sound pressure at the plate's edge). Moreover, Fig. 5.19 demonstrates that the reduction in pressure is fairly uniform as one moves toward the back of the cavity, with the amount of reduction actually increasing near the back wall. As indicated by the results recorded at the noncentered point $c2$, similar reductions were recorded off the central axis.

The controlling voltage is plotted in Fig. 5.20. As expected, it reflects both the transient behavior of the system as it moves toward steady state as well as the periodicity of the driving force. It is noted that the voltage has a maximum magnitude of 41 V and an rms level of 14.5 V. In practice, it has been observed that the patches can be used for extended periods without damage or degradation of performance if the voltage levels are maintained below 8–10 rms V/mil (Metcalf, 1994) (the patches simulated here and used in the apparatus being modeled are 7 mil thick). Hence the control observed here is obtained with physically reasonable voltages to the patches.

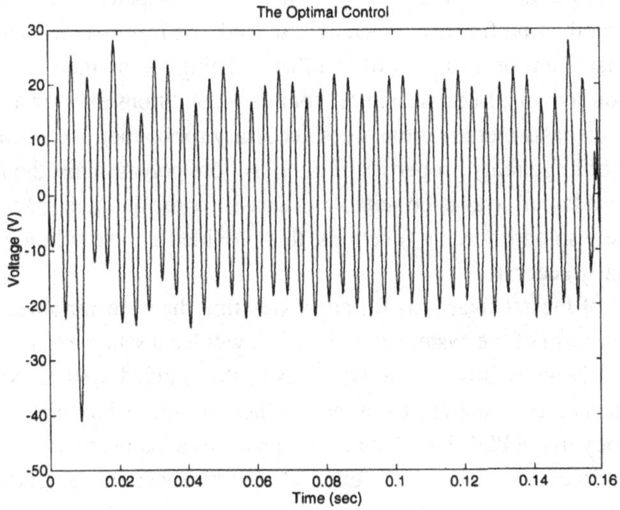

Fig. 5.20. The controlling voltage $u(t)$.

5.7 Conclusion

In this work, the 2-D structural acoustics model developed in Banks, Fang, Silcox, and Smith (1993) and Banks, Silcox, and Smith (1994) has been extended to a 3-D cylindrical domain with a flexible plate at one end. This geometry was chosen since it models the experimental apparatus that is being constructed to test the control methodology developed in Banks, Fang, Silcox, and Smith (1993). The final system of equations is considered in variational form because this avoids the problems associated with the differentiation of the discontinuous plate material parameters as well as the Heaviside function and Dirac delta that arise in the control input term.

A Fourier–Galerkin scheme utilizing modified cubic splines in the radial direction was chosen for the plate expansion. The choice of the splines resulted from the fact that they fit the smoothness criteria, were easily adapted to the clamped boundary conditions, and were suitable for approximations involving the discontinuous plate parameters. As noted in the plate examples, this expansion provides accurate results when approximating the dynamics of an undamped plate, which are then translated into accurate approximations of the plate motion in the fully coupled system.

A Fourier–Galerkin expansion was also chosen for discretizing the wave equation in the cylindrical domain, but in this case, translated Legendre functions were used in the radial and axial directions. The results in Banks, Ito, and Wang (1991) concerning the decay of stability margins under approximation with finite elements and finite differences in weakly damped hyperbolic systems were one motivation for this choice and, indeed, the results in the last section indicate that a uniform margin of stability is being maintained with this approximation scheme. Because natural boundary conditions exist on all walls of the cavity, no basis modifications were necessary for boundary conditions, and due to the orthogonality properties of the Legendre polynomials, the functions were easily adapted so as to be suitable as a basis for the quotient space. Finally, for relatively smooth forcing functions, this provided a wave solution that was exponentially accurate.

The use of these expansions for approximating the state variables w and ϕ in the weak form of the system equations has yielded a scheme that accurately and efficiently approximates the dynamics of the coupled system. As demonstrated in a detailed numerical example, the incorporation of this approximation methodology into a PDE-based controller provides a viable means of reducing structure-borne noise through the excitation of piezoceramic actuators on the structure. Such techniques are currently being experimentally implemented for isolated structures (Banks, Brown, Metcalf, Silcox, Smith, and Y. Wang, 1994; Banks, Smith, Brown, Silcox, and Metcalf, 1995) as a first step toward the implementation of PDE-based controllers for reducing noise in structural acoustic systems.

Acknowledgments

The research of H. T. Banks was supported in part by the Air Force Office of Scientific Research under grants AFOSR-90-0091 and AFOSR-F49620-93-1-0198. This research was also supported by the National Aeronautics and Space Administration under NASA grant NAG-1-1600 and NASA contract numbers NAS1-18605 and NAS1-19480 while H. T. Banks was a visiting scientist and R. C. Smith was in residence at the Institute for Computer Applications in Science and Engineering (ICASE), NASA Langley Research Center, Hampton, VA 23681.

The authors also express their sincere appreciation to R. J. Silcox of the Acoustics Division, NASA Langley Research Center, for numerous consultations concerning the modeling of this problem.

References

Abramowitz, M., and Stegan, I. A. (eds.), 1972, *Handbook of Mathematical Functions with Formulas, Graphs, and Mathematical Tables*, Dover Publications, New York.

Banks, H. T., Smith, R. C., Brown, D. E., Silcox, R. J., and Metcalf, V., 1997, "Experimental Confirmation of a PDE-Based Approach to Design of Feedback Controls," *SIAM Journal of Optimization and Control*, Vol. 35, No. 4, pp. 1263–96.

Banks, H. T., Brown, D. E., Metcalf, V., Silcox, R. J., Smith, R. C., and Wang, Y., 1994. "A PDE-Based Methodology for Modeling, Parameter Estimation and Feedback Control in Structural and Structural Acoustic Systems," *Proceedings of the 1994 North American Conference on Smart Structures and Materials*, Orlando, FL, pp. 311–20.

Banks, H. T., Fang, W., Silcox, R. J., and Smith, R. C., 1993, "Approximation Methods for Control of Acoustic/Structure Models with Piezoceramic Actuators," *Journal of Intelligent Material Systems and Structures*, Vol. 4, No. 1, pp. 98–116.

Banks, H. T., Ito, K., and Wang, C., 1991, "Exponentially Stable Approximations of Weakly Damped Wave Equations," *International Series in Numerical Mathematics*, Birkhäuser, Vol. 100, pp. 1–33.

Banks, H. T., Ito, K., and Wang, Y., 1995, "Well-Posedness for Damped Second Order Systems with Unbounded Input Operators," *Differential and Integral Equations*, Vol. 8, No. 3, pp. 587–606.

Banks, H. T., Silcox, R. J., and Smith, R. C., 1993, "Numerical Simulations of a Coupled 3-D Structural Acoustics System," *Proceedings of the Second Conference on Recent Advances in Active Control of Sound and Vibration*, Blacksburg, VA, pp. 85–97.

Banks, H. T., Silcox, R. J., and Smith, R. C., 1994, "The Modeling and Control of Acoustic/Structure Interaction Problems Via Piezoceramic Actuators: 2-D Numerical Examples," *Transactions of the ASME Journal of Vibration and Acoustics*, Vol. 116, No. 3, pp. 386–96.

Banks, H. T., and Smith, R. C., 1994, "The Modeling and Approximation of a Structural Acoustics Problem in a Hard-Walled Cylindrical Domain," Center for Research in Scientific Computation Technical Report, CRSC-TR94-26, North Carolina State University, December.

Banks, H. T., and Smith, R. C., 1995a. "Feedback Control of Noise in a 2-D Nonlinear Structural Acoustics Model," *Discrete and Continuous Dynamical Systems*, Vol. 1, No. 1, pp. 119–49.

Banks, H. T., and Smith, R. C., 1995b, "Well-Posedness of a Model for Structural Acoustic Coupling in a Cavity Enclosed by a Thin Cylindrical Shell," *Journal of Mathematical Analysis and Applications*, Vol. 191, pp. 1–25.

Banks, H. T., Smith, R. C., and Wang, Y., 1995, "Modeling Aspects for Piezoceramic Patch Activation of Shells, Plates and Beams," *Quarterly of Applied Mathematics*, Vol. 53, No. 2, pp. 353–81.

Banks, H. T., Wang, Y., Inman, D. J., and Slater, J. C., 1992, "Variable Coefficient Distributed Parameter System Models for Structures with Piezoceramic Actuators and Sensors," *Proceedings of the 31st Conference on Decision and Control*, Tucson, AZ, December 16–18, pp. 1803–8.

Bouaoudia, S., and Marcus, P. S., 1991, "Fast and Accurate Spectral Treatment of Coordinate Singularities," *Journal of Computational Physics*, Vol. 96, pp. 217–23.

Canuto, C., Hussaini, M. Y., Quarteroni, A., and Zang, T. A., 1988. *Spectral Methods in Fluid Dynamics*, Springer-Verlag, New York.

Gottlieb, D., and Orszag, S. A., 1977, *Numerical Analysis of Spectral Methods: Theory and Applications*, SIAM, Philadelphia.

Leissa, A. W., 1969. *Vibration of Plates,* NASA SP-160, Washington, D.C.

Metcalf, V. L., 1994. U.S. Army Research Laboratory, NASA Langley Research Center, personal communication.

Morse, P. M., and Ingard, K. U., 1968, *Theoretical Acoustics,* McGraw-Hill, New York.

Orszag, S. A., 1974. "Fourier Series on Spheres," *Monthly Weather Review*, Vol. 102, pp. 56–75.

Orszag, S. A., and Patera, A. T., 1983, "Secondary Instability of Wall-Bounded Shear Flows," *Journal of Fluid Mechanics*, Vol. 128, pp. 347–85.

Patera, A. T., and Orszag, S. A., 1981, "Finite-Amplitude Stability of Axisymmetric Pipe Flow," *Journal of Fluid Mechanics*, Vol. 112, pp. 467–74.

Zang, T. A., Streett, C. L., and Hussaini, M. Y., 1989, "Spectral Methods for CFD," *ICASE Report 89-13*.

Nomenclature

a	plate radius
B_n^m	cubic spline satisfying pole and boundary conditions
c	speed of sound in air
c_D	Kelvin–Voigt damping coefficient for the plate
D	$D = \frac{Eh^3}{12(1-\nu^2)}$, flexural rigidity for the plate
f	external surface force applied to plate
g	distributed force applied to wave equation
h	plate thickness
H	state space for plate displacement and acoustic potential
I_m	modified Bessel function of the first kind
J_m	Bessel function of the first kind
$\mathcal{K}_i^B$	piezoceramic patch constants

ℓ	cylinder length
M	Fourier index
$\mathcal{M}$	discretization index for acoustic potential
$\hat{m}$	truncated Fourier index
$\mathcal{M}_r, \mathcal{M}_\theta, \mathcal{M}_{r\theta}$	general bending and twisting moments for plate
$M_r, M_\theta, M_{r\theta}$	internal bending and twisting moments for plate
$(M_r)_{pe}, (M_\theta)_{pe}$	external moments generated by the piezoceramic patches
N	cubic spline or Legendre index
$\mathcal{N}$	discretization index for plate displacement
p	fluid (air) pressure
P_n	Legendre polynomial mapped to $(0, a)$
$\tilde{P}_n$	parity-preserving Legendre polynomial on $(-a, a)$
s	number of piezoceramic patches
$u_i(t)$	voltage to the ith piezoceramic patch
V	space of test functions for displacement and potential
w	transverse plate displacement

Greek

Γ	boundary to acoustic domain
Γ_0	equilibrium position for the circular plate
λ_{mn}	eigenvalue for plate or wave
ν	Poisson ratio for the plate
ρ_f	fluid (air) density
ρ_p	plate density
ϕ	acoustic potential
Ω	acoustic domain

6

Distributed Transfer Function Analysis of Stepped and Ring-stiffened Cylindrical Shells

B. YANG and J. ZHOU

Abstract

A new analytical method is presented for modeling and analysis of stepped cylindrical shells and cylindrical shells stiffened by circumferential rings. Through use of the distributed transfer functions of the structural systems, various static and dynamic problems of cylindrical shells are systematically formulated. With this transfer function formulation, the static and dynamic response, natural frequencies and mode shapes, and buckling loads of general stiffened cylindrical shells under arbitrary external excitations and boundary conditions can be determined in exact and closed form. The proposed method is illustrated on a Donnell–Mushtari shell and compared with the finite element method and other modeling techniques.

6.1 Introduction

Cylindrical shells are the basic element in many structures and machines and therefore have been extensively studied in the past; for instance, see Donnell (1933), Soedel (1981), Irie et al. (1984), Yamada et al. (1984), Sheinman and Weissman (1987), Koga (1988), Thangaratnam et al. (1990), Huang and Hsu (1992), Heyliger and Jilani (1993), Birman (1993), and Miyazaki and Hagihara (1993). The static and dynamic problems of cylindrical shells are often complicated by engineering design in which a cylindrical shell is composed of a finite number of serially connected shell segments and/or stiffened by circumferential rings. For such complex structural systems, numerical methods are usually adopted.

In the previous research on cylindrical shells, approximate and asymptotic methods are widely used; exact and closed-form solution methods have not been well developed, with notable exceptions in Chaudhuri and Abu-Arja (1989) and Christoforou and Swanson (1990). Although advanced computer technology and powerful numerical algorithms, such as finite element methods (FEM), make it possible to analyze complicated flexible systems, analytical methods are always desirable, for they yield more accurate results and deeper physical understanding of structures.

The objective of this study is to develop a unified analytical method, namely the distributed transfer function method, for modeling and analysis of general cylindrical shells. The distributed transfer functions of a flexible system are the Laplace transforms of the system Green's functions (Butkoviskiy, 1983). As a new modeling tool for vibration analysis and control of flexible mechanical systems, the distributed transfer function method has attracted great attention lately (Yang, 1989; Burke and Hubbard, 1990; Yang and Mote, 1990; Lee and Kuo, 1992; Yang, 1992a, 1992b; Pang et al., 1992; Tan and Chung, 1993; Yang, 1994). The transfer function method is related to the Green's function method (Bergman and Nicholson, 1985; Bergman and Hyatt, 1989). Most work on the distributed transfer function modeling has been focused on one-dimensional flexible systems; exact and closed-form transfer formulation for complex two-dimensional flexible continua has not been addressed.

The material presented herein is mainly from the authors' recent investigation on the distributed transfer function modeling of general cylindrical shells (Yang and Zhou, 1995; Zhou and Yang, 1995). In the development, the distributed transfer functions of a homogeneous cylindrical shell are first derived. With the shell distributed transfer functions, stepped shells are synthesized; stiffened shell are assembled, with the shell and ring stiffeners treated as individual structural components. This way, various static and dynamic problems of stepped and stiffened shells, including the static and dynamic response, natural frequencies and mode shapes, and buckling loads of general cylindrical shells under arbitrary external excitations and boundary conditions, can be determined in exact and closed form. The proposed method is illustrated on a Donnell–Mushtari shell and compared with the finite element method and other modeling techniques. The numerical results show the accuracy, efficiency and flexibility of the proposed transfer function method.

6.2 Distributed Transfer Functions

The response of the homogeneous cylindrical shell in Fig. 6.1 is governed by

$$\sum_{k=1}^{3}\sum_{i=0}^{n_k}\sum_{j=0}^{i}\left[A_{mkij}\frac{\partial^2}{\partial t^2}+B_{mkij}\frac{\partial}{\partial t}+C_{mkij}\right]\frac{\partial^i u^k(x,\theta,t)}{\partial x^{i-j}\partial\theta^j}=f^m(x,\theta,t),$$

$$m=1,2,3, \qquad (6.1)$$

where u^k ($k=1,2,3$) are the shell displacements in the longitudinal ($x-$),

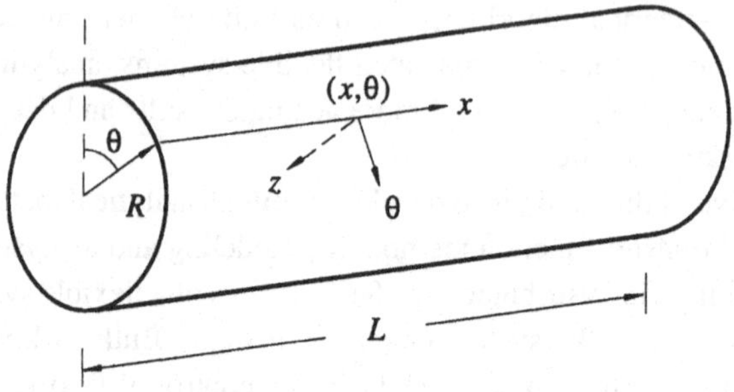

Fig. 6.1. A homogeneous cylindrical shell.

circumferential ($\theta-$), and radial ($z-$) coordinate directions, respectively, n_k is the highest order of differentiation of u^k, A_{mkij}, B_{mkij}, and C_{mkij} are constants related to the shell geometry and material properties, and $f^m(x, \theta, t)$ are the external loads acted on the shell. Here, Eq. (6.1) has been nondimensionalized so that $x = 0$ and $x = 1$ are the left and right boundaries of the shell. The boundary and initial conditions of the shell are

$$\left[\sum_{k=1}^{3}\sum_{i=0}^{n_k-1}\sum_{j=0}^{i}\Gamma_{lkij}\frac{\partial^i u^k(x_\alpha, \theta, t)}{\partial x^{i-j}\partial\theta^j}\right]_{x=x_\alpha} = \lambda^l(\theta, t), \qquad l = 1, 2, \ldots, n_b \tag{6.2a}$$

and

$$u^k(x, \theta, t)\big|_{t=0} = u_0^k(x, \theta), \qquad \frac{\partial}{\partial t}u^k(x, \theta, t)\bigg|_{t=0} = u_0^{k,t}(x, \theta), \qquad k = 1, 2, 3, \tag{6.2b}$$

where $x_\alpha = 0$ and 1 for $\alpha = 1$ and 2, respectively, Γ_{lkij} are constants, and $\lambda^l(\theta, t)$, $u_0^k(x, \theta)$, and $u_0^{k,t}(x, \theta)$, are given functions representing the boundary and initial disturbances. The number of boundary conditions $n_b = n_1 + n_2 + n_3$, which is eight in most situations. Equations (6.1) and (6.2) represent many shell models, such as Donnell–Mushtari, Love–Timoshenko, Flugge–Novozhilov, Reissner, Vlasov, and Sanders. To find the solution of the initial-boundary value problem formed by Eqs. (6.1) and (6.2), the shell displacements and the given forcing/disturbance functions are expanded into Fourier series in the circumferential direction θ:

$$u^k(x, \theta, t) = \sum_{n=0}^{\infty}\left[u_{1,n}^k(x, t)\cos n\theta + u_{2,n}^k(x, t)\sin n\theta\right], \qquad k = 1, 3, \tag{6.3a}$$

$$u^2(x, \theta, t) = \sum_{n=0}^{\infty}\left[u_{1,n}^2(x, t)\sin n\theta + u_{2,n}^2(x, t)\cos n\theta\right], \tag{6.3b}$$

$$f^m(x,\theta,t) = \sum_{n=0}^{\infty}\left[f_{1,n}^m(x,t)\cos n\theta + f_{2,n}^m(x,t)\sin n\theta\right], \qquad m = 1, 2, 3, \tag{6.3c}$$

$$u_0^k(x,\theta) = \sum_{n=0}^{\infty}\left[u_{1,n}^{k,0}(x)\cos n\theta + u_{2,n}^{k,0}(x)\sin n\theta\right], \qquad k = 1, 3, \tag{6.3d}$$

$$u_0^2(x,\theta) = \sum_{n=0}^{\infty}\left[u_{1,n}^{2,0}(x)\sin n\theta + u_{2,n}^{2,0}(x)\cos n\theta\right], \tag{6.3e}$$

$$u_0^{k,t}(x,\theta) = \sum_{n=0}^{\infty}\left[u_{0,1,n}^{k,t}(x)\cos n\theta + u_{0,2,n}^{k,t}(x)\sin n\theta\right], \tag{6.3f}$$

$$u_0^{2,t}(x,\theta) = \sum_{n=0}^{\infty}\left[u_{0,1,n}^{2,t}(x)\sin n\theta + u_{0,2,n}^{2,t}(x)\cos n\theta\right], \tag{6.3g}$$

$$\lambda^l(\theta,t) = \sum_{n=0}^{\infty}\left[\lambda_{1,n}^l(t)\cos n\theta + \lambda_{2,n}^l(t)\sin n\theta\right], \qquad l = 1, 2, \ldots, n_b. \tag{6.3h}$$

Here, the subscripts $(1, n)$ and $(2, n)$ indicate that the terms in Fourier series are related to $\cos n\theta$ or $\sin n\theta$, respectively. If all excitations are symmetric with respect to $\theta = 0$ and the shell is homogeneous and isotropic, each of the above Fourier series will only contain either cosine or sine terms, thus reducing the number of unknowns by half.

Substituting the Fourier series into Eqs. (6.1) and (6.2), conducting Laplace transform with time t, and setting the coefficients of $\cos n\theta$ and $\sin n\theta$ to be equal on both sides of the resulting equations lead to an infinite number of equations.

$$\sum_{k=1}^{3}\sum_{i=0}^{n_k}\left\{\sum_{j=0}^{[\frac{i}{2}]} D_{mki(2j)}(s)(-1)^j \frac{\partial^{i-2j} U_{1,n}^k}{\partial x^{i-2j}} + \sum_{j=1}^{[\frac{i+1}{2}]} D_{mki(2j-1)}(s)(-1)^{j+1} \frac{\partial^{i-(2j-1)} U_{2,n}^k}{\partial x^{i-(2j-1)}}\right\} = \tilde{f}_{1,n}^m(x,s) + \tilde{g}_{1,n}^m(x,s), \tag{6.4a}$$

$$\sum_{k=1}^{3}\sum_{i=0}^{n_k}\left\{\sum_{j=0}^{[\frac{i}{2}]} D_{mki(2j)}(s)(-1)^j \frac{\partial^{i-2j} U_{2,n}^k}{\partial x^{i-2j}} + \sum_{j=1}^{[\frac{i+1}{2}]} D_{mki(2j-1)}(s)(-1)^{j} \frac{\partial^{i-(2j-1)} U_{1,n}^k}{\partial x^{i-(2j-1)}}\right\} = \tilde{f}_{2,n}^m(x,s) + \tilde{g}_{2,n}^m(x,s), \tag{6.4b}$$

where $m = 1, 2$, and 3, $n = 0, 1, 2, \ldots$, $[x]$ denotes the integer part of x, $U^k_{j,n} = \tilde{u}^k_{j,n}$ for $k = 1$ and 3, $U^2_{1,n} = \tilde{u}^2_{2,n}$ and $U^2_{2,n} = \tilde{u}^2_{1,n}$, $\tilde{u}^k_{i,n}$ and $\tilde{f}^m_{j,n}(x,s)$ are the Laplace transforms of $f^m_{j,n}(x,t)$ and $u^k_{i,n}$, respectively, s is the Laplace transform parameter, $\tilde{g}^m_{j,n}(x,s)$ are given in terms of $u^{k,0}_{j,n}(x)$, and the complex numbers $D_{mkij}(s) = (A_{mkij}s^2 + B_{mkij}s + C_{mkij})n^j$. Note that the displacement functions $\tilde{u}^k_{i,n}$ and $\tilde{u}^k_{i,m}$ do not couple if n is not equal to m.

Equations (6.4) are cast in a state-space form

$$\frac{\partial \eta_n(x,s)}{\partial x} = F_n(s)\eta_n(x,s) + \tilde{f}_n(x,s) + \tilde{g}_n(x,s), \qquad n = 0, 1, \ldots, \tag{6.5}$$

where the state-space vector contains the displacement functions $\tilde{u}^k_{i,m}$, i.e.,

$$\eta_n = \left\{ \eta^T_{1,1,n} \quad \eta^T_{2,1,n} \quad \eta^T_{3,1,n} \quad \eta^T_{1,2,n} \quad \eta^T_{2,2,n} \quad \eta^T_{3,2,n} \right\}^T \in C^{2n_b}, \tag{6.6a}$$

$$\eta_{k,i,n} = \left\{ \tilde{u}^k_{1,n} \quad \frac{\partial \tilde{u}^k_{1,n}}{\partial x} \quad \cdots \quad \frac{\partial^{n_k - 1}\tilde{u}^k_{1,n}}{\partial x^{n_k - 1}} \right\}^T \in C^{n_k}, \qquad k = 1, 2, 3, \quad i = 1, 2. \tag{6.6b}$$

$F_n(s)$ is a $2n_b \times 2n_b$ complex matrix containing the coefficients A_{mkij}, B_{mkij}, and C_{mkij} in Eq. (6.1), and the complex $2n_b$-vectors $\tilde{f}_n(x,s)$ and $\tilde{g}_n(x,s)$ are composed of $\tilde{f}^m_{(j,n)}(x,s)$ and $\tilde{g}^m_{j,n}(x,s)$, respectively. Similarly, the boundary conditions (6.2a) can be transformed into the form

$$M_n(s)\eta_n(0,s) + N_n(s)\eta_n(1,s) = \gamma_n(s), \tag{6.7}$$

where the boundary matrices $M_n(s)$ and $N_n(s)(\in C^{2n_b \times 2n_b})$ contain the coefficients Γ_{lkij}, and the vectors $\gamma_n(s)$ contain the Laplace transforms of the boundary excitation functions $\lambda^k_{j,n}$. One example of the above state-space form is given in Section 6.6.

The problem now is to solve the state-space equation (6.5) with the boundary condition (6.7). Following Yang and Tan (1992), the solution to Eqs. (6.5) and (6.7) is obtained as

$$\eta_n(x,s) = \int_0^1 G_n(x,\xi,s)(\tilde{f}_n(\xi,s) + \tilde{g}_n(\xi,s))\,d\xi + H_n(x,s)\gamma_n(s), \tag{6.8a}$$

where the $2n_b \times 2n_b$ complex matrices

$$G_n(x,\xi,s) = \begin{cases} H_n(x,s)M_n(s)e^{-F_n(s)\xi}, & \xi \le x, \\ -H_n(x,s)N_n(s)e^{F_n(s)(L-\xi)}, & \xi > x, \end{cases} \tag{6.8b}$$

$$H_n(x,s) = e^{F_n(s)x}\left(M_n(s) + N_n(s)e^{F_n(s)}\right)^{-1} \tag{6.8c}$$

are the distributed transfer functions of the homogenous cylindrical shell. The displacement functions are found from η_n:

$$u^1(x,s) = \sum_{n=0}^{\infty} \left[\eta_{n,1}(x,s)\cos n\theta + \eta_{n,n_1+n_2+n_3+1}(x,s)\sin n\theta\right], \tag{6.9a}$$

$$u^2(x,s) = \sum_{n=0}^{\infty} \left[\eta_{n,n_1+1}(x,s)\sin n\theta + \eta_{n,2n_1+n_2+n_3+1}(x,s)\cos n\theta\right], \tag{6.9b}$$

$$u^3(x,s) = \sum_{n=0}^{\infty} \left[\eta_{n,n_1+n_2+1}(x,s)\cos n\theta + \eta_{n,2n_1+2n_2+n_3+1}(x,s)\sin n\theta\right], \tag{6.9c}$$

where $\eta_{n,j}$ is the jth element of η_n.

In the above derivation, no approximation has been made; the transfer function formulation provides an exact and closed-form solution. The result here is quite general because different shell models and arbitrary boundary and initial conditions are systematically treated by the same formula.

The transfer function formulation can be applied to the free vibration and buckling problems of cylindrical shells. For free vibration, the characteristic equations of the shell, by Eq. (6.8), are

$$\det\left[M_n(s) + N_n(s)e^{F_n(s)}\right] = 0, \quad n = 1, 2, \ldots \tag{6.10}$$

whose roots are the eigenvalues of the shell. The corresponding eigenfunctions are

$$\eta_n(x,\omega) = e^{F_n(\omega)x}\psi, \qquad x \in (0,1), \tag{6.11a}$$

where ω is a root of Eq. (6.10), and the complex vector $\psi \in C^{2N}$ satisfies

$$\left[M_n(\omega) + N_n(\omega)e^{F_n(\omega)}\right]\psi = 0. \tag{6.11b}$$

Note that the vector function $\eta_n(x,\omega)$ in Eq. (6.11a) simultaneously presents the modal distributions of the shell displacements and internal forces.

In the buckling problem, the matrices M_n, N_n, and F_n are functions of a load parameter p, namely, $M_n = M_n(s;p)$, $N_n = N_n(s;p)$, $F_n = F_n(s;p)$. The shell buckling is analyzed by letting $s = 0$ in Eq. (6.10):

$$\det\left[M_n(0;p) + N_n(0;p)e^{F_n(0;p)}\right] = 0. \tag{6.12}$$

The solutions $p = p_k$ from Eq. (6.12) are the buckling loads; the corresponding buckling mode shapes can be obtained following Eq. (6.11). In addition, for a prestressed shell subjected to a given load $\bar{p}$, its national frequencies in vibration can be determined from the following characteristic equation:

$$\det\left[M_n(s;\,\bar{p}) + N_n(s;\,\bar{p})e^{F_n(s;\,\bar{p})}\right] = 0. \tag{6.13}$$

The frequency response of the shell subjected to harmonic excitations can be expressed by the transfer functions. For instance, under an excitation of frequency ω, the shell longitudinal vibration is

$$u^1(x,t) = \sum_{n=0}^{\infty}\left[\eta_{n,1}(x,\,j\omega)\cos n\theta + \eta_{n,n_1+n_2+n_3+1}(x,\,j\omega)\sin n\theta\right]e^{j\omega t},$$

where $j = \sqrt{-1}$. Moreover, the transfer functions can be directly used to predict the stability of the cylindrical shell under various forcing sources and to design active controllers and smart structure mechanisms. In the subsequent sections, the transfer functions of homogeneous cylindrical shells will be used to synthesize stepped shells and ring-stiffened shells.

6.3 Stepped Cylindrical Shells

A stepped cylindrical shell in Fig. 6.2 is composed of n_s serially connected shell segments of radius R, length l_i, and thickness h_i, respectively. Here, the geometry and material parameters are the same within each shell segment but may change from segment to segment. Stepped shells have various engineering applications and can be used to achieve minimum weight design of structures and machines. Moreover, stepped shells can model cylindrical shells with continuously varying thickness along their longitudinal direction.

The transfer function formulation developed in the previous section is applied to synthesize the stepped shell. For the ith subsystem (shell segment) in Fig. 6.2,

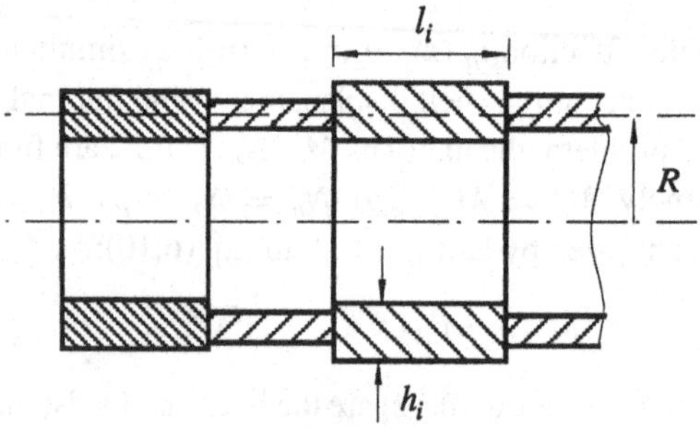

Fig. 6.2. A stepped cylindrical shell.

its response is expressed by

$$\eta_{i,n}(x,s) = e^{F_{i,n}(s)x}\left\{\int_0^x e^{-F_{i,n}(s)\xi}\tilde{f}_{i,n}(\xi,s)\,d\xi + \eta_{i,n}(0,s)\right\}, \tag{6.14}$$

where x is a dimensionless local coordinate, the subscript n is the circumferential wave number, and the subscript i $(1 \le i \le n_s)$ indicates the ith subsystem. At the left and right ends of the stepped shell, $\eta_{1,n}$ and $\eta_{n_s,n}$ must satisfy the boundary condition

$$M_n(s)\eta_{1,n}(0,s) + N_n(s)\eta_{n_s,n}(1,s) = \gamma_n(s). \tag{6.15}$$

The shell is assembled from the n_s subsystems by imposing displacement continuity and force balance at the interconnecting boundaries of the shell segments. Two synthesis techniques are proposed: the connection matrix method and the nodal displacement method.

Connection Matrix Method At the ith connecting boundary, the matching conditions for the two adjacent subsystems in general can be written as

$$R_{i,n}\eta_{i,n}(1,s) - R_{i+1,n}\eta_{i+1,n}(0,s) = \pi_{i,n}(s), \tag{6.16}$$

where $R_{i,n}$ and $R_{i+1,n}$ shall be called the connection matrices, and $\pi_{i,n}(s)$ counts for displacement and/or internal force jump due to the concentrated forces applied at the connecting boundary. By Eqs. (6.14) and (6.15).

$$\eta_{i,n}(0,s) = e^{-F_{i,n}(s)}R_{i,n}^{-1}[R_{i+1}\eta_{i+1,n}(0,s)+\pi_{i,n}(s)] - \int_0^1 e^{-F_{i,n}(s)\xi} f_{i,n}(\xi,s)\,d\xi. \tag{6.17}$$

Apply the above formula recurrently to get

$$\eta_{1,n}(0,s) = P_{n_s}(s)\eta_{n_s,n}(0,s) + \bar{\pi}_n(s) - \bar{f}_n(s), \tag{6.18}$$

where

$$P_{n_s}(s) = \prod_{i=1}^{n_s-1} e^{-F_{i,n}(s)}R_{i,n}^{-1}R_{i+1,n}, \tag{6.19a}$$

$$\bar{\pi}_n(s) = \sum_{i=1}^{n_s-1}\left(\prod_{j=1}^{i-1} e^{-F_{j,n}(s)}R_{j,n}^{-1}R_{j+1,n}\right)e^{-F_{i,n}(s)}R_{i,n}^{-1}\pi_{i,n}(s), \tag{6.19b}$$

and

$$\bar{f}_n(s) = \sum_{i=1}^{n_s-1}\left(\prod_{j=1}^{i-1} e^{-F_{j,n}(s)}R_{j,n}^{-1}R_{j+1,n}\right)\int_0^1 e^{-F_{i,n}(s)\xi} f_{i,n}(\xi,s)\,d\xi. \tag{6.19c}$$

Substituting Eq. (6.18) into Eq. (6.15) yields

$$\bar{M}_n(s)\eta_{n_s,n}(0,s) + N_n(s)\eta_{n_s,n}(1,s) = \bar{\gamma}_n(s) + M_n(s)\bar{f}_n(s), \qquad (6.20)$$

where $\bar{M}_n(s) = M_n(s)P_{n_s}(s)$ and $\bar{\gamma}_n(s) = \gamma_n(s) - M_n\bar{\pi}_n(s)$. From Eqs. (6.14) and (6.20), it is found that

$$\eta_{n_s,n}(0,s) = \left[\bar{M}_n(s) + N_n(s)e^{F_{n_s,n}(s)}\right]^{-1} \times \left\{\bar{\gamma}_n + M_n(s)\bar{f}_n(s) - N_n(s)e^{F_{n_s,n}(s)}\int_0^1 e^{-F_{n_s,n}(s)\xi} f_{n_s,n}(\xi,s)\,d\xi\right\}. \qquad (6.21)$$

Finally, plugging the above into Eq. (6.17), we arrive at

$$\begin{aligned}\eta_{i,n}(0,s) = {} & \left[\prod_{k=i}^{n_s-1} e^{-F_{k,n}(s)}R_{k,n}^{-1}R_{k+1,n}\right]\eta_{n_s,n}(0,s) \\ & + \sum_{k=i}^{n_s-1}\left(\prod_{j=i}^{k-1} e^{-F_{j,n}(s)}R_{j,n}^{-1}R_{j+1,n}\right)e^{-F_{k,n}(s)}R_k^{-1}\prod_{k,n}(s) \\ & - \sum_{k=i}^{n_s-1}\left(\prod_{j=i}^{k-1} e^{-F_{j,n}(s)}R_{j,n}^{-1}R_{j+1,n}\right)\int_0^1 e^{-F_{k,n}(s)\xi} f_{k,n}(\xi,s)\,d\xi.\end{aligned} \qquad (6.22)$$

With $\eta_{i,n}(0,s)$ given in Eq. (6.22), the response $\eta_{i,n}(x,s)$ of each shell segment is completely determined by Eq. (6.14).

The eigenvalue problem (for both free vibration and buckling) of the stepped shell can be formulated by letting $\bar{\gamma}_n(s)$ and $\bar{f}_n(s)$ in Eqs. (6.14) and (6.20) be zero, leading to the characteristic equation

$$\det\left[M_n(s)P_{n_s}(s) + N_n(s)e^{F_{n_s,n}(s)}\right] = 0. \qquad (6.23)$$

With the eigenvalues calculated from the above equation, the mode shapes (eigenfunctions) of each shell segment can be evaluated by Eq. (6.11).

Nodal Displacement Method In this method, the finite element concept is introduced. To this end, the interconnecting boundaries between the adjacent subsystems (shell segments) shall be called nodes and denoted by $0, 1, 2, \ldots, n_s$; see Fig. 6.3. For the ith subsystem, its left and right nodes are $i-1$ and i, respectively. The displacements at node i are $\bar{u}_i$, $\bar{v}_i$, $\bar{w}_i$, the three displacements in the x, θ, and z directions, and $\bar{w}_{x,i}$, the slope of the transverse displacement in the x direction, respectively.

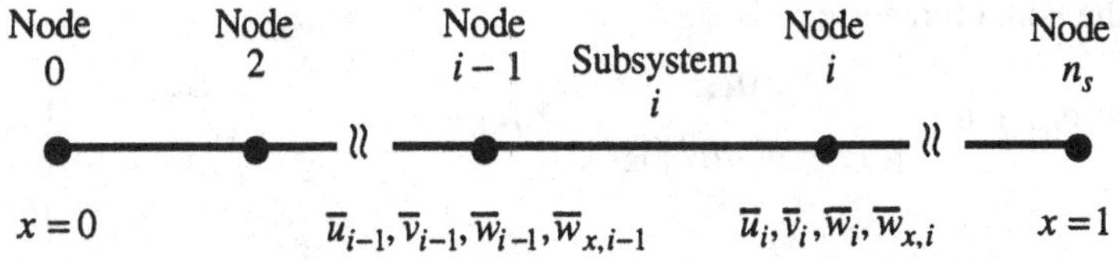

Fig. 6.3. The nodes and nodal displacements of the stepped shell.

Define

$$u_i = \begin{Bmatrix} \bar{u}_i \\ \bar{v}_i \\ \bar{w}_i \\ \bar{w}_{x,i} \end{Bmatrix}, \quad \begin{Bmatrix} f_{i,1} \\ f_{i,2} \end{Bmatrix} = \int_0^1 e^{-F_{i,n}(s)\xi}\, \tilde{f}_{i,n}(\xi, s)\, d\xi, \quad \begin{bmatrix} F_{i,11} & F_{i,12} \\ F_{i,21} & F_{i,22} \end{bmatrix} = e^{F_{i,n}(s)}. \tag{6.24}$$

By letting $x = 0$ and 1 in Eq. (6.14) and making use of Eq. (6.24), the state-space vector $\eta_{i,n}(0, s)$ is expressed by the nodal displacements:

$$\eta_{i,n}(0, s) = \begin{bmatrix} I_{4\times4} & 0_{4\times4} \\ -F_{i,12}^{-1}F_{11} & F_{i,12}^{-1} \end{bmatrix} \begin{Bmatrix} u_{i-1} \\ u_i \end{Bmatrix} - \begin{bmatrix} 0_{4\times4} & 0_{4\times4} \\ -F_{i,12}^{-1}F_{11} & -F_{i,12}^{-1} \end{bmatrix} \begin{Bmatrix} f_{i,1} \\ f_{i,2} \end{Bmatrix}. \tag{6.25}$$

The vector of the internal forces, $\sigma_{i,n}(x, s)$, is related to the state-space vector by

$$\sigma_{i,n}(x, s) = \bar{K}_{i,n}(s)\eta_{i,n}(x, s), \tag{6.26}$$

where $\bar{K}_{i,n}(s) \in C^{8\times8}$ is a constitutive matrix. Force balance at node i reads

$$\sigma_{i,n}(1, s) + \sigma_{i+1,n}(0, s) = q_{i,n}(s), \tag{6.27}$$

where $\sigma_{i,n}(1, s)$ and $\sigma_{i+1,n}(0, s)$ represent the internal forces at node i applied by the ith and $(i + 1)$th subsystems, respectively, and $q_{i,n}(s)$ is the vector of the external forces at node i. By Eqs. (6.14), and (6.24) to (6.27), it is found that

$$K_{i,n} \begin{Bmatrix} u_{i-1} \\ u_i \\ u_{i+1} \end{Bmatrix} = Q_{i,n}, \tag{6.28}$$

where the stiffness matrix is

$$\begin{aligned} K_{i,n} = \bar{K}_{i,n} &\begin{bmatrix} 0_{4\times4} & I_{4\times4} & 0_{4\times4} \\ F_{i,21} - F_{i,22}F_{i+1,12}^{-1}F_{i+1,11} & F_{i,22}^{-1}F_{i+1,12} & 0_{4\times4} \end{bmatrix} \\ + \bar{K}_{i+1,n} &\begin{bmatrix} 0_{4\times4} & 0_{4\times4} & I_{4\times4} \\ 0_{4\times4} & -F_{i+1,12}^{-1}F_{i+1,11} & F_{i+1,12}^{-1}F_{i+1,11} \end{bmatrix} \end{aligned} \tag{6.29a}$$

and the nodal force vector is

$$Q_{i,n} = q_{i,n} + \bar{K}_{i,n} \begin{bmatrix} 0_{4\times 4} \\ F_{i,21} - F_{i,22}F_{i,12}^{-1}F_{i,11} \end{bmatrix} f_{i,1} + \bar{K}_{i+1,n} \begin{bmatrix} I_{4\times 4} \\ -F_{i+1,12}^{-1}F_{i+1,11} \end{bmatrix} f_{i+1,1}. \tag{6.29b}$$

Substituting Eq. (6.25) into Eq. (6.15) gives the boundary conditions of the stepped shell:

$$M_n \begin{bmatrix} I_{4\times 4} & 0_{4\times 4} \\ -F_{1,12}^{-1}F_{i,11} & F_{i,12}^{-1} \end{bmatrix} \begin{Bmatrix} u_0 \\ u_1 \end{Bmatrix} + N_n \begin{bmatrix} I_{4\times 4} & 0_{4\times 4} \\ -F_{n_s,12}^{-1}F_{n_s,11} & F_{n_s,12}^{-1} \end{bmatrix} \begin{Bmatrix} u_{n_s-1} \\ u_{n_s} \end{Bmatrix} = Q_{0,n} + Q_{n_s,n}, \tag{6.30}$$

where

$$Q_{0,n} = M_n \begin{bmatrix} F_{1,12}^{-1}F_{1,11} & 0_{4\times 4} \\ 0_{4\times 4} & I_{4\times 4} \end{bmatrix} \begin{Bmatrix} f_{1,1} \\ f_{1,2} \end{Bmatrix} + \gamma_{n,1}(s), \tag{6.31a}$$

$$Q_{n_s,n} = -N_n \begin{bmatrix} I_{4\times 4} \\ -F_{n_s,12}^{-1}F_{n_s,11} \end{bmatrix} f_{n_s,1} + \gamma_{n,2}, \qquad \gamma_n = \begin{Bmatrix} \gamma_{n,1} \\ \gamma_{n,2} \end{Bmatrix}. \tag{6.31b}$$

Applying Eq. (6.28) to all nodes and combining the boundary condition (6.30) yield a global dynamic equilibrium equation:

$$K_n(s)u(s) = Q_n(s), \tag{6.32}$$

where the subscript n is the shell circumferential wave number, the nodal displacement vector $u(s) = \{u_1^T \cdots u_{n_s}^T\}^T$, the nodal force vector $Q(s) = \{Q_{0,n}^T \cdots Q_{n_s,n}^T\}^T$, and the global dynamic stiffness matrix K_n is constructed in the same way as in the finite element method. Solve Eq. (6.32) for the node displacements and substitute the result back into Eqs. (6.14) and (6.25) to give $\eta_{i,n}(x, s)$. As such, the response of the stepped shell in every segment can be determined by Eq. (6.9).

The eigenvalue problem of the stepped shell becomes ($Q_n = 0$)

$$K_n(s)u(s) = 0. \tag{6.33}$$

The eigenvalues are determined from det $K_n(\omega) = 0$. For an eigenvalue ω, its corresponding eigenfunction is obtained in two steps: (i) determine the nontrivial solution $u(\omega)$ from $K_n(\omega)u(\omega) = 0$; and (ii) substitute $u(\omega)$ into Eqs. (6.25) and (6.14) to calculate $\eta_{i,n}(x, \omega)$.

In the above synthesis methods, no approximation or discretization has been made. Thus, the transfer function method provides accurate prediction of the static and dynamic behavior of stepped cylindrical shells, as shall be seen in

Section 6.6. Also unlike many other analytical methods, the proposed transfer function formulation is convenient for computer coding because matrix manipulation and the FEM concept have been adopted.

6.4 Ring-stiffened Cylindrical Shells

6.4.1 Background

The modeling techniques for ring-stiffened cylindrical shells fall into two categories. The first category smears the stiffness of the stiffeners on the shell and treats the stiffened shell as an orthotropic one; for instance, see Wang (1970) and the references therein. Such treatment is fine when the stiffeners are densely placed. However, if stiffeners are sparsely spaced so that there are not enough stiffeners in each half wavelength in the circumferential and longitudinal directions of the shell, or if the actual deformation and stress distributions of the stiffened shell are needed, orthotropic approximation will lose accuracy. In this case, it is necessary to consider stiffeners as individual structural components, which gives rise to the second category of modeling techniques, namely, the analysis of combined shell-stiffener component systems.

Modeling stiffened cylindrical shells as combined component systems has been studied by many authors. Of various static and dynamic problems, free vibration of different stiffened shells has received great attention. Garnet and Goldberg (1962) and Godzevich and Ivanova (1965) considered ring-stiffened cylindrical shells. Schnell and Heinrichsbauer (1964) studied longitudinally stiffened thin-walled cylindrical shells. Egle and Sewall (1968) analyzed a ring-and-stringer-stiffened circular cylindrical shell by a Rayleigh–Ritz procedure. Wang and Rinehart (1974) modeled longitudinally stiffened cylindrical shells with arbitrary edge boundary conditions. Wang and Hsu (1985) proposed a model for stiffened composite cylindrical shells. Mead and Bardell (1987) investigated periodically stiffened cylindrical shells.

Among other problems, Wang and Lin (1973) examined the stability of stringer-stiffened cylindrical shells under axial pressure and simply supported boundary conditions; Reddy (1980) presented a bifurcation analysis for stringer-stiffened cylinders sustaining elastic-plastic deformation; and Sridharan et al. (1992) studied post-buckling of stiffened composite shells.

Several analytical and numerical methods for stiffened cylindrical shells have been developed. Forsberg (1969) obtained an exact solution for natural frequencies of ring-stiffened cylinders subjected to different boundary conditions. Al-Najafi and Warburton (1970) applied a finite element method to evalute the natural frequencies and mode shapes of ring-stiffened cylindrical shells with each stiffening ring treated as a discrete element. Wang (1970) yielded a Fourier series solution of ring-and-stringer-stiffened cylindrical shells by treating the shell, rings, and stringers as structural components and imposing deformation compatibility among those components. Wilken and Soedel (1976) used a receptance method to describe the modal characteristics of ring-stiffened shells. By Laplace transform and numerical inverse Laplace transform, Deskos and Oates (1981) predicted the dynamic response of ring-stiffened circular cylindrical shells subjected to axis symmetric loading and boundary restriction conditions. Rigo (1992) developed the stiffened sheathing method and LBR-3 software for computing the response of orthotropic cylindrical shells. Combining the wave propagation method and transfer matrix method, Huntington and Lyrintzis (1992) developed a modified wave method for analysis of skin-stringer panels that improves the numerical stability of the transfer matrix method.

In this section, we apply the distributed transfer function formulation to develop a new analytic and numerical method for constrained/combined, ring-stiffened cylindrical shells, which falls in the second category of modeling techniques. In the analysis, the shell and stiffeners are modeled as individual structural components. Through use of the distributed transfer functions of the shell and stiffening rings, the shell and stiffeners are assembled by imposing displacement continuity and force balance on the shell-stiffener contact surface, leading to a global dynamic equilibrium equation. Solution of the equilibrium equation gives accurate estimation of the static deflection, natural frequencies and mode shapes, buckling loads, and forced response of stiffened cylindrical shells under arbitrary boundary conditions and external excitations.

6.4.2 Analysis of Ring-stiffened Cylindrical Shells

In Fig. 6.4, a homogeneous cylindrical shell is stiffened by N^r circumferential rings at $x = x_i$, $i = 1, \ldots, N^r$. Assume that the influence of the ring width is

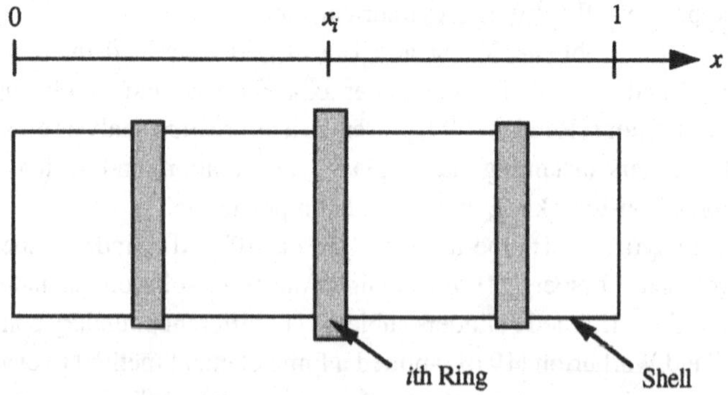

Fig. 6.4. A cylindrical shell stiffened by circumferential rings.

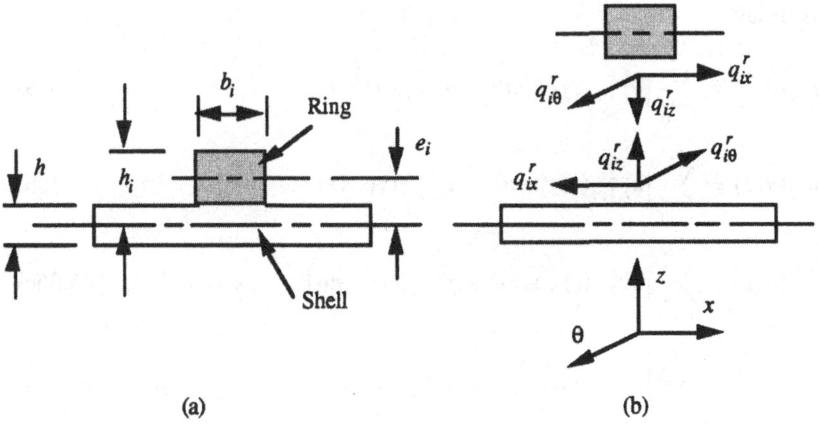

Fig. 6.5. Interaction between the shell and the ith ring: (a) geometry; (b) tractions.

negligible. Denote the middle surface displacement of the ith ring by $u^r_{ik}(\theta, t)$, where the subscript refers to the ith stiffening ring, and the superscript r indicates that the parameters are related to the rings. The dynamic equations of the stiffened ring in general are in the form

$$\sum_{k=1}^{3}\sum_{j=0}^{n_k}\left(\alpha_{imkj}\frac{\partial^2}{\partial t^2}+\beta_{imkj}\frac{\partial}{\partial t}+\gamma_{imkj}\right)\frac{\partial^j u^r_{ik}(\theta,t)}{\partial\theta^j}=q^r_{im}+q^e_{im}, \qquad (6.34)$$

where $m = 1, 2, 3, i = 1, \ldots, N^r$, the constants α_{imkj}, β_{imkj}, and γ_{imkj} are related to the geometry and material parameters of the rings, and q^e_{im} are the external loads applied to the ith ring. Here q^r_{im} are the tractions between the shell and the ith ring; see Fig. 6.5, where $q^r_{i1} = q^r_{ix}$, $q^r_{i2} = q^r_{i\theta}$ and $q^r_{i3} = q^r_{iz}$. Like Eq. (6.1), Eq. (6.34) is quite general and represents many different curved beam models.

For thin-walled cylindrical shells, the Kirchhoff hypothesis holds and the displacements of the shell and rings satisfy the following matching conditions:

$$u^r_{i1}(\theta,t)=\left[u^1-e_i u,^3_x\right]_{x=x_i}, \qquad (6.35a)$$

$$u^r_{i2}(\theta,t)=\left[u^2-e_i\frac{1}{R}u,^3_\theta\right]_{x=x_i}, \qquad (6.35b)$$

$$u^r_{i3}(\theta,t)=u^3(x_i,\theta,t), \qquad (6.35c)$$

where $u,^3_x = \partial u^3/\partial x$, $u,^3_\theta = \partial u^3/\partial\theta$, and e_i is the eccentricity of the centroid of the ith stiffening ring from the middle surface of the shell; see Fig. 6.5(a). The e_i is positive (negative or zero) if the ring is inward (outward or symmetric).

The analysis of the stiffened shell follows Fourier expansion, Laplace transform, and state-space formulation. The Fourier series of the displacements and loads are

$$u_{ik}^{r}(\theta,t)=\sum_{n=0}^{\infty}\left[u_{r,1,n}^{i,k}(t)\cos n\theta+u_{r,2,n}^{i,k}(t)\sin n\theta\right],\qquad k=1,3, \tag{6.36a}$$

$$u_{i2}^{r}(\theta,t)=\sum_{n=0}^{\infty}\left[u_{r,1,n}^{i,2}(t)\sin n\theta+u_{r,2,n}^{i,2}(t)\cos n\theta\right], \tag{6.36b}$$

$$q_{ik}^{r}(\theta,t)=\sum_{n=0}^{\infty}\left[q_{r,1,n}^{i,k}(t)\cos n\theta+q_{r,2,n}^{i,k}(t)\sin n\theta\right],\qquad k=1,2,3, \tag{6.36c}$$

$$q_{ik}^{e}(\theta,t)=\sum_{n=0}^{\infty}\left[q_{e,1,n}^{i,k}(t)\cos n\theta+q_{e,2,n}^{i,k}(t)\sin n\theta\right],\qquad k=1,2,3. \tag{6.36d}$$

By substituting Eqs. (6.3a,b) and (6.36a,b) into Eqs. (6.34) and (6.35), one obtains

$$\begin{aligned}
q_{r,1,n}^{i,m}\cos n\theta+q_{r,2,n}^{i,m}\sin n\theta &= -q_{r,1,n}^{i,m}\cos n\theta-q_{r,2,n}^{i,m}\sin n\theta\\
&+\sum_{k=1}^{3}\sum_{j=0}^{n_k}\left(\alpha_{imkj}\frac{\partial^2}{\partial t^2}+\beta_{imkj}\frac{\partial}{\partial t}+\gamma_{imkj}\right)\frac{\partial^j u_n^k}{\partial\theta^j}\bigg|_{x=x_i}\\
&-\frac{ne_i}{R}\sum_{j=0}^{n_2}\left(\alpha_{im2j}\frac{\partial^2}{\partial t^2}+\beta_{im2j}\frac{\partial}{\partial t}+\gamma_{im2j}\right)\frac{\partial^j}{\partial\theta^j}\left(u_{2,n}^3\cos n\theta-u_{1,n}^3\sin n\theta\right)\bigg|_{x=x_i}\\
&-e_i\sum_{j=0}^{n_1}\left(\alpha_{im1j}\frac{\partial^2}{\partial t^2}+\beta_{im1j}\frac{\partial}{\partial t}+\gamma_{im1j}\right)\frac{\partial}{\partial x}\frac{\partial^j}{\partial\theta^j}\left(u_{1,n}^3\cos n\theta+u_{2,n}^3\sin n\theta\right)\bigg|_{x=x_i}
\end{aligned} \tag{6.37}$$

for $m=1$, 2, and 3, where n is the circumferential wave number of the shell. Laplace transform of Eq. (6.37) and application of the state space form (6.5) lead to

$$\tilde{q}_i^{r,n}(s)=L^{i,n}(s)\eta_n(x_i,s)-\tilde{q}_i^{e,n}(s), \tag{6.38}$$

where $\tilde{q}_i^{r,n}(s)$ is the vector of the tractions between the shell and rings, $\tilde{q}_i^{e,n}(s)$ is the vector of the external and initial disturbances, and the matrix $L^{i,n}(s)$ consists of the parameters of the ith ring; see the appendix.

Since the traction forces $\tilde{q}_i^{r,n}$ are pointwise along the x direction, the total forces π_n applied to the shell can be expressed by

$$\pi_n(x,s)=\hat{f}_n(x,s)+\sum_{i=1}^{N^r}T_i^r L^{i,n}(s)\eta_n(x_i,s)\delta(x-x_i), \tag{6.39}$$

where $\delta(x-x_i)$ is the delta function $\hat{f}_n(x,s)$ is the resultant of the external loads on the shell and rings and the equivalent forces due to the initial disturbances of the rings, T_i^r are scaling matrices guaranteeing the same dimension of $\tilde{q}_i^{r,n}$ in Eq. (6.38) and $\tilde{f}_n$ in Eq. (6.8a). By Eq. (6.8), we have

$$\eta_n(x,s) = e^{F_n(s)x}\left\{\int_0^x e^{-F_n(s)\xi}[\pi_n(\xi,s)+\tilde{g}_n(\xi,s)]\,d\xi + \eta_n(0,s)\right\}. \quad (6.40)$$

Substituting Eq. (6.39) into Eq. (6.40) gives

$$\eta_n(x,s) = A_n^{(0)}(x,s) + \sum_{j=1}^{N^r} B_{n,j}(x,s)\eta_n(x_j,s) + C_n^{(0)}(x,s)\eta_n(0,s), \quad (6.41)$$

where

$$A_n^{(0)}(x,s) = e^{F_n(s)x}\int_0^x e^{-F_n(s)\xi}\left[\hat{f}_n(\xi,s)+\tilde{g}_n(\xi,s)\right]d\xi,$$

$$B_{n,j}(x,s) = e^{F_n(s)(x-x_j)}T_i^r L^{j,n}(s)u(x-x_j), \qquad j=1,\ldots,N^r,$$

$$C_n^{(0)}(x,s) = e^{F_n(s)x},$$

and $u(x-x_j)$ is the unit step function. Eliminating $\eta_n(x_j,s)$ from Eq. (6.41) by setting $x = x_i$, $i = 1,\ldots,N^r$, and solving the N^r resulting equations yield

$$\eta_n(x,s) = \bar{A}_n(x,s) + \bar{C}_n(x,s)\eta_n(0,s), \quad (6.42)$$

where the matrices $\bar{A}_n(x,s)$ and $\bar{C}_n(x,s)$ are obtained from the following N^r-step recurrence procedure:

$$A_n^{(i)}(x,s) = A_n^{(i-1)}(x,s) + B_{n,i}(x,s)[I - B_{n,i}(x_i,s)]^{-1}A_n^{(i-1)}(x_i,s), \quad (6.43a)$$

$$C_n^{(i)}(x,s) = C_n^{(i-1)}(x,s) + B_{n,i}(x,s)[I - B_{n,i}(x_i,s)]^{-1}C_n^{(i-1)}(x_i,s) \quad (6.43b)$$

for $i = 1, 2, \ldots, N^r$, with $\bar{A}_n(x,s) = A_n^{(N^r)}(x,s)$ and $\bar{C}_n(x,s) = C_n^{(N^r)}(x,s)$.

Finally, plug Eq. (6.42) into the boundary conditions (6.7) to obtain the response of the ring-stiffened cylindrical shell under various external loads and initial and boundary conditions:

$$\eta_n(x,s) = \bar{A}_n(x,s) + \bar{C}_n(x,s)[M_n(s) + N_n(s)\bar{C}_n(1,s)]^{-1}[\gamma_n(s) - N_n(s)\bar{A}_n(1,s)]. \quad (6.44)$$

For the eigenvalue problem of the stiffened shell, the characteristic equation by Eq. (6.44) is

$$\det[M_n(s;\bar{p}) + N_n(s;\bar{p})\bar{C}_n(1,s;\bar{p})] = 0 \quad (6.45)$$

and the discussion after Eq. (6.10) applies.

Constrained cylindrical shells can be viewed as the degenerate cases of stiffened shells. Assume that the shell is constrained by N^s pairs of uniformly distributed springs and dampers, located at $x = x_i$ ($i = 1, \ldots, N^s$) and along the circle $0 \leq \theta \leq 2\pi$. At $x = x_i$, the constraint forces by the springs and dampers are described by

$$\begin{Bmatrix} q_{i1}^s \\ q_{i2}^s \\ q_{i3}^s \end{Bmatrix} = -\left(K_i + D_i \frac{\partial}{\partial t} \right) \begin{Bmatrix} u^1 - \bar{u}^1 \\ u^2 - \bar{u}^2 \\ u^3 - \bar{u}^3 \end{Bmatrix}_{x=x_i}, \tag{6.46}$$

where the matrices K_i and D_i contain the spring and damper coefficients, respectively, and the $\bar{u}^j$ count for the unstretched status of the springs. It can be shown that the matrix $L^{i,n}(s)$ in Eq. (6.38) is expressed in terms of the elements of $K_i + sD_i$. The response of the constrained shell can be determined based on Eqs. (6.43) and (6.44).

In summary, the transfer function analysis of the ring-stiffened shell takes three steps: (a) For given shell and ring parameters and boundary conditions, from the state-space matrices $F_n(s)$, $M_n(s)$, $N_n(s)$, and $L^{i,n}(s)$; (b) determine $\bar{A}_n(x, s)$ and $\bar{C}_n(x, s)$ by the recurrence procedure (6.43); and (c) evaluate the response of the shell by (6.44).

6.5 Application to Donnell–Mushtari Shells

6.5.1 Equations of Motion

The transfer function method is applied to the Donnell–Mushtari model of cylindrical shells. The nondimensional equations of motion are (Donnell, 1933; Markus, 1989)

$$\frac{1}{\gamma_1^2}\frac{\partial^2 u}{\partial x^2} + \frac{1}{2}(1-\upsilon)\frac{\partial^2 u}{\partial \theta^2} + \frac{1+\upsilon}{2\gamma_1}\frac{\partial^2 v}{\partial x\,\partial \theta} - \frac{\upsilon}{\gamma_1}\frac{\partial w}{\partial x} = \bar{\rho}\frac{\partial^2 u}{\partial t^2} - \bar{q}_x, \tag{6.47a}$$

$$\frac{1+\upsilon}{2\gamma_1}\frac{\partial^2 u}{\partial x \partial \theta} + \frac{1-\upsilon}{2\gamma_1^2}\frac{\partial^2 v}{\partial x^2} + \frac{\partial^2 v}{\partial \theta^2} - \frac{\partial w}{\partial \theta} = \bar{\rho}\frac{\partial^2 v}{\partial t^2} - \bar{q}_\theta, \tag{6.47b}$$

$$\begin{aligned} &\frac{\upsilon}{\gamma_1}\frac{\partial u}{\partial x} + \frac{\partial v}{\partial \theta} - w - k\left(\frac{1}{\gamma_1^4}\frac{\partial^4 w}{\partial x^4} + \frac{2}{\gamma_1^2}\frac{\partial^4 w}{\partial x^2\,\partial \theta^2} + \frac{\partial^4 w}{\partial \theta^4}\right) \\ &+ \frac{1}{J}\left(N_{x0}\frac{\partial^2 w}{\gamma_1^2 \partial x^2} + 2N_{x\theta 0}\frac{\partial^2 w}{\gamma_1\,\partial x\,\partial \theta} + N_{\theta 0}\frac{\partial^2 w}{\partial \theta^2}\right) = \bar{\rho}\frac{\partial^2 w}{\partial t^2} - q_z, \end{aligned} \tag{6.47c}$$

with

$$(u, v, w) = \frac{1}{h}(u_0, v_0, w_0), \qquad (\bar{q}_x, \bar{q}_\theta, \bar{q}_z) = \frac{(1-\upsilon^2)R^2}{Eh^2}(q_x, q_\theta, q_z),$$

$$\bar{\rho} = \frac{\rho(1-\upsilon^2)R^2}{E}, \qquad \gamma_1 = \frac{L}{R}, \qquad J = \frac{Eh}{1-\upsilon^2}, \qquad \text{and } k = \frac{1}{12}\left(\frac{h}{R}\right)^2.$$

Here u_0, v_0, and w_0 are the displacements of the shell middle surface in the coordinate directions x, θ, and z, respectively; R, L, and h are the radius, length, and thickness of the shell, respectively; E and υ are Young's modulus and Poison's ratio; ρ is the density per unit shell surface; N_{x0}, $N_{\theta 0}$, and $N_{x\theta 0}$ are the membrane stresses; and q_x, q_θ, and q_z are the loads on the shell.

There are four pairs of boundary conditions:

(i) $u(x_i, \theta, t) = \bar{u}_i(\theta, t)$ or

$$\frac{E}{1-\upsilon^2}\frac{h^2}{R}\left[\frac{1}{\gamma_1}u_{,x} + \upsilon(v_{,\theta} - w)\right]_{x=x_i} = \bar{N}_{xi}(\theta, t), \tag{6.48a}$$

(ii) $v(x_i, \theta, t) = \bar{v}_i(\theta, t)$ or

$$\frac{E}{2(1+\upsilon)}\frac{h^2}{R}\left[u_{,\theta} + \frac{1}{\gamma_1}v_{,x} + \frac{2k}{\gamma_1}w_{,x\theta}\right]_{x=x_i} = \bar{N}^e_{x\theta i}(\theta, t), \tag{6.48b}$$

(iii) $w(x_i, \theta, t) = \bar{w}_i(\theta, t)$ or

$$-\frac{E}{12(1-\upsilon^2)}\frac{h^4}{R^3}\left[\frac{1}{\gamma_1^3}w_{,xxx} + \frac{2-\upsilon}{\gamma_1}w_{,x\theta\theta}\right]_{x=x_i} = \bar{Q}^e_{x\theta i}(\theta, t), \tag{6.48c}$$

(iv) $w_{,x}\big|_{x=x_i} = \bar{w}_{x,i}(\theta, t)$ or

$$-\frac{E}{12(1-\upsilon^2)}\frac{h^4}{R^2}\left[\frac{1}{\gamma_1^2}w_{,xx} + \upsilon w_{,\theta\theta}\right]_{x=x_i} = \bar{M}_i(\theta, t), \tag{6.48d}$$

where $x_i = 0$ and 1 for $i = 1$ and 2, for the left and right boundaries of the shell, respectively, and $u_{,x} = \partial u/\partial x$, etc.

6.5.2 Transfer Function Formulation

Although the transfer function method is valid for arbitrary disturbances, in this demonstrative example, it is assumed that all disturbances are symmetric with respect to $\theta = 0$. Thus, the Fourier expansions of the displacements and loads become

$$\begin{Bmatrix} u \\ v \\ w \end{Bmatrix} = \sum_{n=0}^{\infty} \begin{Bmatrix} u_n(x,t)\cos n\theta \\ v_n(x,t)\sin n\theta \\ w_n(x,t)\cos n\theta \end{Bmatrix}, \qquad \begin{Bmatrix} \bar{q}_x \\ \bar{q}_\theta \\ \bar{q}_z \end{Bmatrix} = \sum_{n=0}^{\infty} \begin{Bmatrix} q_{xn}(x,t)\cos n\theta \\ q_{\theta n}(x,t)\sin n\theta \\ q_{zn}(x,t)\cos n\theta \end{Bmatrix}. \tag{6.49}$$

The η_n, F_n, and $\bar{f}_n$ in Eq. (6.5) have the form

$$\eta_n(x,s) = \left\{ \tilde{u}_n \quad \frac{\partial \tilde{u}_n}{\partial x} \quad \tilde{v}_n \quad \frac{\partial \tilde{v}_n}{\partial x} \quad \tilde{w}_n \quad \frac{\partial \tilde{w}_n}{\partial x} \quad \frac{\partial^2 \tilde{w}_n}{\partial x^2} \quad \frac{\partial^3 \tilde{w}_n}{\partial x^3} \right\}^T, \tag{6.50a}$$

$$\begin{aligned} \tilde{f}_n(x,s) + \tilde{g}_n(x,s) = \Big\{ 0 \quad & -\gamma_1^2[\bar{\rho}(s u_{0,n} + \dot{u}_{0,n}) + \tilde{q}_{xn}] \quad 0 \\ & -\frac{2\gamma_1^2}{1-\upsilon}[\bar{\rho}(s v_{0,n} + \dot{v}_{0,n}) + \tilde{q}_{\theta n}] \quad 0 \quad 0 \quad 0 \\ & \frac{\gamma_1^4}{k}[\bar{\rho}(s w_{0,n} + \dot{w}_{0,n}) + \tilde{q}_{zn}] \Big\}^T, \end{aligned} \tag{6.50b}$$

$$F_n(s) = \begin{bmatrix} F_{11} & F_{12} \\ F_{21} & F_{22} \end{bmatrix} \in C^{8\times 8}, \tag{6.50c}$$

where the tilde denotes Laplace transformation, the over dot stands for time derivative, and

$$F_{11} = \begin{bmatrix} 0 & 1 & 0 & 0 \\ \gamma_1^2\left(\frac{1-\upsilon}{2}n^2 + s^2\bar{\rho}\right) & 0 & 0 & -\frac{1+\upsilon}{2}n\gamma_1 \\ 0 & 0 & 0 & 1 \\ 0 & \gamma_1 n\frac{1+\upsilon}{1-\upsilon} & \frac{2n^2\gamma_1^2}{1-\upsilon} + \frac{2\bar{\rho}s^2\gamma_1^2}{1-\upsilon} & 0 \end{bmatrix},$$

$$F_{12} = \begin{bmatrix} 0 & 0 & 0 & 0 \\ 0 & \upsilon\gamma_1 & 0 & 0 \\ 0 & 0 & 0 & 0 \\ -\frac{2n\gamma_1^2}{1-\upsilon} & 0 & 0 & 0 \end{bmatrix}, \qquad F_{21} = \begin{bmatrix} 0 & 0 & 0 & 0 \\ 0 & 0 & 0 & 0 \\ 0 & 0 & 0 & 0 \\ 0 & \gamma_3^1\frac{\upsilon}{k} & \frac{n\gamma_1^4}{k} & 0 \end{bmatrix},$$

$$F_{22} = \begin{bmatrix} 0 & 1 & 0 & 0 \\ 0 & 0 & 1 & 0 \\ 0 & 0 & 0 & 1 \\ -\gamma_4^1\left(\frac{1}{k} + n^4 + \frac{\bar{\rho}}{k}s^2\right) - \frac{n^2\gamma_1^4 N_{\theta 0}}{kJ} & -\frac{2n\gamma_1^3 N_{x\theta 0}}{kJ} & 2n^2\gamma_1^2 + \frac{\gamma_1^2 N_{x0}}{kJ} & 0 \end{bmatrix}.$$

Similarly, Fourier expansions of the boundary disturbances in Eqs. (6.48) are

$$\bar{u}_i(\theta,t) = \sum_{n=0}^{\infty} \bar{u}_n^i(t)\cos n\theta, \qquad \bar{v}_i(\theta,t) = \sum_{n=1}^{\infty} \bar{v}_n^i(t)\sin n\theta,$$
$$\bar{w}_i(\theta,t) = \sum_{n=0}^{\infty} \bar{w}_n^i(t)\cos n\theta, \tag{6.51a,b,c}$$

$$\bar{w}_{x,i}(\theta,t) = \sum_{n=0}^{\infty} \bar{w}_{x_i,n}(t)\cos n\theta, \tag{6.51d}$$

$$\begin{aligned} \bar{N}_{xi}(\theta,t) &= \frac{(1-\upsilon^2)R}{Eh^2}\sum_{n=0}^{\infty} \bar{N}_{x,n}^{i}(t)\cos n\theta, \\ \bar{N}_{x\theta i}^{e}(\theta,t) &= \frac{2(1+\upsilon)R}{Eh^2}\sum_{n=1}^{\infty} \bar{N}_{x\theta,n}^{ei}(t)\sin n\theta, \end{aligned} \tag{6.51e,f}$$

$$\bar{Q}_{xi}^{e}(\theta,t) = -\frac{12(1-\upsilon^2)R^3}{Eh^4}\sum_{n=0}^{\infty} \bar{Q}_{x,n}^{ei}(t)\cos n\theta, \tag{6.51g}$$

$$\bar{M}_i(\theta,t) = -\frac{12(1-\upsilon^2)R^2}{Eh^4}\sum_{n=0}^{\infty} \bar{M}_n^{i}(t)\cos n\theta. \tag{6.51h}$$

By Eqs. (6.48), (6.49), and (6.50), η_n must satisfy the boundary conditons

$$B_i\Gamma_n\eta_n(\bar{x}_i,s) = 0, \qquad i = 1,2, \tag{6.52}$$

where

$$\Gamma_n = \begin{bmatrix} 1 & 0 & 0 & 0 & 0 & 0 & 0 & 0 \\ 0 & 1/\gamma_1 & n\upsilon & 0 & -\upsilon & 0 & 0 & 0 \\ 0 & 0 & 1 & 0 & 0 & 0 & 0 & 0 \\ -n & 0 & 0 & 1/\gamma_1 & 0 & -kn/(6\gamma_1) & 0 & 0 \\ 0 & 0 & 0 & 0 & 1 & 0 & 0 & 0 \\ 0 & 0 & 0 & 0 & 0 & -(2-\upsilon)n^2/\gamma_1 & 0 & 1/\gamma_1^3 \\ 0 & 0 & 0 & 0 & 0 & 1 & 0 & 0 \\ 0 & 0 & 0 & 0 & -\upsilon n^2 & 0 & 1/\gamma_1^2 & 0 \end{bmatrix}$$

and B_i is a 4×8 matrix containing 1 or 0. Write $B_i = [B_i(j,k)]$. Define the Kronecker delta by $\delta_m^k = 1$ for $k = m$ and 0 for $k \neq m$. If the kth boundary conditon is of displacement type ($u = 0$, $v = 0$, $w = 0$, or $w_{,x} = 0$), the nonzero elements of the kth row of B_i are

$$B_i(j, 2j-1) = \delta_{2j-1}^k. \tag{6.53a}$$

If the kth boundary condition is of force type ($N_x = 0$, $N_{x\theta} = 0$, $Q_x = 0$, or $M_x = 0$), then the nonzero element of the kth row of B_i is given by

$$B_i(j,k) = \delta_{2j}^k. \tag{6.53b}$$

The boundary matrices in Eq. (6.7) take the form

$$M_n(s) = \begin{bmatrix} B_1L_n \\ 0_{4\times8} \end{bmatrix}, \quad N_n(s) = \begin{bmatrix} 0_{4\times8} \\ B_2L_n \end{bmatrix}. \tag{6.54}$$

With the state-space matrix F_n and the boundary matrices M_n and N_n, the transfer functions of the shell can be evaluated by Eqs. (6.8b,c) and the response of the shell subject to various external excitations and boundary conditons can be determined by Eq. (6.9).

6.5.3 Connection Matrix

The connection matraix is needed when stepped cylindrical shells are analyzed using the connection matrix method; see Section 6.3. Without loss of generality, assume that all segments of a stepped shell are completely boned such that the displacements and internal forces are continuous at the interconnecting boundaries of the shell segments. Under the assumption, the connection matrix $R_{i,n}$ is found as

$$R_{i,n} = \begin{bmatrix} \bar{h}_i & 0 & 0 & 0 & 0 & 0 & 0 & 0 \\ 0 & 0 & \bar{h}_i & 0 & 0 & 0 & 0 & 0 \\ 0 & 0 & 0 & 0 & \bar{h}_i & 0 & 0 & 0 \\ 0 & 0 & 0 & 0 & 0 & \bar{h}_i/\bar{L}_i & 0 & 0 \\ 0 & \bar{k}_{i,1}\bar{\gamma}_2 & \bar{k}_{i,1}\bar{\upsilon}_i n & 0 & -\bar{k}_{i,1}\bar{\upsilon}_i & 0 & 0 & 0 \\ -\bar{k}_{i,2}n & 0 & 0 & \bar{k}_{i,2}\bar{\gamma}_2 & 0 & -\bar{k}_{i,2}n\bar{h}_i^2\bar{\gamma}_2/(6R^2) & 0 & 0 \\ 0 & 0 & 0 & 0 & 0 & \bar{k}_{i,3}n^2(2-\bar{\upsilon}_i)\bar{\gamma}_2 & 0 & -\bar{k}_{i,3}\bar{\gamma}_2^3 \\ 0 & 0 & 0 & 0 & 0 & \bar{k}_{i,3}\upsilon_i n^2 R & -\bar{k}_{i,3}\bar{\gamma}_2^3 R & 0 \end{bmatrix}, \tag{6.55}$$

where

$$\bar{k}_{i,1} = \frac{\bar{E}_i\bar{h}_i^2}{(1-\bar{\upsilon}_i^2)R}, \qquad \bar{k}_{i,2} = \frac{\bar{E}_i\bar{h}_i^2}{2(1+\bar{\upsilon}_i)R},$$

$$\bar{k}_{i,3} = \frac{\bar{E}_i\bar{h}_i^4}{12(1-\bar{\upsilon}_i^2)R^3}, \qquad \bar{\gamma}_i = \frac{R}{\bar{L}_i},$$

and $\bar{E}_i$, $\bar{\upsilon}_i$, $\bar{L}_i$, and $\bar{h}_i$ are the elastic module, Poison's ratio, and length, and height of the ith shell, respectively.

6.5.4 Stiffened Shells

The displacement u_i^r, v_i^r, and w_i^r of the ith stiffering ring are governed by

$$q_{ix}^r + q_{ix}^e = \rho_i A_i \frac{\partial^2 u_i^r}{\partial t^2}, \tag{6.56a}$$

$$\frac{1}{R}\frac{\partial N_{i\theta}^r}{\partial \theta} + q_{i\theta}^r + q_{i\theta}^e = \rho_i A_i \frac{\partial^2 v_i^r}{\partial t^2}, \tag{6.56b}$$

$$\frac{1}{R^2}\frac{\partial^2 M_{i\theta}^r}{\partial \theta^2} + \frac{N_{i\theta}^r}{R} + N_{\theta 0}\frac{\partial^2 w_i^r}{\partial \theta^2} + q_{iz}^r = \rho_i A_i \frac{\partial^2 w_i^r}{\partial t^2}, \tag{6.56c}$$

where ρ_i and A_i are the density and the cross sectional area of the ith ring, respectively, and the superscript r indicates that the physical parameters are related to the ring. Here the middle surface of the shell has been chosen as the reference plane. The internal forces of the ring are related to the shell displacements by

$$N_{i\theta}^r = k_{11}^i\left(-\frac{w}{R}+\frac{1}{R}\frac{\partial v}{\partial \theta}\right) - k_{12}^i\frac{\partial^2 w}{R^2\partial\theta^2}, \tag{6.57a}$$

$$M_{i\theta}^i = k_{12}^i\left(-\frac{w}{R}+\frac{1}{R}\frac{\partial v}{\partial \theta}\right) - k_{22}^i\frac{\partial^2 w}{R^2\partial\theta^2}, \tag{6.57b}$$

where the coordinate z is measured from the middle surface of the shell,

$$k_{11}^i = E_i A_i, \qquad k_{12}^i = E_i A_i e_i, \qquad k_{22}^i = E_i I_{i\rho}, \tag{6.58}$$

e_i is the eccentricity of the ring centroid shown in Fig. 6.5(a) and $I_{i\rho}$ is the second-order moment of the ring cross sections with respect to the middle surface of the shell.

The key in estimating the response of the stiffened shell is to determined the matrix $L^{i,n}(s)$ in Eq. (6.38), whose nonzero elements $L_{j,l}^{i,n}$ (defined in the appendix) are obtained as follows:

$$L_{2,1}^{i,n} = \bar{k}R^2\rho_i A_i\gamma_1^2 s^2, \qquad L_{2,6}^{i,n} = -\bar{k}R^2\rho_i A_i\gamma_1^2 e_i s^2,$$

$$L_{4,3}^{i,n} = \left(n^2 k_{11}^i + \rho_i A_i R^2 s^2\right)\frac{2\bar{k}\gamma_1^2}{1-\upsilon},$$

$$L_{4,5}^{i,n} = -\left(nk_{11}^i - \frac{n^3 k_{12}^i}{R} - n\rho_i A_i R s^2 e_i\right)\frac{2\bar{k}\gamma_1^2}{1-\upsilon},$$

$$L_{8,3}^{i,n} = \left(nk_{11}^i - \frac{n^3 k_{12}^i}{R}\right)\frac{\bar{k}\gamma_1^4}{k},$$

$$L_{8,5}^{i,n} = -\left(k_{11}^i - \frac{2}{R}k_{12}^i n^2 + \frac{n^4 k_{22}^i}{R^2} + n^2 N_{\theta 0} + R^2\rho_i A_i s^2\right)\frac{\bar{k}\gamma_1^4}{k}$$

with $\bar{k} = (1-\upsilon^2)/(Eh^2)$

For the comparison purpose, the stiffened cylindrical shell is also approximated as an orthotropic shell by smearing the tension and bending stiffness of the rings on the shell. The internal force–strain relations in this case are given by

$$\begin{Bmatrix} N_x \\ N_\theta \\ N_{x\theta} \end{Bmatrix} = \frac{Eh}{1-\upsilon^2}\begin{bmatrix} 1 & \upsilon & 0 \\ \upsilon & 1+\Delta D_t & 0 \\ 0 & 0 & \frac{1-\upsilon}{2} \end{bmatrix}\begin{Bmatrix} \varepsilon_{xo} \\ \varepsilon_{\theta 0} \\ \varepsilon_{x\theta 0} \end{Bmatrix} + \Delta B_{tb}\begin{Bmatrix} 0 \\ \kappa_\theta \\ 0 \end{Bmatrix}, \tag{6.59a}$$

$$\begin{Bmatrix} M_x \\ M_\theta \\ M_{x\theta} \end{Bmatrix} = \frac{Eh^3}{12(1-\upsilon^2)}\begin{bmatrix} 1 & \upsilon & 0 \\ \upsilon & 1+\Delta D_b & 0 \\ 0 & 0 & 1-\upsilon \end{bmatrix}\begin{Bmatrix} \kappa_x \\ \kappa_\theta \\ \kappa_{x\theta} \end{Bmatrix} + \Delta B_{tb}\begin{Bmatrix} 0 \\ \varepsilon_\theta \\ 0 \end{Bmatrix}, \tag{6.59b}$$

where

$$\Delta D_t = \frac{1}{EhL}(1-\upsilon^2)\sum_{i=1}^{N^r} k_{11}^i, \qquad \Delta D_b = \frac{12}{Eh^3L}(1-\upsilon^2)\sum_{i=1}^{N^r} k_{22}^i,$$

$$\Delta B_{tb} = \frac{1}{L}\sum_{i=1}^{N^r} k_{12}^i.$$

For the shell simply supported ($N_x = v = w = M_x = 0$) at its both ends, the frequency equation is

$$\det(A + \omega^2 I) = 0, \tag{6.60}$$

where

$$A = \beta \begin{bmatrix} -\left(\frac{m^2\pi^2}{\gamma_1^2} + \frac{1-\upsilon}{2}n^2\right) & \frac{1+\upsilon}{2\gamma_1}nm\pi & -\frac{\upsilon}{\gamma_1}m\pi \\ \frac{1+\upsilon}{2\gamma_1}nm\pi & -\left(\frac{1-\upsilon}{2\gamma_1^2}m^2\pi^2 + (1+\Delta D_t)n^2\right) & n(1+\Delta D_t) - \Delta\bar{B}_{tb}n^3 \\ -\frac{\upsilon}{\gamma_1}m\pi & n(1+\Delta D_t) - \Delta\bar{B}_{tb}n^3 & \bar{A}_{mn}^{33} \end{bmatrix}$$

with

$$\beta = \frac{E}{\bar{\rho}(1-\upsilon^2)R^2}, \qquad \Delta\bar{B}_{tb} = \frac{\Delta B_{tb}(1-\upsilon^2)}{EhRL}, \qquad \bar{\rho} = \rho + \frac{1}{Lh}\sum_{i=1}^{N^r}\rho_i b_i h_i,$$

$$\bar{A}_{mn}^{33} = -\left[1 + \Delta D_t - 2n^2\Delta\bar{B}_{tb} + k\left(\frac{m^4\pi^4}{\gamma_1^4} + \frac{2n^2m^2\pi^2}{\gamma_1^2} + n^4(1+\Delta D_b)\right)\right].$$

6.6 Numerical Examples

Having established the distributed transfer function formulation, we are ready to address various static and dynamic problems for homogeneous, stepped, and ring-stiffened cylindrical shells. In the following numerical examples, the Donnell–Mushtari shell model (Section 6.5) is used.

6.6.1 Free Vibration of a Homogenous Shell

Consider the following dimensionaless geometry and material parameters:

$$E = 100, \qquad \upsilon = 0.3, \qquad R = L = 100, \qquad h = 1, \qquad \rho = 1. \tag{6.61}$$

Assume simply supported boundary conditions

$$v = 0, \qquad w = 0, \qquad N_x = 0, \qquad M_x = 0 \tag{6.62}$$

Table 6.1. *Natural frequencies ω_{mn} of the cylindrical shell (simply supported, $n = 7$).*

Mode no. m of ω_{m7}	Transfer function method	Boundary value approach
1	0.0242	0.0242
2	0.0516	0.0516
3	0.0764	0.0764
4	0.0985	0.0985
5	0.1222	0.1222
6	0.1505	0.1505
7	0.1849	0.1849
8	0.2259	0.2259
9	0.2743	0.2743
10	0.3276	0.3276

at both ends of the shell. In this case, the classical boundary value approach gives the exact frequency equation, which is (Donnell, 1993)

$$\det\left(\frac{E}{\rho(1-\upsilon^2)R^2}A + \omega^2 I\right) = 0, \tag{6.63}$$

with

$$A = \begin{bmatrix} -\left(\frac{m^2\pi^2}{\gamma_1^2} + \frac{1-\upsilon}{2}n^2\right) & \frac{1+\upsilon}{2\gamma_1}nm\pi & -\frac{\upsilon}{\gamma_1}m\pi \\ \frac{1+\upsilon}{2\gamma_1}nm\pi & -\left(\frac{1-\upsilon}{2\gamma_1^2}m^2\pi^2 + n^2\right) & n \\ -\frac{\upsilon}{\gamma_1}m\pi & n & -1 - k\left(\frac{m^4\pi^4}{\gamma_1^4} + \frac{2n^2m^2\pi^2}{\gamma_1^2} + n^4\right) \end{bmatrix}.$$

Shown in Table 6.1 are the first ten natural frequencies ω_{mn} of the cylindrical shell, where the subscript m is the shell wave number along its longitudinal direction x. The circumferential wave number $n = 7$ is chosen such that the natural frequency ω_{17} is the fundamental frequency (lowest) of the shell. Both the transfer function method (6.10) and the boundary value approach (6.63) have been used in the calculation, and they produce the same result.

6.6.2 Buckling of a Homogenous Shell

The shell is under both axial compression N_{x0} and hydrostatic pressure $N_{\theta 0}$; see Eq. (6.47c). Consider the same shell parameters as given in Eq. (6.61), and the simly supported boundary conditions (6.62). Neglect the prebuckling effects. The buckling loads, calculated by the transfer function method (6.10),

Table 6.2. *Buckling loads N_{mn} of the shell under axial compression (simply supported, $n = 6$).*

Mode no. m of N_{m6}	Transfer function method	Boundary value approach
1	0.6053	0.6054
2	0.6267	0.6267
3	0.6380	0.6380
4	0.6643	0.6643
5	0.6904	0.6904
6	0.7307	0.7307
7	0.7880	0.7880
8	0.8251	0.8251
9	0.9140	0.9140
10	1.0653	1.0653

Table 6.3. *Buckling loads N_{mn} of the shell under hydrostatic pressure (simply supported, $n = 8$).*

Mode no. m of N_{m8}	Transfer function method	Boundary value approach
1	0.1060	0.1060
2	0.3806	0.3806
3	0.8620	0.8620
4	1.4958	1.4958
5	2.3668	2.3668
6	3.6376	3.6376
7	5.5094	5.5094
8	8.2127	8.2127
9	12.0068	12.0069
10	17.1820	17.1821

are shown in Tables 6.2 and 6.3 for two cases: axial compression and hydrostatic pressure, loading. Here the circumferential wave number $n = 6$ ($n = 8$) is chosen to reach the lowest buckling load N_{16}(N_{18}) in the axial compression (hydrostatic pressure) case. The transfer function method is compared to the boundary value approach, which gives exact estimation of the buckling loads by (Yamaki, 1984)

$$\det\begin{bmatrix} -\left(\frac{m^2\pi^2}{\gamma_1^2}+\frac{1-\upsilon}{2}n^2\right) & \frac{1+\upsilon}{2\gamma_1}nm\pi & -\frac{\upsilon}{\gamma_1}m\pi \\ \frac{1+\upsilon}{2\gamma_1}nm\pi & -\left(\frac{1-\upsilon}{2\gamma_1^2}m^2\pi^2+n^2\right) & n \\ -\frac{\upsilon}{\gamma_1}m\pi & n & \phi \end{bmatrix} = 0, \tag{6.64}$$

Table 6.4. *Natural frequencies ω_{mn} of the stepped shell (simply supported SS3, $n = 5$).*

Mode no. m of ω_{m5}	Transfer function method	Finite element method
1	0.02944	0.02909
2	0.07115	0.07099
3	0.08471	0.08464
4	0.11111	0.11080
5	0.13660	0.13659
6	0.15720	0.15701
7	0.21960	0.21958
8	0.24255	0.24240
9	0.28263	0.28396
10	0.33358	0.33315

where

$$\phi = -1 - k\left(\frac{m^4\pi^4}{\gamma_1^4} + \frac{2n^2m^2\pi^2}{\gamma_1^2} + n^4\right) - \frac{1}{J}\left(\frac{m^2\pi^2 N_{x0}}{\gamma_1^2} + n^2 N_{\theta 0}\right).$$

6.6.3 Free Vibration of a Stepped Shell

Consider a stepped shell of three shell segments. The geometry and material parameters of the shell segment are as follows:

Subsystem 1.	$l_1 = 40,$	$h_1 = 1,$	$E_1 = 100,$	$\upsilon_1 = 0.3.$
Subsystem 2.	$l_2 = 20,$	$h_2 = 3,$	$E_2 = 100,$	$\upsilon_2 = 0.3.$
Subsystem 3.	$l_3 = 40,$	$h_3 = 1,$	$E_3 = 100,$	$\upsilon_3 = 0.3.$

The radius of the stepped shell $R = 100$. This is a shell of varying thickness. The transfer function synthesis of the stepped beam is based on the connection matrix method presented in Section 6.3. To show the flexibility of the transfer function method, the following nine sets of boundary conditions are considered in the simulation:

four sets of simple–simple boundary conditions:

$$\text{(SS1)} \quad \begin{aligned} &w^{(1)}(0) = M_x^{(1)}(0) = u^{(1)}(0) = v^{(1)}(0) = 0, \\ &w^{(3)}(1) = M_x^{(3)}(1) = u^{(3)}(1) = v^{(3)}(1) = 0; \end{aligned} \tag{6.65a}$$

$$\text{(SS2)} \quad \begin{aligned} &w^{(1)}(0) = M_x^{(1)}(0) = u^{(1)}(0) = N_{x\theta}^{(1)}(0) = 0, \\ &w^{(3)}(1) = M_x^{(3)}(1) = u^{(3)}(1) = N_{x\theta}^{(3)}(1) = 0; \end{aligned} \tag{6.65b}$$

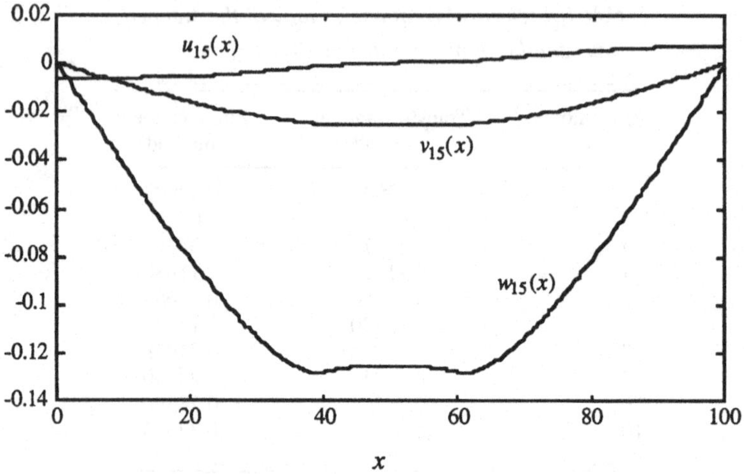

(a)

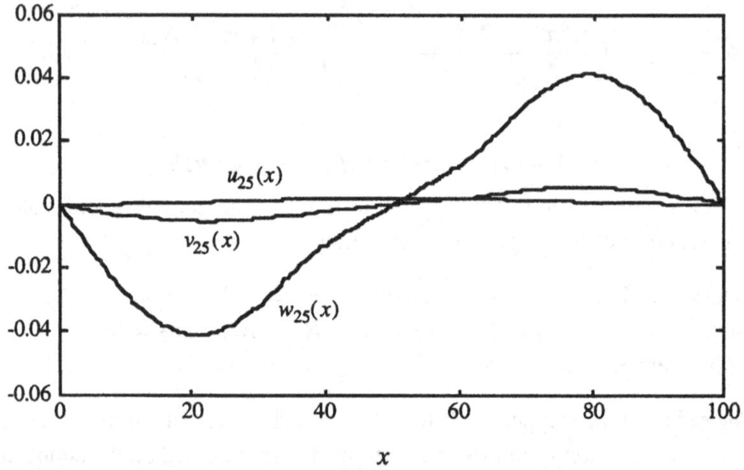

(b)

Fig. 6.6. (a) The mode shapes of the shell (simply supported SS3): $\omega_{15} = 0.02944$, (b) $\omega_{25} = 0.07115$, (c) $\omega_{35} = 0.08471$, and (d) $\omega_{45} = 0.11111$.

$$\text{(SS3)} \quad \begin{aligned} &w^{(1)}(0) = M_x^{(1)}(0) = N_x^{(1)}(0) = v^{(1)}(0) = 0, \\ &w^{(3)}(1) = M_x^{(3)}(1) = N_x^{(3)}(1) = v^{(3)}(1) = 0; \end{aligned} \tag{6.65c}$$

$$\text{(SS4)} \quad \begin{aligned} &w^{(1)}(0) = M_x^{(1)}(0) = N_x^{(1)}(0) = N_{x\theta}^{(1)}(0) = 0, \\ &w^{(3)}(1) = M_x^{(3)}(1) = N_x^{(3)}(1) = N_{x\theta}^{(3)}(1) = 0. \end{aligned} \tag{6.65d}$$

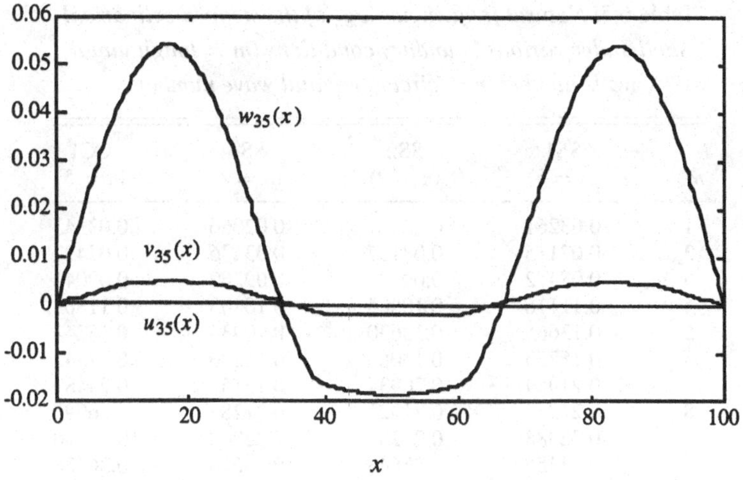

(c)

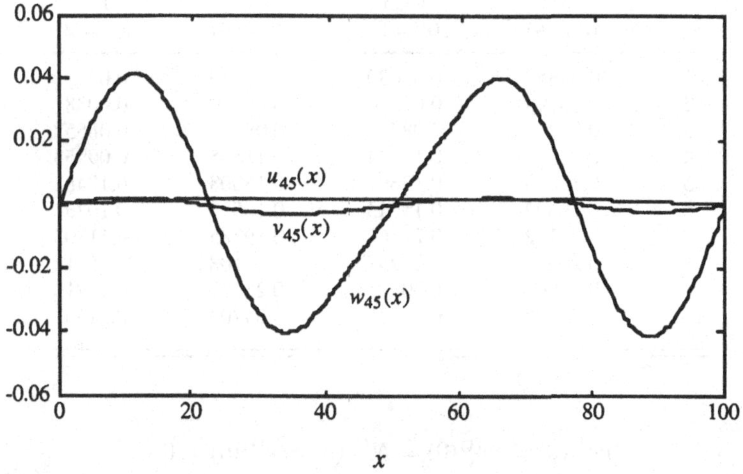

(d)

Fig. 6.6. (*cont.*)

four sets of clamped–clamped boundary conditions:

$$\text{(CC1)} \quad \begin{aligned} &w^{(1)}(0) = w^{(1)}_{,x}(0) = u^{(1)}(0) = v^{(1)}(0) = 0, \\ &w^{(3)}(1) = w^{(3)}_{,x}(1) = u^{(3)}(1) = v^{(3)}(1) = 0; \end{aligned} \tag{6.65e}$$

$$\text{(CC2)} \quad \begin{aligned} &w^{(1)}(0) = w^{(1)}_{,x}(0) = u^{(1)}(0) = N^{(1)}_{x\theta}(0) = 0, \\ &w^{(3)}(1) = w^{(3)}_{,x}(1) = u^{(3)}(1) = N^{(3)}_{x\theta}(1) = 0; \end{aligned} \tag{6.65f}$$

Table 6.5. *Natural frequencies ω_{mn} of the stepped cylindrical shell under various boundary conditions (m = longitudinal wave number, n = circumferential wave number).*

m	SS1 ($n = 5$)	SS2 ($n = 2$)	SS4 ($n = 2$)	CC1 ($n = 5$)
1	0.03261	0.02081	0.02066	0.03343
2	0.07115	0.08137	0.03276	0.07477
3	0.08472	0.09374	0.09260	0.09046
4	0.11116	0.10885	0.10403	0.11869
5	0.13665	0.13630	0.10934	0.15241
6	0.15753	0.15088	0.13630	0.17342
7	0.21964	0.20934	0.15152	0.24489
8	0.24258	0.21327	0.21285	0.26996
9	0.32388	0.23286	0.22008	0.32390
10	0.33356	0.23643	0.23618	0.36426

m	CC2 ($n = 4$)	CC3 ($n = 5$)	CC4 ($n = 4$)	FF1 ($n = 2$)
1	0.02850	0.03030	0.02709	0.00216
2	0.06795	0.07115	0.06149	0.00381
3	0.08715	0.08471	0.08634	0.08557
4	0.11539	0.11114	0.11535	0.09953
5	0.15025	0.13662	0.15008	0.10482
6	0.16892	0.15743	0.16530	0.10737
7	0.24142	0.21962	0.19785	0.11803
8	0.26322	0.24257	0.24108	0.15312
9	0.26902	0.30005	0.26687	0.16913
10	0.35974	0.33356	0.34702	0.21901

$$\text{(CC3)} \quad \begin{aligned} &w^{(1)}(0) = w^{(1)}_{,x}(0) = N^{(1)}_x(0) = v^{(1)}(0) = 0, \\ &w^{(3)}(1) = w^{(3)}_{,x}(1) = N^{(3)}_x(1) = v^{(3)}(1) = 0; \end{aligned} \tag{6.65g}$$

$$\text{(CC4)} \quad \begin{aligned} &w^{(1)}(0) = w^{(1)}_{,x}(0) = N^{(1)}_x(0) = N^{(1)}_{x\theta}(0) = 0, \\ &w^{(3)}(1) = w^{(3)}_{,x}(1) = N^{(3)}_x(1) = N^{(3)}_{x\theta}(1) = 0. \end{aligned} \tag{6.65h}$$

one set of free–free boundary conditions:

$$\text{(FF1)} \quad \begin{aligned} &Q^{(1)}_x(0) = M^{(1)}_x(0) = N^{(1)}_x(0) = N^{(1)}_{x\theta}(0) = 0, \\ &Q^{(3)}_x(1) = M^{(3)}_x(1) = N^{(3)}_x(1) = N^{(3)}_{x\theta}(1) = 0. \end{aligned} \tag{6.65i}$$

Here the superscripts (1) and (3) designate the left and right shell segments respectively.

Table 6.6. *The fundamental frequency of the cylindrical shell with inward stiffeners.*

N^r	Proposed method	Orthotropic approximation
1	0.2699(6)	0.2622(6)
2	0.2777(6)	0.2711(6)
3	0.2849(6)	0.2791(6)
4	0.2914(6)	0.2863(6)
5	0.2974(6)	0.2927(6)
6	0.3028(6)	0.2985(6)
7	0.3048(5)	0.3038(6)
8	0.3053(5)	0.3051(5)
9	0.3058(5)	0.3056(5)
10	0.3061(5)	0.3060(5)

Table 6.7. *The fundamental frequency of the cylindrical shell with outward stiffeners.*

N^r	Proposed method	Orthotropic approximation
1	0.2699(6)	0.2622(6)
2	0.2777(6)	0.2711(6)
3	0.2849(6)	0.2791(6)
4	0.2914(6)	0.2863(6)
5	0.2974(6)	0.2927(6)
6	0.3028(6)	0.2985(6)
7	0.3048(5)	0.3038(6)
8	0.3053(5)	0.3051(5)
9	0.3058(5)	0.3056(5)
10	0.3061(5)	0.3060(5)

The proposed transfer function method is first compared to the finite element method (FEM) since the exact results by other analytical methods are not available. Table 6.4 shows the first ten lowest natural frequencies of the shell under the simply supported boundary conditions (SS3), when m and n are the longitudinal and circumferential wave numbers, respectively. Good agreement between the proposed method and FEM is seen. It is found that the fundamental frequency of the shell is ω_{15}. Figure 6.6 plot the mode shapes of the shell, where $u_{mn}(x)$, $v_{mn}(x)$, and $w_{mn}(x)$ are the eigenfunctions corresponding to the frequency ω_{mn} listed in Table 6.4.

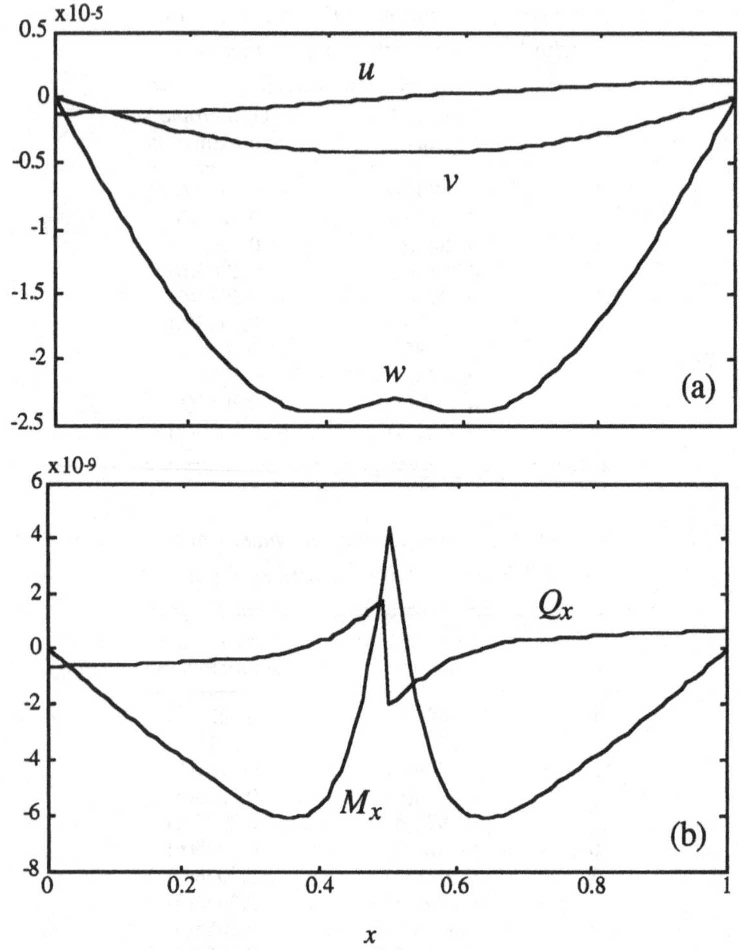

Fig. 6.7. Eigenfunctions of the shell with one symmetric stiffener ($N^r = 1$, fundamental frequency = 0.2567): (*a*) mode shapes; (*b*) moment and shear force.

The natural frequencies of the stepped shell under the other eight sets of boundary conditions are also calculated by the transfer function method, and they are presented in Table 6.5. Here the circumferential wave number n is selected so that ω_{1n} is the fundamental frequency of the shell under the specified boundary conditions. In treating each set of boundary conditions, the transfer function method simply changes the entries of the matrices M_n and N_n and does not need different derivations.

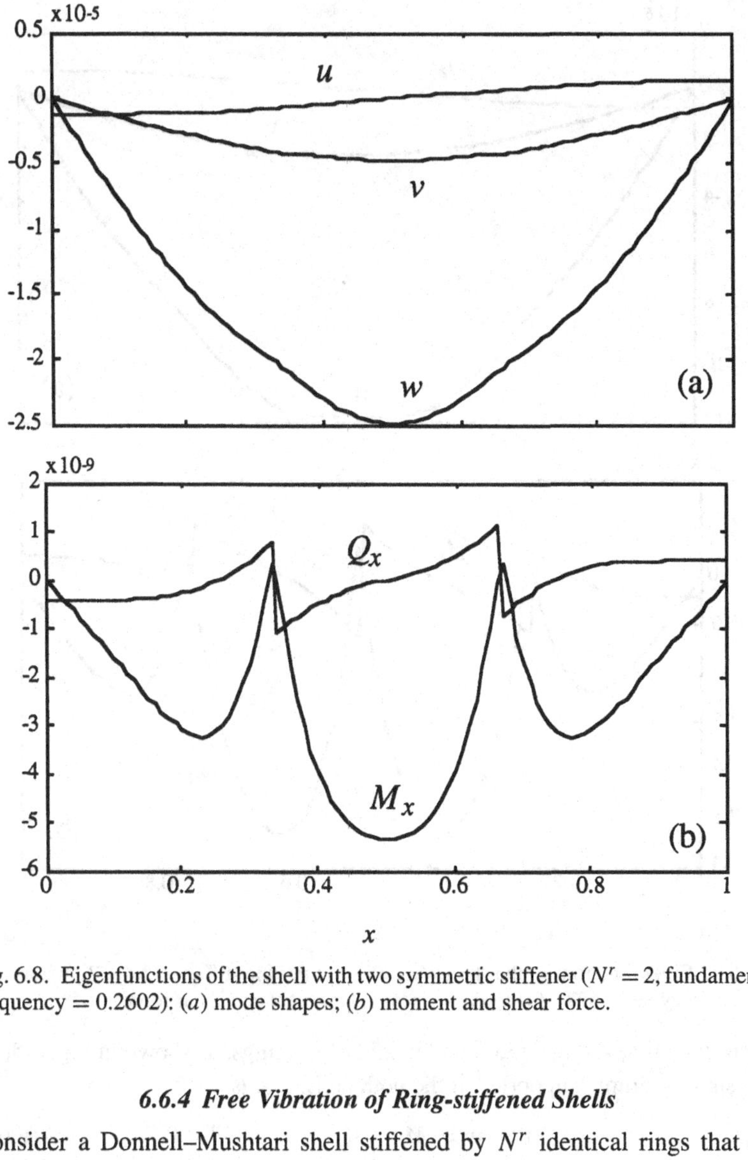

Fig. 6.8. Eigenfunctions of the shell with two symmetric stiffener ($N^r = 2$, fundamental frequency = 0.2602): (*a*) mode shapes; (*b*) moment and shear force.

6.6.4 Free Vibration of Ring-stiffened Shells

Consider a Donnell–Mushtari shell stiffened by N^r identical rings that are equally spaced along the shell longitudinal direction. The parameters of the shell and stiffening rings are chosen as:

$$R = L = 100, \quad h = 1, \quad E = 10^4, \quad \upsilon = 0.3 \qquad \text{(shell)} \qquad (6.66a)$$

$$E_i = 10^4, \quad \upsilon_i = 0.3, \quad b_i = 1, \quad h_i = 2 \qquad \text{(ring),} \qquad (6.66b)$$

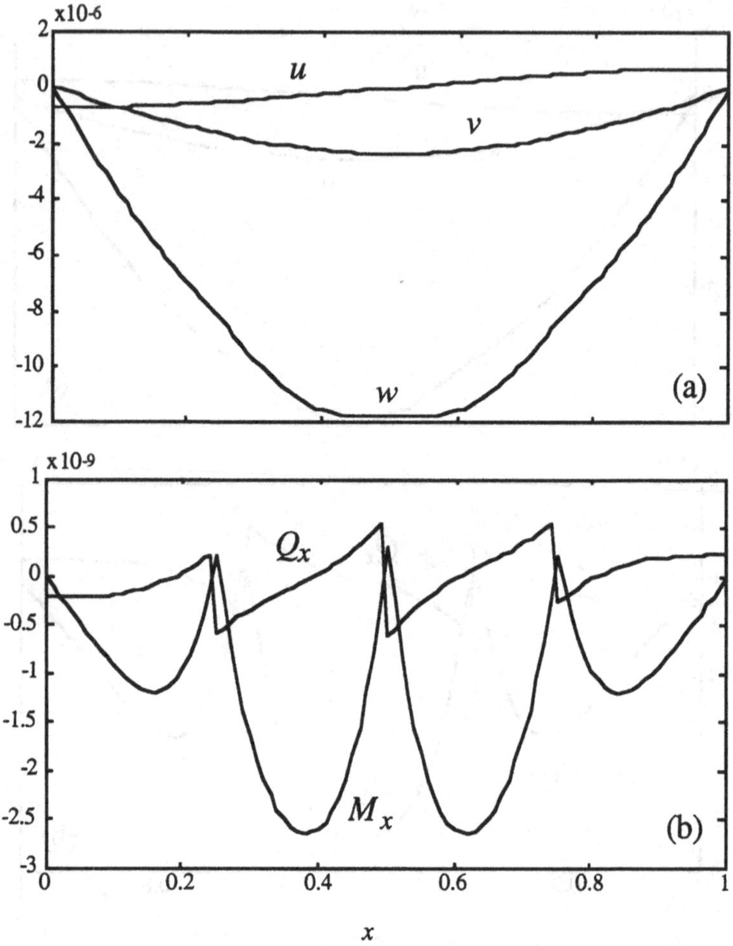

Fig. 6.9. Eigenfunctions of the shell with three symmetric stiffener ($N^r = 3$, fundamental frequency $= 0.2627$): (*a*) mode shapes; (*b*) moment and shear force.

where b_i and h_i are the width and height of the rings, as shown in Fig. 6.5(a). The shell is simply supported at the both ends, that is,

$$N_x = v = w = M_x = 0 \text{ at } x = 1 \text{ and } 100. \tag{6.67}$$

The natural frequencies of the ring-stiffened shell are determined by four methods: (i) the proposed transfer function method (Eq. (6.45)) for ring-stiffened shells; (ii) the orthotropic approximation (Eq. (6.60)); (iii) the stepped shell modeling given in Scction 6.3; and (iv) the finite element method. It should be noted that the last two methods are only valid for shells with symmetric stiffeners (i.e., the stiffener eccentricity e_i in Fig. 6.5(a) is zero). In this case, with

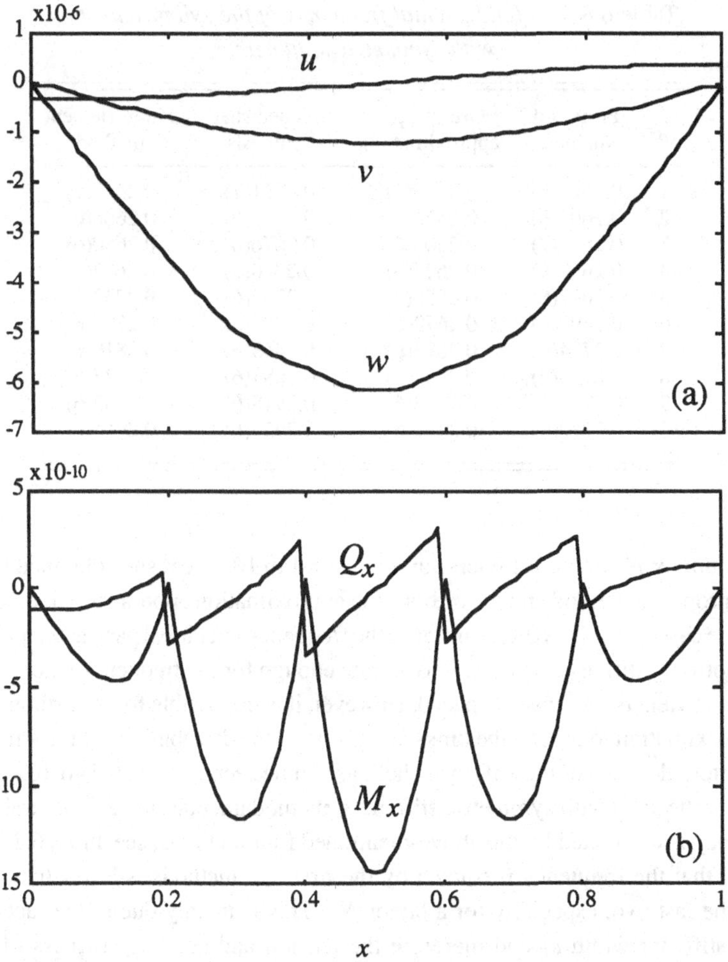

Fig. 6.10. Eigenfunctions of the shell with four symmetric stiffener ($N^r = 4$, fundamental frequency = 0.2650): (*a*) mode shapes; (*b*) moment and shear force.

each ring and the connecting shell segment taken as a short cylindrical shell, the whole stiffened shell becomes a stepped shell, which, accordingly, can be analyzed by the stepped shell synthesis and the finite element method. However, if inward or outward stiffeners are used, the shell can be modeled as a stepped shell because the middle surfaces of those short shells do not coicide; the stepped shell assumption in methods (iii) and (iv) may lead to inaccurate results.

Shown in Tables 6.6 and 6.7 is the fundamental (lowest) natural frequency of the cylindrical shell with inward and outward stiffening rings, respectively, with

Table 6.8. *The fundamental frequency of the cylindrical shell with symmetric stiffeners.*

N^r	Proposed method	Orthotropic approximation	Stepped shell synthesis	Finite element method
1	0.2567(7)	0.2497(7)	0.2571(7)	0.2542(7)
2	0.2602(6)	0.2570(7)	0.2631(6)	0.2606(6)
3	0.2627(7)	0.2603(6)	0.2676(6)	0.2650(6)
4	0.2650(6)	0.2627(6)	0.2720(6)	0.2692(6)
5	0.2672(6)	0.2650(6)	0.2762(6)	0.2733(6)
6	0.2693(6)	0.2672(6)	0.2804(6)	0.2772(6)
7	0.2714(6)	0.2693(6)	0.2843(6)	0.2810(6)
8	0.2733(6)	0.2714(6)	0.2881(6)	0.2847(6)
9	0.2752(6)	0.2733(6)	0.2918(6)	0.2882(6)
10	0.2770(6)	0.2752(6)	0.2953(6)	0.2915(6)

the number N^r of the stiffeners varying from 1 to 10. The results obtained from the proposed method and the orthotropic approximation approach each other in all the cases. This is expected because the frequency is a global parameter and the orthotropic stiffness treatment is accurate enough for the frequency calculation if the stiffeners are densely spaced. However, it is impossible for the orthotropic approximation to predict the kinks and jumps in the distributions of the stresses or internal forces of the stiffened shell, as can be seen in Figs. 6.7–6.10.

For the shell with symmetric stiffeners, its fundamental (lowest) natural frequency is calculated by the above-mentioned four methods; see Table 6.8. It is seen that the frequency predicted by the first two methods is lower than that by the last two, especially for a larger N^r. This is mainly due to the fact that the stiffener width b_i and therefore the tension and bending stiffness of the rings in the longitudinal direction are ignored in the proposed method and in the orthotropic approximation. In all three tables, the digit in brackets is the circumferential wave number n corresponding to the lowest natural frequency.

The mode shapes (u, v, w) and the distributions of the bending moment (M_x) and shear force (Q_x) of the stiffened shell are plotted in Figs. 6.7 to 6.10 by the proposed transfer function method, corresponding to the fundamental frequency in Table 6.8 for $N^r = 1, 2, 3, 4$. The rings have the most significant effect on the transverse displacement w and almost no effect on the longitudinal displacement u. The kinks in the M_x plots and jumps in the Q_x are clearly seen, which can not be obtained by the orthotropic approximation. As N^r increases, more kinks and jumps occur in the moment and shear force curves near the locations of the stiffeners.

6.7 Conclusions

An analytical and numerical method has been developed for stepped and ring-stiffened cylindrical shells. The numerical examples show the accuracy and efficiency of the method. The proposed transfer function method has the following special features:

(i) The transfer function method provides exact and closed-form solutions for static deflection, free vibration, buckling, and frequency response of cylindrical shells under arbitrary boundary conditions and external loads.
(ii) The transfer function formulation represents various linear models of cylindrical shells (Love type, Donnell–Mushtari type, etc.) and characterizes different physical properties, such as gyroscopic effects from a spinning shell, isotropic/anisotropic and elastic/viscoelastic materials, and nonuniformly distributed damping.
(iii) The distributed transfer functions, serving as exact "super elements," can be used to assemble stepped and ring-stiffened cylindrical shells, shells with continuously varying thickness, and other shell structures.
(iv) Being able to take into consideration the ring eccentricity and ring position with respect to the shell middle surface, the transfer function method can easily handle inward, outward, and symmetric ring stiffeners.
(v) The transfer function method is convenient in numerical simulation and computer coding. Different systems, boundary conditions, and external loads are systematically treated in one simple and compact matrix form, just like in finite element analysis.

Acknowledgments

This work was partially supported by the U.S. Army Research Office.

References

Al-Najafi, A. M., and Warburton, G. B., 1970, "Free Vibration of Ring-Stiffened Cylindrical Shells," *Journal of Sound and Vibration*, Vol. 13, No. 1, pp. 9–25.

Bergman, L. A., and Hyatt, J. E., 1989, "Green's Functions for Transversely Vibrating Uniform Euler–Bernoulli Beams Subject to Constant Axial Preload," *Journal of Sound and Vibration*, Vol. 134, No. 1, pp. 175–80.

Bergman, L. A., and Nicholson, J. W., 1985, "Forced Vibration of a Damped Combined Linear System," *ASME Journal of Vibration, Acoustics, Stress, and Reliability in Design*, Vol. 107, pp. 275–81.

Birman, V., 1993, "Axisymmetric Bending of Generally Laminated Cylindrical Shells," *ASME Journal of Applied Mechanics*, Vol. 60, No. 1, pp. 157–62.

Burke, S. E., and Hubbard, J. E., Jr., 1990, "Spatial Filtering Concepts in Distributed Parameter Control," *ASME Journal of Dynamic Systems, Measurement and Control*, Vol. 112, December, pp. 565–73.

Butkoviskiy, A. G., 1983, *Structure Theory of Distributed Systems*, John Wiley & Sons, New York.

Chaudhuri, R. A., and Abu-Arja, K. R., 1989, "Closed-Form Solutions for Arbitrary Laminated Anisotropic Cylindrical Shells (Tubes) Including Shear Deformation," *AIAA Journal*, Vol. 27, No. 11, pp. 1597–605.

Christoforou, A. P., and Swanson, S. R., 1990, "Analysis of Simply-Supported Orthotropic Cylindrical Shells Subjects to Lateral Impact Loads," *ASME Journal of Applied Mechanics*, Vol. 57, No. 2, pp. 376–82.

Deskos, D. E., and Oates, J. B., 1981, "Dynamic Analysis of Ring-Stiffened Circular Cylindrical Shells," *Journal of Sound and Vibration*, Vol. 75, No. 1, pp. 1–15.

Donnell, L. H., 1933, "Stability of Thin-Walled Tubes Under Torsion," NACA-TR479, Washington, D.C.

Egle, D. M., and Sewall, J. L., 1968, "An Analysis of Free Vibration of Orthogonally Stiffened Cylindrical Shells with Stiffeners Treated as Discrete Elements," *AIAA Journal*, Vol. 6, No. 3, pp. 518–26.

Forsberg, K., 1969, "Exact Solution for Natural Frequencies of Ring-Stiffened Cylinders," *AIAA/ASME 10th Structures, Structural Dynamics and Materials Conference*, Volume on Structures and Materials, pp. 18–30.

Garnet, H., Goldberg. M. A., 1962, "Free Vibrations of Ring-Stiffened Shells," RE-156, Grumman Aircraft Research.

Godzevich, V. G., and Ivanova, O. V., 1965, "Free Oscillations of Circular Conical and Cylindrical Shell Reinforced by Rigid Circular Ribs," TTF-291, NASA.

Heyliger, P. R., and Jilani, A., 1993, "Free Vibration of Laminated Anisotropic Cylindrical Shells," *Journal of Engineering Mechanics*, Vol. 119, No. 5, pp. 1062–77.

Huang, S. C., and Hsu, B. S., 1992, "Vibration of Spinning Ring-Stiffened Thin Cylindrical Shells," *AIAA Journal*, Vol. 30, No. 9, pp. 2291–98.

Huntington, D. E., and Lyrintzis, C. S., 1992, "Dynamics of Skin-Stringer Panels Using Modified Wave Methods," *AIAA Journal*, Vol. 30, No. 11, pp. 2765–73.

Irie, T., Yamada, G., and Kudoh, Y., 1984, "Free Vibration of Point-Supported Circular Cylindrical Shell," *Journal of Acoustic Society of America*, Vol. 75, No. 4, pp. 1118–23.

Koga, T., 1988, "Effects of Boundary Conditions on the Free Vibration of Circular Cylindrical Shells," *AIAA Journal*, Vol. 26, No. 11, 1988, pp. 1387–94.

Lee, S. Y., and Kuo, Y. H., 1992, "Exact Solutions for the Analysis of General Elastically Restrained Nonuniform Beams," *ASME Journal of Applied Mechanics*, Vol. 59, No. 2, Pt. 2, pp. S205–12.

Markus, S., 1989, *The Mechanics of Vibrations of Cylindrical Shells*, Elsevier, Amsterdam.

Mead, D. J., and Bardell, N. S., 1987, "Free Vibration of a Thin Cylindrical Shell with Periodic Circumferential Stiffeners," *Journal of Sound and Vibration*, Vol. 115, No. 3, pp. 499–520.

Miyazaki, N., and Hagihara, S., 1993, "Bifurcation Buckling of Circular Cylindrical Shells Subjected to Axial Compression During Creep Deformation," *ASME Journal of Pressure Vessel Technology*, Vol. 115, No. 3, pp. 268–74.

Pang, S. T., Tsao, T-C., and Bergman, L. A., 1992, "Active and Passive Damping of Euler–Bernoulli Beams and Their Interactions," *Proceedings of the 1992 American Control Conference*, Vol. 3, pp. 2144–49.

Reddy, B. D., 1980, "Buckling of Elastic-Plastic Discretely Stiffened Cylinders in Axial Compression," *International Journal of Solid and Structures*, Vol. 16, pp. 313–28.

Rigo, P., 1992, "Stiffened Sheathings of Orthotropic Cylindrical Shells," *Journal of Structural Engineering*, Vol. 118, No. 4, pp. 926–43.

Schnell, W., and Heinrichsbauer, F. J., 1964, "The Determination of Free Vibrations of Longitudinally Stiffened Thin-Walled Circular Shells," TTF-8856, NASA.

Sheinman, I., and Weissman, S., 1987, "Coupling Between Symmetric and Antisymmetric Modes in Shells and Revolution," *Journal of Composite Materials*, Vol. 21, No. 11, pp. 988–1007.

Soedel, W., 1981, *Vibration of Shells and Plates*, Marcel Dekker, New York.

Sridharan, S., Zeggane, M., and Starnes, J. H., Jr., 1992, "Postbuckling Response of Stiffened Composite Cylindrical Shells," *AIAA Journal*, Vol. 30, No. 12, pp. 2897–905.

Tan, C. A., and Chung, C. H., 1993, "Transfer Function Formulation of Constrained Distributed Parameter Systems, Part I: Theory," *ASME Journal of Applied Mechanics*, Vol. 60, No. 4, pp. 1004–11.

Thangaratnam, R. K., Palaninathan, R., and Ramachandran, J., 1990, "Thermal Buckling of Laminated Composite Shells," *AIAA Journal*, Vol. 28, No. 5, pp. 859–60.

Wang, J. T. S., 1970, "Orthogonally Stiffened Cylindrical Shells Subjected to Internal Pressure," *AIAA Journal*, Vol. 8, No. 3, pp. 455–61.

Wang, J. T. S., and Hsu, T. M., 1985, "Discrete Analysis of Stiffened Composite Cylindrical Shell," *AIAA Journal*, Vol. 23, No. 11, pp. 1753–61.

Wang, J. T. S., and Lin, Y. J., 1973, "Stability of Discretely Stringer-Stiffened Cylindrical Shells," *AIAA Journal*, Vol. 11, No. 6, pp. 810–13.

Wang, J. T. S., and Rinehart, S. A., 1974, "Free Vibrations of Longitudinally Stiffened Cylindrical Shells," *ASME Journal of Applied Mechanics*, Vol. 41, No. 4, pp. 1087–93.

Wilken, I. D., and Soedel, W., 1976, "The Receptance Method Applied to Ring-Stiffened Cylindrical Shells: Analysis of Model Characteristics," *Journal of Sound and Vibration*, Vol. 44, No. 4, pp. 563–76.

Yamada, G., Irie, T., and Tsushima, M., 1984, "Vibration and Stability of Orthotropic Circular Cylindrical Shells Subjected to Axial Load," *Journal of Acoustic Society of America*, Vol. 75, No. 3, pp. 842–48.

Yamaki, N., 1984, *Elastic Stability of Circular Cylindrical Shells*, North-Holland, Amsterdam.

Yang, B., 1989, *Active Vibration Control of Axially Moving Materials*, Ph.D. Dissertation, Univ. California, Berkeley, California, pp. 38–41.

Yang, B., 1992a, "Transfer Functions of Constrained/Combined One-Dimensional Continuous Dynamic Systems," *Journal of Sound and Vibration*, Vol. 156, No. 3, pp. 425–43.

Yang B., 1992b, "Eigenvalue Inclusion Principles for Distributed Gyroscopic Systems," *ASME Journal of Applied Mechanics*, Vol. 59, No. 3, pp. 650–56.

Yang, B., 1994, "Distributed Transfer Function Analysis of Complex Distributed Parameter Systems," *ASME Journal of Applied Mechanics*, Vol. 61, No. 1, pp. 84–92.

Yang, B., and Mote, C. D., Jr., 1990, "Vibration Control of Band Saws: Theory and Experiment," *Wood Science and Technology*, Vol. 24, pp. 355–73.

Yang, B., and Tan, C. A., 1992, "Transfer Functions of One-Dimensional Distributed Parameter Systems," *ASME Journal of Applied Mechanics*, Vol. 59, No. 4, pp. 1009–14.

Yang, B., and Zhou, J., 1995, "Analysis of Ring-Stiffened Cylindrical Shells," *ASME Journal of Applied Mechanics*, Vol. 62, No. 4, pp. 1005–14.
Zhou, J., and Yang. B., 1995, "A Distributed Transfer Function Method for Analysis of Cylindrical Shells," *AIAA Journal*, Vol. 33, No. 9, pp. 1698–708.

Nomenclature

$A_{mkij}, B_{mkij}, C_{mkij}$	constants of cylindrical shells
e_i	the eccentricity of the centroid of the ith stiffening ring from the shell middle surface
f^m	external force components
$F_n(s)$	state-space matrix
$G_n(x, \xi, s), H_n(x, s)$	distributed transfer functions of shell
$K_n(s)$	dynamic stiffness matrix
$M_n(s), N_n(s)$	boundary matrices
$R_{i,n}$	connection matrix
u^k, u, v, w	shell displacements
u_0, v_0, w_0	displacements of the shell middle surface
$u_{ik}^r, u_i^r, v_i^r, w_i^r$	displacements of stiffening rings
x	shell longitudinal coordinate
z	shell radial coordinate

Greek

$\alpha_{imkj}, \beta_{imkj}, \gamma_{imkj}$	constants of stiffening rings
Γ_{lkij}	coefficients of shell boundary conditions
$\eta_n(x, s)$	state-space vector
θ	shell circumferential coordinate

Appendix: The Elements of $L^{i,n}(s)$

Define

$$d_{mkij}(s) = \alpha_{mkij}s^2 + \beta_{mkij}s + \gamma_{mkij};$$

$$l_{1m} = \textstyle\sum_{j=1}^{m} n_j, \qquad l_{2m} = n_1 + n_2 + n_3 + \sum_{j=1}^{m} n_j.$$

Let δ_j^i be the Kronecker delta. Let $[y]$ be the integer part of the number y. Denote the elements of $L^{i,n}(s)$ by $L_{j,k}^{i,n}$, $j, k = 1, \dots, 2n_b$. The nonzero elements of $L^{i.n}(s)$ are as follows:

$$L_{l_{1m},l}^{i,n} = \delta_l^{l_{1k}} \sum_{j=0}^{[n_k/2]} d_{imk(2j)} s(-1)^j n^{2j}$$

$$- \delta_l^{2n_1+2n_2+2n_3} \frac{ne_i}{R} \left\{ \sum_{j=0}^{[n_k/2]} d_{im2(2j)}(s)(-1)^j n^{2j} \right\},$$

$$L^{i,n}_{l_{1m},l} = \delta_l^{l_{2k}} \sum_{j=1}^{[(n_k+1)/2]} d_{imk(2j-1)}(s)(-1)^{j+1} n^{2j-1}$$

$$- \delta_l^{n_1+n_2+n_3} \frac{ne_i}{R} \left\{ \sum_{j=1}^{[n_2+1)/2]} d_{im2(2j-1)}(s)(-1)^j n^{2j-1} \right\},$$

$$L^{i,n}_{l_{2m},l} = \delta_l^{l_{1k}} \sum_{j=1}^{[(n_k+1)/2]} d_{imk(2j-1)}(s)(-1)^j n^{2j-1}$$

$$- \delta_l^{2n_1+2n_2+2n_3} \frac{ne_i}{R} \left\{ \sum_{j=1}^{[n_2+1)/2]} d_{im2(2j-1)}(-1)^j n^{2j-1} \right\},$$

$$L^{i,n}_{l_{2m},l} = \delta_l^{l_{2k}} \sum_{j=0}^{[n_k/2]} d_{imk(2j)}(s)(-1)^j n^{2j}$$

$$- \delta_l^{n_1+n_2+n_3} \frac{ne_i}{R} \left\{ \sum_{j=0}^{[n_k/2]} d_{im2(2j)}(-1)^{j+1} n^{2j} \right\},$$

$$L^{i,n}_{l_{2m},n_1+n_2+n_3} = -e_i \sum_{j=1}^{[(n_1+1)/2]} d_{im1(2j-1)}(s)(-1)^j n^{2j-1},$$

$$L^{i,n}_{l_{2m},2n_1+2n_2+2n_3} = -e_i \sum_{j=0}^{[n_1/2]} d_{im1(2j)}(s)(-1)^j n^{2j},$$

$$L^{i,n}_{l_{2m},n_1+n_2+n_3} = -e_i \sum_{j=1}^{[(n_1+1)/2]} d_{im1(2j-1)}(s)(-1)^j n^{2j-1},$$

$$L^{i,n}_{l_{2m},2n_1+2n_2+2n_3} = -e_i \sum_{j=0}^{[n_1/2]} d_{im1(2j)}(s)(-1)^j n^{2j}.$$

7

Orthogonal Sensing and Control of Continua with Distributed Transducers – Distributed Structronic System

H. S. TZOU, V. B. VENKAYYA, and J. J. HOLLKAMP

Abstract

Distributed sensing and control of flexible shells and continua using distributed transducers has posted challenging issues for decades. This chapter focuses on distributed sensing and control of a generic double-curvature elastic shell and its derived geometries laminated with distributed piezoelectric transducers. Generic distributed orthogonal sensing and actuation of shells and continua are proposed. Spatially distributed orthogonal sensors/actuators and self-sensing actuators are presented. Collocated independent modal control with self-sensing orthogonal actuators is demonstrated and its control effectiveness evaluated. Spatially distributed orthogonal piezoelectric sensors/actuators for circular ring shells are designed and their modal sensing and control are investigated. Membrane and bending contributions in sensing and control responses are studied.

7.1 Introduction

Control of distributed parameter systems has posted many challenging problems and issues stimulating sophisticated research for decades (Balas, 1988; Brichkin et al., 1973; Butkovskii, 1962; Lions, 1968; Meirovitch, 1988; Oz and Meirovitch, 1983; Robinson, 1971; Sakawa, 1966; Tzafestas, 1970; Tzou, 1988, 1991, 1993; Vidyasagar, 1988; Wang, 1966; Zimmerman, Inman, and Juang, 1988). However, implementing distributed control of elastic continua, e.g., shells, plates, etc., using distributed devices has continuously been hampered by the practical availability of distributed sensing/actuation devices. Recent development of *smart structures* and *intelligent structural systems* (or *structronic* (*struc*ture-elec*tronic*) systems, in a new generic term) using active electromechanical materials has revealed the missing link of distributed transducers. The original work of intelligent structural systems involved studies of distributed control of flexible beams using spatially distributed piezoelectric layers (Baily and Hubbard, 1985; Crawley and de Luis, 1987; Baz and Poh,

1988; Tzou, 1987). This intelligent structure concept was quickly emerging and has been expanding to include other active electromechanical materials, e.g., shape memory alloys, electrostrictive materials, electromagnetostrictive materials, electro/magnetorheological materials, etc. (Rogers, 1988; Wada, Fanson, and Crawley, 1989; Gandhi and Thompson, 1992; Tzou and Anderson, 1992; Tzou and Fukuda, 1992; Tzou, 1993; Tzou et al., 1998). Today this intelligent structure technology has become part of the mainstream research, having many applications in distributed sensing and control of distributed elastic systems (Tzou and Anderson, 1992; Tzou, 1993), precision manipulation and control, microelectromechanical systems (Tzou and Fukuda, 1992), structronic systems (Gandhi and Thompson, 1992; Tzou et al., 1998), etc.

Various issues of distributed vibration control of one-dimensional (1-D) continua-beams using distributed piezoelectric transducers have been studied over the years. Usov and Surygin (1984) studied damping changes of a semiconductor structure using piezofilm. Hubbard et al. used distributed piezoelectric polyvinylidene fluoride (PVDF) films to enhance the damping of cantilever beams. (Baily and Hubbard, 1985; Plumb, Hubbard, and Baily, 1997). Crawley et al. (1990) investigated the use of piezoceramic actuators as elements of intelligent structures and developed a detailed piezoceramic actuator model for beams. Lee and Moon (1990) proposed piezoelectric modal sensor/actuator designs for flexible plates and applied them to two-dimensional thin beams. Collins et al. (1994) evaluated spatially shaped distributed piezoelectric sensors defined by sinc functions for monitoring beam oscillations. Hanagud and Obal (1988) identified dynamic coupling coefficients of structures using piezoceramic sensors and actuators. Baz and Poh (1988) evaluated the performance of an active control system with piezoceramic actuators. Cudney et al. (1989) proposed a distributed multilayered actuator based on the deep beam theory. Fanson and Garba (1988) experimentally studied active members made of piezoelectric ceramics. Tzou and coworkers also applied PVDF films as active dampers (Tzou, 1987), distributed orthogonal self-sensing/actuation transducers (Tzou and Hollkamp, 1994), and active precision vibration isolators/exciters (Tzou and Gadre, 1988, 1990). Distributed piezoelectric sensors and actuators were applied to distributed sensing and multimode control of flexible robot manipulators (Tzou, 1989b; Tzou, Tseng, and Wan, 1990). Rao and Sunar (1993) studied distributed thermopiezoelectric sensors and actuators. Tzou and Ye (1994) investigated the temperature/pyroelectric/dynamic coupling effect and also combined dynamic and thermal deflection control of beams. Nonlinear boundary transition and control of piezoelectric laminated cantilever beams and plates subjected to large control voltages were studied by Tzou, Johnson, and Liu (1995).

Sensing and control of two-dimensional (2-D) zero-curvature continua-plates using distributed piezoelectric transducers were also investigated. Ricketts studied a piezoelectric polymer flexural plate hydrophone (Ricketts, 1981) and frequencies of completely free composite piezoelectric plates (Ricketts, 1989). Theories of distributed sensing and vibration control of generic plates with distributed transducers have been proposed (Tzou, 1989a, 1993; Tzou and Zhong, 1993). Lee (1990, 1992) proposed a theory for laminated piezoelectric plates with applications to distributed sensor/actuator designs. Lee's formulations suggested that the distributed piezoelectric layers are capable of sensing and controlling bending, shearing, shrinking, and stretching effects of rectangular plates. Burke and Hubbard (1990) studied distributed transducer control designs for thin plates with general boundary conditions. Modal control and observation deficiencies were also exploited. Dimitriadis et al. (1991) investigated modal excitations of thin plates using piezoelectric actuators. Crawley and Lazarus (1991) studied induced strain actuation of anisotropic plates. Tzou and Fu (1994) studied distributed sensing and control (velocity and displacement feedback) of square plates laminated with segmented distributed sensor and actuator patches. Distributed modal voltages and feedback control of plates were also investigated using the finite element method with a newly developed 8-node thin piezoelectric solid element (Tzou and Tseng, 1990, 1991). Detwiler et al. (1994) and Suleman and Venkayya (1994) investigated 2-D plate integrated with piezoelectric sensors/actuators and flutter control using piezoelectric actuators. Birman (1992) proposed a theory of geometrically nonlinear composite plates with piezoelectric stiffeners. Pai et al. (1993) studied a refined nonlinear model of piezoelectric plate laminates. Nonlinear characteristics and static/dynamic control of nonlinear deformations/frequencies of circular plates (Tzou and Zhou, 1995) and opto-thermoelastic characteristics and control effectiveness of distributed opto-mechanical actuators have also been investigated (Liu and Tzou, 1996; Tzou and Liu, 1996).

Distributed sensing and control of 1-D and 2-D nonzero curvature elastic continua (e.g., rings, cylindrical shells, spherical shells, conical shells, shells of revolution, etc.) using distributed piezoelectric transducers contribute an added level of sophistication and complication. In general, for 2-D zero-curvature continua, primary sensing/actuation depends on bending actions. Sensing signals usually depend on bending strains and distributed actuation also depends on counteracting control moments (Tzou, 1993). However, both membrane and bending actions are important for shell-type continua. Sensing signals are contributed by both membrane and bending strains, and both counteracting control moments and in-plane membrane control forces are essential for shell control. A theory of multilayered shell actuators was proposed and evaluated (Tzou and Gadre, 1989; Tzou, 1989a). A generic distributed piezoelectric identification and vibration control theory was proposed for a shell continuum with surface-coupled distributed sensor and actuator layers (Tzou, 1991). Hagood et al. (1990) proposed a constitutive dynamic modeling of piezoelectric actuator and control. Distributed active vibration control of a blade cylindrical shell was investigated (Tzou and Tseng, 1988). Qiu and Tani (1994) studied vibration control of a cylindrical shell using distributed piezoelectric sensors and actuators. Linear and nonlinear theories of piezoelectric or piezothermoelastic shells and shell laminates were proposed and applications to other geometries or materials demonstrated (Tzou, 1991, 1992, 1993; Tzou and Zhong, 1993; Tzou and Howard, 1994; Tzou, Bao, and Ye, 1994; Tzou and Bao, 1994, 1995a). Distributed orthogonal sensing (Tzou, Zhong, and Natori, 1993) and control (Tzou and Hollkamp, 1994) of ring shells and detailed membrane or bending contributions to sensing/control effectiveness were evaluated. Tzou and Bao (1995b) investigated dynamics and control of adaptive shells with curvature transformations. Frequency and damping variations during the transformation process were investigated. In this chapter, detailed electromechanics and sensing/control characteristics of distributed shell transducers are presented. Spatially distributed orthogonal sensing and control of elastic continua are studied. Self-sensing orthogonal actuators are also demonstrated.

7.2 System Definition: Shell with Distributed Transducers

A 2-D double-curvature, deep, flexible, and elastic shell continuum defined in a triorthogonal curvilinear coordinate system ($\alpha_1, \alpha_2, \alpha_3$) is used as the generic distributed (parameter) system discussed in this chapter. The generic shell is sandwiched between two flexible piezoelectric shell layers serving as distributed transducers (Figure 7.1). The top piezoelectric layer serves as a distributed sensor (based on the *direct piezoelectric effect*), and the bottom layer serves as a distributed actuator (the *converse piezoelectric effect*) (Tzou, 1993). With four geometric parameters – two Lamé parameters and two radii of in-plane axes – this generic shell can be reduced to a variety of common geometries, such as shells of revolution, spherical shells, conical shells, cylindrical shells, plates, arches, rings, beams, rods, etc. (Soedel, 1993; Tzou, 1993). Accordingly, derived theories and governing equations for the generic double-curvature shell can be simplified to apply to these common geometries and detailed analysis can be further carried out.

It is assumed that 1. piezoelectric layers are much thinner than the elastic shell and they are perfectly bonded on the elastic shell; 2. physical properties of the bonding material are negligible; 3. the laminated shell is still thin and the Kirchoff–Love thin shell hypothesis is still valid; and 4. the surface electrodes, with negligible physical properties, are perfectly conductive. Localization, discretization, and spatial shaping of the sensor/actuator can be achieved by step functions, Dirac delta functions, and/or other spatial shape functions.

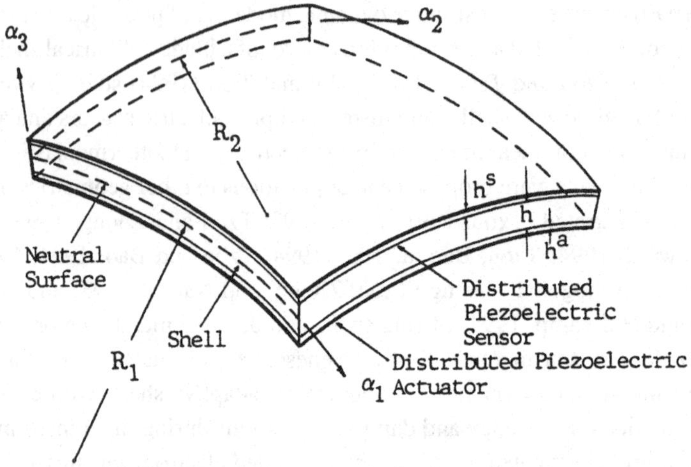

Fig. 7.1. A deep elastic shell with distributed piezoelectric transducers.

Orthogonal shaping of distributed transducers will be discussed later. (Note that the distributed piezoelectric layers can also be embedded in a composite shell laminate.) (Tzou and Bao, 1994, 1995a; Tzou and Gadre, 1989)

As mentioned previously, there are four geometric parameters (i.e., two Lamé parameters and two radii) required to simplify the generic shell to a large number of common geometries. Two radii can be observed from the coordinate system of the common geometry and the two Lamé parameters are determined by a fundamental form related to an infinitesimal distance ds on the neutral surface of the double-curvature shell (Figure 7.1),

$$(ds)^2 = A_1^2(d\alpha_1)^2 + A_2^2(d\alpha_2)^2, \tag{7.1}$$

where A_1 and A_2 are the Lamé parameters and $d\alpha_1$ and $d\alpha_2$ are the projected infinitesimal components in the α_1 and α_2 directions, respectively. (Simplifications of the shell theory to other common geometries using the four parameters will be demonstrated later.) Equations of motion of the elastic shell can be derived from either Hamilton's principle or simplification of the electromechanical equations of thin piezoelectric shells (Tzou, 1993):

$$\begin{aligned}&\left\{-\frac{\partial(N_{11}A_2)}{\partial\alpha_1} - \frac{\partial(N_{21}A_1)}{\partial\alpha_2} - N_{12}\frac{\partial A_1}{\partial\alpha_2} + N_{22}\frac{\partial A_2}{\partial\alpha_1}\right.\\ &\quad\left. - \frac{1}{R_1}\left(\frac{\partial(M_{11}A_2)}{\partial\alpha_1} + \frac{\partial(M_{21}A_1)}{\partial\alpha_2} + M_{12}\frac{\partial A_1}{\partial\alpha_2} - M_{22}\frac{\partial A_2}{\partial\alpha_1}\right)\right\}\\ &\quad + A_1A_2\rho h\ddot{u}_1 = A_1A_2F_1,\end{aligned} \tag{7.2}$$

$$\left\{-\frac{\partial(N_{12}A_2)}{\partial\alpha_1}-\frac{\partial(N_{22}A_1)}{\partial\alpha_2}-N_{21}\frac{\partial A_2}{\partial\alpha_1}+N_{11}\frac{\partial A_1}{\partial\alpha_2}\right.$$
$$\left.-\frac{1}{R_2}\left(\frac{\partial(M_{12}A_2)}{\partial\alpha_1}+\frac{\partial(M_{22}A_1)}{\partial\alpha_2}+M_{21}\frac{\partial A_2}{\partial\alpha_1}-M_{11}\frac{\partial A_1}{\partial\alpha_2}\right)\right\}$$
$$+A_1A_2\rho h\ddot{u}_2=A_1A_2F_2, \tag{7.3}$$

$$\left\{\left(-\frac{\partial}{\partial\alpha_1}\left[\frac{1}{A_1}\left(\frac{\partial(M_{11}A_2)}{\partial\alpha_1}+\frac{\partial(M_{21}A_1)}{\partial\alpha_2}+M_{12}\frac{\partial A_1}{\partial\alpha_2}-M_{22}\frac{\partial A_2}{\partial\alpha_1}\right)\right]\right.\right.$$
$$-\frac{\partial}{\partial\alpha_2}\left[\frac{1}{A_2}\left(\frac{\partial(M_{12}A_2)}{\partial\alpha_1}+\frac{\partial(M_{22}A_1)}{\partial\alpha_2}+M_{21}\frac{\partial A_2}{\partial\alpha_1}-M_{11}\frac{\partial A_1}{\partial\alpha_2}\right)\right]$$
$$\left.+A_1A_2\left(\frac{N_{11}}{R_1}+\frac{N_{22}}{R_2}\right)\right\}+A_1A_2\rho h\ddot{u}_3=A_1A_2F_3, \tag{7.4}$$

where $\ddot{u}_i$ is the acceleration in the ith direction, ρ is the mass density, h is the shell thickness, F_i is the externally applied mechanical load, and R_1 and R_2 are the curvature radii of the α_1 and α_2 axes, respectively. All terms inside the brace are the resultant forces N_{ij} and moments M_{ij} contributed by induced strains; these can be simply represented by a Love's operator $L_i(u_1, u_2, u_3)$, where u_i are displacements. (This Love's operator will be used in Section 7.4.6.) Note that generic damping terms (e.g., viscous damping proportional to velocities $\dot{u}_i$) can be added to the equations or introduced via the force terms.

7.2.1 Membrane Forces and Moments

Resultant forces N_{ij} and moments M_{ij} are defined by generic displacements u_1, u_2, and u_3 on the neutral surface. Note that the subscript ij denotes a component defined on the ith surface and in the jth direction.

Membrane Forces The Membrane Forces are

$$N_{11}=K\left[\left(\frac{1}{A_1}\frac{\partial u_1}{\partial\alpha_1}+\frac{u_2}{A_1A_2}\frac{\partial A_1}{\partial\alpha_2}+\frac{u_3}{R_1}\right)+\mu\left(\frac{1}{A_2}\frac{\partial u_2}{\partial\alpha_2}+\frac{u_1}{A_1A_2}\frac{\partial A_2}{\partial\alpha_1}+\frac{u_3}{R_2}\right)\right], \tag{7.5}$$

$$N_{22}=K\left[\left(\frac{1}{A_2}\frac{\partial u_2}{\partial\alpha_2}+\frac{u_1}{A_1A_2}\frac{\partial A_2}{\partial\alpha_1}+\frac{u_3}{R_2}\right)+\mu\left(\frac{1}{A_1}\frac{\partial u_1}{\partial\alpha_1}+\frac{u_2}{A_1A_2}\frac{\partial A_1}{\partial\alpha_2}+\frac{u_3}{R_1}\right)\right], \tag{7.6}$$

$$N_{12}=\frac{K(1-\mu)}{2}\left[\frac{A_2}{A_1}\frac{\partial}{\partial\alpha_1}\left(\frac{u_2}{A_2}\right)+\frac{A_1}{A_2}\frac{\partial}{\partial\alpha_2}\left(\frac{u_1}{A_1}\right)\right], \tag{7.7}$$

where μ is Poisson's ratio and K is the membrane stiffness, given by $K = [Yh/(1-\mu^2)]$, where Y is Young's modulus of the shell.

Bending Moments The bending moments are

$$M_{11} = D\left\{\left[\frac{1}{A_1}\frac{\partial}{\partial\alpha_1}\left(\frac{u_1}{R_1} - \frac{1}{A_1}\frac{\partial u_3}{\partial\alpha_1}\right) + \frac{1}{A_1A_2}\left(\frac{u_2}{R_2} - \frac{1}{A_2}\frac{\partial u_3}{\partial\alpha_2}\right)\frac{\partial A_1}{\partial\alpha_2}\right]\right.$$
$$\left. +\mu\left[\frac{1}{A_2}\frac{\partial}{\partial\alpha_2}\left(\frac{u_2}{R_2} - \frac{1}{A_2}\frac{\partial u_3}{\partial\alpha_2}\right) + \frac{1}{A_1A_2}\left(\frac{u_1}{R_1} - \frac{1}{A_1}\frac{\partial u_3}{\partial\alpha_1}\right)\frac{\partial A_2}{\partial\alpha_1}\right]\right\}, \tag{7.8}$$

$$M_{22} = D\left\{\left[\frac{1}{A_2}\frac{\partial}{\partial\alpha_2}\left(\frac{u_2}{R_2} - \frac{1}{A_2}\frac{\partial u_3}{\partial\alpha_2}\right) + \frac{1}{A_1A_2}\left(\frac{u_1}{R_1} - \frac{1}{A_1}\frac{\partial u_3}{\partial\alpha_1}\right)\frac{\partial A_2}{\partial\alpha_1}\right]\right.$$
$$\left. +\mu\left[\frac{1}{A_1}\frac{\partial}{\partial\alpha_1}\left(\frac{u_1}{R_1} - \frac{1}{A_1}\frac{\partial u_3}{\partial\alpha_1}\right) + \frac{1}{A_1A_2}\left(\frac{u_2}{R_2} - \frac{1}{A_2}\frac{\partial u_3}{\partial\alpha_2}\right)\frac{\partial A_1}{\partial\alpha_1}\right]\right\}, \tag{7.9}$$

$$M_{12} = \frac{D(1-\mu)}{2}\left\{\frac{A_2}{A_1}\frac{\partial}{\partial\alpha_1}\left[\frac{1}{A_2}\left(\frac{u_2}{R_2} - \frac{1}{A_2}\frac{\partial u_3}{\partial\alpha_2}\right)\right]\right.$$
$$\left. + \frac{A_1}{A_2}\frac{\partial}{\partial\alpha_2}\left[\frac{1}{A_1}\left(\frac{u_1}{R_1} - \frac{1}{A_1}\frac{\partial u_3}{\partial\alpha_1}\right)\right]\right\}, \tag{7.10}$$

where D is the bending stiffness and $D = (Yh^3)/[12(1-\mu^2)]$. (Note that the resultant forces and moments could also include piezoelectricity induced feedback forces and moments, which are to be derived later.)

7.2.2 Modal Expansion Method

Dynamic response of a shell continuum can be determined by the *modal expansion method* in which the total response is a sum of all participating modes U_{ik} weighted with their corresponding modal participation factors η_ks:

$$u_i(\alpha_1, \alpha_2, t) = \sum_{k=1}^{\infty} \eta_k(t) U_{ik}(\alpha_1, \alpha_2), \qquad i = 1, 2, 3, \tag{7.11}$$

where i denotes the α_i direction and $U_{ik}(\alpha_1, \alpha_2)$ is the mode shape function. For a distributed system, the number of modes is infinite (i.e., $k = 1, 2, 3, \ldots, \infty$). This modal expression concept will be used to estimate the modal contribution of distributed piezoelectric sensors/actuators, distributed control, and also orthogonal transducers presented in later sections.

Considering viscous damping, substituting the modal expansion expression into the shell equations, and imposing the modal orthogonality, one can derive a second-order ordinary differential equation – the modal equation – in the kth modal participation factor (Soedel, 1993):

$$\ddot{\eta}_k + \frac{c}{\rho h}\dot{\eta}_k + \omega_k^2 \eta_k = F_k, \tag{7.12}$$

where c is the damping constant, ω_k is the kth natural frequency, and F_k is the modal excitation force defined by

$$F_k = \frac{1}{\rho h N_k} \int_{\alpha_1} \int_{\alpha_2} \left(\sum_{j=1}^{3} F_j(\alpha_1, \alpha_2, t) U_{jk}(\alpha_1, \alpha_2) \right) A_1 A_2 \, d\alpha_1 \, d\alpha_2, \tag{7.13}$$

and

$$N_k = \int_{\alpha_1} \int_{\alpha_2} \left(\sum_{j=1}^{3} U_{jk}^2(\alpha_1, \alpha_2) \right) A_1 A_2 \, d\alpha_1 \, d\alpha_2. \tag{7.14}$$

Note that $F_j(\alpha_1, \alpha_2, t)$ can be either distributed mechanical excitation or control force, depending on the system setup (Tzou, 1993). This modal equation will be used in distributed vibration control in later analysis.

7.3 Sensor Electromechanics

In general, piezoelectric layers respond to strain variations and generate electric signals corresponding to dynamic states of the shell – the *direct piezoelectric effect*. Electromechanics of a generic spatially distributed shell convolving sensor are discussed in this section, and the shell sensor theory can be simplified to a broad class of distributed sensors applied to common geometries based on four geometric parameters (i.e., two Lamé parameters and two radii of in-plane coordinate axes).

In this section, the electrostatic equation of a thin piezoelectric shell continuum is discussed first. Then, the electrostatic equation is extended to apply to distributed shell sensors. Detailed electromechanics of spatially distributed shell sensors are derived, followed by spatial shaping and orthogonal modal sensing, that is, piezoelectric layers are shaped and convolved – change of polarity. Practical applications of the theory are demonstrated in case studies presented in later sections.

7.3.1 Thin Distributed Piezoelectric Shell Continuum

A charge equation of electrostatics, in the transverse direction, of a piezoelectric thin shell is (Tzou, Zhong, and Natori, 1993)

$$\frac{\partial}{\partial \alpha_3}[e_{31}S_{11} + e_{32}S_{22} + \epsilon_{33}E_3]A_1 A_2 \left(1 + \frac{\alpha_3}{R_1}\right)\left(1 + \frac{\alpha_3}{R_2}\right) = 0, \tag{7.15}$$

where e_{31} and e_{32} are the piezoelectric strain constants (for a symmetrical hexagonal piezoelectric material, $e_{31} = e_{32}$); S_{ii} is the normal strain in the ith direction; ϵ_{33} is the dielectric constant; E_3 is the transverse electric field; and α_3/R_i denotes the curvature effect, where α_3 is the transverse distance measured from the neutral surface. It is assumed that the piezoelectric layer is insensitive to in-plane shear deformations and that the material is symmetrical hexagonal. Note that since $\alpha_3 \ll R_i$, the curvature effect can be neglected:

$$\left(1 + \frac{\alpha_3}{R_i}\right) \cong 1. \tag{7.16}$$

In sensor applications, it is assumed that there is no externally applied electric boundary condition in an open-circuit condition. Thus,

$$e_{31} S_{11} + e_{32} S_{22} + \epsilon_{33} E_3 = 0. \tag{7.17}$$

It is also assumed that the effective axes of the piezoelectric layer are aligned with the principal axes of the shell (i.e., the skew angle is zero). Integrating Eq. (7.17) over the piezoelectric layer thickness yields

$$\int_{\alpha_3} (e_{31} S_{11} + e_{32} S_{22})\, d\alpha_3 + \epsilon_{33} \int_{\alpha_3} E_3\, d\alpha_3 = 0. \tag{7.18}$$

Integrating the electric field E_3 gives the electric potential, i.e.,$\int_{\alpha_3} E_3\, d\alpha_3 = \phi_3$. ϕ_3 is the electric potential difference in an open-circuit condition. Thus,

$$\phi_3 = -\frac{1}{\epsilon_{33}} \int_{\alpha_3} (e_{31} S_{11} + e_{32} S_{22})\, d\alpha_3. \tag{7.19}$$

This potential equation of a generic piezoelectric shell continuum will be used in sensor signal calculations presented next.

7.3.2 Spatially Distributed Shell Sensor

For a spatially *distributed* piezoelectric shell sensor continuum with an effective surface electrode area $\mathbf{S}^{\mathrm{e}}$, the total signal output ϕ_3^{s} is

$$\begin{aligned}
&\int_{\mathbf{S}^{\mathrm{e}}} \phi_3^{\mathrm{s}} A_1 A_2 \left(1 + \frac{\alpha_3}{R_1}\right)\left(1 + \frac{\alpha_3}{R_2}\right) d\alpha_1\, d\alpha_2 \\
&\quad = -\frac{1}{\epsilon_{33}} \int_{\mathbf{S}^{\mathrm{e}}} \int_{\alpha_3} (e_{31} S_{11} + e_{32} S_{22}) A_1 A_2 \left(1 + \frac{\alpha_3}{R_1}\right)\left(1 + \frac{\alpha_3}{R_2}\right) d\alpha_1\, d\alpha_2\, d\alpha_3.
\end{aligned} \tag{7.20}$$

Note that $\alpha_3/R_1 \ll 1$ and $\alpha_3/R_2 \ll 1$. Thus, taking the surface average over the entire electrode area $\mathbf{S}^{\mathrm{e}}$, neglecting the curvature effect, and rearranging the equation, one can derive

$$\phi_3^s = -\frac{\int_{S^e}\int_{\alpha_3}(e_{31}S_{11} + e_{32}S_{22})A_1A_2\,d\alpha_1\,d\alpha_2\,d\alpha_3}{\epsilon_{33}S^e}. \tag{7.21}$$

Note that $\int_{S^e} A_1A_2(1+\frac{\alpha_3}{R})(1+\frac{\alpha_3}{R})\,d\alpha_1\,d\alpha_2 = S^e$ – the effective electrode area. For a thin elastic shell continuum, (mechanical) normal strains can be further divided into two strain components: membrane strains S_{ii}^0 and bending strains k_{ii}:

$$S_{11} = S_{11}^0 + \alpha_3 k_{11} \quad \text{and} \quad S_{22} = S_{22}^0 + \alpha_3 k_{22}. \tag{7.22, 7.23}$$

For distributed sensors made of symmetrical hexagonal piezoelectric materials, the piezoelectric constants $e_{31} = e_{32}$. Thus, the sensor output signal becomes

$$\phi_3^s = -\frac{1}{\epsilon_{33}S^e}\int_{S^e}\int_{\alpha_3}^{h^s}\left[e_{31}\left((s_{11}^0 + S_{22}^0) + \alpha_3(k_{11} + k_{22})\right)\right]A_1A_2\,d\alpha_1\,d\alpha_2\,d\alpha_3. \tag{7.24}$$

For a distributed piezoelectric shell sensor layer (laminated or embedded), the thickness integration is with respect to the thickness h^s of the piezoelectric sensor layer, which can be either constant (uniform thickness) or variable (nonuniform thickness). (This issue will be discussed next.) Note that the overall signal output of the shell sensor is contributed by both membrane and bending strains experienced in the piezoelectric sensor layer. The membrane strains S_{ii}^0 and bending strains k_{ii} can be further expressed as a function of three neutral-surface displacements, u_1, u_2, and u_3, in the three axial directions (Tzou and Zhong, 1993):

$$S_{11}^0 = \frac{1}{A_1}\frac{\partial u_1}{\partial \alpha_1} + \frac{u_2}{A_1A_2}\frac{\partial A_1}{\partial \alpha_2} + \frac{u_3}{R_1}, \tag{7.25}$$

$$S_{22}^0 = \frac{1}{A_2}\frac{\partial u_2}{\partial \alpha_2} + \frac{u_1}{A_1A_2}\frac{\partial A_2}{\partial \alpha_1} + \frac{u_3}{R_2}, \tag{7.26}$$

and

$$k_{11} = \frac{1}{A_1}\frac{\partial}{\partial \alpha_1}\left(\frac{u_1}{R_1} - \frac{1}{A_1}\frac{\partial u_3}{\partial \alpha_1}\right) + \frac{1}{A_1A_2}\left(\frac{u_2}{R_2} - \frac{1}{A_2}\frac{\partial u_3}{\partial \alpha_2}\right)\frac{\partial A_1}{\partial \alpha_2}, \tag{7.27}$$

$$k_{22} = \frac{1}{A_2}\frac{\partial}{\partial \alpha_2}\left(\frac{u_2}{R_2} - \frac{1}{A_2}\frac{\partial u_3}{\partial \alpha_2}\right) + \frac{1}{A_1A_2}\left(\frac{u_1}{R_1} - \frac{1}{A_1}\frac{\partial u_3}{\partial \alpha_1}\right)\frac{\partial A_2}{\partial \alpha_1}. \tag{7.28}$$

Note that the Lamé parameters, the A_is, and radii of curvatures, the R_is, are geometry dependent, e.g., $A_1 = A_2 = 1$ and $R_1 = R_2 = \infty$ for a rectangular plate; $A_1 = 1$, $A_2 = R$, $R_1 = \infty$, and $R_2 = R$ for a cylindrical shell, etc. Thus the sensor equation can be further simplified based on these four parameters defined for the geometries and accordingly the shell sensor equation can be applied to a large number of common shell and nonshell continua in practical applications.

7.3.3 Spatially Distributed Piezoelectric Convolving Sensor

Observation spillover can introduce unstable dynamic responses in undamped structural systems (Meirovitch and Baruh, 1983). Thus it is highly desirable that sensors only monitor those modes needed to be controlled such that the observation spillover is prevented. In reality, however, sensors not only respond to controlled modes but also to those uncontrolled residual modes. There are several techniques to reduce the observation spillover. Conventional practice is to place sensors – spatially *discrete* sensors – at modal nodes or nodal lines of the residual modes. The difficulty with this arrangement is that it is very difficult, if not impossible, to avoid all uncontrolled residual modes. Another common approach is to prefilter the sensor data using comb filters with phase-locked loops (Gustafson and Speyer, 1976) which reduces the observation spillover in the frequency domain. This technique requires that 1. the controlled modal frequencies are precisely known, 2. there is a reasonable separation from the nearby residual modes, and 3. the signal-to-noise ratio is sufficiently high. The other method is to use spatially distributed *modal sensors* which respond only to a structural mode or a group of modes.

Lee and Moon (1990) studied distributed piezoelectric modal sensors for a flexible beam and a one-dimensional plate. Collins et al. (1994) proposed spatially shaped distributed piezoelectric sensors using sinc functions for monitoring beam oscillations. Busch-Vishniac (1990) evaluated spatially distributed transducers with sensor and actuator applications. Tzou and coworkers also derived a generic distributed orthogonal sensor/actuator theory and applied it to beams and ring shells (Tzou, Zhong, Natori, 1993; Tzou, Zhong, Hollkamp, 1994; Tzou and Hollkamp, 1994). In this section, detailed electromechanics of spatially distributed convolving shell sensors for orthogonal modal sensing/control are presented.

Distributed piezoelectric shell layers can be either embedded or surface bonded with a flexible elastic shell, and these layers are used as distributed sensors. For a spatially distributed piezoelectric shell convolving sensor, a weighting function $W(\alpha_1, \alpha_2)$ and a polarity function $\mathrm{sgn}[U_3(\alpha_1, \alpha_2)]$ can be added to the generic shell sensor equation. Note that $\mathrm{sgn}(\cdot)$ denotes a signum function used to change the piezoelectric polarity, in which $\mathrm{sgn}(\cdot) = 1$ when $(\cdot) > 0$,

0 when $(\cdot) = 0$, and -1 when $(\cdot) < 0$; $U_3(\alpha_1, \alpha_2)$ denotes a transverse modal function, mode shape function, or any orthogonal function. In later derivations, two weighting functions are discussed. First, in *thickness shaping* it is assumed that the thickness of the piezoelectric shell is a spatial function (i.e., $h^s = W_t(\alpha_1, \alpha_2)$). Second, in *surface shaping* the sensor thickness is constant and the electrode shape is defined as $\mathscr{S}(\alpha_1, \alpha_2) = W_s(\alpha_1, \alpha_2)$. Each of the two shapings are respectively discussed.

7.3.3.1 Spatial Thickness Shaping

In the first case, the sensor thickness varies over the effective sensor area. Figure 7.2 illustrates the thickness shaping of distributed shell sensors.

The piezoelectric shell sensor thickness is a spatial function $W_t(\alpha_1, \alpha_2)$. Thus, the sensor equation becomes

$$\int_{\mathbf{S}^e} \phi_3^s A_1 A_2 \left(1 + \frac{\alpha_3}{R_1}\right)\left(1 + \frac{\alpha_3}{R_2}\right) d\alpha_1\, d\alpha_2$$
$$= -\frac{e_{31}}{\epsilon_{33}} \int_{\alpha_1}\int_{\alpha_2} \operatorname{sgn}[U_3(\alpha_1, \alpha_2)] \cdot \Bigg[W_t(\alpha_1, \alpha_2)\left(S_{11}^0 + S_{22}^0\right)$$
$$+ \int_{r_1}^{W_t(\alpha_1,\alpha_2)+r_1} \alpha_3 (k_{11} + k_{22})\, d\alpha_3 \Bigg] \left(1 + \frac{\alpha_3}{R_1}\right)\left(1 + \frac{\alpha_3}{R_2}\right) A_1 A_2\, d\alpha_1\, d\alpha_2, \tag{7.29}$$

where $\mathbf{S}^e$ denotes the effective sensor area or electrode area and r_1 is the distance measured from the shell neutral surface to the bottom of the piezoelectric

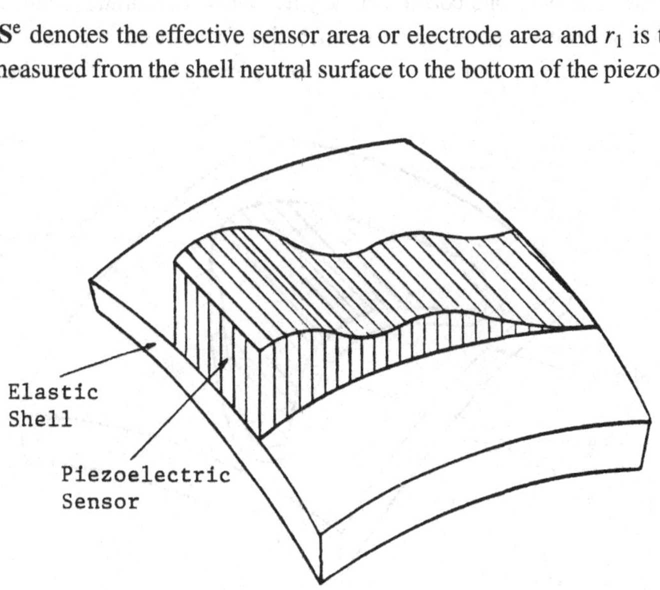

Fig. 7.2. Spatial thickness shaping of distributed shell sensors.

sensor layer. Note that the first term inside the square bracket is contributed by the membrane strains (*membrane sensitivity*) and the second term by the bending strains (*bending* or *transverse sensitivity*). The total output signal is contributed by a summation of membrane and bending effects. Recall that the output voltage is measured in an open-circuit condition. In addition $R_i \gg \alpha_3$, and thus $(1 + \alpha_3/R_i) \cong 1$.

7.3.3.2 Spatial Surface Shaping

In this case, the sensor thickness is constant. Only the sensor shape or effective electrode changes. Figure 7.3 shows the spatial distribution of a shell sensor.

The electrode shape (or sensor shape) can be designed using a spatial (shape) weighting function $W_s(\alpha_1, \alpha_2)$ and the sensor signal can be calculated by

$$\begin{aligned}
&\int_{\alpha_1}\int_{\alpha_2} \phi_3^s \left(1 + \frac{\alpha_3}{R_1}\right)\left(1 + \frac{\alpha_3}{R_2}\right) A_1 A_2 \, d\alpha_1 \, d\alpha_2 \\
&\quad = -\frac{e_{31}}{\epsilon_{33}} \int_{\alpha_1}\int_{\alpha_2} W_s(\alpha_1, \alpha_2) \cdot \text{sgn}[U_3(\alpha_1, \alpha_2)] \int_{\alpha_3} (S_{11} + S_{22}) \\
&\quad \cdot \left(1 + \frac{\alpha_3}{R_1}\right)\left(1 + \frac{\alpha_3}{R_2}\right) A_1 \, A_2 \, d\alpha_1 \, d\alpha_2 \, d\alpha_3. \qquad (7.30)
\end{aligned}$$

Again, the curvature effects can be neglected (i.e., $(1 + \alpha_3/R_i) \cong 1$). For a one-dimensional continuum (e.g., beam, ring, arch, etc.) the mechanical strains are functions of only one coordinate, say α_1. Thus, membrane strains S_{11}^0, S_{22}^0 and bending strains k_{11}, k_{22} can be described as functions of the coordinate α_1, that is, $W_s(\alpha_1, \alpha_2) \equiv W_s(\alpha_1)$. The averaged sensor signal with its membrane

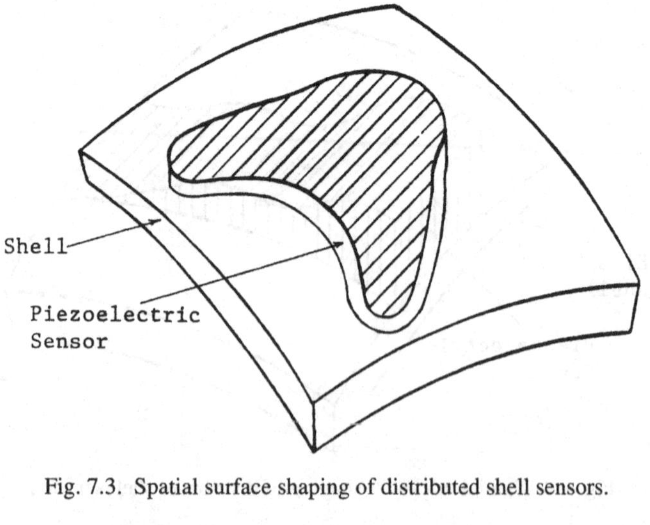

Fig. 7.3. Spatial surface shaping of distributed shell sensors.

and bending contributions is

$$\phi_3^s = -\frac{e_{31}}{\epsilon_{33}\mathbf{S}^e}\int_{\alpha_1} \operatorname{sgn}[U_3(\alpha_1)]\cdot\left(h^s\left(S_{11}^0 + S_{22}^0\right)\right.$$
$$\left.+\frac{1}{2}\left(r_2^2 - r_1^2\right)(k_{11}+k_{22})\right)A_1A_2\int_0^{W_s(\alpha_1)} d\alpha_2\, d\alpha_1$$
$$= -\frac{e_{31}}{\epsilon_{33}\mathbf{S}^e}\int_{\alpha_1} W_s(\alpha_1)\cdot\operatorname{sgn}[U_3(\alpha_1)]\cdot\left(h^s\left(S_{11}^0 + S_{22}^0\right)\right.$$
$$\left.+\frac{1}{2}h^s(h+h^s)(k_{11}+k_{22})\right)A_1A_2\, d\alpha_1, \tag{7.31}$$

where h^s is the thickness of the sensor layer and r_2 is the distance measured from the neutral surface of the shell to the top surface of the distributed piezoelectric sensor layer. Note that $r_2 - r_1 = h^s$ and $(r_2^2 - r_1^2) = h^s(h+h^s)$. Since the surface (or electrode) shaping is easier to achieve in practice, only surface shaping is adopted in later applications.

7.3.4 Modal Contribution

Recall that the uniformly distributed piezoelectric sensor layer is much thinner than the elastic shell continuum. Thus the piezoelectric sensor strains are assumed constant and approximately equal to the outer surface strains of the elastic shell. The sensor equation (without shaping) becomes

$$\phi^s(\alpha_1,\alpha_2)$$
$$= -\frac{e_{31}h^s}{\epsilon_{33}\mathbf{S}^e}\int_{\mathbf{S}^e}\left(e_{31}\left\{\left(\frac{1}{A_1}\frac{\partial u_1}{\partial\alpha_1} + \frac{u_2}{A_1A_2}\frac{\partial A_1}{\partial\alpha_2} + \frac{u_3}{R_1}\right)\right.\right.$$
$$\left.+ r_1^s\left[\frac{1}{A_1}\frac{\partial}{\partial\alpha_1}\left(\frac{u_1}{R_1} - \frac{1}{A_1}\frac{\partial u_3}{\partial\alpha_1}\right) + \frac{1}{A_1A_2}\frac{\partial A_1}{\partial\alpha_2}\left(\frac{u_2}{R_2} - \frac{1}{A_2}\frac{\partial u_3}{\partial\alpha_2}\right)\right]\right\}$$
$$+ e_{32}\left\{\left(\frac{1}{A_2}\frac{\partial u_2}{\partial\alpha_2} + \frac{u_1}{A_1A_2}\frac{\partial A_2}{\partial\alpha_1} + \frac{u_3}{R_2}\right) + r_2^s\left[\frac{1}{A_2}\frac{\partial}{\partial\alpha_2}\left(\frac{u_2}{R_2} - \frac{1}{A_2}\frac{\partial u_3}{\partial\alpha_2}\right)\right.\right.$$
$$\left.\left.\left. + \frac{1}{A_1A_2}\frac{\partial A_2}{\partial\alpha_1}\left(\frac{u_1}{R_1} - \frac{1}{A_1}\frac{\partial u_3}{\partial\alpha_1}\right)\right]\right\}\right] d\mathbf{S}^e, \tag{7.32}$$

where $r_2^s = r_1^s = 0.5(h+h^s) \cong 0.5\,h$, if $h \gg h^s$, for a uniform thickness sensor layer. Note that this equation is valid for both surface-coupled and internally embedded distributed piezoelectric sensors. The total signal is contributed by two strain components, membrane and bending, as discussed previously. Thus, this sensor equation can be used in 1. membrane sensor applications, such as in-plane contraction and expansion, and 2. bending sensor applications, such as bending vibrations.

To examine the modal contribution to the sensor output, one can further substitute the modal expansion expression, Eq. (7.11), into the expanded sensor equation, Eq. (7.32):

$$\begin{aligned}
\phi^s(\alpha_1,\alpha_2) = {} & -\frac{e_{31}h^s}{\epsilon_{33}S^e}\int_S \left[e_{31}\left\{ \left[\frac{1}{A_1}\frac{\partial}{\partial\alpha_1}\sum_{k=1}^{\infty}\eta_k(t)U_{1k}(\alpha_1,\alpha_2) \right.\right.\right. \\
& \left. + \frac{1}{A_1A_2}\sum_{k=1}^{\infty}\eta_k(t)U_{2k}(\alpha_1,\alpha_2)\frac{\partial A_1}{\partial\alpha_2} + \frac{1}{R_1}\sum_{k=1}^{\infty}\eta_k(t)U_{3k}(\alpha_1,\alpha_2) \right] \\
& + r_1^s\left[\frac{1}{A_1}\frac{\partial}{\partial\alpha_1}\left(\frac{1}{R_1}\sum_{k=1}^{\infty}\eta_k(t)U_{1k}(\alpha_1,\alpha_2) - \frac{1}{A_1}\frac{\partial}{\partial\alpha_1} \right.\right. \\
& \left. \times \sum_{k=1}^{\infty}\eta_k(t)U_{3k}(\alpha_1,\alpha_2) \right) + \frac{1}{A_1A_2}\frac{\partial A_1}{\partial\alpha_2}\left(\frac{1}{R_2}\sum_{k=1}^{\infty}\eta_k(t)U_{2k}(\alpha_1,\alpha_2) \right. \\
& \left.\left.\left. - \frac{1}{A_2}\frac{\partial}{\partial\alpha_2}\sum_{k=1}^{\infty}\eta_k(t)U_{3k}(\alpha_1,\alpha_2) \right) \right] \right\} + e_{32}\left\{ \left[\frac{1}{A_2}\frac{\partial}{\partial\alpha_2} \right.\right. \\
& \times \sum_{k=1}^{\infty}\eta_k(t)U_{2k}(\alpha_1,\alpha_2) + \frac{1}{A_1A_2}\sum_{k=1}^{\infty}\eta_k(t)U_{1k}(\alpha_1,\alpha_2)\frac{\partial A_2}{\partial\alpha_1} \\
& \left. + \frac{1}{R_2}\sum_{k=1}^{\infty}\eta_k(t)U_{3k}(\alpha_1,\alpha_2) \right] + r_2^s\left[\frac{1}{A_2}\frac{\partial}{\partial\alpha_2} \right. \\
& \times \left(\frac{1}{R_2}\sum_{k=1}^{\infty}\eta_k(t)U_{2k}(\alpha_1,\alpha_2) - \frac{1}{A_2}\frac{\partial}{\partial\alpha_2}\sum_{k=1}^{\infty}\eta_k(t)U_{3k}(\alpha_1,\alpha_2) \right) \\
& + \frac{1}{A_1A_2}\frac{\partial A_2}{\partial\alpha_1}\left(\frac{1}{R_1}\sum_{k=1}^{\infty}\eta_k(t)U_{1k}(\alpha_1,\alpha_2) \right. \\
& \left.\left.\left.\left. - \frac{1}{A_1}\frac{\partial}{\partial\alpha_1}\sum_{k=1}^{\infty}\eta_k(t)U_{3k}(\alpha_1,\alpha_2) \right) \right] \right\} \right] A_1A_2\,d\alpha_1\,d\alpha_2. \qquad (7.33)
\end{aligned}$$

Eq. (7.33) indicates that the total output of the distributed sensor is a summation of all participating modes and the specific contribution of each mode is determined by the modal participation factor. The modal participation factors η_ks are zeros for those modes not participating in the shell oscillation. Detailed electromechanics of *distributed shell modal sensors* (*shaped* and/or *convolved*) will be presented later. Observation spillover problems (Meirovitch and Baruh, 1983; Balas, 1978) and solutions of distributed sensors are also discussed.

7.3.5 Discrete Voltage and Modal Voltage

It is desirable to identify the mode shapes of a elastic shell continuum via the outputs of distributed/discrete piezoelectric sensor electrodes. Since mode shapes and modal strain distributions are mostly distinct for natural modes, the induced modal potential (spatial) distributions – *modal voltages* – should also be unique. To construct a modal voltage contour of a natural mode, one needs to calculate a number of discrete point signals corresponding to modal strains and graphically connect the discrete signal magnitudes together for the natural mode. The *potential map* or *contour* represents the modal voltage of the natural mode (Tzou and Tseng, 1990, 1991; Tzou, 1993). Note that a finite separation of surface electrode is required to prevent the global voltage average since electric charges are freely moving on the electrode.

It is assumed that there is a finite electrode centering around a discrete location (α_1^*, α_2^*) where the asterisk denotes given or known coordinate. If the surface average and integration is removed from the distributed sensor equation, a discrete voltage amplitude can be estimated at the location (α_1^*, α_2^*). In this case, dynamic states (i.e., a *potential map* or *contour*), of the whole shell continuum can be defined. A discrete voltage amplitude $\phi_3^s(\alpha_1^*, \alpha_2^*)$ at a known location (α_1^*, α_2^*) can be defined as

$$\begin{aligned}\phi_3^s = -\frac{h^s}{\epsilon_{33}}\left[e_{31}\left(\left[S_{11}^0(\alpha_1^*, \alpha_2^*) + S_{22}^0(\alpha_1^*, \alpha_2^*)\right]\right.\right.\\ \left.\left. + \alpha_3\left[k_{11}(\alpha_1^*, \alpha_2^*) + k_{22}(\alpha_1^*, \alpha_2^*)\right]\right)\right],\end{aligned} \tag{7.34}$$

where α_3 denotes the sensor placement normal to the neutral surface and α_3 equals r_i if it is surface bonded. This equation indicates that a local (*discrete*) output signal is a function of local strains. The local output signal $\phi_k^s(\alpha_1^*, \alpha_2^*)$ of the kth natural mode can be expressed as a function of natural modes:

$$\begin{aligned}\phi_k^s(\alpha_1^*, \alpha_2^*) = -\frac{h^s}{\epsilon_{33}}\Bigg\{e_{31}\Bigg\{\Bigg[\frac{1}{A_1}\frac{\partial}{\partial \alpha_1}\eta_k(t)U_{1k}(\alpha_1^*, \alpha_2^*) + \frac{1}{A_1 A_2}\eta_k(t)\\ \cdot U_{2k}(\alpha_1^*, \alpha_2^*)\frac{\partial A_1}{\partial \alpha_2} + \frac{1}{R_1}\eta_k(t)U_{3k}(\alpha_1^*, \alpha_2^*)\Bigg]\\ + r_1^s\Bigg[\frac{1}{A_1}\frac{\partial}{\partial \alpha_1}\left(\frac{1}{R_1}\eta_k(t)U_{1k}(\alpha_1^*, \alpha_2^*) - \frac{1}{A_1}\frac{\partial}{\partial \alpha_1}\eta_k(t)\right.\\ \left.\cdot U_{3k}(\alpha_1^*, \alpha_2^*)\right) + \frac{1}{A_1 A_2}\frac{\partial A_1}{\partial \alpha_2}\left(\frac{1}{R_2}\eta_k(t)U_{2k}(\alpha_1^*, \alpha_2^*)\right.\\ \left.- \frac{1}{A_2}\frac{\partial}{\partial \alpha_2}\eta_k(t)U_{3k}(\alpha_1^*, \alpha_2^*)\right)\Bigg]\Bigg\}\end{aligned}$$

$$+e_{32}\left\{\left[\frac{1}{A_2}\frac{\partial}{\partial\alpha_2}\eta_k(t)U_{2k}(\alpha_1^*,\alpha_2^*)+\frac{1}{A_1A_2}\eta_k(t)\right.\right.$$

$$\left.\cdot U_{1k}(\alpha_1^*,\alpha_2^*)\frac{\partial A_2}{\partial\alpha_1}+\frac{1}{R_2}\eta_k(t)U_{3k}(\alpha_1^*,\alpha_2^*)\right]$$

$$+r_2^s\left[\frac{1}{A_2}\frac{\partial}{\partial\alpha_2}\left(\frac{1}{R_2}\eta_k(t)U_{2k}(\alpha_1^*,\alpha_2^*)-\frac{1}{A_2}\frac{\partial}{\partial\alpha_2}\eta_k(t)\right.\right.$$

$$\left.\cdot U_{3k}(\alpha_1^*,\alpha_2^*)\right)+\frac{1}{A_1A_2}\frac{\partial A_2}{\partial\alpha_1}\left(\frac{1}{R_1}\eta_k(t)U_{1k}(\alpha_1^*,\alpha_2^*)\right.$$

$$\left.\left.\left.-\frac{1}{A_1}\frac{\partial}{\partial\alpha_1}\eta_k(t)U_{3k}(\alpha_1^*,\alpha_2^*)\right)\right]\right\}. \tag{7.35}$$

In summary, the output signal of a distributed piezoelectric sensor is contributed by strains that can be further classified into two components: 1. the in-plane membrane strains and 2. the out-of-plane bending strains. Accordingly, two sensor sensitivities: 1. a *bending* or *transverse sensitivity* and 2. a *membrane sensitivity* can be defined. Furthermore, if these sensitivities are defined for natural modes, the *bending modal sensitivity* and the *membrane modal sensitivity* can be defined for each natural mode. In general, the bending modal sensitivity is defined for the transverse bending natural modes and the membrane modal sensitivity for the in-plane membrane natural modes. (Detailed contribution of each component will be evaluated in case studies.) Besides, depending on the placement of the sensor, the contribution of each strain component could be different. For example, only the bending strain contributes to the output signal if the shell experiences only bending oscillations. On the other hand, the membrane strain contributes to the output if the shell only experiences membrane oscillations.

Note that for a single layer piezoelectric shell continuum, the bending modal sensitivity is zero because the neutral surface of the shell is located in the middle of the shell thickness. The thickness integration from $-h/2$ to $+h/2$ gives a zero output in the bending sensitivity. Thus a piezoelectric shell continuum or a distributed piezoelectric shell layer on the neutral surface of a structure responds to only in-plane contractions and expansions, and consequently it can be used only as a *membrane sensor*. (The membrane sensitivity can exist if the membrane strains are not zero.) Note that the generic piezoelectric shell is fully covered with conducting electrodes on the top and bottom surfaces. These surface electrodes can be spatially shaped and the electric polarization can also be altered in order to design generic distributed orthogonal convolving shell sensors.

7.4 Distributed Actuation and Vibration Control

Recall that the bottom piezoelectric layer laminated on the elastic shell continuum is designated as a distributed actuator for active vibration suppression and control. It is assumed that the distributed piezoelectric layer is made of a biaxially polarized piezoelectric material. Thus, a voltage ϕ^a applied to the distributed actuator layer introduces two in-plane normal strains (α_1 and α_2 directions) on the actuator layer due to the *converse piezoelectric effect*. It is assumed that the piezoelectric actuator layer is unconstrained and is free from external in-plane normal forces. Thus, the stress effects can be neglected in the derivation, (i.e., the stress-free condition). (This stress-free condition implies that the boundaries of the actuator layer can not be fully constrained.) The induced strains due to imposed control voltages, the converse piezoelectric effect, are used to counteract the shell oscillation. It is also assumed that the applied control voltage ϕ^a is much more significant than the self-generated voltage ϕ due to the direct piezoelectric effect in the distributed actuator. Thus the self-generated voltage ϕ is neglected in the active vibration control system. The control voltage ϕ^a can be either a reference voltage in an open-loop control or the distributed sensor signal ϕ^a in a closed-loop control. (Note that the superscript "a" denotes the actuator and "s" the sensor.) In this section, micro-electromechanics of this distributed vibration control mechanism is analyzed and three generic feedback algorithms, namely, 1. the direct proportional feedback, 2. the negative velocity feedback, and 3. the Lyapunov feedback, are discussed. System dynamic equations including all control effects are also derived.

Induced strains S^a_{ii} in the distributed actuator can be calculated by

$$S^a_{11} = (d_{31}\phi^a)/h^a \quad \text{and} \quad S^a_{22} = (d_{32}\phi^a)/h^a, \tag{7.36a,b}$$

where d_{3i} is the piezoelectric strain constant, h^a is the actuator thickness, and the superscript "a" denotes the distributed piezoelectric actuator. (The induced strains and the resultant effect are illustrated in Figure 7.4.) Note that these strains are generated in the distributed actuator layer, which is located a distance away from the shell neutral surface. Thus, besides in-plane membrane control forces, these strains also introduce counteracting control moments for the shell structure.

7.4.1 Control Forces/Moments and System Equations

In-plane control forces N^a_{ii} and counteracting control moments M^a_{ii} induced by the imposed actuator voltage ϕ^a can be expressed as

$$N^a_{11} = d_{31}Y_p\phi^a \quad \text{and} \quad N^a_{22} = d_{32}Y_p\phi^a, \tag{7.37, 7.38}$$

$$M^a_{11} = r^a_1 d_{31}Y_p\phi^a \quad \text{and} \quad M^a_{22} = r^a_2 d_{32}Y_p\phi^a, \tag{7.39, 7.40}$$

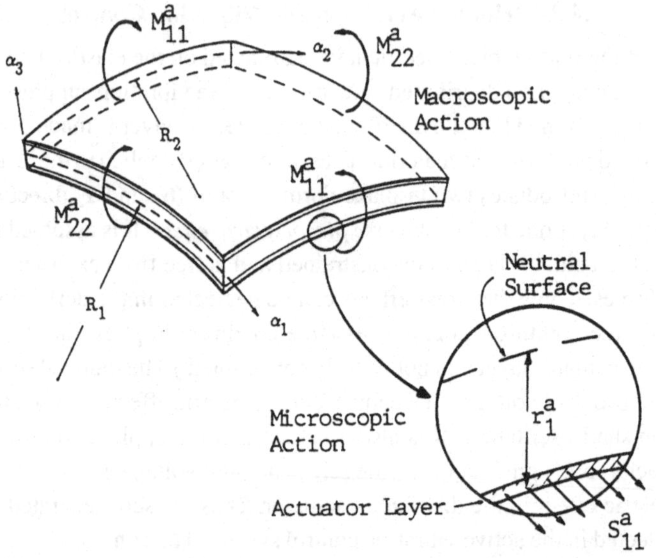

Fig. 7.4. Distributed control by a distributed piezoelectric actuator.

where Y_p is Young's modulus of the piezoelectric actuator and r_i^a is the effective moment arm (distance measured from the neutral surface to the midplane of the piezoelectric actuator). The in-plane twisting (shear) effect is not considered. Note that the imposed actuator voltage ϕ^a is determined by the control schemes (e.g., open- or closed-loop control), which will be discussed later. Positive or negative signs of the feedback voltage should be carefully controlled such that the induced forces and moments counteract the shell oscillation. Substituting these induced normal forces and counteracting moments into the shell's equations of motion yields

$$-\frac{\partial(\tilde{N}_{11}A_2)}{\partial\alpha_1} - \frac{\partial(\tilde{N}_{21}A_1)}{\partial\alpha_2} - \tilde{N}_{12}\frac{\partial A_1}{\partial\alpha_2} + \tilde{N}_{22}\frac{\partial A_2}{\partial\alpha_1}\frac{1}{R_1}\left[\frac{\partial(\tilde{M}_{11}A_2)}{\partial\alpha_1}\right.$$

$$\left. + \frac{\partial(\tilde{M}_{21}A_1)}{\partial\alpha_2} + \tilde{M}_{12}\frac{\partial A_1}{\partial\alpha_2} - \tilde{M}_{22}\frac{\partial A_2}{\partial\alpha_1}\right] + A_1A_2\rho h\ddot{u}_1 = A_1A_2F_1, \quad (7.41)$$

$$-\frac{\partial(\tilde{N}_{12}A_2)}{\partial\alpha_1} - \frac{\partial(\tilde{N}_{22}A_1)}{\partial\alpha_2} - \tilde{N}_{21}\frac{\partial A_2}{\partial\alpha_1} + \tilde{N}_{11}\frac{\partial A_1}{\partial\alpha_2} - \frac{1}{R_2}\left[\frac{\partial(\tilde{M}_{12}A_2)}{\partial\alpha_1}\right.$$

$$\left. + \frac{\partial(\tilde{M}_{22}A_1)}{\partial\alpha_2} + \tilde{M}_{21}\frac{\partial A_2}{\partial\alpha_1} - \tilde{M}_{11}\frac{\partial A_1}{\partial\alpha_2}\right] + A_1A_2\rho h\ddot{u}_2 = A_1A_2F_2, \quad (7.42)$$

$$-\frac{\partial}{\partial\alpha_1}\left\{\frac{1}{A_1}\left[\frac{\partial(\tilde{M}_{11}A_2)}{\partial\alpha_1}+\frac{\partial(\tilde{M}_{21}A_1)}{\partial\alpha_2}+\tilde{M}_{12}\frac{\partial A_1}{\partial\alpha_2}-\tilde{M}_{22}\frac{\partial A_2}{\partial\alpha_1}\right]\right\}$$

$$-\frac{\partial}{\partial\alpha_2}\left\{\frac{1}{A_2}\left[\frac{\partial(\tilde{M}_{12}A_2)}{\partial\alpha_1}+\frac{\partial(\tilde{M}_{22}A_1)}{\partial\alpha_2}+\tilde{M}_{21}\frac{\partial A_2}{\partial\alpha_1}-\tilde{M}_{11}\frac{\partial A_1}{\partial\alpha_2}\right]\right\}$$

$$+A_1A_2\left(\frac{\tilde{N}_{11}}{R_1}+\frac{\tilde{N}_{22}}{R_2}\right)+A_1A_2\rho h\ddot{u}_3=A_1A_2F_3, \qquad (7.43)$$

where the $\tilde{}$ terms include the feedback control effects induced by the converse piezoelectric effect. These resultant forces and moments are modified to include the induced control forces and counteracting moments, that is,

$$\tilde{N}_{ij}=N_{ij}-N_{ij}^{\mathrm{a}} \quad \text{and} \quad \tilde{M}_{ij}=M_{ij}-M_{ij}^{\mathrm{a}}, \qquad (7.44, 7.45)$$

where N_{ij} and M_{ij} are respectively the elastic membrane forces and the bending moments defined in Section 7.2.1. Note that the in-plane twisting effect is assumed negligible (i.e., $\tilde{M}_{12}=M_{12}$ and $\tilde{N}_{12}=N_{12}$). In addition, material properties of the piezoelectric actuator layer is not considered. Otherwise, one needs to consider the multilayer piezoelastic composite lamination theory (Tzou and Bao, 1994, 1995a). The feedback voltage ϕ^{a} is determined by the control algorithms (i.e., the direct proportional feedback, the negative velocity feedback, and the Lyapunov feedback), which will be discussed next. Spatial feedback algorithms and orthogonal modal filtering using spatial functions in the spatial and modal domain are also discussed.

7.4.2 Direct Proportional Feedback Control

In the direct proportional feedback control, the feedback voltage ϕ^{a} is generated by amplifying the sensor output ϕ^{s} directly,

$$\phi^{\mathrm{a}}=\mathcal{G}\phi^{\mathrm{s}}, \qquad (7.46)$$

where $\mathcal{G}$ denotes the voltage amplified ratio – a *feedback control gain* – which can be adjusted depending on the performance requirement. The sensor signal $\phi^{\mathrm{s}}(u_1,u_2,u_3,t)$ is defined in Eq. (7.24). (Note that the indices inside the parenthesis are redefined in terms of displacements and time.) The feedback induced effective control forces and moments are

$$N_{11}^{\mathrm{a}}=\mathcal{G}d_{31}Y_{\mathrm{p}}\phi^{\mathrm{s}} \quad \text{and} \quad N_{22}^{\mathrm{a}}=\mathcal{G}d_{32}Y_{\mathrm{p}}\phi^{\mathrm{s}}, \qquad (7.47\text{a,b})$$

$$M_{11}^{\mathrm{a}}=\mathcal{G}r_1^{\mathrm{a}}d_{31}Y_{\mathrm{p}}\phi^{\mathrm{s}} \quad \text{and} \quad M_{22}^{\mathrm{a}}=\mathcal{G}r_2^{\mathrm{a}}d_{32}Y_{\mathrm{p}}\phi^{\mathrm{s}}. \qquad (7.48\text{a,b})$$

These forces and moments can be substituted into shell equations leading to closed-loop system equations. Since the sensor signal is a function of strains ultimately contributed by displacements (u_1, u_2, u_3), this control scheme usually controls system frequencies. The oscillation amplitude could also change due to the frequency change. Other detailed discussions and examples can be found in Tzou (1993) and Tzou, Johnson, and Liu (1995).

7.4.3 Negative Velocity Feedback Control

The feedback signal can be differentiated so that the strain rate (related to the velocity) information is obtained. The velocity feedback can enhance the system damping and therefore effectively control the oscillation amplitude. Note that, for the velocity feedback, the sensor signal ϕ^s used in control force/moment equations in Section 7.4.2 should be replaced by $\dot{\phi}^s$:

$$N_{11}^a = -\mathcal{G} d_{31} Y_p \dot{\phi}^s \quad \text{and} \quad N_{22}^a = -\mathcal{G} d_{32} Y_p \dot{\phi}^s, \tag{7.49a,b}$$

$$M_{11}^a = -\mathcal{G} r_1^a d_{31} Y_p \dot{\phi}^s \quad \text{and} \quad M_{22}^a = -\mathcal{G} r_2^a d_{32} Y_p \dot{\phi}^s, \tag{7.50a,b}$$

where

$$\dot{\phi}^s \equiv \frac{\partial}{\partial t}[\phi^s(u_1, u_2, u_3, t)] = \phi^s(\dot{u}_1, \dot{u}_2, \dot{u}_3, t). \tag{7.51}$$

7.4.4 Lyapunov Feedback Control

In the Lyapunov feedback control, the feedback voltage amplitude is constant and the sign is opposite to the velocity (Tzou, 1988, 1993). The amplitude of feedback signal can be expressed as

$$\phi^a = -\mathcal{G}\,\mathrm{sgn}\left[\frac{\partial}{\partial t}\phi^s(u_1, u_2, u_3, t)\right], \tag{7.52}$$

where $\mathcal{G}$ is the feedback gain and "sgn" is a signum function (i.e., sgn$[z] = -1$ if $z < 0$, 0 if $z = 0$, and $+1$ if $z > 0$). Control forces and moments can be written as

$$N_{11}^a = -\mathcal{G}\, d_{31} Y_p \cdot \mathrm{sgn}(\dot{\phi}^s) \quad \text{and} \quad N_{22}^a = -\mathcal{G}\, d_{32} Y_p \cdot \mathrm{sgn}(\dot{\phi}^s), \tag{7.53a,b}$$

$$M_{11}^a = -\mathcal{G}\, r_1^a d_{31} Y_p \cdot \mathrm{sgn}(\dot{\phi}^s) \quad \text{and} \quad M_{22}^a = -\mathcal{G}\, r_2^a d_{32} Y_p \cdot \mathrm{sgn}(\dot{\phi}^s). \tag{7.54a,b}$$

Note that when ϕ^s is used in the feedback control, an averaged dynamic state of the shell is considered.

In practical applications, a transverse velocity $\dot{u}_3(\alpha_1^*, \alpha_2^*, t)$ at a reference location (α_1^*, α_2^*) or a reference area can also be used in the feedback control. Thus, the feedback signal becomes

$$\phi^a = -\mathcal{G}\,\mathrm{sgn}\left[\dot{u}_3\left(\alpha_1^*, \alpha_2^*, t\right)\right], \tag{7.55}$$

where (α_1^*, α_2^*) denotes the coordinates of the reference location. Thus, the effective control forces and moments become point-velocity dependent functions:

$$N_{11}^{\mathrm{a}} = -\mathcal{G}\, d_{31} Y_{\mathrm{p}} \cdot \mathrm{sgn}\left[\dot{u}_3\left(t, \alpha_1^*, \alpha_2^*\right)\right],$$
$$N_{22}^{\mathrm{a}} = -\,\mathcal{G}\, d_{32} Y_{\mathrm{p}} \cdot \mathrm{sgn}\left[\dot{u}_3\left(t, \alpha_1^*, \alpha_2^*\right)\right], \qquad (7.56\mathrm{a,b})$$

$$M_{11}^{\mathrm{a}} = -\mathcal{G}\, r_1^{\mathrm{a}} d_{31} Y_{\mathrm{p}} \cdot \mathrm{sgn}\left[\dot{u}_3\left(t, \alpha_1^*, \alpha_2^*\right)\right],$$
$$M_{22}^{\mathrm{a}} = -\mathcal{G}\, r_2^{\mathrm{a}} d_{32} Y_{\mathrm{p}} \cdot \mathrm{sgn}\left[\dot{u}_3\left(t, \alpha_1^*, \alpha_2^*\right)\right]. \qquad (7.57\mathrm{a,b})$$

7.4.5 Modal Feedback Gains

In addition, since the sensor signal can be represented by its modal contribution, the system feedback gain $\mathcal{G}$ can be extended to a *modal feedback gain* $\hat{\mathcal{G}}_k$ from which control of each individual mode can be achieved. Using the proportional feedback as an example, one can write control forces and moments with the modal feedback gains $\hat{\mathcal{G}}_k$s and mode shape functions as

$$\begin{aligned}
N_{ii}^{\mathrm{a}} = {} & d_{3i} Y_{\mathrm{p}} \frac{h^{\mathrm{s}}}{\epsilon_{33} \mathbf{S}^{\mathrm{e}}} \int_{\mathbf{S}^{\mathrm{e}}} \left[e_{31} \left\{ \left[\frac{1}{A_1} \frac{\partial}{\partial \alpha_1} \sum_{k=1}^{\infty} \hat{\mathcal{G}}_k \eta_k U_{1k} \right. \right. \right. \\
& \left. + \frac{1}{A_1 A_2} \sum_{k=1}^{\infty} \hat{\mathcal{G}}_k \eta_k U_{2k} \frac{\partial A_1}{\partial \alpha_2} + \frac{1}{R_1} \sum_{k=1}^{\infty} \hat{\mathcal{G}}_k \eta_k U_{3k} \right] \\
& + r_1^{\mathrm{s}} \left[\frac{1}{A_1} \frac{\partial}{\partial \alpha_1} \left(\frac{1}{R_1} \sum_{k=1}^{\infty} \hat{\mathcal{G}}_k \eta_k U_{1k} - \frac{1}{A_1} \frac{\partial}{\partial \alpha_1} \sum_{k=1}^{\infty} \hat{\mathcal{G}}_k \eta_k U_{3k} \right) \right. \\
& \left. \left. + \frac{1}{A_1 A_2} \frac{\partial A_1}{\partial \alpha_2} \left(\frac{1}{R_2} \sum_{k=1}^{\infty} \hat{\mathcal{G}}_k \eta_k U_{2k} - \frac{1}{A_2} \frac{\partial}{\partial \alpha_2} \sum_{k=1}^{\infty} \hat{\mathcal{G}}_k \eta_k U_{3k} \right) \right] \right\} \\
& + e_{32} \left\{ \left[\frac{1}{A_2} \frac{\partial}{\partial \alpha_2} \sum_{k=1}^{\infty} \hat{\mathcal{G}}_k \eta_k U_{2k} + \frac{1}{A_1 A_2} \sum_{k=1}^{\infty} \hat{\mathcal{G}}_k \eta_k U_{1k} \frac{\partial A_2}{\partial \alpha_1} \right. \right. \\
& \left. + \frac{1}{R_2} \sum_{k=1}^{\infty} \hat{\mathcal{G}}_k \eta_k U_{3k} \right] + r_2^{\mathrm{s}} \left[\frac{1}{A_2} \frac{\partial}{\partial \alpha_2} \left(\frac{1}{R_2} \sum_{k=1}^{\infty} \hat{\mathcal{G}}_k \eta_k U_{2k} \right. \right. \\
& \left. - \frac{1}{A_2} \frac{\partial}{\partial \alpha_2} \sum_{k=1}^{\infty} \hat{\mathcal{G}}_k \eta_k U_{3k} \right) + \frac{1}{A_1 A_2} \frac{\partial A_2}{\partial \alpha_1} \left(\frac{1}{R_1} \sum_{k=1}^{\infty} \hat{\mathcal{G}}_k \eta_k U_{1k} \right. \\
& \left. \left. \left. \left. - \frac{1}{A_1} \frac{\partial}{\partial \alpha_1} \sum_{k=1}^{\infty} \hat{\mathcal{G}}_k \eta_k U_{3k} \right) \right] \right\} \right] A_1 A_2 \, d\alpha_1 \, d\alpha_2
\end{aligned} \qquad (7.58)$$

and

$$
\begin{aligned}
M_{ii}^{\mathrm{a}} = {} & r_i^{\mathrm{a}} d_{3i} Y_{\mathrm{p}} \frac{h^{\mathrm{s}}}{\epsilon_{33} \mathbf{S}^{\mathrm{e}}} \int_{\mathbf{S}^{\mathrm{e}}} \left[e_{31} \left\{ \left[\frac{1}{A_1} \frac{\partial}{\partial \alpha_1} \sum_{k=1}^{\infty} \hat{\mathcal{G}}_k \eta_k U_{1k} \right. \right. \right. \\
& \left. + \frac{1}{A_1 A_2} \sum_{k=1}^{\infty} \hat{\mathcal{G}}_k \eta_k U_{2k} \frac{\partial A_1}{\partial \alpha_2} + \frac{1}{R_1} \sum_{k=1}^{\infty} \hat{\mathcal{G}}_k \eta_k U_{3k} \right] \\
& + r_1^{\mathrm{s}} \left[\frac{1}{A_1} \frac{\partial}{\partial \alpha_1} \left(\frac{1}{R_1} \sum_{k=1}^{\infty} \hat{\mathcal{G}}_k \eta_k U_{1k} - \frac{1}{A_1} \frac{\partial}{\partial \alpha_1} \sum_{k=1}^{\infty} \hat{\mathcal{G}}_k \eta_k U_{3k} \right) \right. \\
& \left. \left. + \frac{1}{A_1 A_2} \frac{\partial A_1}{\partial \alpha_2} \left(\frac{1}{R_2} \sum_{k=1}^{\infty} \hat{\mathcal{G}}_k \eta_k U_{2k} - \frac{1}{A_2} \frac{\partial}{\partial \alpha_2} \sum_{k=1}^{\infty} \hat{\mathcal{G}}_k \eta_k U_{3k} \right) \right] \right\} \\
& + e_{32} \left\{ \left[\frac{1}{A_2} \frac{\partial}{\partial \alpha_2} \sum_{k=1}^{\infty} \hat{\mathcal{G}}_k \eta_k U_{2k} + \frac{1}{A_1 A_2} \sum_{k=1}^{\infty} \hat{\mathcal{G}}_k \eta_k U_{1k} \frac{\partial A_2}{\partial \alpha_1} \right. \right. \\
& \left. + \frac{1}{R_2} \sum_{k=1}^{\infty} \hat{\mathcal{G}}_k \eta_k U_{3k} \right] + r_2^{\mathrm{s}} \left[\frac{1}{A_2} \frac{\partial}{\partial \alpha_2} \left(\frac{1}{R_2} \sum_{k=1}^{\infty} \hat{\mathcal{G}}_k \eta_k U_{2k} \right. \right. \\
& \left. - \frac{1}{A_2} \frac{\partial}{\partial \alpha_2} \sum_{k=1}^{\infty} \hat{\mathcal{G}}_k \eta_k U_{3k} \right) + \frac{1}{A_1 A_2} \frac{\partial A_2}{\partial \alpha_1} \left(\frac{1}{R_1} \sum_{k=1}^{\infty} \hat{\mathcal{G}}_k \eta_k U_{1k} \right. \\
& \left. \left. \left. \left. - \frac{1}{A_1} \frac{\partial}{\partial \alpha_1} \sum_{k=1}^{\infty} \hat{\mathcal{G}}_k \eta_k U_{3k} \right) \right] \right\} \right] A_1 A_2 \, d\alpha_1 \, d\alpha_2. \qquad (7.59)
\end{aligned}
$$

Similarly, a new set of system equations can be derived. It can be observed that all three actuator parameters (i.e., the piezoelectric constant d_{ij}, modulus of elasticity Y_{p}, and moment arm r_j^{a}) are of importance to the overall control effects. In general, a higher piezoelectric constant, stiffer piezoelectric actuator, and longer moment arm contribute better control effects. For example, piezoceramic actuators are more effective than polymeric piezoelectric actuators, such as polyvinylidene fluoride (PVDF) (Tzou, 1993).

It should be noted that control spillover could occur in the above formulations since they include cross coupling terms (Meirovitch and Baruh, 1981; Balas, 1978). Control spillover and design of distributed shaped and convolved actuators are discussed in Sections 7.4.7 and 7.4.8.

7.4.6 State Equation

Dynamic equations of the distributed shell/sensor/actuator system can be transferred into the state equation. Define the elastic terms associated with the elastic shell by generic Love's operators $L_i(u_1, u_2, u_1)$ $(i = 1, 2, 3)$ and the feedback control terms as H_is $(i = 1, 2, 3)$:

$$L_1 = \frac{1}{\rho h A_1 A_2}\left[\frac{\partial(N_{11}A_2)}{\partial\alpha_1} + \frac{\partial(N_{21}A_1)}{\partial\alpha_2} + N_{12}\frac{\partial A_1}{\partial\alpha_2} - N_{22}\frac{\partial A_2}{\partial\alpha_1} + \frac{1}{R_1}\left(\frac{\partial(M_{11}A_2)}{\partial\alpha_1} + \frac{\partial(M_{21}A_1)}{\partial\alpha_2} + M_{12}\frac{\partial A_1}{\partial\alpha_2} - M_{22}\frac{\partial A_2}{\partial\alpha_1}\right)\right], \quad (7.60)$$

$$L_2 = \frac{1}{\rho h A_1 A_2}\left[\frac{\partial(N_{12}A_2)}{\partial\alpha_1} + \frac{\partial(N_{22}A_1)}{\partial\alpha_2} + N_{21}\frac{\partial A_2}{\partial\alpha_1} - N_{11}\frac{\partial A_1}{\partial\alpha_2} + \frac{1}{R_2}\left(\frac{\partial(M_{12}A_2)}{\partial\alpha_1} + \frac{\partial(M_{22}A_1)}{\partial\alpha_2} + M_{21}\frac{\partial A_2}{\partial\alpha_1} - M_{11}\frac{\partial A_1}{\partial\alpha_2}\right)\right], \quad (7.61)$$

$$L_3 = \frac{1}{\rho h A_1 A_2}\left\{\left[\frac{\partial}{\partial\alpha_2}\left(\frac{1}{A_1}\right)\left(\frac{\partial(M_{11}A_2)}{\partial\alpha_1} + \frac{\partial(M_{21}A_1)}{\partial\alpha_2} + M_{12}\frac{\partial A_1}{\partial\alpha_2} - M_{22}\frac{\partial A_2}{\partial\alpha_1}\right) + \frac{\partial}{\partial\alpha_2}\left(\frac{1}{A_2}\right)\left(\frac{\partial(M_{12}A_2)}{\partial\alpha_1} + \frac{\partial(M_{22}A_1)}{\partial\alpha_2} + M_{21}\frac{\partial A_2}{\partial\alpha_1} - M_{11}\frac{\partial A_1}{\partial\alpha_2}\right)\right] - A_1A_2\left[\frac{N_{11}}{R_1} + \frac{N_{22}}{R_2}\right]\right\} \quad (7.62)$$

and

$$H_1 = \frac{1}{\rho h A_1 A_2}\left[\frac{\partial\left(N^{a}_{11}A_2\right)}{\partial\alpha_1} + \frac{\partial\left(N^{a}_{21}A_1\right)}{\partial\alpha_2} + N^{a}_{12}\frac{\partial A_1}{\partial\alpha_2} - N^{a}_{22}\frac{\partial A_2}{\partial\alpha_1} + \frac{1}{R_1}\left(\frac{\partial\left(M^{a}_{11}A_2\right)}{\partial\alpha_1} + \frac{\partial\left(M^{a}_{21}A_1\right)}{\partial\alpha_2} + M^{a}_{12}\frac{\partial A_1}{\partial\alpha_2} - M^{a}_{22}\frac{\partial A_2}{\partial\alpha_1}\right)\right], \quad (7.63)$$

$$H_2 = \frac{1}{\rho h A_1 A_2}\left[\frac{\partial\left(N^{a}_{12}A_2\right)}{\partial\alpha_1} + \frac{\partial\left(N^{a}_{22}A_1\right)}{\partial\alpha_2} + N^{a}_{21}\frac{\partial A_2}{\partial\alpha_1} - N^{a}_{11}\frac{\partial A_1}{\partial\alpha_2} + \frac{1}{R_2}\left(\frac{\partial\left(M^{a}_{12}A_2\right)}{\partial\alpha_1} + \frac{\partial\left(M^{a}_{22}A_1\right)}{\partial\alpha_2} + M^{a}_{21}\frac{\partial A_2}{\partial\alpha_1} - M^{a}_{11}\frac{\partial A_1}{\partial\alpha_2}\right)\right], \quad (7.64)$$

$$H_3 = \frac{1}{\rho h A_1 A_2}\left\{\left[\frac{\partial}{\partial\alpha_1}\left(\frac{1}{A_1}\right)\left(\frac{\partial\left(M^{a}_{11}A_2\right)}{\partial\alpha_1} + \frac{\partial\left(M^{a}_{21}A_1\right)}{\partial\alpha_2} + M^{a}_{12}\frac{\partial A_1}{\partial\alpha_2} - M^{a}_{22}\frac{\partial A_2}{\partial\alpha_1}\right) + \frac{\partial}{\partial\alpha_2}\left(\frac{1}{A_2}\right)\left(\frac{\partial\left(M^{a}_{12}A_2\right)}{\partial\alpha_1} + \frac{\partial\left(M^{a}_{22}A_1\right)}{\partial\alpha_2} + M^{a}_{21}\frac{\partial A_2}{\partial\alpha_1} - M^{a}_{11}\frac{\partial A_1}{\partial\alpha_2}\right)\right] - A_1A_2\left[\frac{N^{a}_{11}}{R_1} + \frac{N^{a}_{22}}{R_2}\right]\right\}. \quad (7.65)$$

Note that the in-plane control shear forces and moments can be neglected (i.e., $N^{a}_{12} = N^{a}_{21} = 0$ and $M^{a}_{12} = M^{a}_{21} = 0$), as discussed previously. Substituting L_is and H_is into the system equation and transferring them into the state space gives

$$\frac{\partial}{\partial t}\begin{bmatrix} u_1 \\ u_2 \\ u_3 \\ \dot{u}_1 \\ \dot{u}_2 \\ \dot{u}_3 \end{bmatrix} = \begin{bmatrix} 0 & 0 & 0 & 1 & 0 & 0 \\ 0 & 0 & 0 & 0 & 1 & 0 \\ 0 & 0 & 0 & 0 & 0 & 1 \\ L_1 & 0 & 0 & 0 & 0 & 0 \\ 0 & L_2 & 0 & 0 & 0 & 0 \\ 0 & 0 & L_3 & 0 & 0 & 0 \end{bmatrix} \begin{bmatrix} u_1 \\ u_2 \\ u_3 \\ \dot{u}_1 \\ \dot{u}_2 \\ \dot{u}_3 \end{bmatrix} + [I] \begin{bmatrix} 0 \\ 0 \\ 0 \\ F_1/\rho h + H_1 \\ F_2/\rho h + H_2 \\ F_3/\rho h + H_3 \end{bmatrix}, \tag{7.66}$$

where $[I]$ is an identity matrix. Symbolically,

$$\frac{\partial \mathbf{U}}{\partial t} = \mathbf{AU} + \mathbf{Bm}, \tag{7.67}$$

where **U**, **A**, and **B** are defined in the above state equation and

$$\mathbf{m} = [0 \quad 0 \quad 0 \quad F_1/\rho h + H_1 \quad F_2/\rho h + H_2 \quad F_3/\rho h + H_3]^t, \tag{7.68}$$

where $[\cdot]^t$ denotes the vector or matrix transpose. The state equation, Eqs. (7.66) or (7.67), is for a generic shell continuum (distributed parameter system) with distributed piezoelectric sensors/actuators. The state equation is defined in a generic form that includes the in-plane control normal and shear forces and out-of-plane control moments. Depending on the four system parameters (i.e., two Lamé parameters and two radii) this equation can be simplified to apply to many other common geometries. Simplification procedures are outlined in case studies.

7.4.7 Spatially Distributed Feedback and Modal Feedback Functions

In general, there are two kinds of forces for the piezoelectric laminated shell continuum. One is the conventional mechanical force and the other is the electric (control) force introduced by the converse piezoelectric effect. (Note that this "force" is a generic term that includes both membrane forces and bending moments.) If a generic Love's operator $L_i^c(\phi_3)$ representing the converse piezoelectric effect is designated as a function of mechanical motions, such as displacement, velocity, or acceleration, a closed-loop feedback control system can be established. In this section, spatial characteristics, distributed modal feedback functions, and modal filtering of distributed actuators are discussed. Three generic distributed feedback algorithms: 1. the displacement feedback, 2. the velocity feedback, and 3. the acceleration feedback are proposed and their corresponding governing equations are derived. Combination of the displacement and velocity feedback gives the conventional proportional plus derivative feedback. Note that only the transverse vibration (α_3 direction) is considered as the primary motion of interest. The general procedures can be applied to all directions.

7.4.7.1 Distributed Displacement Feedback

In the distributed displacement feedback, it is assumed that the Love's operator $L_3^c\{\phi_3\}$ contributed by the converse piezoelectric effect is a function of the transverse displacement u_3, that is,

$$L_3^c\{\phi_3\} = \mathscr{F}_1^*\{u_3(\alpha_1, \alpha_2, t)\}, \tag{7.69}$$

where $\mathscr{F}_1^*\{u_3(\alpha_1, \alpha_2, t)\}$ denotes a generic spatial and time function. Since the shell response is contributed by all participating modes, the feedback can be further expressed in a modal participation form and a generic spatial function:

$$L_3^c\{\phi_3\} = \sum_{m=1}^{\infty} \mathcal{G}_{3m}^{\mathrm{df}}(\alpha_1, \alpha_2)\eta_m(t), \tag{7.70}$$

where $\mathcal{G}_{3m}^{\mathrm{df}}(\alpha_1, \alpha_2)$ is the *displacement modal feedback gain function*, a generic spatially distributed function, denoting the spatial actuation to the shell continuum. The modal control force defined for the kth mode is

$$\hat{F}_k = \frac{1}{\rho h N_k} \int_{\alpha_1}\int_{\alpha_2} \left(\left[\sum_{m=1}^{\infty} \mathcal{G}_{3m}^{\mathrm{df}}(\alpha_1, \alpha_2)\eta_m(t) \right] U_{3k}(\alpha_1, \alpha_2) \right) A_1 A_2 \, d\alpha_1 d\alpha_2. \tag{7.71}$$

Substituting the modal control force into the modal equation discussed in Section 7.2.2 gives

$$\ddot{\eta}_k + \frac{c}{\rho h}\dot{\eta}_k + \omega_k^2 \eta_k$$
$$= \frac{1}{\rho h N_k} \sum_{m=1}^{\infty} \int_{\alpha_1}\int_{\alpha_2} (\mathcal{G}_{3m}^{\mathrm{df}}(\alpha_1, \alpha_2)\eta_m(t) U_{3k}(\alpha_1, \alpha_2)) A_1 A_2 \, d\alpha_1 d\alpha_2, \tag{7.72}$$

where the external mechanical forces are not considered (i.e., $F_3 = 0$) and

$$N_k = \int_{\alpha_1}\int_{\alpha_2} U_{3k}^2 A_1 A_2 \, d\alpha_1 d\alpha_2. \tag{7.73}$$

7.4.7.2 Distributed Velocity Feedback

Following the same procedures, one can further define the governing modal equations for the distributed velocity feedback:

$$L_3^c\{\phi_3\} = \mathscr{F}_2^*\{\dot{u}_3(\alpha_1, \alpha_2, t)\}, \tag{7.74}$$

$$L_3^c\{\phi_3\} = \sum_{m=1}^{\infty} \mathcal{G}_{3m}^{\mathrm{vf}}(\alpha_1, \alpha_2)\dot{\eta}_m(t), \tag{7.75}$$

and furthermore

$$\ddot{\eta}_k + \frac{c}{\rho h}\dot{\eta}_k + \omega_k^2 \eta_k = \frac{1}{\rho h N_k} \sum_{m=1}^{\infty} \int_{\alpha_1}\int_{\alpha_2} \left(\mathcal{G}_{3m}^{\mathrm{vf}}(\alpha_1, \alpha_2)\dot{\eta}_m(t) U_{3k}(\alpha_1, \alpha_2)\right) A_1 A_2 \, d\alpha_1 \, d\alpha_2, \quad (7.76)$$

where the modal feedback control force for the velocity feedback is defined as

$$\hat{F}_k = \frac{1}{\rho h N_k} \int_{\alpha_1}\int_{\alpha_2} \left(\sum_{m=1}^{\infty} \mathcal{G}_{3m}^{\mathrm{vf}}(\alpha_1, \alpha_2)\dot{\eta}_m(t) U_{3k}(\alpha_1, \alpha_2)\right) A_1 A_2 \, d\alpha_1 \, d\alpha_2. \quad (7.77)$$

7.4.7.3 *Distributed Acceleration Feedback*

The distributed acceleration feedback can be defined in a similar manner:

$$L_3^c\{\phi_3\} = \mathcal{F}_3^*\{\ddot{u}_3(\alpha_1, \alpha_2, t)\}, \quad (7.78)$$

$$L_3^c\{\phi_3\} = \sum_{m=1}^{\infty} \mathcal{G}_{im}^{\mathrm{af}}(\alpha_1, \alpha_2)\ddot{\eta}_m(t), \quad (7.79)$$

and furthermore

$$\ddot{\eta}_k + \frac{c}{\rho h}\dot{\eta}_k + \omega_k^2 \eta_k = \frac{1}{\rho h N_k} \sum_{m=1}^{\infty} \int_{\alpha_1}\int_{\alpha_2} \left(\mathcal{G}_{3m}^{\mathrm{af}}(\alpha_1, \alpha_2)\ddot{\eta}_m(t) U_{3k}(\alpha_1, \alpha_2)\right) A_1 A_2 \, d\alpha_1 \, d\alpha_2, \quad (7.80)$$

where

$$\hat{F}_k = \frac{1}{\rho h N_k} \int_{\alpha_1}\int_{\alpha_2} \left(\sum_{m=1}^{\infty} \mathcal{G}_{3m}^{\mathrm{af}}(\alpha_1, \alpha_2)\ddot{\eta}_m(t) U_{3k}(\alpha_1, \alpha_2)\right) A_1 A_2 \, d\alpha_1 d\alpha_2. \quad (7.81)$$

Note that a generic shell control system can be described by an infinite number of second-order modal equations. The overall system response is contributed by all participating modes. However, only a finite number of modes is considered in practical applications. As mentioned previously, control spillovers could occur when feedback control forces for controlled modes appear in the modal equations of other uncontrolled modes due to modal coupling and interaction among all participating modes. Generic distributed orthogonal control concepts based on distributed convolving piezoelectric actuators are presented next. In addition, the *modal feedback gain functions* $\mathcal{G}_{3m}^{\mathrm{df}}(\alpha_1, \alpha_2)$, $\mathcal{G}_{3m}^{\mathrm{vf}}(\alpha_1, \alpha_2)$, and $\mathcal{G}_{3m}^{\mathrm{af}}(\alpha_1, \alpha_2)$ are spatially distributed functions. If these modal feedback gain functions are defined so that they are orthogonal to a mode or a group of modes, these modes can be filtered out from being activated or controlled in the feedback control system. This concept will be discussed next.

7.4.8 Distributed Orthogonal Actuators and Control

Note that control spillovers resulting from the cross-coupling control forces can be observed in the modal equations of all three distributed feedback algorithms. Based on the modal orthogonality concept, one can design the modal feedback gain functions, $\mathcal{G}_{3n}^{\mathrm{df}}(\alpha_1, \alpha_2)$, $\mathcal{G}_{3n}^{\mathrm{vf}}(\alpha_1, \alpha_2)$, and $\mathcal{G}_{3n}^{\mathrm{af}}(\alpha_1, \alpha_2)$, with spatially distributed characteristics such that they are orthogonal to other natural modes (i.e., $k \neq p$). Thus, the modal feedback gain functions can be defined as a product of a constant modal gain $\mathcal{G}_{3n}^{f}$ and the nth mode shape function $U_{3n}(\alpha_1, \alpha_2)$,

$$\mathcal{G}_{3n}^{\mathrm{df}}(\alpha_1, \alpha_2) = \mathcal{G}_{3n}^{\mathrm{df}} U_{3n}(\alpha_1, \alpha_2), \qquad \mathcal{G}_{3n}^{\mathrm{vf}}(\alpha_1, \alpha_2) = \mathcal{G}_{3n}^{\mathrm{vf}} U_{3n}(\alpha_1, \alpha_2),$$
$$\mathcal{G}_{3n}^{\mathrm{af}}(\alpha_1, \alpha_2) = \mathcal{G}_{3n}^{\mathrm{af}} U_{3n}(\alpha_1, \alpha_2), \qquad (7.82\text{a,b,c})$$

where $\mathcal{G}_{3n}^{\mathrm{df}}$, $\mathcal{G}_{3n}^{\mathrm{vf}}$, and $\mathcal{G}_{3n}^{\mathrm{af}}$ are weighting factors (gain constants). Imposing the modal orthogonality to the generic feedback functions can make the control forces appear only in the modal equations of controlled modes, but not in the equations of uncontrolled modes. Based on this modal orthogonality concept, one can further define modal control forces and modal equations for the double-curvature shell with distributed modal actuators. Note that in practice this mode shape function represents the shape or spatial distribution of distributed actuators.

7.4.8.1 Distributed Displacement Feedback

Using the modal orthogonality concept and Eq. (7.82a), one can define the modal control force

$$\hat{F}_k = \frac{\mathcal{G}_3^{\mathrm{df}}}{\rho h N_k} \eta_n(t) \int_{\alpha_1}\int_{\alpha_2} U_{3n}(\alpha_1, \alpha_2) U_{3k}(\alpha_1, \alpha_2) A_1 A_2 \, d\alpha_1 \, d\alpha_2, \qquad (7.83)$$

such that

$$\text{(i)} \qquad \hat{F}_k = \frac{\mathcal{G}_3^{\mathrm{df}}}{\rho h N_n} \eta_n(t) \int_{\alpha_1}\int_{\alpha_2} U_{3n}^2 A_1 A_2 \, d\alpha_1 \, d\alpha_2, \quad (\text{for } k = n), \qquad (7.84\text{a})$$

$$\text{(ii)} \qquad \hat{F}_k = 0 \quad (\text{for } k \neq n). \qquad (7.84\text{b})$$

Thus, the modal equation is decoupled from the residual modes, that is,

$$\ddot{\eta}_n + \frac{c}{\rho h}\dot{\eta}_n + \left(\omega_n^2 - \frac{\mathcal{G}_3^{\mathrm{df}}}{\rho h N_n} \int_{\alpha_1}\int_{\alpha_2} U_{3n}^2 A_1 A_2 \, d\alpha_1 \, d\alpha_2\right)\eta_n = 0. \qquad (7.85)$$

Decoupled modal equations for the velocity and acceleration feedback controls can be derived in the same manner.

7.4.8.2 Distributed Velocity Feedback

In this case the modal control force is

$$\hat{F}_k = \frac{\mathcal{G}_3^{\mathrm{vf}}}{\rho h N_k} \dot{\eta}_n(t) \int_{\alpha_1} \int_{\alpha_2} U_{3n}(\alpha_1, \alpha_2) U_{3k}(\alpha_1, \alpha_2) A_1 A_2 \, d\alpha_1 \, d\alpha_2, \tag{7.86}$$

such that

$$\text{(i)} \qquad \hat{F}_k = \frac{\mathcal{G}_3^{\mathrm{vf}}}{\rho h N_n} \dot{\eta}_n(t) \int_{\alpha_1} \int_{\alpha_2} U_{3n}^2 A_1 A_2 \, d\alpha_1 \, d\alpha_2 \quad (\text{for } k = n), \tag{7.87a}$$

$$\text{(ii)} \qquad \hat{F}_k = 0 \quad (\text{for } k \neq n), \tag{7.87b}$$

which implies

$$\ddot{\eta}_n + \frac{1}{\rho h}\left(c - \frac{\mathcal{G}_3^{\mathrm{vf}}}{N_n} \int_{\alpha_1} \int_{\alpha_2} U_{3n}^2 A_1 A_2 \, d\alpha_1 \, d\alpha_2\right) \dot{\eta}_n + \omega_n^2 \eta_n = 0. \tag{7.88}$$

7.4.8.3 Distributed Acceleration Feedback

Here the modal control force is

$$\hat{F}_k = \frac{\mathcal{G}_3^{\mathrm{af}}}{\rho h N_k} \ddot{\eta}_n(t) \int_{\alpha_1} \int_{\alpha_2} U_{3n}(\alpha_1, \alpha_2) U_{3k}(\alpha_1, \alpha_2) A_1 A_2 \, d\alpha_1 \, d\alpha_2, \tag{7.89}$$

and

$$\text{(i)} \qquad \hat{F}_k = \frac{\mathcal{G}_3^{\mathrm{af}}}{\rho h N_n} \ddot{\eta}_n(t) \int_{\alpha_1} \int_{\alpha_2} U_{3n}^2 A_1 A_2 \, d\alpha_1 \, d\alpha_2 \quad (\text{for } k = n), \tag{7.90a}$$

$$\text{(ii)} \qquad \hat{F}_k = 0 \quad (\text{for } k \neq n), \tag{7.90b}$$

which implies

$$\left(1 - \frac{\mathcal{G}_3^{\mathrm{af}}}{\rho h N_n} \int_{\alpha_1} \int_{\alpha_2} U_{3n}^2 A_1 A_2 \, d\alpha_1 \, d\alpha_2\right) \ddot{\eta}_n + \frac{c}{\rho h} \dot{\eta}_n + \omega_n^2 \eta_n = 0. \tag{7.91}$$

Again, G_i^{df}, G_i^{vf}, and G_i^{af} are weighting factors (gain constants) that are independent of spatial coordinates and time. $U_{3n}(\alpha_1, \alpha_2)$ denotes the transverse mode shape function in the transverse direction. (Note that only the transverse oscillation is considered in the formulation, although the procedures can be extended to encompass all three directions.) As a demonstration of distributed modal actuators, spatially distributed convolving actuators are presented in case studies.

7.5 Applications of Generic Theory to Common Continua

As discussed previously, distributed sensing and actuation theories developed for a generic deep shell can be applied to a number of common geometries using four geometric parameters: two Lamé parameters and two radii of curvatures. In this section, detailed reduction procedures and examples are presented.

There are four geometric parameters, two Lamé parameters (A_1 and A_2) and two radii (R_1 and R_2) of curvatures, required to carry out the simplification and reduction of generic shell sensing and actuation theories. The general procedures, a step-by-step approach, are presented as follows.

1. *Select a Coordinate System:* The original generic shell continuum was defined in a triorthogonal curvilinear coordinate system $(\alpha_1, \alpha_2, \alpha_3)$. Depending on the given geometries (host elastic structure) and/or sensor/actuator shapes, these coordinates can be redefined to best represent the problem, e.g., $\alpha_1 = x$ and $\alpha_2 = y$ for a *rectangular plate*; $\alpha_1 = x$ and $\alpha_2 = \theta$ for a *cylindrical shell* and *thin cylindrical tube shell*; $\alpha_1 = \psi$ and $\alpha_2 = \theta$ for a *spherical shell*; etc.
2. *Determine the Radii of Curvatures:* The radii of curvature R_1 and R_2 of the two in-plane coordinate axes α_1 and α_2 can be easily observed from the coordinate system defined in Step 1. For example, the radii of x and y axes in a *rectangular plate* are $R_1 = \infty$ and $R_2 = \infty$. In a *cylindrical shell*, $R_1 = \infty$ and $R_2 = R$. In a *spherical shell*, radii are $R_1 = R_2 = \mathbb{R}$.
3. *Derive a Fundamental Form:* A *fundamental form* represents an infinitesimal distance ds on the neutral surface of the shell continuum; the distance is the hypotenuse of a (or an approximate) right-angle triangle defined by the infinitesimal distances ($d\alpha_1$ and $d\alpha_2$) of the two in-plane coordinates; see Figure 7.1. From the *fundamental form*, two Lamé parameters A_1 and A_2 and the two selected coordinates α_1 and α_2 can be defined:

$$(ds)^2 = (A_1)^2(d\alpha_1)^2 + (A_2)^2(d\alpha_2)^2. \tag{7.92}$$

For example, the fundamental form for a *rectangular plate* (defined in a Cartesian coordinate system x and y) is

$$(ds)^2 = (1)^2(dx)^2 + (1)^2(dy)^2, \tag{7.93}$$

where the Lamé parameters are $A_1 = 1$ and $A_2 = 1$ (Tzou, 1993).
For a *cylindrical shell* defined by the x and θ axes, the fundamental form is defined as

$$(ds)^2 = (1)^2(dx)^2 + (R)^2(d\theta)^2, \tag{7.94}$$

where $A_1 = 1$ and $A_2 = R$ (radius of the cylinder).

The fundamental form defining the *spherical shell* is

$$(ds)^2 = \mathbb{R}^2(d\psi)^2 + \mathbb{R}^2 \sin^2 \psi (d\theta)^2, \tag{7.95}$$

where $\mathbb{R}$ is the radius of the spherical shell. Thus, the Lamé parameters are $A_1 = \mathbb{R}$ and $A_2 = \mathbb{R} \sin \psi$. Other geometries can be defined accordingly (Tzou, 1993; Soedel, 1993).

4. *Simplify the Shell Sensor/Actuator Equations:* By substituting the four parameters A_1, A_2, R_1, and R_2 into the generic shell sensor/actuator and/or system equations, one can easily derive the governing equations for that geometry or sensor/actuator, e.g., shells (cylindrical shells, conical shells, spherical shells, shells of revolution, etc.) or nonshell continua (rings, arches, beams, etc.). Detailed sensing/actuation and shell equations can be found in Tzou (1993).

7.6 Self-sensing Orthogonal Actuators

A perfect sensor/actuator collocation usually provides a stable control performance in closed-loop feedback systems. A self-sensing piezoelectric actuator is a single piece of piezoelectric device simultaneously used for both sensing and control. (The sensor signal is separated from the control signal by using a difference-amplifier; this signal is then amplified and fed back to induce control actions.) Dosch, Inman, and Garcia (1992) proposed a self-sensing piezoelectric actuator for collocated control of a cantilever beam. Anderson, Hagood, and Goodliffe (1992) presented an analytical modeling of the self-sensing actuator system and studied its applications to beam and truss structures. Small rectangular piezoelectric devices attached near the fixed end were used in both studies.

It is known that the spatially distributed orthogonal sensors and actuators are sensitive to a mode or a group of natural modes (Tzou, 1993; Lee, 1992). Spatially distributed piezoelectric sensors and actuators were investigated in a number of recent studies, such as beams, plates, rings, shells, etc. (Lee and Moon, 1990; Lee, 1992; Hubbard and Burke, 1992; Collins, Miller, and von Flotow, 1994; Gu et al., 1994; Tzou and Fu, 1994; Tzou, Zhong, and Natori, 1993; Tzou, 1993; Tzou, Zhong, and Hollkamp, 1994). Based on the modal orthogonality, a spatially shaped self-sensing orthogonal modal actuator is effective to only a single mode; consequently, each vibration mode can be independently controlled, often referred to as the *independent modal control* (Meirovitch, 1988), while the feedback control system is kept simple.

This section presents an investigation of sensing and control characteristics of self-sensing orthogonal modal actuators. A simplified one-dimensional (1-D) theory for a spatially distributed self-sensing orthogonal actuator is derived first, followed by an experimental study of self-sensing orthogonal piezoelectric actuators. Independent modal control with the self-sensing orthogonal actuators are demonstrated (Tzou and Hollkamp, 1994). Emphasis is placed on experimental demonstrations.

7.6.1 1-D Orthogonal Sensor/Actuator Theory

It is assumed that a spatially distributed piezoelectric layer is laminated on a 1-D structure, such as arches, rings, beams, rods, etc. (Figure 7.5). The shape of the piezoelectric layer is defined by a 1-D shape function $W_s(\alpha_1)$. Both the piezoelectric layer and the elastic continuum have constant thickness (i.e., constant h^s and h, respectively). It is assumed that the piezoelectric material is hexagonal symmetrical such that the piezoelectric constants $e_{31} = e_{32}$. An open-circuit sensor signal from a 1-D spatially distributed orthogonal sensor can be estimated from its strains, which can be further separated into two components: a membrane component and a bending component; see Section 7.3 (Tzou, 1993).

Thus the sensor signal ϕ_3^s equation of a 1-D spatially distributed orthogonal sensor with a shape function of $W_s(\alpha_1)$ becomes

$$\begin{aligned}\phi_3^s &= -\frac{e_{31}}{\epsilon_{33}\mathbf{S}^e}\int_{\alpha_1} \operatorname{sgn}[U_3(\alpha_1)]\cdot\big(h^s\big(S_{11}^0 + S_{22}^0\big)\\ &\quad + 0.5h^s(h+h^s)(k_{11}+k_{22})\big)\int_0^{W_s(\alpha_1)} A_1A_2\,d\alpha_2\,d\alpha_1\\ &= -\frac{e_{31}}{\epsilon_{33}\mathbf{S}^e}\int_{\alpha_1} W_s(\alpha_1)\cdot\operatorname{sgn}[U_3(\alpha_1)]\cdot\big[h^s\big(S_{11}^0+S_{22}^0\big)\\ &\quad + 0.5h^s(h+h^s)(k_{11}+k_{22})\big]A_1A_2\,d\alpha_1,\end{aligned} \tag{7.96}$$

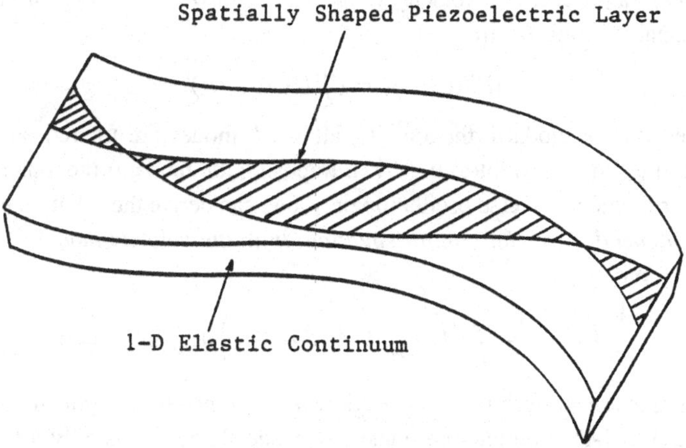

Fig. 7.5. A 1-D spatially distributed orthogonal sensor/actuator.

where e_{31} is the piezoelectric constant, ϵ_{33} is the dielectric constant, $\mathbf{S}^e$ is the effective electrode area, $W_s(\alpha_1)$ is a 1-D shape function, $U_3(\alpha_1)$ is an orthogonal function, and sgn[$\cdot$] is a signum function that defines the polarity changes of the orthogonal sensor. Note that modal functions (mode shape functions) are used to define the shape functions and orthogonal functions discussed later. The terms S_{22}^0 and k_{22} are usually neglected since $\partial(\cdot)/\partial\alpha_2 = 0$. In addition, the first term (leading by h^s) denotes the membrane strain contribution to the sensor output, and the second term (leading by $0.5h^s$) denotes the bending strain contribution. The total output signal is contributed by a summation of membrane and bending strain components. In a 1-D elastic continuum with finite radius of curvatures (e.g., arches and rings) both membrane and bending components contribute to the output signal. However, for flat 1-D continua with infinite radius of curvature (e.g., beams and rods) the output signal is contributed either by the membrane component (e.g., rods) or the bending component (e.g., beams) (Tzou, 1993).

The distributed velocity (strain-rate) feedback can be derived using the modal expansion method and a spatially distributed modal feedback force (Tzou, Zhong, Hollkamp, 1994). As discussed previously, the kth modal equation can be written as

$$\ddot{\eta}_k + \frac{c}{\rho h}\dot{\eta}_k + \omega_k^2\eta_k = \frac{1}{\rho h N_k}\sum_{j=1}^{3}\sum_{m=1}^{\infty}\int_{\alpha_1}\int_{\alpha_2}\left(\mathcal{G}_{jm}^{\mathrm{vf}}(\alpha_1,\alpha_2)\dot{\eta}_m U_{jk}\right)A_1A_2\,d\alpha_1\,d\alpha_2, \quad (7.97)$$

where $N_k = \int_{\alpha_1}\int_{\alpha_2}[\sum_{j=1}^{\infty}(U_{jk})^2]A_1A_2\,d\alpha_1\,d\alpha_2$ and $\mathcal{G}_{jm}^{\mathrm{vf}}(\alpha_1,\alpha_2)$ is the distributed *velocity modal feedback gain function*, see Sections 7.4.7 and 7.4.8. Using the modal orthogonality, one can write the distributed *velocity feedback gain function* as a product of a velocity weighting factor $\mathcal{G}_i^{\mathrm{vf}}$ (gain constant) and an orthogonal function $U_{3n}(\alpha_1,\alpha_2)$:

$$\mathcal{G}_{jm}^{\mathrm{vf}}(\alpha_1,\alpha_2) = \mathcal{G}_i^{\mathrm{vf}}U_{3n}(\alpha_1,\alpha_2). \quad (7.98)$$

Based on the modal orthogonality, all $n \neq k$ modes are filtered out when substituting Eq. (7.98) into Eq. (7.97). Imposing the modal orthogonality and considering the transverse oscillation only, one can derive the nth modal equation (*independent modal control equation*) with the *velocity modal feedback control force* as

$$\ddot{\eta}_n + \frac{1}{\rho h}\left(c - \frac{\mathcal{G}_3^{\mathrm{vf}}}{N_n}\int_{\alpha_1}\int_{\alpha_2}U_{3n}^2A_1A_2\,d\alpha_1\,d\alpha_2\right)\dot{\eta}_n + \omega_n^2\eta_n = 0. \quad (7.99)$$

Note that the subscript k is replaced by n when modal orthogonality is considered. For 1-D continua, the transverse mode shape U_{3n} is only a function of one coordinate α_1, for example, the circumferential direction in rings and arches, the longitudinal direction in beams and rods, etc. If electrode areas of the actuators are designed as 1-D shape functions of $W_a(\alpha_1)$, the modal control force for the 1-D spatially shaped actuator can be redefined and the modal equation becomes

$$\ddot{\eta}_n + \frac{1}{\rho h}\left(c - \frac{\mathcal{G}_3^{vf}}{N_n}\int_{\alpha_1} W_a(\alpha_1)\cdot U_{3n}^2 A_1 A_2\, d\alpha_1\right)\dot{\eta}_n + \omega_n^2 \eta_n = 0. \quad (7.100)$$

Note that the modal coupling and spillover from all other natural modes are eliminated. The shape functions can be defined by the modal functions. This modal filtering characteristic will be demonstrated in an experimental study of a cantilever beam laminated with self-sensing orthogonal actuators presented next.

7.6.2 Orthogonal Sensor/Actuator for a Cantilever Beam

A 1-D cantilever Bernoulli–Euler beam usually exhibits transverse oscillations only. (The in-plane longitudinal oscillation is neglected.) The Lamé parameters for a flat uniform beam are $A_1 = 1$, $A_2 = 1$; the radii are $R_1 = \infty$ and $R_2 = \infty$. In addition, $\partial(\cdot)/\partial\alpha_2 = 0$. Accordingly, the closed-loop equation of motion of a cantilever beam can be derived:

$$\rho h\ddot{u}_3 + YI\frac{\partial^4 u_3}{\partial x^4} - b\frac{\partial^2(M_{11}^a)}{\partial x^2} = bF_3, \quad (7.101)$$

where Y is Young's modulus, I is the area-moment of inertia, b is the beam width; M_{11}^a is the induced control moment, and F_3 is the external mechanical force. As discussed previously, the orthogonal modal sensors/actuators are designed based on the *modal functions* (or mode shape function) of $U_{3m}(x)$:

$$U_{3m}(x) = \frac{1}{\lambda_m^2}\frac{d^2U_{3m}(0)}{dx^2}\left(C(\lambda_m x) - \frac{A(\lambda_m L)}{B(\lambda_m L)}D(\lambda_m x)\right), \quad (7.102)$$

where the coefficients are

$$A(\lambda_m x) = 0.5[\cosh(\lambda x) + \cos(\lambda x)], \quad (7.103a)$$

$$B(\lambda_m x) = 0.5[\sinh(\lambda x) + \sin(\lambda x)], \quad (7.103b)$$

$$C(\lambda_m x) = 0.5[\cosh(\lambda x) - \cos(\lambda x)], \quad (7.103c)$$

$$D(\lambda_m x) = 0.5[\sinh(\lambda x) - \sin(\lambda x)], \quad (7.103d)$$

where x defines the distance measured from the fixed end. The eigenvalue λ_m is determined by its characteristic equation:

$$\cos(\lambda L)\cosh(\lambda L) + 1 = 0, \quad (7.104)$$

where $\lambda_1 L = 1.875$, $\lambda_2 L = 4.694$, $\lambda_3 L = 7.855$, $\lambda_4 L = 10.996$, $\lambda_5 L = 14.137$, etc. and L is the beam length. The first derivative $d[U_{3m}(x)]/dx$ is the *modal slope function* and the second derivative $d^2[U_{3m}(x)]/dx^2$ is the *modal strain function*. The modal strain function is used to define the shapes of orthogonal modal sensors/actuators:

$$\begin{aligned} U''_{3m}(x) = {} & [(e^{(\lambda L - \lambda x)}[e^{\lambda L} + \cos(\lambda L) + \sin(\lambda L)]/\{2[e^{2\lambda L} + 2e^{\lambda L}\sin(\lambda L) - 1]\}) \\ & + (-e^{\lambda x}\{e^{\lambda L}[0.5\cos(\lambda L) - 0.5\sin(\lambda L)] + 0.5\} + e^{2\lambda L}[0.5\cos(\lambda x) \\ & - 0.5\sin(\lambda x)] + e^{\lambda L}[\cos(\lambda x)\sin(\lambda L) - \sin(\lambda x)\cos(\lambda L)] \\ & - 0.5\cos(\lambda x) - 0.5\sin(\lambda x))/[e^{2\lambda L} + 2e^{\lambda L}\sin(\lambda L) - 1]]/(\lambda L)^2. \end{aligned} \tag{7.105}$$

Note that each orthogonal modal sensor/actuator has a distinct shape based on its modal strain function and eigenvalue. Detailed layouts of the spatially shaped orthogonal sensors/actuators are presented next.

7.6.3 Model Fabrication

The shapes of distributed orthogonal sensors/actuators follow the definitions of modal strain functions defined by their eigenvalues. The first four modal functions are plotted in Figure 7.6, and their modal strain functions are plotted in Figure 7.7. Note that the effective regions are from zero to one, since they are normalized in the length direction. Polymeric piezoelectric polyvinylidene fluoride (PVDF) is flexible and easy to cut into various shapes in a laboratory

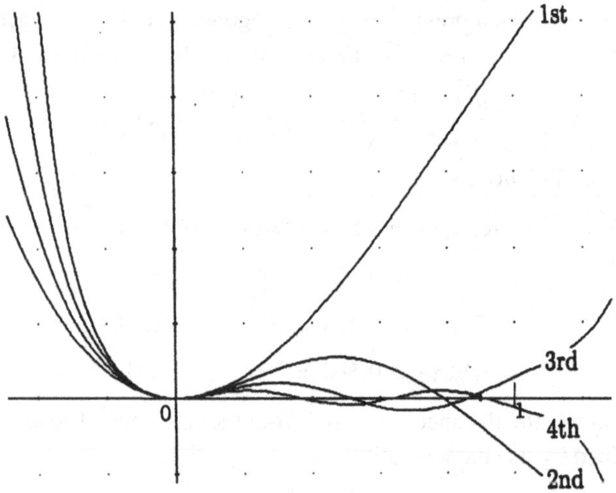

Fig. 7.6. Mode shape functions of the cantilever beam.

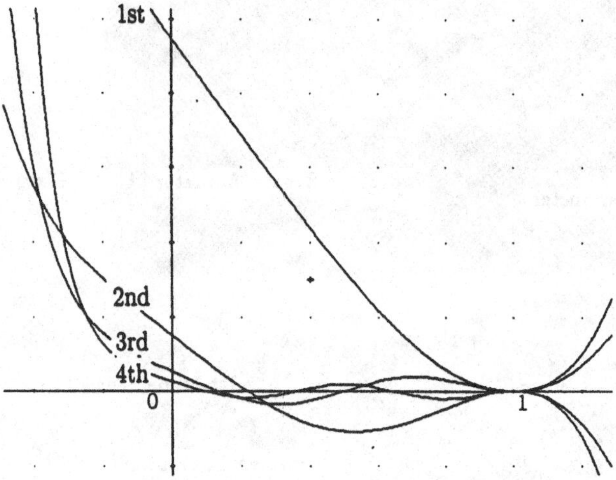

Fig. 7.7. Modal strain functions of the cantilever beam.

environment. In this study, a 40 μm PVDF sheet is used for the orthogonal sensors/actuators and the effective axis is aligned with the beam length. These sensor/actuator layers are cut according to their strain functions and then glued on a plexiglas beam ($15 \times 1 \times 1/8$ in). Patterns of surface electrodes are first laid out on the plexiglas beam using a thin-film silver paste to ensure good electrical conductivity. Individual silver electrodes are connected by either thin silver-paste lines (internal connections) or 0.5-mm Teflon-coated surgical wires (external connections). Polarity changes are achieved by reversing the cut PVDF sheets. The finished PVDF/plexiglas beam is shown in Figure 7.8.

7.6.4 Apparatus and Experimental Setup

A self-sensing feedback control circuit is set up with two current amplifiers and a difference-circuit (Anderson et al., 1992; Dosch et al., 1992). Three operational amplifiers (AD-711JN), six resistors (24.9 kΩ, 8 MΩ, and 16 MΩ), and a capacitor (14 nF) are used to build the circuits for the first and second orthogonal modal sensors/actuators. Figure 7.9 shows the circuit. The capacitor is used to match the capacitance of the orthogonal piezoelectric sensor/actuator. A power amplifier (BK-1651) supplies 30 V to the operational amplifiers, and a signal amplifier is used to amplify the sensor signal to induce control actions in the piezoelectric layers. A reference accelerometer (Kistler 5205) is mounted at the free end to provide a reference signal. All signals are input into an HP data acquisition system (HP3566A) for signal processing and recording.

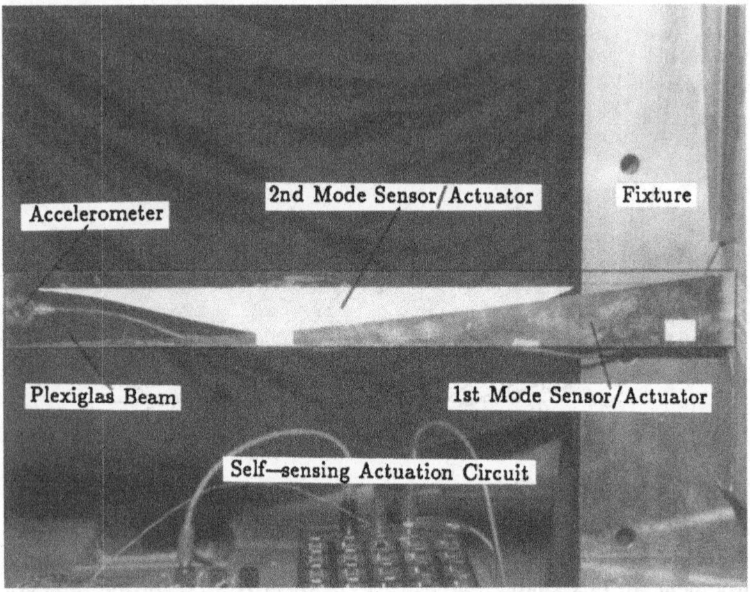

Fig. 7.8. Experimental beam model with orthogonal PVDF transducers.

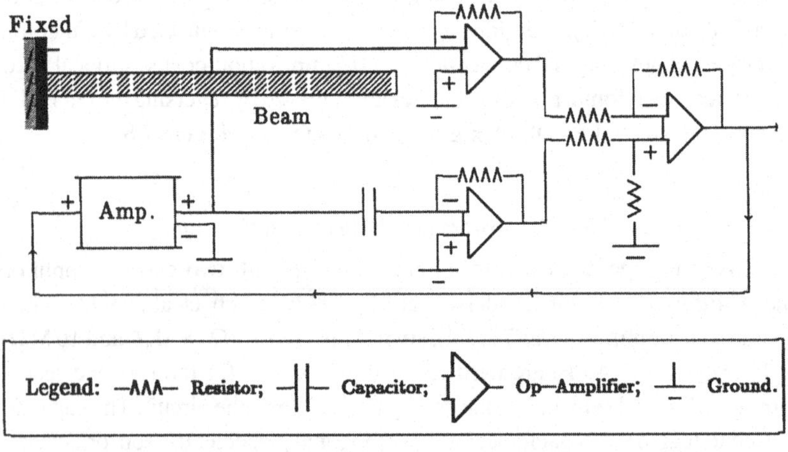

Fig. 7.9. A self-sensing feedback control circuit.

7.6.5 Experimental Procedures

There are two sets of experiments carried out in this study. The first set is to test the modal orthogonality of orthogonal modal sensors; the second set is to evaluate the control effectiveness of the self-sensing orthogonal actuators.

The first set involves two tests: 1. strain signals and 2. strain-rate signals. The strain signal is contributed by elastic bending strains of the cantilever beam; it is ultimately related to the transverse deflection u_3. Thus, the strain signal is often referred to as a "displacement" signal; the strain-rate can be regarded as a "velocity" signal. The signs of these signals are individually checked to ensure correct feedback signals in the self-sensing feedback control.

A self-sensing feedback control circuit, discussed previously, is used in the second set of experiments. Controlled time histories from the accelerometer are acquired; modal damping ratios are calculated using the eigensystem realization algorithm (ERA) method (Juang and Pappa, 1985).

7.6.6 Results and Discussion

Test results acquired from the modal signals of orthogonal modal sensors and the modal control effectiveness of the self-sensing orthogonal piezoelectric actuators are presented in this section.

7.6.6.1 Modal Sensing

To evaluate the sensing effectiveness of orthogonal modal sensors, the spatially shaped piezoelectric layers were subjected to external excitations and their dynamic responses recorded. Strain and strain-rate responses were also tested using a strain-rate circuit (Lee, 1992). Figure 7.10 shows the spectra of the accelerometer (top), the first modal sensor (bottom-1), and the second modal sensor (bottom-2). It is observed that the accelerometer senses multiple modes of the cantilever beam, and the modal sensors only respond to their respective modes. Sixty Hertz line noises also appear in these spectra.

7.6.6.2 Self-sensing and Feedback Control

In this section, free oscillation and controlled time histories are presented and their respective damping ratios are calculated.

Free Oscillations For the first mode, an initial displacement was applied to the free end and the snap-back response recorded. The free oscillation time histories of strain and strain-rate signals of the first modal sensor/actuator are plotted in Figures 7.11 and 7.12, and those of the second sensor/actuator are plotted in Figures 7.13 and 7.14, respectively. Note that those time histories of

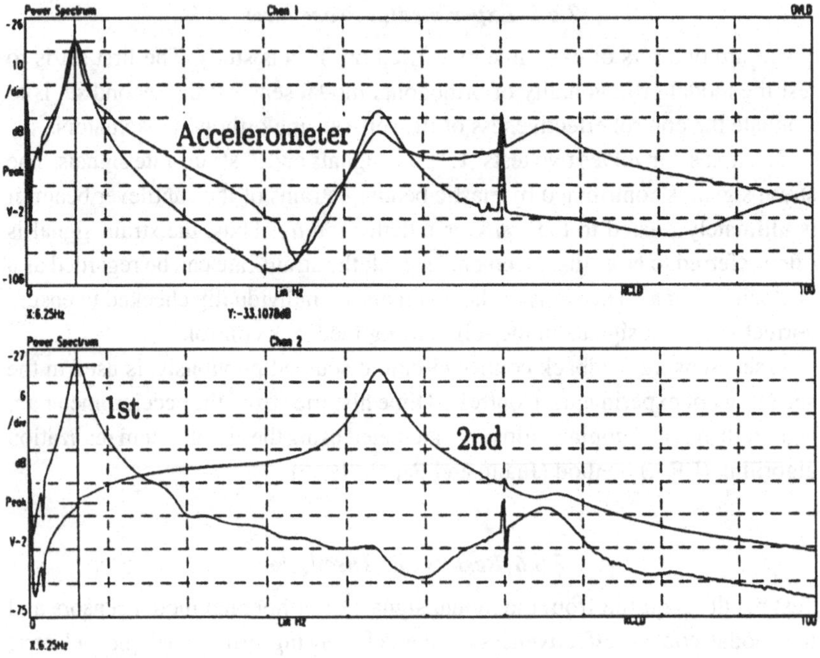

Fig. 7.10. Spectra of the orthogonal modal sensors and the accelerometer.

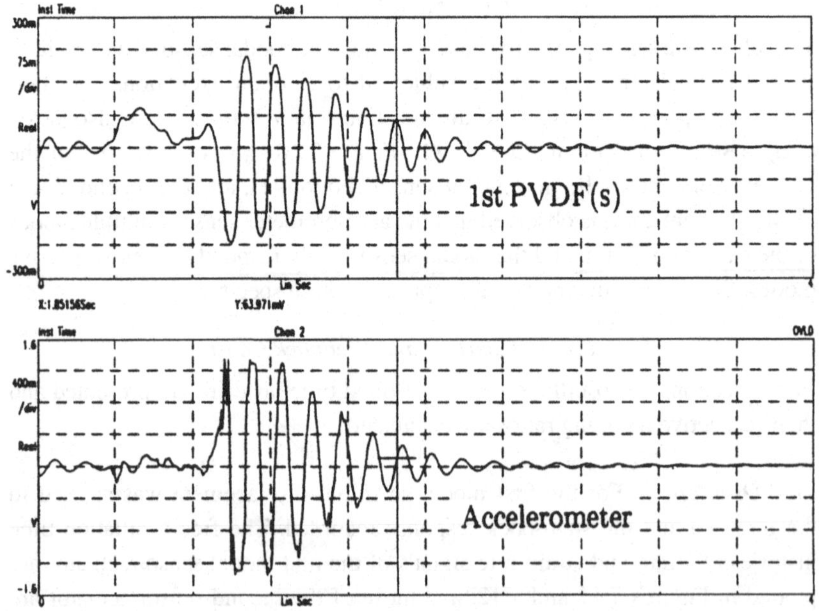

Fig. 7.11. Free oscillation of the plexiglas beam (1st strain).

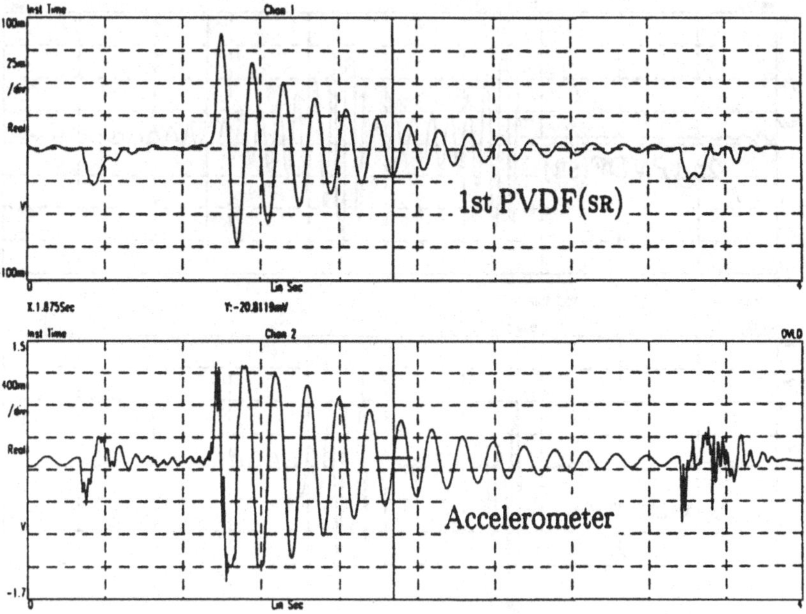

Fig. 7.12. Free oscillation of the plexiglas beam (1st strain rate).

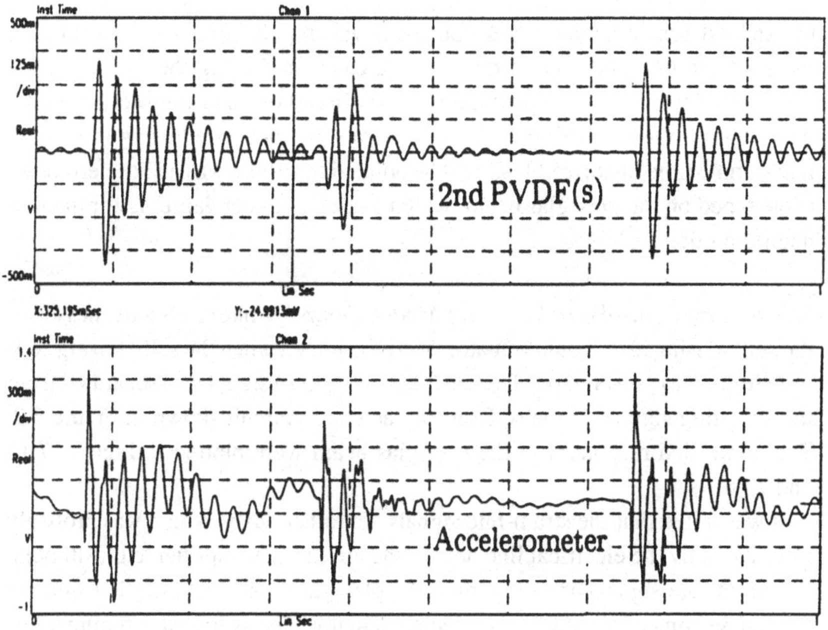

Fig. 7.13. Free oscillation of the plexiglas beam (2nd strain).

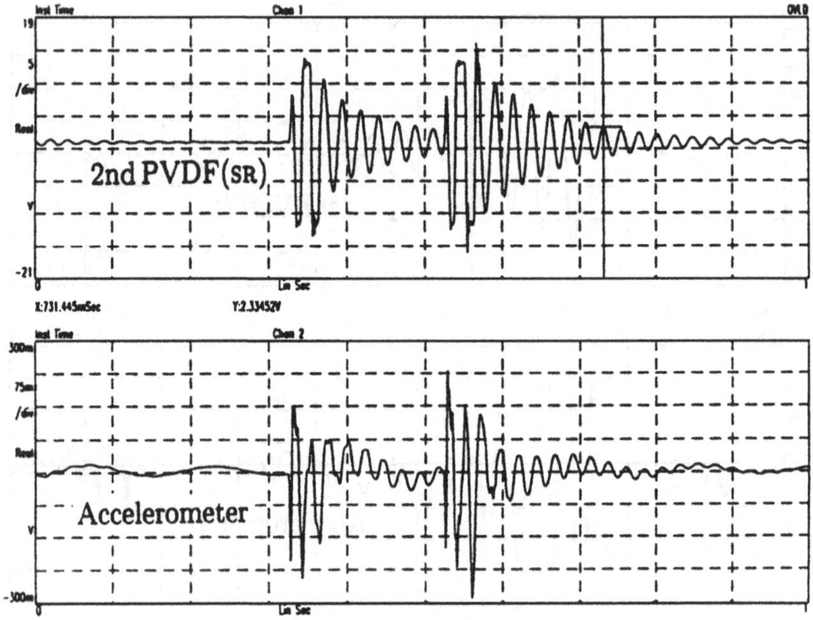

Fig. 7.14. Free oscillation of the plexiglas beam (2nd strain rate).

the second sensor/actuator were obtained via impulse excitations. Damping ratios of the first and second modes were calculated using the free vibration time histories. The first modal damping ratio is 4.0% and the second modal damping ratio is 3.4%. (Note that these data were calculated using more than five sample time histories.) It should be pointed that there was an accelerometer cable taped on the plexiglas beam, which caused a higher damping for the first natural mode.

Self-sensing Control – Independent Modal Control Control effectivenesses of the self-sensing orthogonal actuators were evaluated when the self-sensing control circuit was powered on. The sensing (strain-rate) signal was separated from the actuating signal via the self-sensing actuation circuit shown in Figure 7.9. The controlled responses of the plexiglas beam were plotted in Figures 7.15 and 7.16.

It was noted that the strain-rate signals are rather noisy, which was probably introduced by the electrical line noises, the circuit, the amplifier, etc. Although the strain-rate signals are very noisy, the plexiglas beam is controlled well via the self-sensing orthogonal piezoelectric actuator. The averaged damping ratios of the controlled responses are 7.1% for the first mode and 4.2% for the second mode. (Note that since the second mode was relatively difficult to excite by

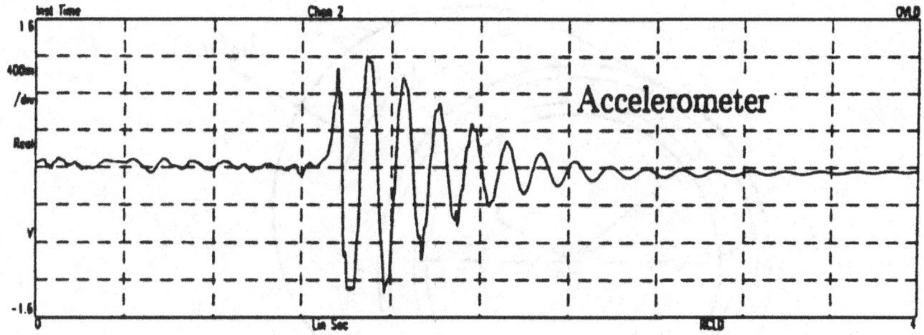

Fig. 7.15. Controlled time history of the plexiglas beam (1st).

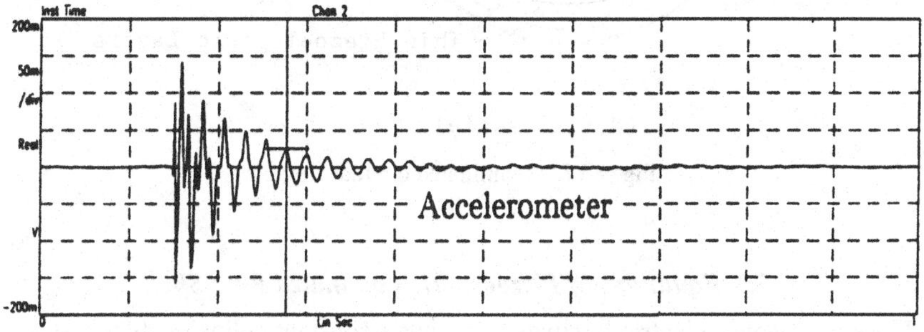

Fig. 7.16. Controlled time history of the plexiglas beam (2nd).

an impulse excitation, twelve samples were used to obtain the averaged data.) Control of the second mode could be improved by changing the resistors in the circuit. It should be pointed out that the convergence of a modal response is determined by the product of the damping ratio and the modal frequency (i.e., $e^{-\zeta_n \omega_n t}$), consequently, responses of the higher modes usually converge much faster than those of the lower modes.

7.7 Orthogonal Sensing and Control of Circular Ring Shells

In this section, distributed convolving sensors and actuators of flexible ring shells are discussed and detailed micro-electromechanics and numerical simulations of sensor/actuator effectiveness are evaluated (Tzou, Zhong, and Natori, 1993; Tzou, Zhong, and Hollkamp, 1994). As discussed in Section 7.2, an elastic ring shell is laminated with piezoelectric sensor and actuator layers on the inner and outer surfaces. Since the piezoelectric layers are thin, mechanical (i.e., mass and stiffness) properties of the piezoelectric layers are neglected. It is assumed that the piezoelectric sensor layer has a constant thickness. Figure 7.17 illustrates the piezoelectric laminated circular ring.

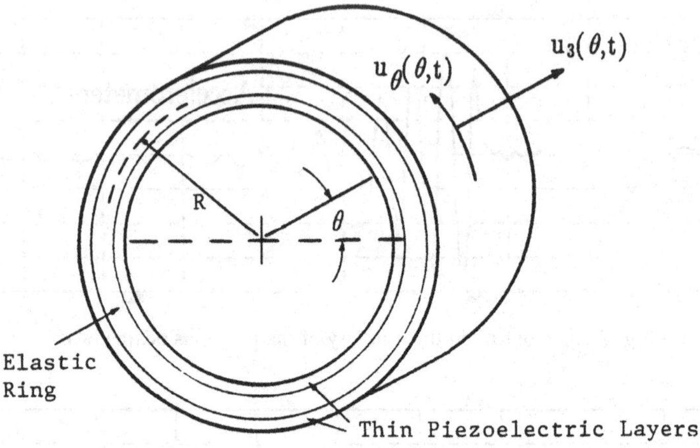

Fig. 7.17. A laminated circular ring.

7.7.1 Equations of Piezoelectric Laminated Ring Shell

The circular piezoelectric laminated ring has a constant radius R, thickness h, and width b. The fundamental form of a circular ring is

$$(ds)^2 = (R)^2(d\theta)^2 + (1)^2(dx)^2, \tag{7.106}$$

where θ defines the circumferential direction and x the width direction. α_3 is normal to the neutral surface defined by the x and θ axes. Thus the Lamé parameters are $A_1 = R$, $A_2 = 1$, where $\alpha_1 = \theta$ and $\alpha_2 = x$. The radii are $R_1 = R$ and $R_2 = \infty$. Substituting the four parameters into the thin-shell equations and simplifying them, one can derive the ring shell equations respectively in the circumferential and transverse directions:

$$\frac{D}{R^4}\left(\frac{\partial^2 u_\theta}{\partial\theta^2} - \frac{\partial^3 u_3}{\partial\theta^3}\right) + \frac{K}{R^2}\left(\frac{\partial^2 u_\theta}{\partial\theta^2} + \frac{\partial u_3}{\partial\theta}\right) = \rho h \frac{\partial^2 u_\theta}{\partial t^2}, \tag{7.107}$$

$$\frac{D}{R^4}\left(\frac{\partial^3 u_\theta}{\partial\theta^3} - \frac{\partial^4 u_3}{\partial\theta^4}\right) - \frac{K}{R^2}\left(\frac{\partial u_\theta}{\partial\theta} + u_3\right) = \rho h \frac{\partial^2 u_3}{\partial t^2}, \tag{7.108}$$

where the membrane stiffness K and bending stiffness D are defined in Section 7.2.

7.7.2 Distributed Orthogonal Ring Sensors

A closed elastic circular ring is a special case of an open arch, which has a constant radius. (Note that a beam is a special case of an open ring with an infinite curvature.) Ring-type structures are very common in mechanical, aeronautical, and astronautical applications. In this section, spatially distributed convolving orthogonal sensors for ring structures are proposed and their analytical solutions are derived. Based on the modal expansion technique (Section 7.2) the circumferential and transverse displacements for a free-floating circular ring can be expressed as

$$u_\theta(\theta, t) = \sum_{n=0}^{\infty} \eta_n(t) A_n \sin(n\theta - \varphi), \tag{7.109}$$

$$u_3(\theta, t) = \sum_{n=0}^{\infty} \eta_n(t) B_n \cos(n\theta - \varphi), \tag{7.110}$$

where, as defined previously, $\eta_n(t)$ denotes the modal participation factor, or modal coordinate; φ is an arbitrary phase angle; and A_n and B_n are constants. It is assumed that a reference point is defined so that the phase angle $\varphi = 0$.

For a circular ring with free boundary conditions, the first mode $n = 0$ is a breathing mode and the second mode $n = 1$ is a translational rigid-body mode. For $n \geq 2$, sets of a *transverse component mode* and a *circumferential component mode* appear. It should be noted that for a given integer n, there are two *component* natural frequencies ω_{n_1} and ω_{n_2} in which the former corresponds to a *transverse component mode* (i.e., the transverse oscillation dominates) and the latter to a *circumferential component mode* (i.e., the circumferential oscillation dominates) (Soedel, 1993; Tzou, 1993). Also, for low transverse component modes, the modal oscillation amplitudes of the transverse and circumferential oscillation components are coupled by a factor:

$$B_{n_1}/A_{n_1} \cong -n \quad \text{or} \quad B_{n_1} \cong -nA_{n_1}. \tag{7.111}$$

Similarly, for low circumferential component modes, the ratio between the transverse oscillation component and the circumferential component is

$$B_{n_2}/A_{n_2} \cong \frac{1}{n} \quad \text{or} \quad A_{n_2} \cong nB_{n_2}, \tag{7.112}$$

where A_{n_2} and B_{n_2} denote the modal amplitudes of circumferential component modes and A_{n_1} and B_{n_1} denote the modal amplitudes of transverse component modes. Substituting the modal equations and strain expressions into the sensor equation (Section 7.3) gives

$$\phi_3^s = -\frac{e_{31}}{\epsilon_{33} S^e} \sum_{n=0}^{\infty} \eta_n(t) \cdot \left[h^s (n A_{n_2} + B_{n_2}) + \frac{1}{2R} (r_2^2 - r_1^2)(n A_{n_1} + B_{n_1} n^2) \right]$$

$$\cdot \int_0^{2\pi} \mathrm{sgn}(W_s(\theta)) \cdot W_s(\theta) \cdot \cos(n\theta)\, d\theta. \quad (7.113)$$

Thus, the first part, membrane strains, is primarily contributed by the circumferential component modes with amplitudes A_{n_2} and B_{n_2}; the second part, bending strains, is contributed by the transverse component modes with amplitudes A_{n_1} and B_{n_1}. The modal amplitudes, either A_{n_1}/B_{n_1} or A_{n_2}/B_{n_2}, are coupled by a constant as discussed previously. Eq. (7.113) is the distributed sensor equation for spatially shaped piezoelectric ring sensors with spatially convolving electrodes.

Based on the modal orthogonality, appropriate shaped and convolved sensors only respond to the modal amplitudes of selected modes without the spillover of other residual modes. Thus, the observation spillover problem can be prevented. Since the kth mode shape is orthogonal to all other mode shapes, the shape function $W_{s^k}(\alpha_1)$ of the distributed sensor can be designed as a cosine function, that is,

$$W_{s^k}(\theta) = b' \cos(k\theta), \quad (7.114)$$

where b' is a weighting factor. For convenience, it is usually assumed that $b' = b$ – the ring width. Substituting $W_{s^k}(\theta)$ into the sensor equation and using $(r_2^2 - r_1^2) = h^s(h + h^s)$, one can derive

$$\phi_3^s = -\frac{e_{31}\pi b'}{\epsilon_{33} S_k^e} \mathrm{sgn}(\cos k\theta) \cdot \Big(h^s (k A_{k_2} + B_{k_2})$$

$$+ \frac{1}{2R} h^s (h + h^s)(k A_{k_1} + k^2 B_{k_1}) \Big) \eta_k(t), \quad (7.115)$$

where k_1 denotes the transverse component mode and k_2 the circumferential component mode for the $n = k$ natural mode. A_{k_1} and B_{k_1} are respectively the circumferential and transverse modal oscillation amplitudes of the kth transverse natural mode with a component natural frequency ω_{k_1}. A_{k_2} and B_{k_2} are the circumferential and transverse modal amplitudes of the kth circumferential natural mode with a component natural frequency ω_{k_2}.

Thus one can define two modal sensitivities: 1. a *transverse modal sensitivity* and 2. a *circumferential modal sensitivity*. (Note that the membrane modal sensitivity used for a piezoelectric shell continuum is defined as the circumferential modal sensitivity in ring sensors.) Each of the sensitivities can be further divided into two component sensitivities respectively defined in terms of either the transverse or the circumferential oscillation amplitudes. Note that these two amplitudes are coupled by a constant discussed above. The transverse modal sensitivity S_t^t defined by the transverse oscillation amplitude B_{k_1} of the distributed cosine shaped convolving sensor is

$$S_t^t = \frac{\phi_3^s}{B_{k_1}\eta_k(t)} = \frac{-e_{31}\pi b'}{\epsilon_{33}\mathbf{S}_k^e}\left(\frac{1}{2R}h^s(h+h^s)(k^2-1)\right). \qquad (7.116a)$$

The transverse modal sensitivity S_c^t defined by the circumferential oscillation amplitude A_{k_1} of the distributed sensor is

$$S_c^t = \frac{\phi_3^s}{A_{k_1}\eta_k(t)} = \frac{e_{31}\pi b'}{\epsilon_{33}\mathbf{S}_k^e}\left(\frac{1}{2R}h^s(h+h^s)(k^3-k)\right). \qquad (7.116b)$$

The circumferential modal sensitivity S_c^t defined by the transverse oscillation amplitude B_{k_2} is

$$S_t^c = \frac{\phi_3^s}{B_{k_2}\eta_k(t)} = \frac{-e_{31}\pi b'}{\epsilon_{33}\mathbf{S}_k^e}(h^s(k^2+1)), \qquad (7.117a)$$

and the circumferential modal sensitivity S_c^c defined by the circumferential oscillation amplitude A_{k_2} is

$$S_c^c = \frac{\phi_3^s}{A_{k_2}\eta_k(t)} = \frac{-e_{31}\pi b'}{\epsilon_{33}\mathbf{S}_k^e}\left(h^s\left(k+\frac{1}{k}\right)\right). \qquad (7.117b)$$

Note that $h^s = r_2 - r_1$, and $(r_2^2 - r_1^2) = h^s(h+h^s)$. These two sensor sensitivities for the distributed convolving ring sensors are analyzed. Thus, for bending oscillations where the transverse modes dominate, S^t should be used to estimate the oscillation amplitudes. In contrast, for circumferential oscillations, S^c is used. Detailed evaluation of orthogonal sensors will be presented later.

7.7.3 Distributed Orthogonal Ring Actuators

In this section, distributed modal actuators designed for flexible ring structures are proposed and their performances evaluated. Since the radius of curvature R_θ, i.e., $R_\theta = R$, is not equal to infinity, the electric bending moment and the membrane force induced by the converse piezoelectric effect are preserved in the system equation and they can be used for structural excitation and control. In closed-loop feedback control, these forces/moments can be related to the velocity or displacement signals such that damping ratios or frequencies can be manipulated. The modal participation equation of the ring structure is (Tzou, Zhong, and Hollkamp, 1994)

$$\ddot{\eta}_k + \frac{c}{\rho h}\dot{\eta}_k + \omega_k^2\eta_k = \frac{1}{\rho h N_k}\int_\theta\left(\mathrm{W_a}(\theta)\cdot\frac{N_{\theta\theta}^e}{R} - \frac{1}{R^2}\frac{\partial^2\left[\mathrm{W_a}(\theta)\cdot M_{\theta\theta}^e\right]}{\partial\theta^2}\right)U_{3k}R\,d\theta, \qquad (7.118)$$

where $N^{\mathrm{e}}_{\theta\theta}$ is the electric control force, $M^{\mathrm{e}}_{\theta\theta}$ is the control moment, $W_{\mathrm{a}}(\theta)$ is an actuator shape function defined by the modal orthogonality, and N_k is defined by the mode shape function

$$N_k = Rb \int_0^{2\pi} \cos^2(n\theta)\, d\theta = \pi Rb. \tag{7.119}$$

For spatially convolving actuators, these force/moment components can be expressed as a product of a spatially shaped weighing function $W_{\mathrm{a}}(\theta)$ and the electric forces or moments. Since the actuator is surface laminated, the electric membrane force $N^{\mathrm{e}}_{\theta\theta}$ and moment $M^{\mathrm{e}}_{\theta\theta}$ can be derived as

$$N^{\mathrm{e}}_{\theta\theta} = \int_{\alpha_3} e_{31} E_3\, d\alpha_3 = e_{31} \int_{\alpha_3} E_3\, d\alpha_3 = e_{31}\phi_3, \tag{7.120}$$

$$M^{\mathrm{e}}_{\theta\theta} = e_{31}\phi_3 \cdot r^{\mathrm{a}}_{\theta} = \frac{e_{31}(h + h^{\mathrm{a}})}{2}\phi_3, \tag{7.121}$$

where h^{a} is the actuator thickness. Substituting Eqs. (7.119)–(7.121) as well as the mode shape function into the ring modal equation gives

$$\ddot{\eta}_k + \frac{c}{\rho h}\dot{\eta}_k + \omega_k^2 \eta_k = \frac{e_{31}\phi_3}{\pi R b \rho h} \int_0^{2\pi} \left(W_{\mathrm{a}}(\theta) - \frac{h + h^{\mathrm{a}}}{2R} \frac{\partial^2 W_{\mathrm{a}}(\theta)}{\partial \theta^2} \right) \cos(n\theta)\, d\theta. \tag{7.122}$$

Note that the first term in the bracket comes from the in-plane membrane force and the second is due to the moment effect. The actuator weighting function $W_{\mathrm{a}^k}(\theta)$ (spatially convolving electrode area) is designed based on the kth mode shape function

$$W_{\mathrm{a}^k}(\theta) = b'' \cos(k\theta), \tag{7.123}$$

where b'' is a weighting constant. For convenience, let $b'' = b$, the ring width. In this case, only the kth mode can be controlled. The kth modal equation of the ring with cosine-shaped actuator can be written as

$$\begin{aligned} \ddot{\eta}_k + \frac{c}{\rho h}\dot{\eta}_k + \omega_k^2 \eta_k &= \frac{e_{31}\phi_3}{\pi R \rho h} \left(1 + \frac{k^2(h + h^{\mathrm{a}})}{2R} \right) \int_0^{2\pi} \cos(k\theta) \cdot \cos(n\theta)\, d\theta \\ &= \frac{e_{31}\phi_3}{R \rho h} \left(1 + \frac{k^2(h + h^{\mathrm{a}})}{2R} \right). \end{aligned} \tag{7.124}$$

Again, the first term is related to the in-plane membrane control effect, and the second is due to the bending control effect. The velocity feedback can be set up via the input voltage ϕ_3. For the velocity feedback, the signal taken from a sensor is differentiated, amplified, and then applied to the piezoelectric actuator, that is,

$$\phi_3 = -\mathcal{G}_r^{\mathrm{vf}} \dot{\eta}_k(t), \tag{7.125}$$

where $\mathcal{G}_r^{\text{vf}}$ is the total gain of the feedback loop, which can be further expressed as

$$\mathcal{G}_r^{\text{vf}} = S_r^{tr} \cdot \mathcal{G}_{\text{am}} \cdot \mathcal{G}_{\text{de}}, \tag{7.126}$$

where S_r^{tr} is the sensitivity of the sensor layer and $\mathcal{G}_{\text{de}}$ and $\mathcal{G}_{\text{am}}$ are differentiator and amplifier gains, respectively. Substituting the control voltage into the ring modal equation yields

$$\ddot{\eta}_k + \frac{1}{\rho h}\left[c + \frac{e_{31}}{R}\left(1 + \frac{k^2(h + h^{\text{a}})}{2R}\right)\mathcal{G}_r^{\text{vf}}\right]\dot{\eta}_k + \omega_k^2 \eta_k = 0. \tag{7.127}$$

The equivalent controlled modal damping ratio ζ_k' for the ring shell is

$$\zeta_k' = \frac{1}{2\rho h \omega_k}\left\{c + \frac{e_{31}}{R}\left(1 + \frac{k^2(h + h^{\text{a}})}{2R}\right)\mathcal{G}_r^{\text{vf}}\right\}. \tag{7.128}$$

Recall that the damping ratio $\zeta_k = c/(2\rho h \omega_k)$. The velocity feedback can be used to enhance the modal damping.

7.7.4 Evaluation of Orthogonal Ring Sensors

A steel ring structure with a radius of 50 mm, width 10 mm, and thickness 1 mm is used in this study. Piezoelectric polymeric polyvinylidene fluoride material (25 μm) is spatially shaped as distributed modal sensors. Material and geometric properties are summarized in Table 7.1. Detailed sensor electromechanics and parametric studies of transverse and circumferential component modes are analyzed and results presented in this section. Contributions from two modal oscillation amplitudes (e.g., the in-plane circumferential component and the out-of-plane transverse component) are analyzed and compared. Variations of ring and sensor thicknesses are also investigated.

As discussed previously, for a free-floating ring, the first mode, $k = 0$, is a breathing mode and the second mode, $k = 1$, is a translational rigid-body mode. For $k \geq 2$, sets of transverse and circumferential component modes, with distinct component natural frequencies, appear.

Thus, cosine-shaped distributed piezoelectric sensors are primarily designed for $k \geq 2$ modes and sensitivity analyses of distributed orthogonal sensors are also evaluated for $k \geq 2$ modes. Spatially distributed cosine-shaped convolving sensors for $k = 2$, 3, and 4 modes are shown in Figures 7.18a,b,c. Note that the sensors are defined from 0 to 2π and it is cut at $\theta = 0$. The polarity changes are also illustrated.

Transverse modal sensitivity and circumferential modal sensitivity of the distributed cosine-shaped convolving sensor are calculated and plotted for $k \geq 2$ modes. As discussed previously, there are two component natural modes for

Table 7.1. *Material and geometric properties*

Properties	Steel	PVDF	Units
Radius R	5.00×10^{-2}		m
Width b	1.00×10^{-2}	1.00×10^{-2}	m
Thickness h	1.00×10^{-3}	2.50×10^{-5}	m
Density ρ	7.80×10^{3}	1.80×10^{3}	kg/m^3
Modulus Y	2.10×10^{11}	1.60×10^{9}	N/mm^2
Poisson μ	0.300	0.29	
Piezo d_{31}/d_{32}		6.00×10^{-12}	C/N
d_{33}		13.0×10^{-12}	C/N
ϵ/ϵ_0		10	
ϵ_0		8.85×10^{-10}	F/m

each k value: a transverse component mode and a circumferential component mode, with distinct natural frequencies. Each mode consists of two modal oscillation amplitudes: 1. an in-plane circumferential oscillation component and 2. an out-of-plane transverse oscillation component. These two oscillation components are coupled by a constant. Thus, for an output signal measured, the voltage can be used to estimate both the transverse oscillation amplitude and the circumferential oscillation amplitude for a specific component natural mode at its natural frequency.

Transverse modal sensitivities, with two component sensitivities, for natural modes $k = 2, 3, \ldots, 10$ are plotted in Figure 7.19a. Circumferential modal sensitivities, with two component sensitivities, for the same modes are plotted in Figure 7.19b. In each figure, sensitivities defined by two oscillation components are plotted respectively. It is assumed that the output sensitivities are estimated at the same strain amplitude for all natural modes.

Figure 7.19a suggests that for a measured signal, the inferred transverse oscillation amplitude (bending effect) is larger than the in-plane circumferential oscillation (membrane effect) for the transverse component natural modes. This is true because the transverse oscillation amplitude has to be much more significant in transverse natural modes. Figure 7.19b also leads to a similar conclusion for the circumferential component natural modes.

Transverse and circumferential modal sensitivities are also evaluated when the thickness of the ring and the sensor layer are changed. Figures 7.20a and 7.20b show the sensitivities of a ring (5-mm thick) with a 25-μm sensor layer. Figures 7.21a and 7.21b show the sensitivities of a ring (1-mm thick) with a 40-μm piezoelectric sensor layer. (Note that the original ring was 1-mm thick coupled with a 25-μm piezoelectric sensor layer.)

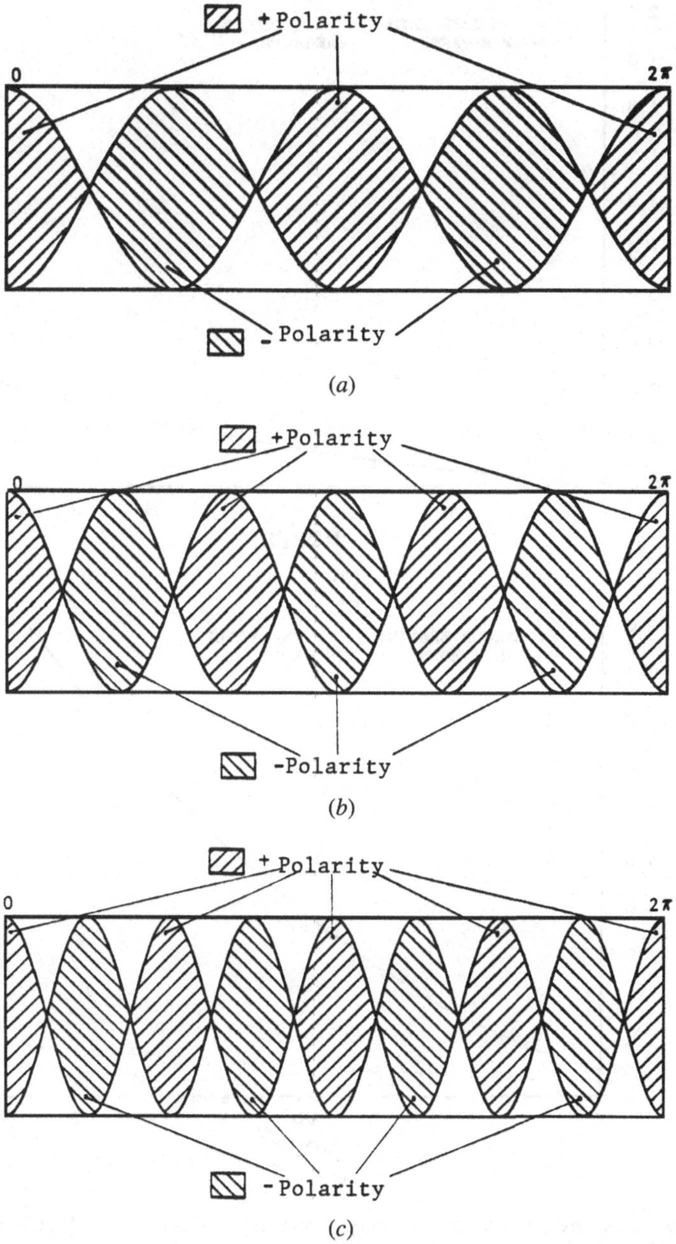

Fig. 7.18. Spatially distributed cosine shaped sensors. (*a*) Cosine shaped sensor for $k = 2$. (*b*) Cosine shaped sensor for $k = 3$. (*c*) Cosine shaped sensor for $k = 4$.

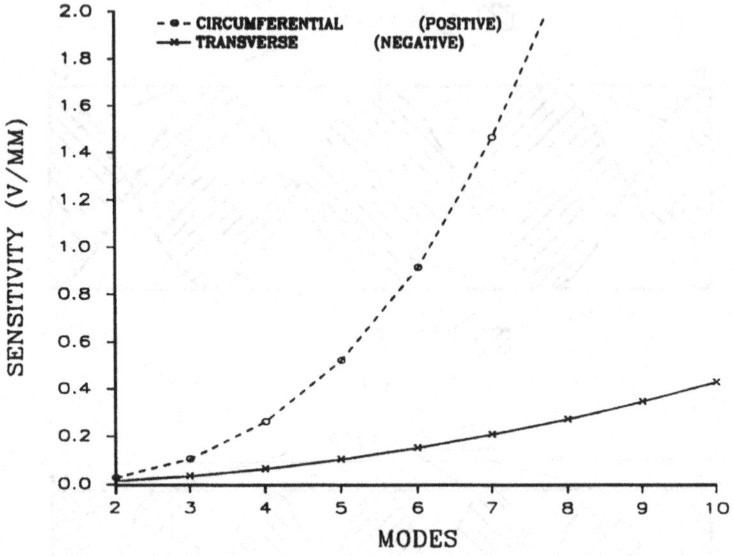

(*a*)

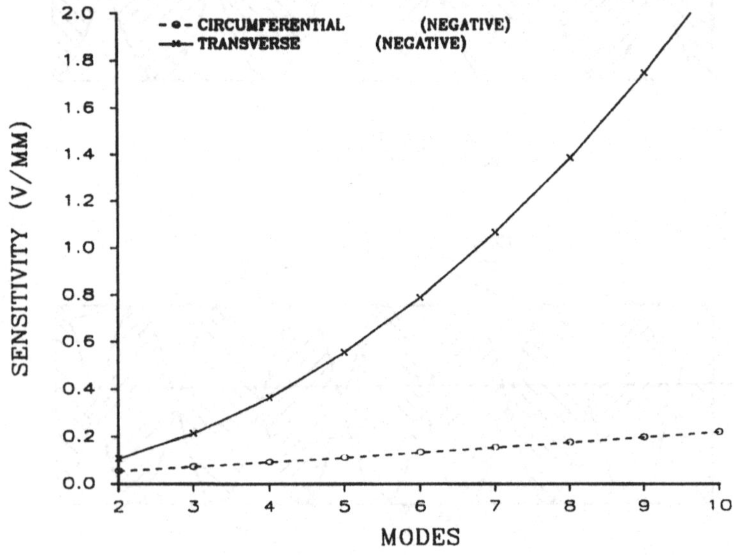

(*b*)

Fig. 7.19. (*a*) Transverse modal sensitivity (1 mm ring, 25 μm sensor). (*b*) Circumferential modal sensitivities (1 mm ring, 25 μm sensor).

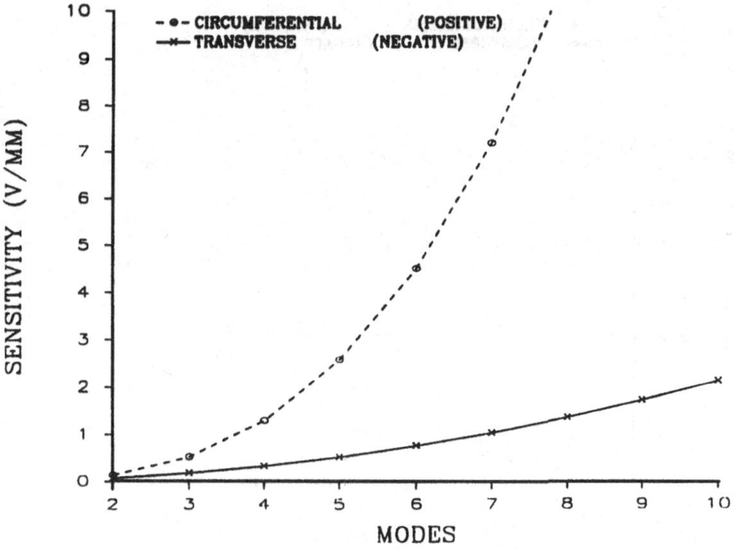

(*a*)

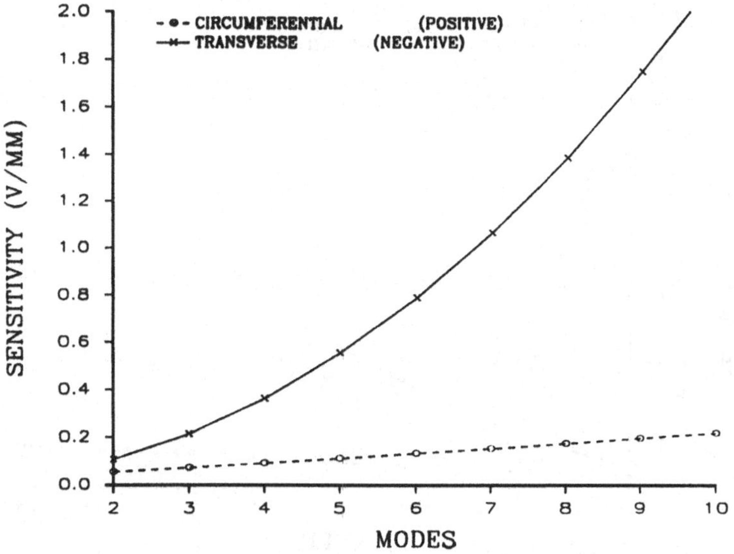

(*b*)

Fig. 7.20. (*a*) Transverse modal sensitivity (5 mm ring, 25 μm sensor). (*b*) Circumferential modal sensitivities (5 mm ring, 25 μm sensor).

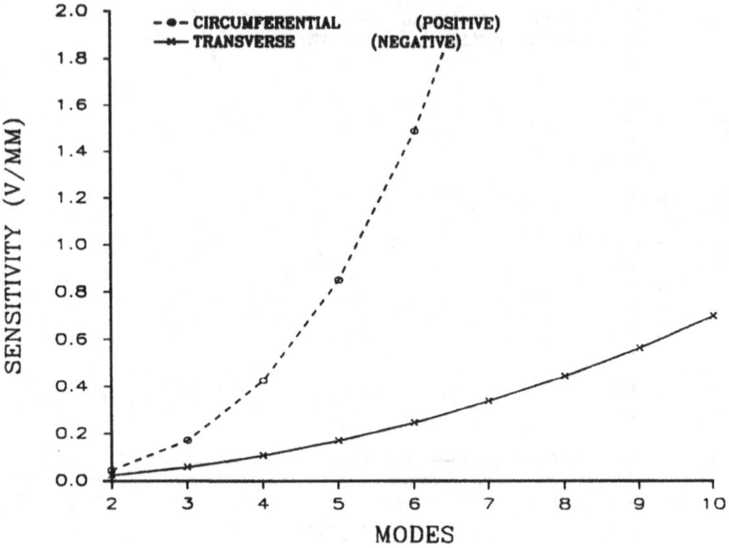

(*a*)

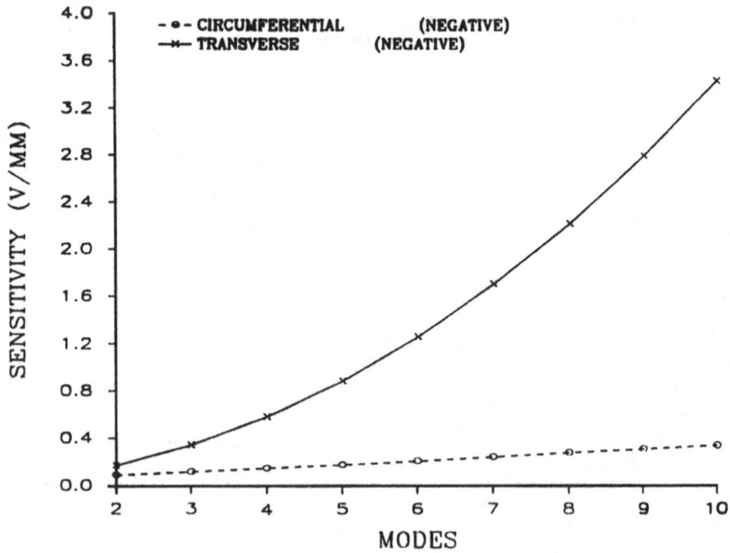

(*b*)

Fig. 7.21. (*a*) Transverse modal sensitivity (1 mm ring, 40 μm sensor). (*b*) Circumferential modal sensitivities (1 mm ring, 40 μm sensor).

In general, membrane strains (the circumferential component) in the distributed sensor should be the same regardless the ring thickness. However, bending strains (the transverse component) increase when the ring becomes thicker. Higher strains in the sensor layer generate higher output signals. Comparing Figures 7.20a and 7.19a, one can observe that the transverse sensitivity increases when the ring thickness increases. However, the circumferential sensitivities in Figures 7.20b and 7.19b are identical because the membrane strains remained unchanged. These results suggested that the membrane (circumferential) modal sensitivity is independent of ring thickness and the transverse modal sensitivity is a linear function of ring thickness. Figures 7.21a and 7.21b show that the sensitivities increase when the sensor layer becomes thicker, for the same oscillation amplitudes.

Distributed sensor electromechanics and distributed convolving sensors have been presented. Design and performance of distributed convolving cosine sensors for circular ring structures have been studied. Next, distributed convolving actuators for a flexible circular ring structure are demonstrated and their vibration control effectiveness evaluated.

7.7.5 Evaluation of Orthogonal Ring Actuators

In this section, control effectiveness of distributed cosine-shaped actuators for ring structures is evaluated. The laminated ring has a radius $R = 50$ mm, ring thickness $h = 1$ mm, and piezoelectric actuator thickness $h^{\mathrm{a}} = 0.025$ mm; see Figure 7.17 and Table 7.1. The physical model was presented in Section 7.6 and the material properties were listed in Table 7.1. The shape of the orthogonal convolving ring actuators are described by cosine functions, i.e., $W_{\mathrm{a}_k}(\theta) = b\cos(k\theta)$, which have the same shapes as those plotted in Figures 7.18a,b,c for natural modes $k = 2, 3, 4$. Note that the mode number $k = n$ in the distributed orthogonal sensor/actuator.

7.7.5.1 Comparison of Bending and Membrane Effects

The ratio of bending and membrane control is defined by

$$\Delta_r = \frac{\text{Bending}}{\text{Membrane}} = \frac{k^2(h + h^{\mathrm{a}})}{2R} = \frac{(1 + 0.025)k^2}{2 \times 50} = 0.01025k^2. \quad (7.129)$$

In order to compare the effective control effects respectively induced by the electric membrane force and the control moment, ratios of bending to membrane contributions are calculated and summarized in Table 7.2. Note that the membrane effect is assumed to be unity.

It is observed that the ratio contributed by the bending control effect varies from 1.025% for the first mode up to 25.625% for the fifth mode. The

Table 7.2. *Comparison of membrane and bending control effects*

Mode no. k	0	1	2	3	4	5
Rat i 0 Δ_r	0	1.025%	4.100%	9.225%	16.400%	25.625%

contribution of bending effects increases as the mode number increases. These calculations suggest that the membrane control effect is the dominating control action for the low natural modes when $k \leq 10$, and the bending control becomes significant when $k > 10$. Note that both membrane and bending control effects are considered in the later analyses.

7.7.5.2 Modal Damping Control

It is assumed that the circular ring is made of steel and the distributed orthogonal convolving actuators are made of polymeric piezoelectric polyvinylidene fluoride (PVDF); see Table 7.1. Note that the actuator layer is bonded on the top surface and the sensor layer on the bottom surface. Material properties of the PVDF and the bonding material are neglected. It is also assumed that the velocity signals are acquired from the distributed orthogonal convolving sensors such that observation spillover is prevented. In this section, only the velocity feedback is considered and the controlled damping ratios are studied.

The controlled damping ratios of the first three transverse modes ($k = 2, 3, 4$) are plotted versus the (normalized) system feedback gains with units of V/mm (voltage per unit displacement) in Figures 7.22a,b,c. The initial damping ratio is set at 1%. Recall that the $k = 0$ mode is a breathing mode and the $k = 1$ mode is a rigid-body mode. The distributed actuators are designed for $k \geq 2$ modes.

It is observed that the control damping ratios linearly increase with the control gains and they decrease as the mode number increases. Note that the convergence of a modal time history is determined by a combined effect of the modal natural frequency and the damping ratio (i.e., $e^{-\zeta_k \omega_k}$). Higher mode oscillations usually converge much faster than those of lower modes. It is also observed that the electric membrane force contributes the majority of total control effects and that this effect decreases as the mode number increases. The resultant control effect is a summation of the membrane control effect and the bending control effect. The control effect from the electric bending moments is relatively less sensitive than that from the electric membrane forces for different modes.

Damping controls of a thinner (0.5-mm) circular ring with the 25-μm actuator and a 1-mm ring with a thicker (40-μm) piezoelectric actuator are respectively plotted in Figures 7.23 and 7.24.

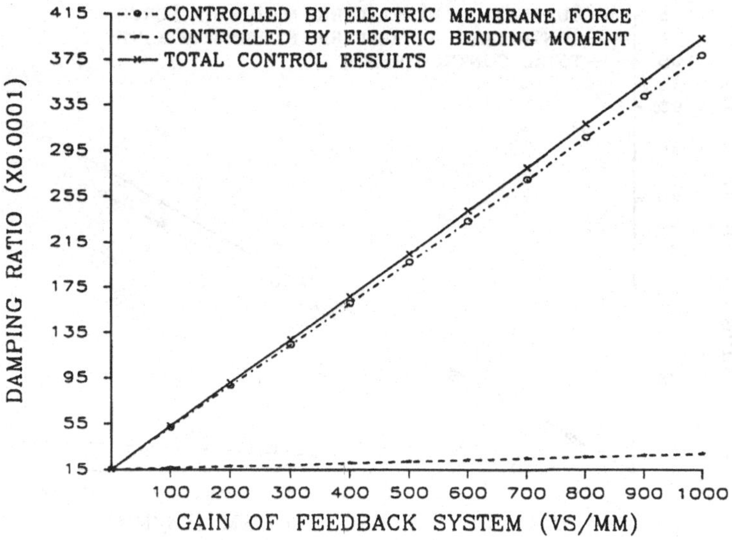

(*a*)

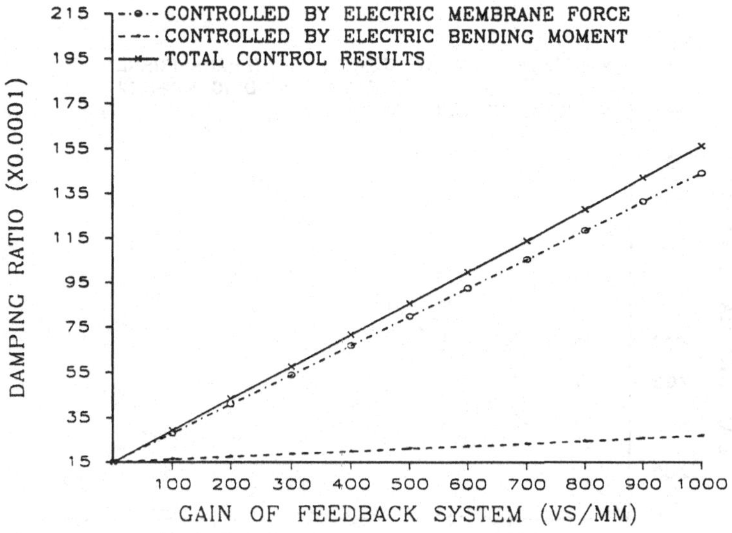

(*b*)

Fig. 7.22. Damping control of a circular ring (1 mm, 25 μm) with (*a*) $k = 2$; (*b*) $k = 3$; (*c*) $k = 4$.

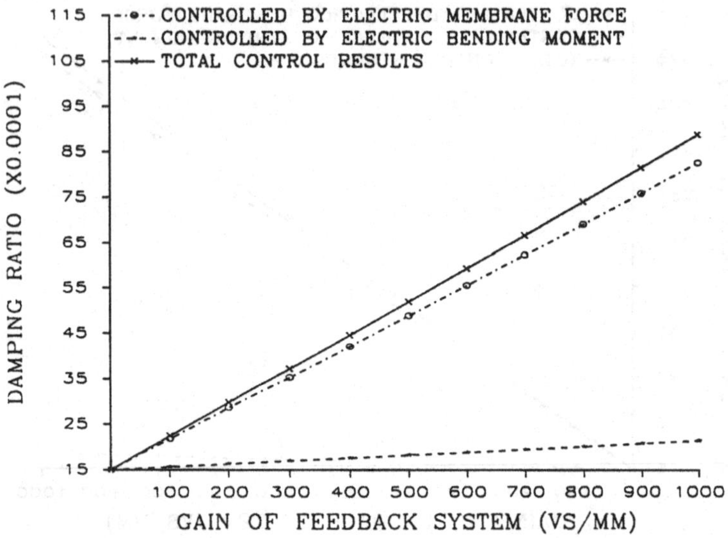

(*c*)

Fig. 7.22. (*cont.*)

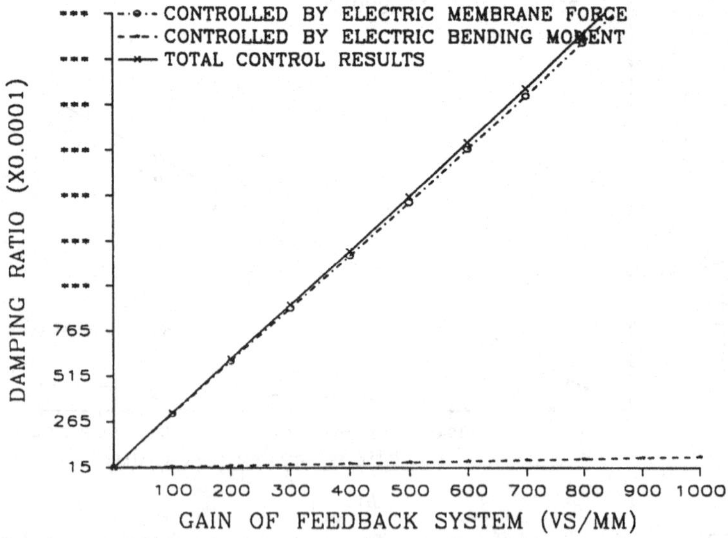

Fig. 7.23. Damping control of a circular ring (0.5 mm, 25 μm) with $k = 2$.

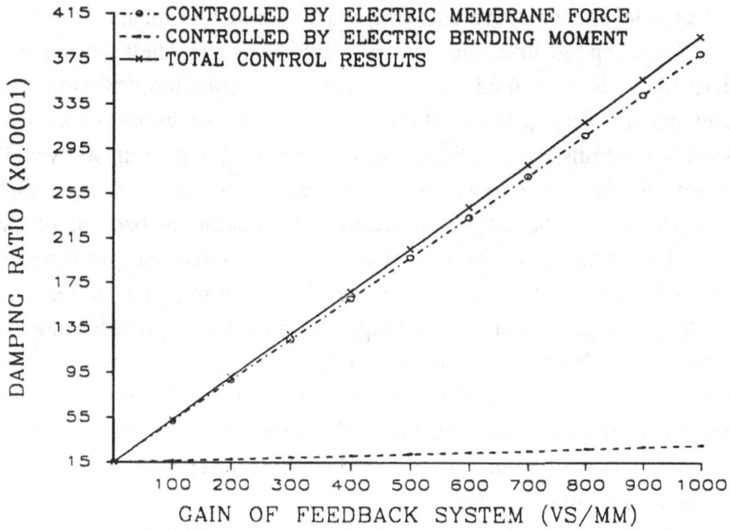

Fig. 7.24. Damping control of a circular ring (1 mm, 40 μm) with $k = 2$.

When the elastic ring becomes thinner, the ring becomes more flexible and the control effect increases rather significantly, about five times. It is observed that the reduction of control moment arm, from $(1.025/2)$ mm to $(0.525/2)$ mm, is insignificant because the control action primarily comes from the in-plane circumferential control forces. Consequently, the control effect to the flexible ring becomes much more prominent. The control contribution from the 40-μm actuator is relatively insignificant. (Note that this may not be the case when the primary control action comes from the electric bending moments, e.g., for $k > 10$ modes.) In general, for the control effectiveness, the change of structural flexibility outweighs the moment arm change if the distributed actuator is made of PVDF materials. Using piezoceramic materials usually increases the control effect. However, mass and stiffness of piezoceramic actuators need to be considered (Tzou and Bao, 1994, 1995a). It should be noted that the gain used in these figures is the *system gain* of the total control system. In a real control system design, it includes the sensitivity of the distributed modal sensor, the gain of the amplifier, and the gain of the derivative circuit (to acquire the velocity information).

7.8 Summary and Conclusions

Distributed sensing and control of a generic flexible elastic shell continua – a generic distributed parameter system – using distributed transducers were investigated in this study. The generic deep elastic shell was laminated with piezoelectric layers respectively serving as a distributed sensor and a distributed actuator. Based on the direct piezoelectric effect, a generic shell sensing theory was derived, and it was extended to include modal contribution, orthogonal sensing, and spatial filtering. It was observed that the sensor output is contributed by membrane strains and bending strains experienced in the sensor layer. Two sensor sensitivities; a *bending* or *transverse modal sensitivity* and a *membrane modal sensitivity* can be defined accordingly. In general, the bending or transverse modal sensitivity is defined for out-of-plane transverse (bending-type) natural modes and the membrane modal sensitivity for in-plane natural modes. Proper design of distributed sensor shape and convolution provides modal filtering to prevent observation spillover in distributed structural control systems. Spatial distributed and shaped orthogonal shell sensors based on the modal orthogonality concept were proposed and associated theory derived. Voltage distribution contours (*modal voltages* or *potential maps*) of shell sensors were studied and limitations discussed.

Distributed actuations induced by the distributed piezoelectric actuator, due to the converse piezoelectric effect, were investigated. Distributed membrane control forces and control moments were formulated and integrated with the elastic shell equations. Depending on the reference sensor signal, these shell system equations can be either open loop or closed loop. Generic displacement, velocity, and acceleration feedback algorithms were discussed and their respective merits evaluated. In addition, based on the modal orthogonality and spatial shaping, distributed orthogonal shell actuators were proposed and their spatial modal filtering characteristics discussed. Based on four geometric parameters (i.e., two Lamé parameters and two radii of curvatures) the original shell sensor, control, and closed-loop system equations can be simplified to apply to a variety of common geometries, for example, plates, rings, cylinders, spheres, beams, cylindrical shells, conical shells, shells of revolution, etc. Procedures and examples were also provided to show the utilities.

Self-sensing actuation provides a perfect collocation of sensors and actuators in closed-loop control systems. To separate control actions for different natural modes (independent modal control), spatially distributed self-sensing orthogonal piezoelectric actuators were proposed and evaluated. Spatially distributed orthogonal sensors/actuators were designed based on the modal strain functions of a cantilever beam. A physical model was fabricated and its self-sensing control effectiveness tested. Polymeric piezoelectric PVDF sheets (40 μm) were cut into shapes and laminated on a plexiglas beam. Surface electrodes were connected by either silver pastes or surgical wires. A self-sensing feedback control circuit was set up and tested. Experimental results showed that the orthogonal modal sensors are sensitive to their respective modes. Free and controlled (via the self-sensing feedback control circuit) time histories were recorded and their modal damping ratios calculated. The calculated results suggested that the modal damping ratios were enhanced by 77.5% for the first mode and by 23.5% for the second mode. Note that the convergence of modal responses is determined by the product of the modal damping and the modal frequency. The independent modal control of continua can be effectively achieved by using the spatially distributed self-sensing orthogonal piezoelectric actuators.

Cosine-shaped piezoelectric convolving modal sensors and actuators were designed and analyzed for flexible circular rings. A *transverse modal sensitivity* for transverse *component* natural modes and a *circumferential modal sensitivity* (equivalent to the *membrane modal sensitivity* in shell sensors) for circumferential *component* natural modes were defined and results studied. Parametric studies suggested that the transverse modal sensitivity increases when the ring structure becomes thicker because the bending strains in the sensor layer increase. However, the circumferential modal sensitivity remained unchanged because the membrane strains were independent of the ring thickness. It was also observed that both modal sensitivities increase when the piezoelectric sensor layer becomes thicker.

Control effectiveness of distributed orthogonal modal actuators for ring structures was also evaluated. Analyses suggested that the primary control action comes from the in-plane membrane control forces; the contribution from the electric bending moment was relatively insignificant for natural modes $k < 10$. In addition, structural flexibility was observed to be of importance in feedback controls. It was observed that control effect increases as the structural stiffness decreases. Increasing control moment arm (as the ring becomes thicker) of the actuator layer seems insignificant in overall control effects for lower natural modes. Theoretical and parametric studies of distributed piezoelectric transducers presented in this chapter provide a better understanding of fundamental sensing/actuation electromechanics and functionalities of spatially distributed piezoelectric shell transducers. Proper designs and selections of sensor/actuator thickness, shape, and convolution can provide spatial modal filtering and prevent observation and control spillovers in distributed control systems.

In recent years, significant research and development activities focus on integrating active electromechanical materials (such as piezoelectrics, shape-memory alloys, electrostrictive materials, magnetostrictive materials, electrorheological fluids, etc.) and control electronics with elastic continua such that the elastic continua transform from completely passive systems to active and adaptive *structronic* (*struc*ture-elec*tronic*) *systems* – *smart structures* or *intelligent structural systems*. In addition, with the rapid development of VLSI technologies and miniature computer systems, adding artificial "intelligence" to the continua could also become a reality in the near future. Accordingly, distributed sensing and control of distributed systems can be automatically and optimally carried out to guarantee optimal performance and stability of high-performance distributed systems.

Acknowledgment

This research was supported by grants from the Wright Laboratory (Flight Dynamics Directorate)/AFOSR, Army Research Office, National Science Foundation, and the Commonwealth of Kentucky. These support sources are gratefully acknowledged. Information presented here does not necessarily reflect the position or the policy of the government, nor should official endorsement be inferred.

References

Anderson, E. H., Hagood, N. W., and Goodliffe, J. M., 1992, "Self-Sensing Piezoelectric Actuation: Analysis and Application to Controlled Structures," *AIAA Paper AIAA-92-2465-CP*, 33rd Structures, Structural Dynamics and Materials Conference.

Baily, T., and Hubbard, J. E., 1985, "Distributed Piezoelectric Polymer Active Vibration Control of a Cantilever Beam," *J. Guidance, Control, and Dynamics*, Vol. 8, No. 5, pp. 605–11.

Balas, M. J., 1978, "Active Control of Flexible Systems," *Journal of Optimization Theory and Applications*, Vol. 25, No. 3, pp. 415–36.

Balas, M. J., 1988, "Nonlinear Finite-Dimensional Control of a Class of Nonlinear Distributed Parameter Systems Using Residual Mode Filters," *Recent Development in Control of Nonlinear and Distributed Parameter Systems*, ASME-DSC Vol. 10, pp. 19–22.

Baz, A., and Poh, S., 1988, "Performance of an Active Control System with Piezoelectric Actuators," *Journal of Sound and Vibration*, Vol. 126, No. 8, pp. 327–43.

Birman, V., 1992, "Theory of Geometrically Nonlinear Composite Plates with Piezoelectric Stiffeners," *Active Control of Vibration and Noise*, ASME-DSC Vol. 38, pp. 231–37.

Brichkin, L. A., Butkovskii, A. G., and Pustyl'nikou, L. M., 1973, "Application of Finite Integral Transformation to Optimal Control Problems," *Automatika Telemekhanika*, Vol. 7, pp. 13–24.

Burke, S., and Hubbard, J. E., 1990, "Distributed Transducer Control Design for Thin Plates," *Electro-Optical Materials for Switches, Coatings, Sensor Optics, and Detectors – 1990*, SPIE Vol. 1307, pp. 222–31.

Busch-Vishniac, I. J., 1990, "Spatially Distributed Transducers: Part 2," *ASME Journal of Dynamic Systems, Measurements, and Control*, Vol. 112, pp. 381–90.

Butkovskii, A. G., 1962, "The Maximum Principle for Optimum Systems with Distributed Parameters," *Automation and Remote Control*, Vol. 22, pp. 1429–38.

Collins, S. A., Miller, D. W., von Flotow, A. H., 1994, "Distributed Sensors as Spatial Filters for Active Structural Control," *Journal of Sound and Vibration*, Vol. 173, No. 4, pp. 471–501.

Crawley, E. F., and Anderson, E. H., 1990, "Detailed Models of Piezoceramic Actuation of Beams," *Journal of Intelligent Material Systems*," Vol. 1.1, pp. 4–25.

Crawley, E. F., and de Luis, J., 1987, "Use of Piezoelectric Actuators as Elements of Intelligent Structures," *AIAA Journal*, Vol. 25, No. 10, pp. 1373–85.

Crawley, E. F., and Lazarus, K. S., 1991, "Induced Strain Actuation of Anisotropic Plates," *AIAA Journal*, Vol. 29, No. 6, pp. 944–51.

Cudney, H. H., Inman, D. J., and Oshman, Y., 1989, "Distributed Parameter Actuators for Structural Control," *Proceedings of 1989 American Control Conference*, pp. 1189–94.

Detwiler, D. T., Shen, M.-H., and Venkayya, V. B., 1994, "Two-dimensional Finite Element Analysis of Laminated Composite Plates Containing Distributed Piezoelectric Actuators and Sensors," *Proc. 35th AIAA/ASME Adaptive Structures Forum*, pp. 451–60, Hilton Head, SC, April 21–22, 1994.

Dimitriadis, E. K., Fuller, C. R., and Rogers, C. A., 1991, "Piezoelectric Actuators for Distributed Vibration Excitation of Thin Plates," *ASME Journal of Vibration and Acoustics*, Vol. 113, No. 1, pp. 100–7.

Dosch, J. J., Inman, D., and Garcia, E., 1992, "A Self-Sensing Piezoelectric Actuator for Collocated Control," *J. of Intelligent Material Systems and Structures*, Vol. 3, pp. 166–85.

Fanson, J. L., and Garba, J. A., 1988, "Experimental Studies of Active Members in Control of Large Space Structures," *AIAA Paper 88-2207, Proc. of AIAA/ASME/AHS 29th Structures, Structural Dynamics, and Materials Conference*, pp. 9–17.

Gandhi, M. V., and Thompson, B. S., 1992, *Smart Materials and Structures*, Chapman and Hall, New York.

Gu, Y, Clark, R. L., Fuller, C. R., and Zander, A. C., 1994, "Experiments on Active Control of Plate Vibration Using Piezoelectric Actuators and Polyvinylidene Fluoride Modal Sensors, *ASME Journal of Vibration and Acoustics*, Vol. 116, pp. 303–8.

Gustafson, D., and Speyer, J., 1976, "Linear Minimum Variable Filters Applied to Carrier Tracking," *IEEE Transactions on Automatic Control*, Vol. AC-21, pp. 65–73.

Hagood, N., Chung, W., and von Flotow, A., 1990, "Modeling of Piezoelectric Actuator Dynamics for Active Structural Control," *AIAA Paper No. 90-1087, 31st Structures, Structural Dynamics and Materials Conference*, Long Beach, CA, April 2–4, 1990.

Hanagud, S., and Obal, M. W., 1988, "Identification of Dynamic Coupling Coefficients in a Structure with Piezoelectric Sensors and Actuators," *Proc. of AIAA/ASME/AHS 29th Structures, Structural Dynamics, and Materials Conference*, (Paper No. 88-2418), Part 3, pp. 1611–20.

Hubbard, J. E., and Burke, S. E., 1992, "Distributed Transducer Design for Intelligent Structural Components," in *Intelligent Structural Systems*, eds. Tzou, H. S., and Anderson, G. L., Kluwer Academic Publishers, Dordrecht/Boston/London, pp. 305–24.

Juang, J. N., and Pappa, R. S., 1985, "An Eigensystem Realization Algorithm for Modal Parameter Identification and Model Reduction," *J. of Guidance and Control*, Vol. 8, No. 5, pp. 620–27.

Lee, C. K., 1990, "Theory of Laminated Piezoelectric Plates for the Design of Distributed Sensors/Actuators, Part 1: Governing Equations and Reciprocal Relationships," *Journal of Acoustic Society of America*, 87(3), 1144–58.

Lee, C. K., 1992, "Piezoelectric Laminates: Theory and Experimentation for Distributed Sensors and Actuators," in *Intelligent Structural Systems*, eds. H. S., Tzou, and Anderson, G. L., Kluwer Academic Publishers, Dordrecht/Boston/London, pp. 75–167.

Lee, C. K., and Moon, F. C., 1990, "Modal Sensors/Actuators," *ASME Journal of Applied Mechanics*, Vol. 57, pp. 434–41.

Lions, J. L., 1968, *Optimal Control of Systems Governed by Partial Differential Equations*, Dunod and Gauthier-Villars, Paris.

Liu, B., and Tzou, H. S., 1996, "Photodeformation and Light–Temperature–Electric

Coupling of Optical Actuators: Part 1: Parameter Calibration and Part 2: Vibration Control," *Proceedings of ASME Aerospace Division*, AD Vol. 52, pp. 663–678, 1996 ASME International Congress.
Meirovitch, L, 1988, "Control of Distributed Systems," in *Large Space Structures: Dynamics and Control*, eds. Atluri, S. N., and Amos, A. K., Springer-Verlag, Berlin, Heidelberg, pp. 193–212.
Meirovitch, L., and Baruh, 1981, "Effect of Damping on Observation Spillover Instability," *Journal of Optimization Theory and Applications*, Vol. 35, No. 1, Sept. pp. 31–44.
Meirovitch, L., and Baruh, H., 1983, "On the Problem of Observation Spillover in Self-Adjoint Distributed-Parameter Systems," *Journal of Optimization Theory and Applications*, Vol. 39, No. 2, pp. 269–91.
Oz, H., and Meirovitch, 1983, "Stochastic Independent Modal-Space Control of Distributed-Parameter Systems," *Journal of Optimization Theory and Applications*, Vol. 40, No. 1, pp. 121–54.
Pai, P. F., Nafeh, A. H., Oh, K., and Mook, D. T., 1993, "A Refined Nonlinear Model of Piezoelectric Plate with Integrated Piezoelectric Actuators and Sensors," *Intl. J. Solids and Structures*, Vol. 30, pp. 1603–30.
Plump, J. M., Hubbard, J. E., and Baily, T., 1987, "Nonlinear Control of a Distributed System: Simulation and Experimental Results," *ASME J. Dynamic Systems, Measurement, and Control*, pp. 133–39.
Qiu, J., and Tani, J., 1994, "Vibration Control of a Cylindrical Shell Using Distributed Piezoelectric Sensors and Actuators," *Proceedings of the Second International Symposium on Intelligent Materials*, pp. 1003–14.
Rao, S. S., and Sunar, M., 1993, "Analysis of Distributed Thermopiezoelectric Sensors and Actuators in Advanced Intelligent Structures," *AIAA Journal*, Vol. 31, No. 7, pp. 1280–86.
Ricketts, D., 1981, "Model for a Piezoelectric Polymer Flexural Plate Hydrophone," *Journal of Acoustic Society of America*, Vol. 70, No. 4, pp. 929–35.
Ricketts, D., 1989, "The Frequency of Flexible Vibration of Completely Free Composite Piezoelectric Polymer Plates," *Journal of Acoustic Society of America*, Vol. 80, No. 3, pp. 2432–39.
Robinson, A. C., 1971, "A Survey of Optimal Control of Distributed Parameter Systems," *Automatica*, Vol. 7, pp. 371–88.
Rogers, C. A. (ed.), 1988, *Smart Materials, Structures, and Mathematical Issues*, Technomic Publ.
Sakawa, Y., 1966, "Optimal Control of Certain Type of Linear Distributed-Parameter Systems," *IEEE Trans. on Automatic Control,* Vol. AC-11, pp. 35–41.
Soedel, W., 1993, *Vibrations of Shells and Plates*, Marcel Dekker, New York.
Suleman, A., and Venkayya, V. B., 1994, "Flutter Control of an Adaptive Composite Panel," *Proc. 35th AIAA/ASME Adaptive Structures Forum*, pp. 118–26, Hilton Head, SC, April 21–22, 1994.
Tzafestas, S. G., 1970, "Optimal Distributed-Parameter Control Using Classical Variational Theory," *Int. J. Control*, 12, pp. 593–608.
Tzou, H. S., 1987, "Active Vibration Control of Flexible Structures Via Converse Piezoelectricity," *Developments in Mechanics*, Vol. 14-C, pp. 1201–6.
Tzou, H. S., 1988, "Integrated Sensing and Adaptive Vibration Suppression of Distributed Systems," *Recent Development in Control of Nonlinear and Distributed Parameter Systems*, ASME-DSC Vol. 10, pp. 51–58, 1988 ASME Winter Annual Meetings.
Tzou, H. S., 1989a, "Theoretical Development of a Layered Thin Shell with Internal Distributed Controllers," *Failure Prevention and Reliability – 1989*, pp. 17–20, 1989 ASME Technical Design Conference, Montreal, Canada, Sept. 17–20.

Tzou, H. S., 1989b, "Integrated Distributed Sensing and Active Vibration Suppression of Flexible Manipulators Using Distributed Piezoelectrics," *Journal of Robotic Systems*, Vol. 6, No. 6, pp. 745–67.

Tzou, H. S., 1991, "Distributed Modal Identification and Vibration Control of Continua: Theory and Applications," *ASME Journal of Dynamic Systems, Measurements, and Control*, Vol. 113, No. 3, pp. 494–99.

Tzou, H. S., 1992, "A New Distributed Sensor and Actuator Theory for "Intelligent" Shells," *Journal of Sound and Vibration*, Vol. 153, No. 2, pp. 335–50.

Tzou, H. S., 1993, *Piezoelectric Shells (Distributed Sensing and Control of Continua)*, Kluwer Academic Publ., Dordrecht/Boston/London.

Tzou, H. S., and Anderson, G. L. (eds.), 1992. *Intelligent Structural Systems*, Kluwer Academic Publishers, Dordrecht/Boston.

Tzou, H. S., and Bao, Y., 1994, "Modeling of Thick Anisotropic Triclinic Piezoelectric Shell Transducer Laminates," *Journal of Smart Materials and Structures*, Vol. 3, pp. 285–92.

Tzou, H. S., and Bao, Y., 1995a, "A Theory on Anisotropic Piezothermoelastic Shell Laminae with Sensor and Actuator Applications," *Journal of Sound and Vibration*, Vol. 184, No. 3, pp. 453–73.

Tzou, H. S., and Bao, Y., 1995b, "Dynamics and Control of Adaptive Shells with Curvature Transformations," *Shock and Vibration Journal*, Vol. 2, No. 2, pp. 143–54.

Tzou, H. S., Bao, Y., and Ye, R., 1994, "A Theory on Nonlinear Piezothermoelastic Shell Laminates," SPIE Paper No. 2190-19, *Proceedings of SPIE Smart Structures and Materials Conference*, Orlando, FL, February 1994.

Tzou, H. S., and Fu, H. Q., 1994, "A Study of Segmentation of Distributed Piezoelectric Sensors and Actuators, Parts 1 and 2," *Journal of Sound and Vibration*, Vol. 172, No. 2, pp. 247–76.

Tzou, H. S., and Fukuda T. (eds.), 1992, *Precision Sensors, Actuators, and Systems*, Kluwer Academic Publishers, Dordrecht/Boston/London.

Tzou, H. S., and Gadre, M., 1988, "Active Vibration Suppression by Piezoelectric Polymer with Variable Feedback Gain," *AIAA Journal*, Vol. 26, No. 8, pp. 1014–17.

Tzou, H. S., and Gadre, M., 1989, "Theoretical Analysis of a Multi-Layered Thin Shell Coupled with Piezoelectric Shell Actuators for Distributed Vibration Controls," *Journal of Sound and Vibration*, Vol. 132, No. 3, pp. 433–50.

Tzou H. S., and Gadre, M., 1990, "Active Vibration Isolation and Excitation by a Piezoelectric Slab with Constant Feedback Gains," *Journal of Sound and Vibration* Vol. 136, No. 3, pp. 477–90.

Tzou, H. S., Guran, A., Anderson, G. L., Natori, M. C., Gabbert, U., Tani, J., and Breitbach, E., (eds.), 1998, *Structronic Systems: Smart Structures, Devices, and Systems*, World Scientific Publishers, River Edge, N.J., and Singapore.

Tzou, H. S., and Hollkamp, J. J., 1994, "Collocated Independent Modal Control with Self-Sensing Orthogonal Piezoelectric Actuators (Theory and Experiments)," *Journal of Smart Materials and Structures*, Vol. 3, pp. 277–84.

Tzou, H. S., and Howard, R. V., 1994, "A Piezothermoelastic Thin Shell Theory Applied to Active Structures," *ASME Journal of Vibration and Acoustics*, Vol. 116, No. 3, pp. 295–302.

Tzou, H. S., Johnson, D., and Liu, K. J., 1995, "Boundary Transition and Nonlinear Control of Distributed Systems," *in Stability, Vibration, and Control of Structures, Vol. 1: Wave Motion, Intelligent Structures, and Nonlinear Mechanics*, eds. Guran, A., and Inman, D. J., World Scientific, pp. 163–93, Singapore.

Tzou, H. S., and Liu, B., 1996, "Study of Spatially Distributed Opto-Mechanical Shell Actuators," *Active Control of Vibration and Noise*, DE Vol. 93, pp. 305–14, 1996 ASME International Congress.

Tzou, H. S., and Tseng, C. I., 1988, "Active Vibration Control of Distributed Parameter Systems by Finite Element Method," *ASME Computers in Engineering 1988*, Vol. 3, pp. 599–604.

Tzou, H. S., and Tseng, C. I., 1990, "Distributed Piezoelectric Sensor/Actuator Design for Dynamic Measurement/Control of Distributed Parameter Systems: A Piezoelectric Finite Element Approach," *Journal of Sound and Vibration*, Vol. 138, No. 1, pp. 17–34.

Tzou, H. S., and Tseng, C. I., 1991, "Distributed Modal Identification and Vibration Control of Continua: Piezoelectric Finite Element Formulation and Analysis," *ASME Journal of Dynamic Systems, Measurements, and Control*, Vol. 113, No. 3, 500–5.

Tzou, H. S., Tseng, C. I., and Wan, G. C., 1990, "Distributed Structural Dynamics Control of Flexible Manipulators, Part 2: Distributed Sensor and Active Electromechanical Actuator," *Journal of Computers and Structures*, Vol. 35, No. 6, pp. 679–87.

Tzou, H. S., and Ye, R., 1994, "Piezothermoelasticity and Precision Control of Piezoelectric Laminates: Theory and Finite Element Analysis," *ASME Journal of Vibration and Acoustics*, Vol. 116, No. 4, pp. 489–95.

Tzou, H. S., and Zhong, J. P., 1993, "Electromechanics and Vibrations of Piezoelectric Shell Distributed Systems: Theory and Applications," *ASME Journal of Dynamic Systems, Measurements, and Control*, Vol. 115, No. 3, pp. 506–17.

Tzou, H. S., Zhong, J. P., and Hollkamp, J. J., 1994, "Spatially Distributed Orthogonal Piezoelectric Shell Actuators (Theory and Applications)," *Journal of Sound and Vibration*, Vol. 177, No. 3, pp. 363–78.

Tzou, H. S., Zhong, J. P., and Natori, M. C., 1993, "Sensor Mechanics of Distributed Shell Convolving Sensors Applied to Flexible Rings," *ASME Journal of Vibration and Acoustics*, Vol. 115, No. 1, pp. 40–6.

Tzou, H. S., and Zhou, Y.-H., 1995, "Dynamics and Control of Nonlinear Circular Plates with Piezoelectric Actuators," *Journal of Sound and Vibration*, Vol. 188, No. 2, pp. 189–207.

Usov, V. S., and Surygin, A. I., 1984, "Variation of SAW Velocity and Damping in a Piezofilm–Semiconductor Structure Under the Effect of a Constant Transverse Electric Field," *Radioelektronika* (ISSN 0021-3470), Vol. 27. (Nov. 1984).

Vidyasagar, M., 1988, "Control of Distributed Parameter System Using the Coprime Factorization Approach," *Recent Development in Control of Nonlinear and Distributed Parameter Systems*, ASME-DSC Vol. 10, pp. 1–10.

Wada, B. K., Fanson, J. L., and Crawley, E. F., 1989, "Adaptive Structures," *in Adaptive Structures*, ed. B. K. Wada, AD Vol. 15, pp. 1–8, 1989 ASME WAM.

Wang, P. K. C., 1966, "On the Feedback Control of Distributed Parameter Systems," *Int. J. on Control*, Vol. 3, No. 3, pp. 255–73.

Zimmerman, D. C., Inman, D. J., and Juang, J. N., 1988, "Low Authority-Threshold Control for Large Flexible Structures," AIAA Paper 88-2270, *Proc. of AIAA/ASME/AHS 29th Structures, Structural Dynamics, and Materials Conference*, pp. 459–69.

Nomenclature

A_1, A_2	Lamé parameter
A_n, B_n	modal amplitudes
b	width
c	damping constant
d_{ij}	piezoelectric strain constant

D	bending stiffness, $D = [Yh^3/12(1-\mu^2)]$
ds	infinitesimal distance on the neutral surface
e_{ij}	piezoelectric stress coefficient
E_i	electric field
F_i	excitation
$\mathcal{G}$	gain
$\hat{\mathcal{G}}_k$	modal feedback gain
h	thickness
I	area moment of inertia
k	kth mode
k_{ij}	bending strains
K	membrane stiffness, $K = [Yh/(1-\mu^2)]$
L	length
$L_i(\alpha_1, \alpha_2)$	Love's operator
M_{ij}	moment on the ith surface and in the jth direction
N_{ij}	membrane force on the ith surface and in the jth direction
Q_{ij}	shear force effect
r_i	distance
R, R_i	radius or radius of curvature of the ith axis
S_{ij}	strain components
S_{ij}^0	membrane strain components
$\mathbf{S}^e$	electrode surface
S_{mn}^s	sensitivity
sgn	signum function: $\mathrm{sgn}(x) = -1$ if $x < 0$; 1 if $x = 0$; $+1$ if $x > 0$.
t	time
$u_i, \dot{u}_i, \ddot{u}_i$	displacement, velocity, acceleration components
U_n	modal function or mode shape function
W_s, W_a	shape function or weighting function
Y_i	Young's modulus
$[\cdot]^t$	vector or matrix transpose
$[\cdot]^{-1}$	matrix inverse

Greek

α_i	coordinate of the α_i axis
$\alpha_1, \alpha_2, \alpha_3$	curvilinear coordinates, α_1 and α_2 for the reference surface and α_3 the normal axis
ϵ	dielectric constant or permittivity
ζ_n	damping ratio of the nth mode
λ_n	eigenvalue or characteristic roots
μ	Poison's ratio
ρ	material density
ϕ	electric potential
η_n	modal participation factor or modal coordinate
φ	phase angle
ω_n	nth natural frequency

Superscripts

a	actuator
af	acceleration feedback

c control
df displacement feedback
e electric component
k kth mode
m mechanical component
s sensor
vf velocity feedback

Subscripts

a actuator
i, j indices
ij on the ith surface, in the jth direction
k kth mode
n nth mode
p piezoelectric related properties
s sensor

Appendix

A.1 Piezoelectricity Theory

Two fundamental equations are used in the derivation of distributed sensor theory:

$$\{T\} = [c^D]\{S\} - [h]^t\{D\}$$

and

$$\{E\} = [\beta^s]\{D\} - [h]\{S\},$$

where $\{T\}$ is the stress vector (i.e., $\{T\} = \{T_{11}\ T_{22}\ T_{33}\ T_{23}\ T_{31}\ T_{12}\}^t$); $[c^D]$ is the elasticity matrix evaluated at constant dielectric displacement; $\{S\}$ is the strain vector (i.e., $\{S\} = \{S_{11}\ S_{22}\ S_{33}\ S_{23}\ S_{31}\ S_{12}\}^t$); $[h]$ is the piezoelectric constant matrix; $\{D\}$ is the electric displacement vector; $[.]^t$ indicates the matrix transpose; $\{E\}$ is the electric field vector; and $[\beta^s]$ is the dielectric impermeability matrix evaluated at constant strain.

A.2 Piezoelectric Matrix of Polyvinylidene Fluoride (PVDF)

Polymeric polyvinylidene fluoride (PVDF) has a mm2 structure. The piezoelectric matrix $[d]$ of a PVDF polymer can be expressed

$$[d_{ij}] = \begin{bmatrix} 0 & 0 & 0 & 0 & d_{15} & 0 \\ 0 & 0 & 0 & d_{24} & 0 & 0 \\ d_{31} & d_{32} & d_{33} & 0 & 0 & 0 \end{bmatrix}.$$

Note that the piezoelectric coefficient d_{24} is equal to d_{15} for a PVDF electrically polarized and not mechanicaly stretched.

Index